Infrared Detectors and Systems

WILEY SERIES IN PURE AND APPLIED OPTICS

Founded by Stanley S. Ballard, University of Florida

EDITOR: Joseph W. Goodman, Stanford University

Infrared Detectors and Systems

E. L. DERENIAK
University of Arizona

G. D. BOREMAN
University of Central Florida

A Wiley-Interscience Publication

JOHN WILEY & SONS, INC.

New York / Chichester / Brisbane / Toronto / Singapore

Library of Congress Cataloging in Publication Data:
Dereniak, Eustace L.
   Infrared detectors and systems / E.L. Dereniak, G.D. Boreman.
     p.   cm.—(Wiley series in pure and applied optics)
   "A Wiley-Interscience publication."
   Includes index.
   ISBN 0-471-12209-2 (cloth : alk. paper)
   1. Infrared detectors. I. Boreman, G. D. (Glenn D.) II. Title.
III. Series.
TA1570.D47   1996
621.36′2—dc20   96-212

To our wives, Barbara and Maggie
Our children, Teresa, Andreana, and Edward
Our fathers, Peter and Lyell
And in memory of our mothers, Julia and Miriam

Contents

Preface

This textbook covers the range of subjects necessary for the understanding of modern infrared-imaging systems at a level appropriate for seniors or first-year graduate students in physics or electrical engineering. It should also be valuable as a reference for the practicing engineer involved in the use, analysis, design, or testing of infrared-detector-based systems.

The first six chapters are devoted to fundamental background issues for radiation detection. We begin in Chapter 1 with the basics of geometrical optics, concentrating on issues of image formation and image quality. Chapter 2 is a discussion of radiometry and flux-transfer issues needed for systems analysis. Chapter 3 introduces the basic radiation-detector mechanisms. Chapter 4 reviews random-process mathematics, so that the fundamentals of detector noise can be considered in Chapter 5. This development culminates in Chapter 6 with a discussion of the figures of merit used for describing the signal-to-noise performance of a detector system, chiefly noise-equivalent power and D^*. We then proceed to five chapters that deal with specific detector technologies. In each chapter we concentrate on the fundamental mechanism of detection, with special attention to issues of responsivity and noise performance. Chapter 7 describes photovoltaic detectors, Chapter 8 photoconductive detectors, Chapter 9 thermal detectors, Chapter 10 Schottky-barrier detectors, and Chapter 11 describes bandgap-engineered structures used in quantum-well and superlattice detectors. Chapters 12 to 14 have a system orientation rather than a device orientation. Chapter 12 considers infrared search systems and develops the range equation in terms of the optical and detector parameters of the system. Chapter 13 considers the modulation transfer function (MTF), a spatial-frequency-domain description of image quality. Chapter 14 uses this material to describe the design equations for thermal-imager systems in terms of noise-equivalent temperature differenc (NETD) and minimum resolvable temperature difference (MRTD). In each chapter we strive for an up-to-date viewpoint on a rapidly moving field.

Material contained in this book has been used for several years in graduate courses on radiometry and infrared detectors, both at the University of Arizona, Optical Sciences Center, and at the University of Central Florida, Center for Research and Education in Optics and Lasers. Consequently numerous students have contributed by asking probing questions and finding errors in drafts of this material. Some of these, but surely not all, are the following: (at Arizona) Dave Kaplan, Geoff Walter, Mike Brown, Steve Stuut, Lee Dettman, Paul Spyak, Debra Kuo, Carl Robinson, Linda Lingg, John Stacy, John Hayes, Scott McNown, Steve Dziuban, Scott Devore, Jackson He, Ed Dubiel, Thorston Graeve, Lee Hudson, Arvi Jeffrey, Mel Benedict, Michael Wiedman, Steve Kurtz, Mary Turner, Chong Im Kim, James Fahnestock, Joan Moll, Mary Ann Moscynski, Ted Bailey, Dorothy Tinkler, Mark Voelker, Mike Salcido, Louie Rosiek, Marsha Bilmont, and Jack Shaffer; (at Central Florida) Ron Driggers, Ken Barnard, Buck Burns, Arnold Daniels, Al Ducharme, Martin Sensiper, Sidney Yang, Patrick Thompson, and Jamie Rhead. Special thanks for technical discussions go to Dr. Bill Horter, Greg Herman, Larry C. Guy, Jon Mooney, Dave Perry, John Garcia, Sean Ross, Dot Dorman, and Mike Nofziger. Also to John "IR" Hubbs, whose down-to-earth questions and diligence made a number of passages clearer.

A substantial portion of this book was finished while one of the authors (GDB) was on sabbatical leave with the Applied Optics group at Imperial College, London. Thanks to Prof. Chris Dainty for providing a delightful and quiet place to work with excellent facilities.

The authors appreciate the cheerfulness and forebearance of their families during the writing process. Finally, we would like to say a very special thank you to Maggie Boreman, whose many volunteer hours of technical editing have made this book a reality. We greatly appreciate her generosity.

ELD
GDB

Tucson
London

Nomenclature

A	area
$A_{\Delta T_d}$	amplitude of time-harmonic detector temperature difference
A_{wire}	cross-sectional area of a wire
$\mathfrak{A}$	separation constant
$\mathcal{A}$	bolometric resistance-temperature exponent
A_s	source area
A_d	detector area
$A_{\mathrm{footprint}}$	footprint area
A_{lens}	lens area
A_{enp}	entrance-pupil area
A_{obj}	area of the object
A_n	cross-sectional area of a detector normal to current flow
A_{img}	area of the image
a	intermediate variable
α	absorptance
a	absorption coefficient
$B_{1/f}$	$1/f$ constant of proportionality
B_{max}	maximum brightness in an object or image waveform
B_{min}	minimum brightness in an object or image waveform
b_{Bose}	Bose factor
B	magnetic field magnitude
$\mathcal{B}$	ratio of T_d to T_{sink}
b	nonideality factor for diode i-v equation
$\mathfrak{b}$	ratio of electron to hole mobility
β	semiconductor material constant
$\mathfrak{B}$	separation constant
c	speed of light in vacuum
C	capacitance
C_1	quantum-yield coefficient for Schottky-barrier diode
C_d	capacitance of a detector
C_{th}	thermal capacitance
c_n	speed of light in a medium of refractive index n
D^*	specific detectivity of a detector
D^*_{BLIP}	D^* under conditions of background-limited operation
D^{**}	angle-normalized specific detectivity
DOS_x	1-dimensional density of states
DOS	3-dimensional density of states

D_{lens}	diameter of a lens
D_{input}	diameter of input marginal ray bundle for telescope
D_{output}	diameter of output marginal ray bundle for telescope
D_{enp}	diameter of entrance pupil
D_{exp}	diameter of exit pupil
D_e	diffusion constant for electrons
D_h	diffusion constant for holes
d	separation between capacitor plates; width of p-n junction/ depletion region
d_{spot}	diameter of target spot on SPRITE detector
D	electric displacement field magnitude
$d_{\text{diffraction}}$	diameter of the diffraction blur spot
e	subscript for radiometric units denoting energy-based unit
E_{image}	image irradiance
$E_{q,\text{sig}}$	signal photon irradiance
$E_{q,\text{bkg}}$	background photon irradiance
$\mathcal{E}$	energy of a photon or of a particle in a given energy state
$\mathcal{E}_m$	a series of m energy levels
$\mathcal{E}_c$	energy of the conduction-band edge
$\mathcal{E}_v$	energy of the valence-band edge
$\mathcal{E}_c^B$	energy of the conduction-band edge in the barrier
$\mathcal{E}_c^w$	energy of the conduction-band edge in the well
$\mathcal{E}_v^B$	energy of the valence-band edge in the barrier
$\mathcal{E}_v^w$	energy of the valence-band edge in the well
$\Delta\mathcal{E}_v$	well-to-barrier valence-energy difference
$\Delta\mathcal{E}_c$	well-to-barrier conduction-energy difference
$\mathcal{E}_g$	energy gap of a material
$\mathcal{E}_f$	Fermi energy level
$\mathcal{E}_n$	energy level in an n-type material
$\mathcal{E}_p$	energy level in a p-type material
E	electric field magnitude
$\mathbf{E}$	electric field vector
E_{ox}, E_{oy}, E_{oz}	components of electric field at $t = 0$
E_x, E_y, E_z	components of electric field at general time
$e(x)$	edge response
F	focal length of a lens
F_F	chopper form factor, for conversion of waveform from peak-to-peak to rms
f	electrical frequency
$f(x, y)$	object function
$F(\xi, \eta)$	object spectrum
f_0	center electrical frequency of a filter
$\mathfrak{f}$	intermediate variable
$F_1, F_2, \cdots$	focal length of lens 1, 2, $\cdots$

$F_{\text{combination}}$	aggregate focal length of a system of two or more thin lenses
FOV	full-angle field of view
$\text{FOV}_{\text{half-angle}}$	half-angle field of view
f	electrical frequency
f_1	front focal point
f_2	rear focal point
$\mathcal{f}(G)$	APD excess noise factor
f_{drift}	3-dB response frequency caused by electron drift
f_{RC}	3-dB response frequency caused by RC time constant
f_c	3-dB corner frequency, aggregate of effects
$F/\#$	f-number
$F/\#_{\text{object-space}}$	f-number in object space
$F/\#_{\text{image-space}}$	f-number in image space
$g(x, y)$	image function
$G(\xi, \eta)$	image spectrum
$\mathcal{G}$	electronic filter gain
G	photoconductive gain or APD gain
G_s	shunt conductance
G_{OL}	open-loop gain for op amp
g	electron-hole generation rate
$\mathcal{g}(t)$	Fourier transform of the filter function $\mathcal{G}(f)$
g_{th}	thermal generation rate
gb	superscript for radiometric variable pertaining to a graybody
$h(x, y)$	impulse response
$H(\xi, \eta)$	spectrum of impulse response
h	Planck's constant
H	heat capacity
HIFOV	horizontal instantaneous field of view
HFOV	horizontal field of view
h_{obj}	object height
h_{img}	image height
$h_{\text{obj},x}$	horizontal object height
$h_{\text{obj},y}$	vertical object height
$h_{\text{img},x}$	horizontal image height
$h_{\text{img},y}$	vertical image height
i	current
i'	operating point on an i-v curve
i_h	hole current
i_e	electron current
i_r	reverse-saturation current
$\bar{i}$	average current
i_J	rms Johnson noise current
$i_{1/f}$	rms $1/f$-noise current
i_{sc}	short circuit current

i_s	shot noise rms current
i_{pa}	rms preamplifier current noise
i_{eq}	equivalent current in a PIN diode
i_d	dark current
i_{bkg}	background-generated current
i_{sig}	signal current
i_{gr}	rms generation-recombination noise current
i_0	reverse saturation current
i_f^2	current noise power spectral density
i_g	photogenerated current
i_n	rms noise current
i_s	signal current
J	current density magnitude
J'	surface-leakage current density
J_{gr}	gr current density
J_s	saturation current density
J_e	electron current density from aggregate contributions
$J_{d,h}$	hole diffusion current density
$J_{d,e}$	electron diffusion current density
$J_{f,h}$	hole field–current density
$J_{d,e}$	electron field–current density
J^+	hole current density
J^-	electron current density
J_x, J_y, J_z	Cartesian components of the current density vector
j	square root of -1
k	Boltzmann's constant
$K(\xi_f)$	spatial-frequency-dependent proportionality factor for objective MRTD
K	thermal conductance for differential temperatures
K_0	integrated thermal conductance for finite temperature difference
K_{eff}	effective thermal conductance, including the effects of Joule heating
K_t	thermal conductivity
$\mathcal{K}_r$	recombination rate constant
k	wave vector
k	wave-vector magnitude
$\mathit{k}_x, \mathit{k}_y, \mathit{k}_z$	wave-vector components
k_n	normal component of the wave vector
$\mathcal{L}$	Lorentz number
L_d	ambipolar diffusion length
L_h	diffusion length for holes
L_e	diffusion length for electrons
L_{bkg}	background radiance

L_t	transport length
L_{mfp}	mean free path length
ℓ	length of a detector along current flow direction
ℓ_W	width of a well
ℓ_B	width of a barrier
$\ell(x)$	line response
ℓ_x, ℓ_y, ℓ_z	dimensions of a cavity or of a detector
$\ell_{z,n}$	z-dimension of n-type material
M_{meas}	measured exitance
M_{obj}	exitance of an object
$\mathfrak{M}$	modulation depth
$\mathfrak{M}$	magnification
$\mathfrak{M}_{angular}$	angular magnification of a telescope
$\mathfrak{M}_{relay}$	magnification of a relay system
m	counting parameter, an integer
m_e^*	effective mass of an electron
m_h^*	effective mass of a hole
n	number of photons or number of states
n_{cy}	number of cycles required to satisfy the Johnson criterion
n_d	number of detectors
n_T	total number of possible hole states
n_L	number of lines
n_e	number of electrons
$\overline{n_e}$	average number of electrons
n_m	number of photons in the state m
$\overline{n}$	average number of photons
n_i	intrinsic carrier concentration
$\underline{\mathsf{n}}$	electron concentration
$\Delta\mathsf{n}$	mean excess electron concentration
n_{bkg}	concentration of background-generated carriers
n_a	acceptor concentration
n_b	bulk impurity concentration
n_d	donor concentration
n_p	electron concentration in p-type material
$\overline{n}_k$	average number of photons per k-state
N	number of photons per cavity volume
NA	numerical aperture
NEP	noise equivalent power
N_k	spectral density of N, (N per unit wave vector)
n	refractive index
$\mathit{n}_1, \mathit{n}_2, \cdots$	refractive index of medium #1, #2, $\cdots$
OPD	optical path difference, a measure of wavefront aberration
$\mathfrak{P}$	pressure
$\mathfrak{p}$	momentum of electron

P	probability of an event
P_{av}	average power
P_L	power delivered to load
P	polarization magnitude
P_s	spontaneous polarization
p	hole concentration
p_p	hole concentration in p-type material
p_n	hole concentration in n-type material
$\wp$	pyroelectric coefficient
p	object distance
p_1	front principal point
p_2	rear principal point
$p(n)$	probability that n photons exist in a given energy state
q	image distance
q	unit of electric charge on one electron or one hole
Q	amount of charge in a charge packet
q	subscript for units based on photons
R	resistance
R_0	steady-state resistance (i.e., at zero voltage, or for a zero flux input)
$R_1, R_2, \cdots$	resistance #1, #2, $\cdots$
R_d	detector resistance
R_L	load resistance
R_s	shunt resistance
R_{eq}	equivalent resistance
R_{out}	output impedance
R_{th}	thermal resistance
$\mathfrak{R}$	autocorrelation
$\mathcal{R}$	responsivity
$\mathcal{R}_i$	current responsivity
$\mathcal{R}_v$	voltage responsivity
r	radial variable
r_{12}	radial variable between two area elements
$\mathbf{r}$	vector in direction of propagation
$\imath$	reflectance
$\imath_1, \imath_2, \cdots$	reflectance of interface 1, 2, $\cdots$
$\mathfrak{r}$	Hall scattering coefficient
s	arc length
s_o	radius of mirror obscuration
s_m	radius of mirror aperture
S	surface recombination velocity
SNR	signal-to-noise ratio
$\mathfrak{S}$	power spectrum
t	time
T	kinetic temperature, in kelvins

T_{target}	target temperature
T_s	critical temperature for a superconductor
T_d	detector temperature
T_L	load resistor temperature
T_{bkg}	background temperature
T_{source}	source temperature
T_0	path temperature
T_0'	effective path temperature
$T, T_o, T(t)$	intermediate quantity for separation of variables
t	transmittance
$t_1, t_2, \cdots$	transmittance of interface 1, 2, $\cdots$
$t_{internal}$	internal transmittance
$t_{external}$	external transmittance
T_R	radiation temperature
T_B	brightness temperature
T_C	color temperature
u	energy of an ensemble of molecules
u	conversion factor from blackbody to spectral responsivity or $D*$
V	volume
v	voltage
VIFOV	vertical instantaneous field of view
VFOV	vertical field of view
v_d	voltage across diode or across detector
v_d'	operating voltage point on i-v curve
v_d''	tangent intercept point on i-v curve
v_i	input voltage
v_{sig}	signal voltage
v_{bi}	built-in voltage in a photovoltaic
v_{pa}	preamplifier rms voltage noise
v_f	forward bias voltage
v_r	reverse bias voltage
v_n	rms noise voltage
v_B	bias voltage
v_o	output voltage
v_{oc}	open-circuit voltage
v_{max}	maximum voltage
v_j	Johnson-noise rms voltage
v_s	shot-noise rms voltage
$v_{t, rms}$	total rms noise voltage
$\mathbf{v}_s$	saturation velocity
$\mathbf{v}_{scan}$	scan velocity
$\mathbf{v}_d$	drift velocity
$\mathbf{v}_{d, e}$	drift velocity for electrons
$\mathbf{v}_{d, h}$	drift velocity for holes

$v_{n,\text{in}}^2(f)$	input noise power to a filter
$v_{n,\text{out}}^2(f)$	output noise power from a filter
v_{oc}	open-circuit voltage
v_x	x-component of velocity vector
$\bar{v}_x$	average x-component of velocity vector
x	distance variable in Cartesian coordinates
$x_{\min}$	minimum dimension of target, for Johnson criterion
x_p	distance from center of p-n junction to edge of depletion region on p side
x_n	distance from center of p-n junction to edge of depletion region on n side
$\hat{x}$	unit vector in Cartesian coordinates
$X, X_o, X(x)$	intermediate quantity for separation of variables
$\mathcal{X}$	spatial period of a bar target
x	x-parameter, equal to $h\nu/kT$
$\propto$	mixing ratio for semiconductor alloys
y	distance variable in Cartesian coordinates
y_d	vertical dimension of detector
$y_{\text{footprint}}$	vertical dimension of detector footprint
y	mixing ratio for semiconductor alloys
$\hat{y}$	unit vector in Cartesian coordinates
$Y, Y_o, Y(y)$	intermediate quantity for separation of variables
z	distance variable in Cartesian coordinates
z	intermediate variable
$z_{o,i}$	axial distance from object to image
$z_{o,\text{enp}}$	axial distance from object to entrance pupil
$z_{\text{exp},i}$	axial distance from exit pupil to image
$z_{\text{exp},f2}$	axial distance from exit pupil to rear focal point
$z_{\text{separation}}$	axial separation of two lenses
$\hat{z}$	unit vector in Cartesian coordinates
$Z, Z_o, Z(z)$	intermediate quantity for separation of variables

Greek Letters

α	fractional resistance change per unit temperature for a bolometer
α_i	impact ionization coefficient for electrons
α_{taper}	readout taper coefficient for SPRITE detector
$\alpha_{1/f}$	current exponent for $1/f$ noise
α_s	Seebeck coefficient
β	blur angle caused by diffraction
β_i	impact ionization coefficient for holes
β_d	diode nonideality factor

β_f	op-amp feedback factor
β_0	$1/f$ proportionality factor
$\beta_{1/f}$	negative f exponent function
β_{img}	blur angle caused by diffraction, in image space
β_{obj}	blur angle caused by diffraction, in object space
γ	gain exponent for impact ionization
$\Delta\lambda$	wavelength interval
Δp	change in hole concentration
Δn	change in electron concentration
Δt	time interval
ΔT	temperature difference
ΔT_d	change in detector temperature
Δv	voltage range
Δf	electronic frequency bandwidth
$\Delta f_{3\,dB}$	3-dB electronic frequency bandwidth
δt	time interval
δ_d	degeneracy factor
ε	emissivity
ϵ	dielectric constant
ϵ_0	dielectric constant of free space
θ	angle variable
θ'	angle after refraction
θ_s	angle between source normal and line of centers
θ_d	angle between detector normal and line of centers
θ_{input}	angle of parallel ray bundle entering into a telescope
θ_{output}	angle of parallel ray bundle exiting from a telescope
θ_{max}	limiting planar angular subtense
θ_i	angle of incidence
θ_r	angle of reflection
η	quantum efficiency
η_{sc}	scan efficiency
λ	subscript for spectral radiometric quantities, having units of per wavelength interval
λ	wavelength
Λ	deBroglie wavelength
λ_c	cutoff wavelength of a detector
λ_p	peak-response wavelength of a detector
λ_o	particular fixed wavelength
λ_{max}	wavelength of maximum exitance
$\lambda_{max\,contrast}$	wavelength of maximum exitance contrast
μ_e	electron mobility
μ_h	hole mobility
ν	optical frequency
ρ	resistivity

ρ_c	charge density
σ_c	conductivity
$\sigma_{c,i}$	conductivity arising from intrinsic-carrier concentration
$\sigma_{c,\text{bkg}}$	conductivity arising from background-induced carriers
σ	standard deviation
σ^2	variance
σ_v	rms voltage noise
σ_{pe}^2	photoelectron variance
σ_e	Stefan–Boltzmann constant for radiant exitance
σ_q	Stefan–Boltzmann constant for photon exitance
σ_n^2	variance of number of photons per mode
τ	carrier lifetime, time constant
τ_{RC}	RC time constant
τ_{int}	integration time for a detector
τ_{th}	thermal time constant
τ_f	frame time
τ_d	dwell time
τ_L	line time
τ_ρ	dielectric relaxation time
τ_h	lifetime for holes
τ_e	lifetime for electrons
τ_f	transit time
φ	angle variable
ϕ_d	flux reaching the detector
$\phi_{q,\text{sig}}$	signal photon flux
$\phi_{q,\text{bkg}}$	background photon flux
ϕ_{img}	flux reaching the image
ϕ_{obj}	flux radiated by the object
ϕ_{incident}	flux incident on a body or surface
ϕ_{absorbed}	flux absorbed by a body or surface
$\phi_{\text{reflected}}$	flux reflected by a body or surface
$\phi_{\text{transmitted}}$	flux transmitted by a body or surface
ϕ_0	flux at origin of coordinates for an absorbing medium
$\psi_{\text{wf},m}$	work function for metal
$\psi_{\text{wf},s}$	work function for semiconductor
Ψ	Schottky-barrier height
ξ	x-direction spatial frequency
ξ_f	fundamental spatial frequency of a bar target
ζ	y-direction spatial frequency
Ω_{enp}	entrance pupil solid angle
Ω_{exp}	exit pupil solid angle
Ω_s	source solid angle
Ω_d	detector solid angle
Ω	solid angle

Ω_{bkg}	background solid angle
Ω_{lens}	lens solid angle
Ω_{img}	image solid angle
Ω_{obj}	object solid angle

Basic Radiometric Quantities

Q	energy in joules
ϕ	flux or power in watts
I	intensity in W/sr
E	irradiance in W/cm^2
M	exitance in W/cm^2
L	radiance in W/cm^2-sr
Q_λ	spectral energy in joules per μm
ϕ_λ	spectral flux or power in watts per μm
I_λ	spectral intensity in W/sr-μm
E_λ	spectral irradiance in W/cm^2-μm
M_λ	spectral exitance in W/cm^2-μm
L_λ	spectral radiance in W/cm^2-sr-μm
Q_q	number of photons
ϕ_q	photon flux in photon/s
I_q	photon intensity in photon/s-sr
M_q	photon exitance in photon/s-cm^2
L_q	photon radiance in photon/s-cm^2-sr
$Q_{q,\lambda}$	number of photons per μm
$\phi_{q,\lambda}$	spectral photon flux in photon/s-μm
$I_{q,\lambda}$	spectral photon intensity in photon/s-sr-μm
$M_{q,\lambda}$	spectral photon exitance in photon/s-cm^2-μm
$L_{q,\lambda}$	spectral photon radiance in photon/s-cm^2-sr-μm
$M_{q,k}$	spectral density of M_p (M_p per unit wave vector)
$L_{q,k}$	spectral density of L_p (L_p per unit wave vector)

1

Geometrical Optics

1.1. INTRODUCTION

Geometrical optics is the study of image formation. Investigating geometrical optics first will allow us to conveniently introduce concepts and definitions we will use throughout the rest of the book to describe various aspects of propagation and detection of radiation. Geometrical optics is an appropriate starting point for our consideration of infrared detectors and systems because the imaging properties of lens and mirror systems determine the flux distribution sensed by the detectors. Issues related to the location and magnification of images are considered first. The configuration of the apertures in the imaging system defines both the field of view and the flux-collection efficiency. The amount of flux collected determines the signal-to-noise ratio of the detected image. We discuss the relation of the sizes and locations of apertures to the flux-collection efficiency in terms of the area-solid-angle product. Scanning systems are often necessary to cover a given field of view (FOV) using a limited number of detectors. The interaction of the scanning system with the image-forming components will affect the radiometric accuracy of the images formed. The specific configuration of the optical system, along with the aperture and field requirements, will impact the fidelity of the resulting image. Image quality is discussed in terms of the smallest spot size that is formed by the system, and is compared with the spot size that would be formed by a system limited only by the effects of diffraction. We end this first chapter with a short discussion of material properties pertinent to their use as optical elements in the infrared portion of the spectrum, specifically reflectance and transmittance as a function of wavelength.

This chapter introduces the geometrical-optics concepts needed throughout the succeeding chapters. The concepts explained include image formation; stops and pupils; lens combinations; scanning; image quality; and optical materials.

1.2. ELECTROMAGNETIC SPECTRUM

An important parameter for electromagnetic radiation is wavelength (λ). Wavelength enables us to classify the radiation into one of several categories:

for example, infrared, visible, or microwave. The divisions between these categories are based on the different source and detector technologies used in each region. The most important region for our purposes is the infrared region, extending on the short-wavelength side from the longer limit of human vision around 0.77 μm, to around 1-mm wavelength on the long-wave side. Figure 1.1 is an illustration of the electromagnetic spectrum.

The wavelength parameter relates to two other parameters, frequency (ν) and photon energy ($\mathcal{E}$). They are related by

$$\nu = \frac{c}{\lambda} \qquad (1.1)$$

where $c \cong 3 \times 10^8$ m/s is the speed of light in free space, and

$$\mathcal{E} = h\nu \qquad (1.2)$$

where $h \cong 6.67 \times 10^{-34}$ Js is Planck's constant. Since the electromagnetic wave travels in different materials, the speed of propagation and the wavelength will change proportionally to keep the frequency constant. For infrared radiation at $\lambda = 1$ μm, Eqs. (1.1) and (1.2) predict a frequency of approximately 3×10^{14} Hz and a photon energy of approximately 2×10^{-19} J. At a longer wavelength (10 μm), the frequency and the photon energy both decrease by a factor of 10, to yield $\nu = 3 \times 10^{13}$ Hz and $\mathcal{E} = 2 \times 10^{-20}$ J.

1.3. IMAGING CONCEPTS

Our approach to imaging concentrates on first-order optics, an approximate analysis technique that predicts image size and image location. In first-order optics, any angles made by rays with the horizontal axis are assumed to be small enough that sines of the angles are replaced by the angles themselves in radians. This approximation, known as the *paraxial (small-angle) approximation*, linearizes the raytrace equations that determine the ray paths through the optical system. More detailed analysis that does not involve the paraxial approximation is needed to predict image quality, particularly for the case of larger angles.

An object is a collection of source points, each of which emits light rays into all forward directions (see Fig. 1.2). Each source point has a different strength. The spatial distribution of brightness across the object is the information we want to reproduce in the image. The lens causes rays that diverge from an object point to intersect again at a corresponding point in the image plane. The image is thus built up on a point-by-point basis. The brightness at any point in the image is proportional to the strength of the corresponding object point, resulting in a geometrically similar distribution of flux. To simplify illustrations, we draw only a few rays from a few object points. A point

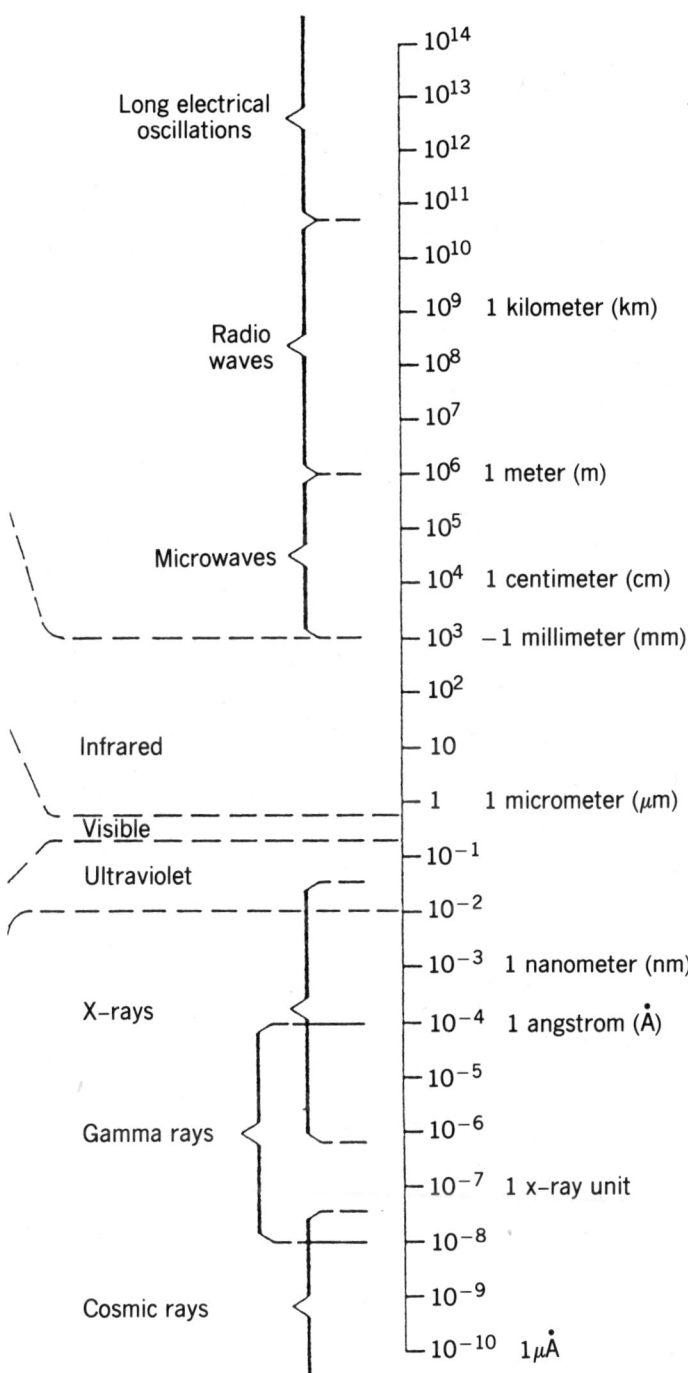

Figure 1.1. Electromagnetic spectrum. (Originally appeared in D. R. Carter, *Burle Electro-Optics Handbook, 1974.* Reprinted by permission of Burle Industries Inc.)

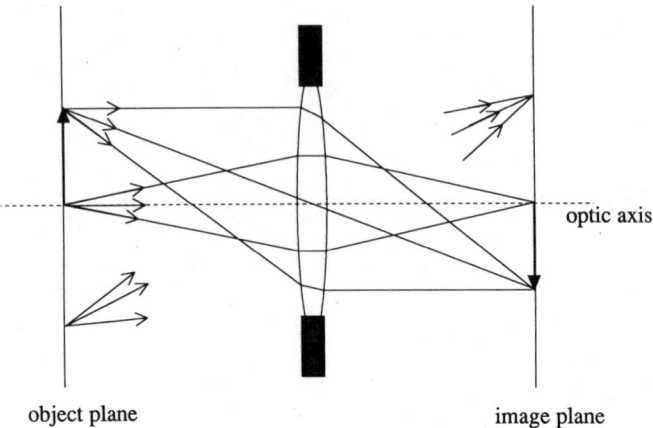

Figure 1.2.   Basic image formation.

of nomenclature is that the optic axis is the axis of symmetry for the lens, containing the centers of curvature of the lens surfaces.

We can choose the rays we want to draw with the aid of the graphical ray-trace rules for thin lenses. These rules, illustrated in Fig. 1.3, allow us to get the maximum amount of insight from a just a few rays. The rules are as follows:

1. Rays entering the lens parallel to the optic axis will exit through the focal point.
2. Rays entering the lens through the focal point will exit parallel to the optic axis.
3. Rays that pass through the center of the lens do not change direction.

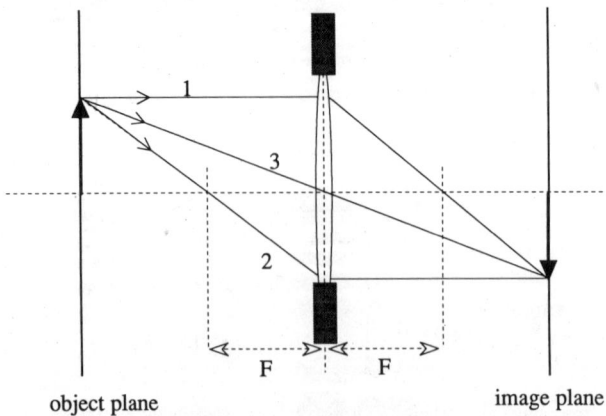

Figure 1.3.   Graphical raytrace rules.

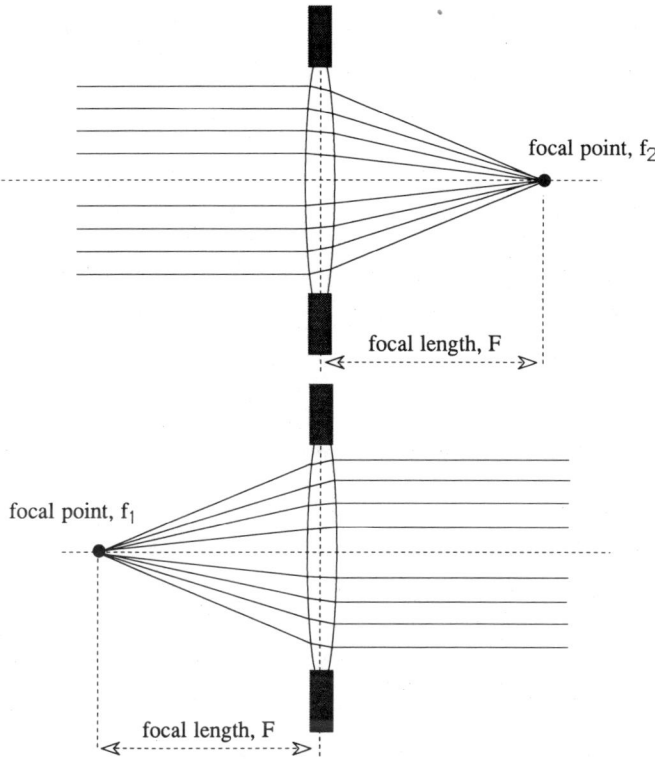

Figure 1.4. Focal point—both sides of the lens.

Raytrace rules 1 and 2 can be justified by the definition of focal point of a lens, seen in Fig. 1.4. Rays that originate from the on-axis point at infinity enter the lens parallel to the optic axis, and will converge at the focal point of the lens. There is a focal point (denoted f_1 on the left side of the lens and f_2 on the right side of the lens) on each side of the lens, because parallel rays can enter from either side. The light rays are reversible and do not have an inherent directionality. The focal length F is the distance from the center of the lens to the focal point for a thin lens. The focal length is a positive distance on either side of the lens when the lens is positive, that is, capable of bringing rays to a focus. A lens is thin if its thickness is small compared to its focal length. Analysis is simplified for thin lenses. Regarding rule 3, we see that a ray passing through the center of the lens will be refracted according to Snell's law, as seen in Fig. 1.5. The region of the lens near the axis is a plane-parallel plate. The ray passing through this region will experience a parallel displacement, but will not change direction. In the limit of an infinitessimally thin lens, the ray through the center is continuous and undeviated.

Figure 1.6 illustrates an application of the graphical raytrace rules; a point

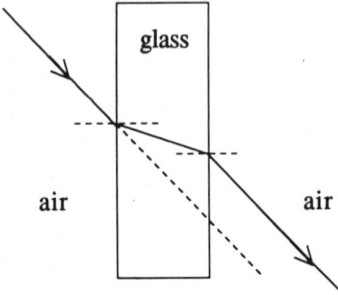

Figure 1.5. Illustration of Snell's law: Rays passing through the center of the lens do not change direction.

source is at an infinite distance from the lens, and hence produces parallel rays, but is not located on the optic axis of the lens. There is an angle θ between the direction of the rays and the optic axis. For small θ, the rays focus at a distance F behind the lens, and at a distance θF away from the axis.

These graphical rules can be used to formulate algebraic relations for thin-lens imaging. The results of using geometrical constructions to develop such formulas is (Fig. 1.7)

$$\frac{1}{p} + \frac{1}{q} = \frac{1}{F} \qquad (1.3)$$

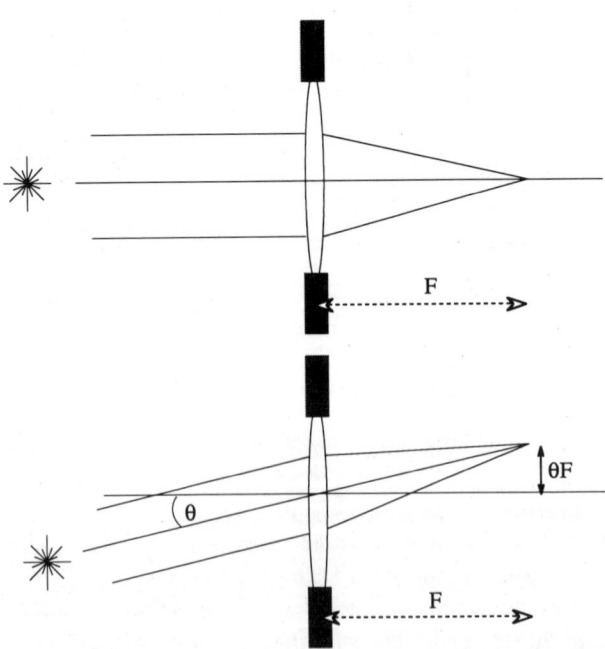

Figure 1.6. Off-axis focusing of a parallel bundle of rays.

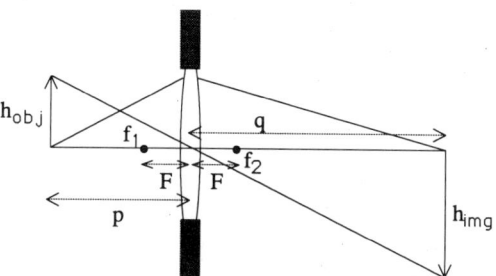

Figure 1.7. Imaging equation for single thin lens.

and

$$\mathfrak{M} \cong \frac{h_{\text{img}}}{h_{\text{obj}}} = -\frac{q}{p} \tag{1.4}$$

where p is the distance from the object to the lens (defined to be positive to the left of the lens) and q is the distance from the lens to the image (defined to be positive to the right of the lens). The magnification $\mathfrak{M}$ is defined as the ratio of the image height, h_{img}, to object height, h_{obj} (both heights defined to be positive above the optic axis). Because of the signs involved in the definitions of these quantities, and because a single lens produces an inverted image, there is a minus sign in Eq. (1.4)

Equations (1.3) and (1.4) answer the questions: Given an object and a lens, where is the image formed, and how big is the image? For example, given a lens with a focal length of 5 cm and an object located 20 cm to the left of the lens, we find for the image distance and the magnification:

$$q = \frac{1}{((1/F) - (1/p))} = 6.67 \text{ cm} \tag{1.5}$$

$$\mathfrak{M} = -\frac{6.67}{20} = -0.33 \tag{1.6}$$

For single-lens imaging, Eqs. (1.3) and (1.4) are sufficient to determine the first-order image size and location for a thin lens. We can examine some limiting cases of these equations:

as $p \to \infty$, $q \to F$; object at infinity, image at the focus

as $p \to F$, $q \to \infty$; object at the front focus; image at infinity

We can also verify from Eq. (1.3) that the minimum distance between real object and real image (p and q both positive) is $4F$. In this symmetrical case,

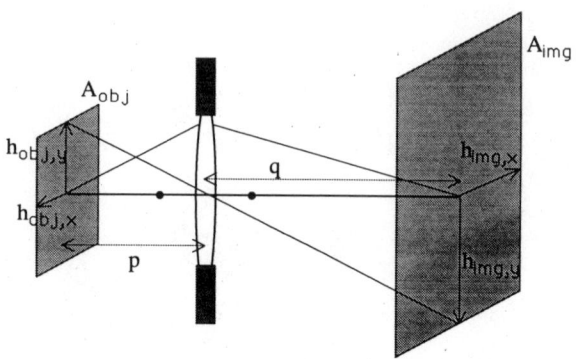

Figure 1.8.   Area magnification.

the object and image distance are both $2F$, and the magnification is -1. We can also verify from Eq. (1.3) that two possible pairs of p and q exist for a given distance $(p + q) = z_{o,i} > 4F$ between object and image. From the quadratic equation

$$p = \frac{z_{o,i}}{2} \pm \frac{\sqrt{z_{o,i}^2 - 4z_{o,i}F}}{2} \tag{1.7}$$

These two object distances will produce configurations of reciprocal magnification. In the previous example, the total length $z_{o,i} = 26.67$ cm. Another solution with the same total length is $p = 6.67$ and $q = 20$, which yields a magnification $\mathfrak{M} = -3$.

It should be noted that the magnification factor $\mathfrak{M}$ defined in Eq. (1.4) is a linear magnification, and that if one wants to compare the areas of the object and image as seen in Fig. 1.8, the ratio used is the square of the linear magnification:

$$\frac{A_{img}}{A_{obj}} = \frac{(2h_{img,y})(2h_{img,x})}{(2h_{obj,y})(2h_{obj,x})} = \mathfrak{M}^2 = \left(\frac{q}{p}\right)^2 \tag{1.8}$$

1.4. THICK LENSES AND LENS COMBINATIONS

Thick lenses are individual lenses with thickness not small compared to the focal length, or combinations of thin lenses. For thick lenses, the focal length can no longer be measured from the focal point to the center of the lens, but must now be measured from the focal point to the principal point. To locate the principal points of a thick lens (Fig. 1.9):

1. Use a ray that enters parallel to the optic axis from the right.

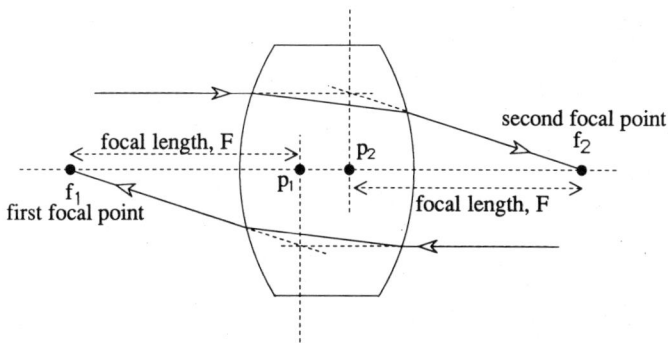

Figure 1.9. Principal planes.

2. Project the entering ray and the exiting ray until they intersect.
3. Construct a perpendicular line to intersect the optic axis.
4. The intersection of this line and the optic axis is the first principal point p_i.
5. The second principal point p_2 can be found in a similar fashion, by using a parallel ray entering from the left.

For thin lenses, both principal points coincide at the lens. The focal length $F \equiv \overline{p_1 f_1} = \overline{p_2 f_2}$ is the same on either side of the lens, given that the medium on each side has the same refractive index.

To calculate the aggregate focal length of two thin lenses, with focal lengths f_1 and f_2 and a separation $z_{\text{separation}}$, we can use the following formula:

$$F_{\text{combination}} = \left[\frac{1}{F_1} + \frac{1}{F_2} - \frac{z_{\text{separation}}}{F_1 F_2} \right]^{-1} \qquad (1.9)$$

As seen in Fig. 1.10, the focal length is measured from the appropriate principal plane, not from either of the individual lenses. As we let $z_{\text{separation}} \to 0$, we obtain the formula for two lenses in contact:

$$\frac{1}{F_{\text{combination}}} = \frac{1}{F_1} + \frac{1}{F_2} \qquad (1.10)$$

For example, two thin lenses in contact each with a focal length of 3 cm will result in a combination focal length of 1.5 cm.

A useful special case of the two-lens combination is the relay lens pair seen in Fig. 1.11, where the object is placed one focal length to the left of the first lens, and the image is formed at one focal length to the right of the second lens. The separation of the lenses will affect both the location of the principal planes and the focal length of the combination. However, because the object

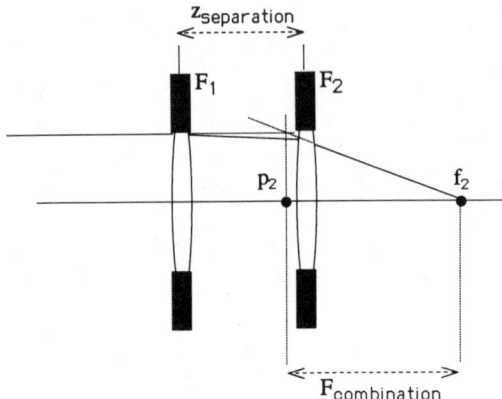

Figure 1.10. Focal length of two-lens combination.

is located at the front focal point of the lens 1, rays from the on-axis object point will be parallel to the optic axis in the space between the lenses. The lens separation will thus not affect the magnification of the relay, which is

$$\mathfrak{M}_{\text{relay}} = \frac{h_{\text{img}}}{h_{\text{obj}}} = -\frac{F_2}{F_1} \tag{1.11}$$

For increasing separation, the diameter of the second lens must increase to allow for an object of finite size to be imaged without losing the edges (vignetting). Consider a ray from the top of the object to the bottom of the image in Fig. 1.11. As the lenses are moved apart, the size of lens 2 must increase to pass this ray.

Afocal telescopes are two-lens systems that are intended to be used with the object approximately at infinity. The lenses are separated by the sum of their focal lengths, so that focal length is indeterminate using Eq. (1.9). This means that no single-lens equivalent to an afocal telescope exists. Parallel rays are

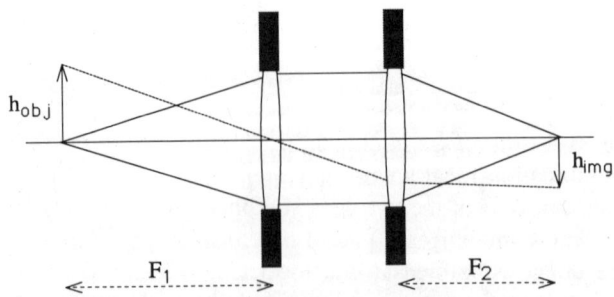

Figure 1.11. Two-lens combination: Relay lens pair.

input, and parallel rays are output. No single lens can accomplish this. Afocal telescopes do not produce an image by themselves, but provide an angular magnification, $\mathfrak{M}_{angular}$:

$$\mathfrak{M}_{angular} \equiv \frac{\theta_{output}}{\theta_{input}} = -\frac{F_1}{F_2} \tag{1.12a}$$

$$|\mathfrak{M}_{angular}| = \frac{D_{input}}{D_{output}} = \left|\frac{F_1}{F_2}\right| \tag{1.12b}$$

where Eq. (1.12b) follows from Eq. (1.12a) by similar triangles.

Figure 1.12 illustrates two basic afocal telescopes. The first, called the *Keplerian telescope*, combines two lenses with positive focal length. The angular magnification for the astronomical telescope is negative, because both focal lengths are positive, as illustrated in Eq. (1.12a). The second is known as the *Galilean telescope*, which has a positive-focal-length first lens and a negative-focal-length second lens. Thus, by Eq. (1.12a), the angular magni-

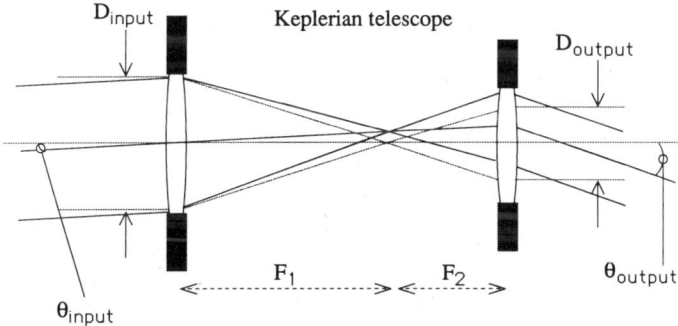

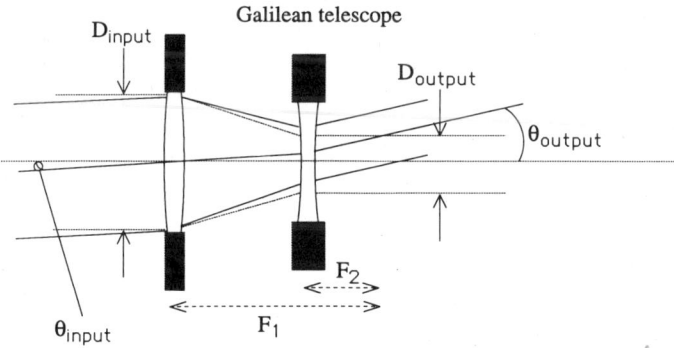

Figure 1.12. Two basic afocal telescopes.

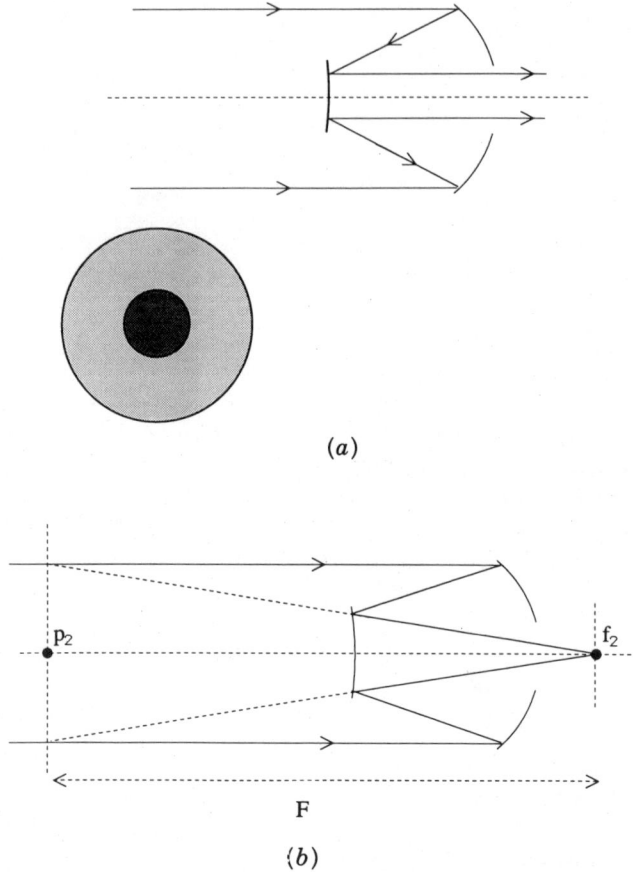

Figure 1.13. (*a*): Reflective Galilean afocal telescope; (*b*) Cassegrain objective.

fication for the Galilean telescope is positive. Galilean telescopes are more compact for a given angular magnification than are astronomical telescopes.

A variant of the Galilean telescope that uses two reflective elements is seen in Fig. 1.13*a*. This two-mirror afocal telescope has an annular obscured aperture because of the obstruction caused by the secondary mirror. If the spacing between the elements or the power distribution between the elements is adjusted, the two-mirror combination can be configured to have positive overall power (capability to focus parallel rays), in which case the combination is termed a *Cassegrain system* (Fig. 1.13*b*). A Cassegrain system is compact; the physical length of the system is shorter than the focal length.

Afocal telescopes often used in infrared systems to change the FOV. The afocal telescope acts in combination with a positive detector lens, shown in Fig. 1.14, to form the final image on the detectors. The focal length of the

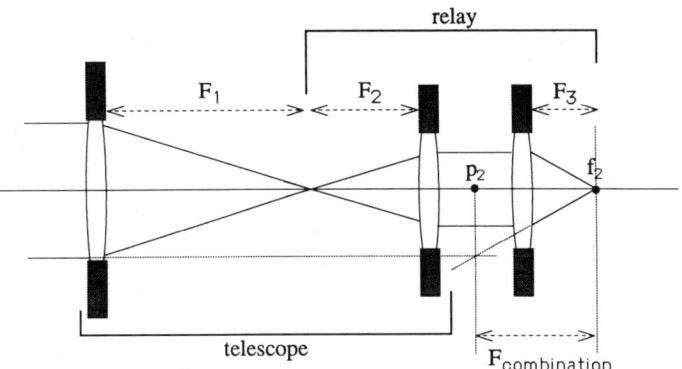

Figure 1.14. Afocal telescope with detector lens.

three-lens system (measured from the principal plane to the focal plane) is given by

$$F_{\text{combination}} = \frac{F_1 \times F_3}{F_2}$$

$$= |\mathfrak{M}_{\text{angular}}| \times F_3$$

$$= F_1 \times |\mathfrak{M}_{\text{relay}}| \tag{1.13}$$

Once the system is built, the detector lens remains fixed, but the afocal telescope can be changed.

1.5. STOPS AND PUPILS

The size and location of the various apertures in an optical system impact both the image quality and the radiometry. We consider here a single lens, with an aperture located at the lens, as seen in Fig. 1.15.

An axial ray begins at the on-axis point in the object, and ends at the on-axis point in the image. The aperture stop is the surface in the optical system that limits the angle over which the lens will accept rays from the on-axis object point. The marginal ray is a particular axial ray that starts at the on-axis point in the object, goes through the edge of the aperture stop, and ends at the on-axis point in the image.

The marginal ray is used to define an important parameter, the *f*-number (*F/#*). For the case of a single lens with a stop at the lens, seen in Fig. 1.15, the *F/#* describes the amount of aperture used compared to one of the following distances: object distance, image distance, or focal length. In this case, *F/#* is

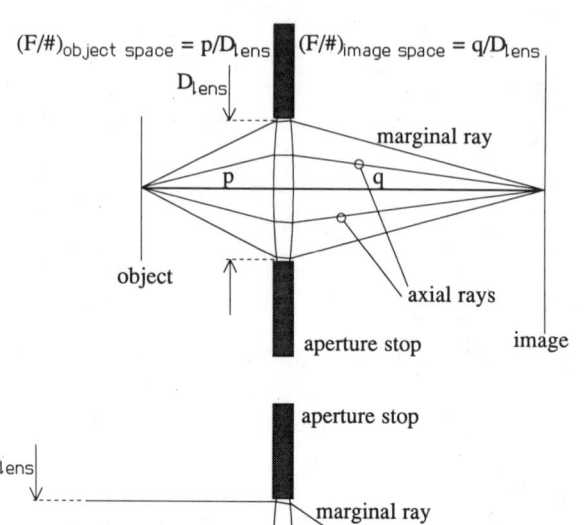

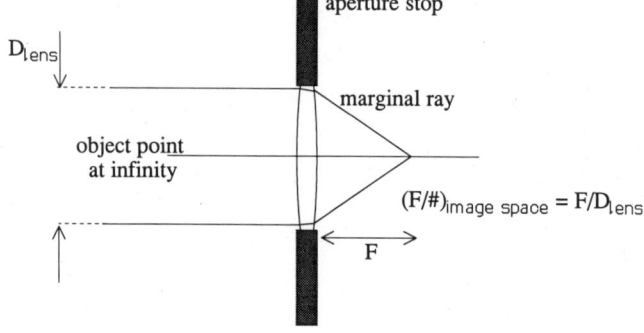

Figure 1.15.   *F/#* for single lens with stop at lens.

defined in one of three ways:

$$(F/\#)_{\text{object space, object not at } \infty} = \frac{p}{D_{\text{lens}}} \tag{1.14a}$$

$$(F/\#)_{\text{image space, object not at } \infty} = \frac{q}{D_{\text{lens}}} \tag{1.14b}$$

$$(F/\#)_{\text{image space, object at } \infty} = \frac{F}{D_{\text{lens}}} \tag{1.14c}$$

The larger the *F/#*, the less relative aperture is used. As seen in Fig. 1.16, an *F/#* = 1 implies that $F = D_{\text{lens}}$. An *F/#* = 3 (written as *F*/3) implies that $D_{\text{lens}} = (1/3) \times F$.

A more general configuration, which has the aperture stop buried inside of a compound-lens system, is seen in Fig. 1.17. The marginal ray is still the particular axial ray that goes through the edge of the aperture stop. Looking in from the object side of the system, we do not see the aperture stop directly, but rather the entrance pupil (enp), which is the image of the aperture stop

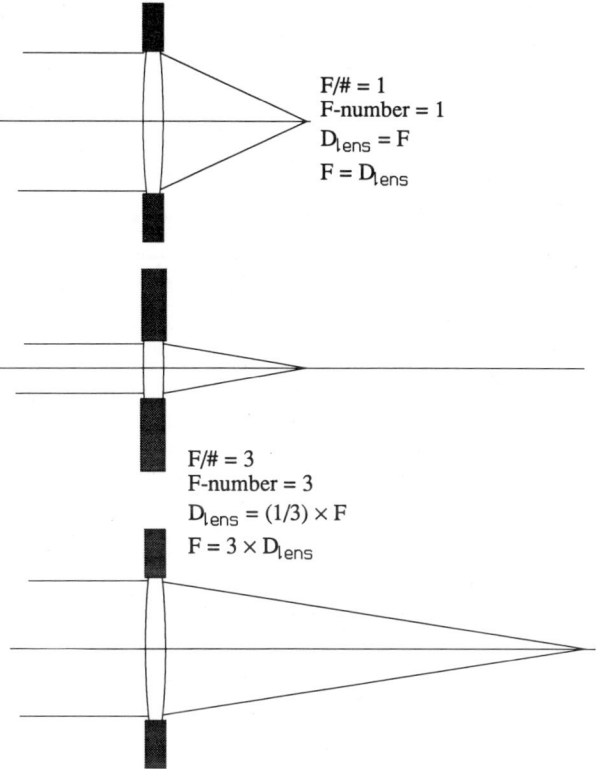

Figure 1.16. Comparison of *F*/3 and *F*/1.

formed by the lenses that precede it in the system. Similarly, the exit pupil (exp) is the image of the aperture stop formed by the lenses that follow it in the system, and is what we would see if we looked back into the system from the image side. The sizes and locations of the entrance pupil and exit pupil can be found using the imaging equations, Eqs. (1.3) and (1.4). The projection of the marginal ray appears to go through the edge of the entrance pupil and exit pupil because the pupils are images of the aperture stop. For this situation, the *F*/# is similarly defined by the marginal ray, using the diameters of the entrance and exit pupils, and the corresponding distances.

$$(F/\#)_{\text{object space, object not at }\infty} = \frac{z_{o,\text{enp}}}{D_{\text{enp}}} \tag{1.15a}$$

$$(F/\#)_{\text{image space, object not at }\infty} = \frac{z_{\text{exp},i}}{D_{\text{exp}}} \tag{1.15b}$$

$$(F/\#)_{\text{image space, object at }\infty} = \frac{z_{\text{exp},f2}}{D_{\text{exp}}} = \frac{F}{D_{\text{enp}}} \tag{1.15c}$$

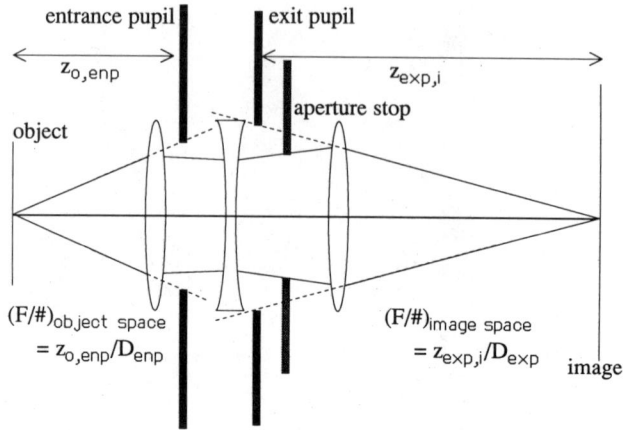

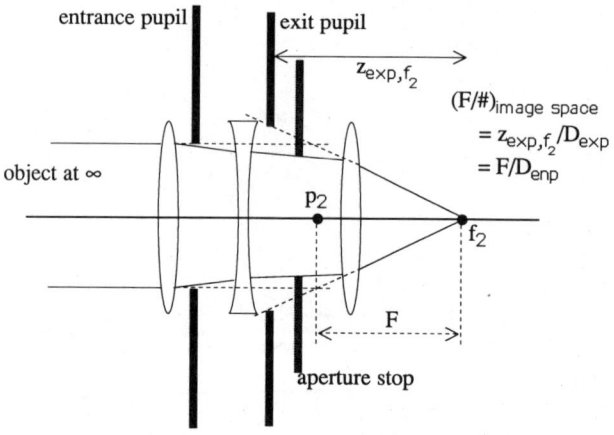

Figure 1.17. *F/#* for a compound-lens system.

Equation (1.15c) uses the definition of the focal length for a compound-lens system as the distance from the principal point to the focal point. The image-space *F/#*s of Eqs. (1.14) and (1.15) are related to another parameter called the *numerical aperture* (NA) as: NA $= 1/(2F/\#_{\text{image space}})$.

To obtain the best performance from infrared detectors, they must be cooled to cryogenic temperatures (typically 77 K). Only a portion of the optical train can be cooled to such temperatures because of limited cooling capacity of the refrigeration system. The detector is housed in a vacuum enclosure called a *Dewar*, which is usually the only part of the system to be cooled. Within the Dewar a cooled aperture called a *cold shield* is placed adjacent to the detector plane, to limit the angle over which the detectors receive radiation. The cold shield passes radiation from scene sources and blocks radiation from internal sources.

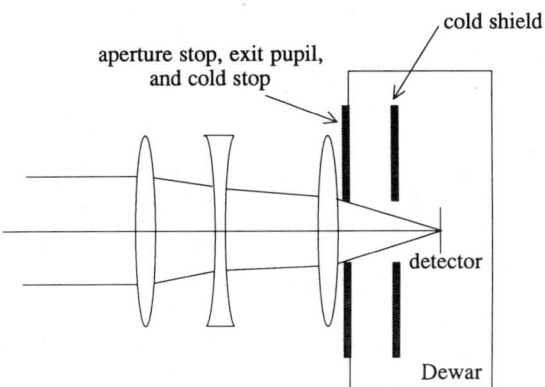

Figure 1.18. Cold shield and cold stop.

The degree to which the cold shield accomplishes this task is quantified by a parameter called the *cold-stop efficiency*. Cold-stop efficiency is the percentage of total flux reaching the detector that comes from sources in the scene. If the cold shield is placed at a plane that is an image of the aperture stop, then the detectors will receive flux only from scene sources (100% cold-stop efficiency, assuming no scattered radiation), and the cold shield is called a *cold stop* (Fig. 1.18).

The FOV is the angular coverage of an optical system. FOV can be defined as full angle or half-angle. Using the half-angle FOV convention,

$$\text{FOV}_{\text{half-angle}} = \left| \tan^{-1}\left(\frac{h_{\text{obj}}}{p}\right) \right| = \left| \tan^{-1}\left(\frac{h_{\text{img}}}{q}\right) \right| \tag{1.16}$$

The angles in Eq. (1.16) are equal because of graphical raytrace rule 3. As seen in Fig. 1.19, the FOV is usually determined by the finite size of the

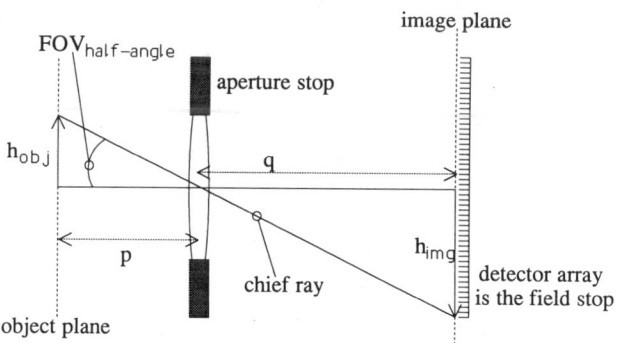

Figure 1.19. Illustration of field of view.

image-recording medium (a detector array, a piece of film, or an individual detector). The surface that limits the FOV is called the *field stop*. One special ray associated with the FOV is the chief ray, which starts at the maximum extent of the object and goes through the center of the aperture stop, and ends at the maximum extent of the image.

The $F/\#$ and the FOV affect both flux transfer and image quality. A small $F/\#$ and large FOV both promote high flux-transfer efficiency, but the image quality suffers (large aberrations). A system with a large $F/\#$ and a small FOV will tend to have better image quality (smaller aberrations), but would tend to have less flux reaching the image plane.

This relationship is defined by the so-called $A\Omega$ *product*, also called the *optical invariant* or the *throughput*. The $A\Omega$ product has the same value at any location in the optical system, and is the product of an area and a solid angle. A solid angle (measured in steradians) is defined as an area divided by the square of the distance to that area. The marginal ray and the chief ray are the special rays used to define $A\Omega$, as seen in Fig. 1.20.

$$A\Omega = A_{obj}\Omega_{enp} = A_{enp}\Omega_{obj} = A_{img}\Omega_{exp} = A_{exp}\Omega_{img} \qquad (1.17a)$$

From the definition of solid angle

$$A\Omega = A_{obj}\frac{A_{enp}}{(z_{o,enp})^2} = A_{enp}\frac{A_{obj}}{(z_{o,enp})^2} = A_{img}\frac{A_{exp}}{(z_{exp,i})^2} = A_{exp}\frac{A_{img}}{(z_{exp,i})^2} \qquad (1.17b)$$

which becomes

$$A\Omega = \frac{A_{enp}A_{obj}}{(z_{o,enp})^2} = \frac{A_{img}A_{exp}}{(z_{exp,i})^2} \qquad (1.17c)$$

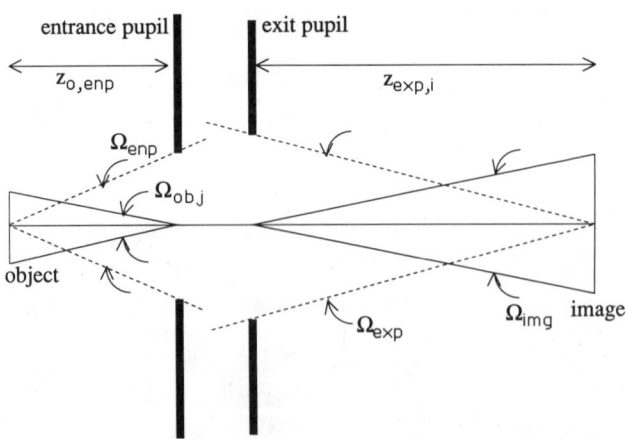

Figure 1.20. $A\Omega$ product.

To show the equivalence of the two expressions in Eq. (1.17c), we demonstrate, for the special case of the aperture stop located at the lens, that the preceding expression reduces to

$$A\Omega = \frac{A_{\text{lens}}A_{\text{obj}}}{p^2} = \frac{A_{\text{img}}A_{\text{lens}}}{q^2} \tag{1.18}$$

which can be seen to be true by the relationship in Eq. (1.18) and defines area magnification. The invariance of $A\Omega$ is true for the more general case of Eq. (1.17c) also. The amount of flux (W) transferred from object to image is related to the $A\Omega$ product as follows:

$$\phi = L \times A\Omega \tag{1.19}$$

where L is the radiance (Wcm^{-2} sr^{-1}) of the source, a quantity proportional to the source brightness.

Frequently, we think of the object as being imaged onto the detector or detector array; however, there is a useful alternative viewpoint. Consider the imaging of the detector onto the object plane. The image of the detector in the object (the footprint of the detector on the object) is the region of the object that contributes flux to the detector. The detector acts as the field stop in this situation. For the case of Fig. 1.21, where the object is far enough away that $q \approx F$, the linear dimension of the footprint $y_{\text{footprint}}$ is

$$y_{\text{footprint}} = \frac{y_d p}{F} \tag{1.20}$$

where y_d is the linear dimension of the detector (or detector array).

It is also of interest to relate FOV and magnification for the case of object approximately at infinity. As seen in Fig. 1.22, for this case the FOV $= y_d/F$. For a given detector dimension, a long focal length lens gives a narrow FOV

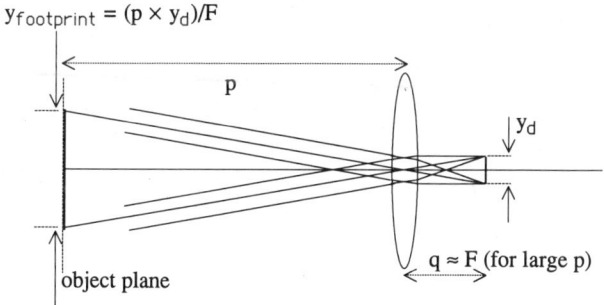

Figure 1.21.   Geometry for calculation of the footprint where $q \approx F$.

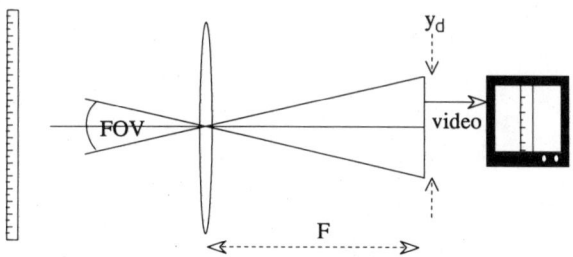

Long F gives narrow FOV and relatively high magnification.

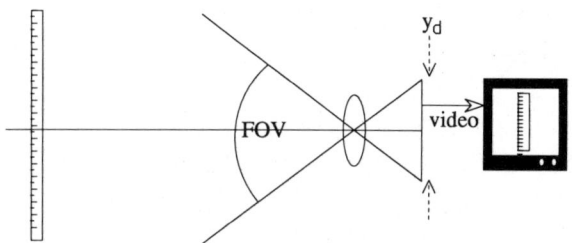

Short F gives wide FOV and relatively low magnification.

Figure 1.22.   Relationship of magnification to FOV.

and a relatively high magnification, while a short focal length lens gives a wide FOV and a relatively low magnification. Because magnification $\mathfrak{M} = -q/p$, a longer focal length implies a longer image distance, which yields a higher magnification. Another way of thinking of this is that whatever portion of object space is imaged onto the detectors will appear as the full width of the system display.

Field lenses increase the FOV of a system, typically a single-detector system. For a detector operating at a fixed noise level (e.g., amplifier noise), if the detector can collect a larger amount of flux, the signal-to-noise ratio increases. A field lens increases system FOV by imaging the aperture stop of the system onto the detector, redirecting rays that would otherwise miss the detector. As seen in Fig. 1.23, this increases the FOV without making the detector larger. The detector is imaged at the aperture stop, not at the object plane. This configuration is good for flux-collection systems (like radiometers), but not for imaging systems because the object is not imaged onto the detector, but rather into the field lens.

If the field lens is moved all the way to the image plane it becomes an immersion lens, as seen in Fig. 1.24, which also increases the FOV for a given

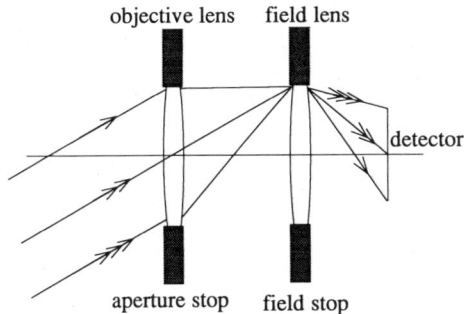

Figure 1.23.   Field lens used in a single-detector system to increase FOV.

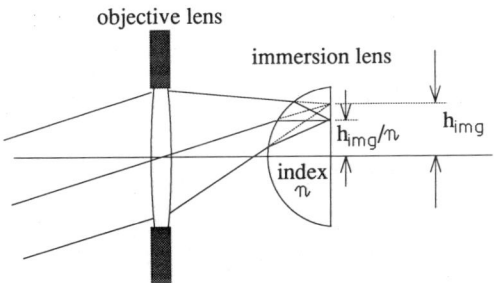

Figure 1.24.   Immersion lens.

detector size, but now the object is imaged onto the detector, so this system can form an image.

1.6. SCANNING

Usually, a two-dimensional object FOV must be covered. Because infrared detectors are more expensive in two-dimensional array formats than in one-dimensional array formats or single-element formats, scanning techniques have been developed that allow the sharing of detectors over the FOV in a time-multiplexed manner. Scanned systems are more complex mechanically than are staring systems, because of the moving mirrors required to accomplish the scan motions.

Three basic scan formats apply for a two-dimensional FOV, as seen in Fig. 1.25. In a raster scan, a single detector is swept in a two-dimensional raster. The raster motion requires at least two separate scan mirror motions. The moving footprint of the individual detector on the object at any instant is called the instantaneous field of view (IFOV). A parallel scan uses a sufficient number of detectors to cover the vertical FOV. The entire array is scanned in the hor-

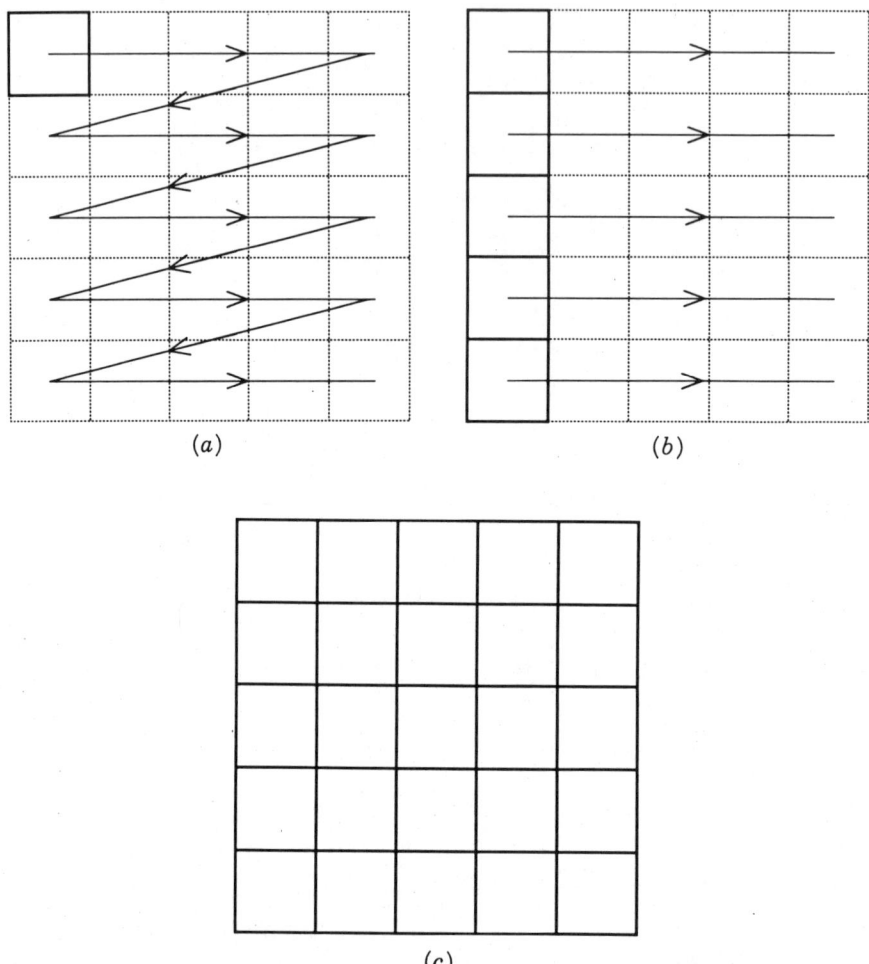

Figure 1.25.   Scan formats: (a) serial; (b) parallel; (c) staring system.

izontal direction, requiring only one scan mirror motion. A staring system has enough detectors to cover all portions of the FOV simultaneously. We discuss the systems-level implications of the choice of scan formats in Chapter 14.

The movement of the IFOV is accomplished by a moving mirror or mirrors in the optical path of the imager system. As a conceptual example, recall that, for a plane mirror, the angle of incidence equals the angle of reflection; it is then possible to verify the positions of the IFOVs in Fig. 1.26a. More complex scan patterns require more complex mirror arrangements. For instance, a two-dimensional motion of the IFOV would require an additional orthogonal mirror motion. In practice, the question arises of scan mirror location. If the scan mirror is placed on the object side of the system, as in Fig. 1.26a, then the

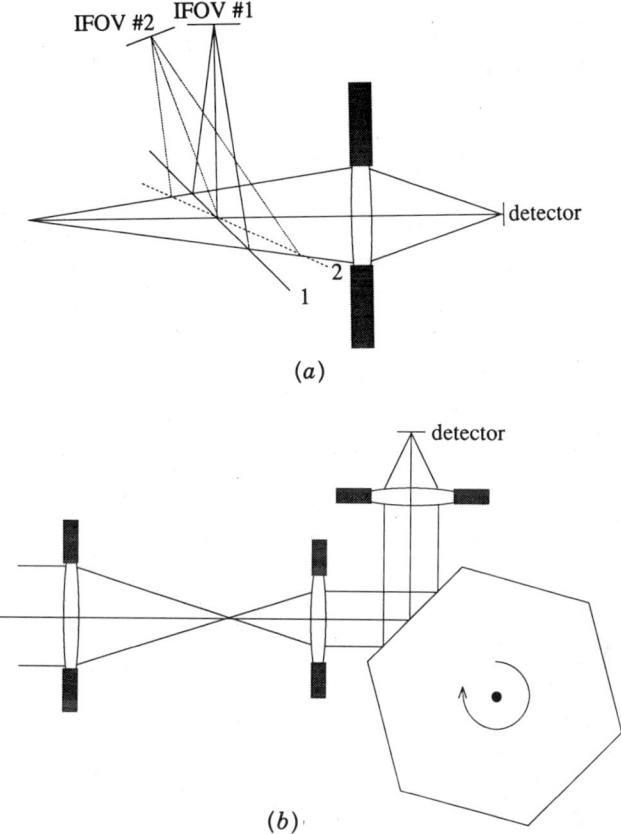

Figure 1.26. (*a*) Object-side scanner; (*b*) image-side scanner.

scan mirror is relatively large. If the scan mirror is placed on the image side of the system, then the scan mirror can generally be made smaller, with less mass to move. A disadvantage of image-side scanning is that the size of the front-end optics grows. The reason for this is seen in Fig. 1.26*b*, where the scan mirror is placed in collimated space on the image-plane side of the system. To cover the required FOV, the size of the rear lens of the afocal telescope must allow the off-axis ray bundles to pass without vignetting (loss of rays at the edge of an off-axis ray bundle). If the diameter of this lens is increased, then the front lens diameter must increase proportionally, to keep the same telescope magnification, consistent with Eq. (1.12b).

Three common image cosmetic artifacts are associated with infrared systems: scan noise, shading, and narcissus. Scan noise is caused by the variation of the amount of internal radiation received by the detectors as a function of scan position. Scan noise is caused by the interaction of the scan system and the self-radiation from room-temperature internal housings or self-radiation from room-temperature optical components (lenses or mirrors). In Fig. 1.26a, position 2 has more mirror area contribution than does position 1. Also, the different scan positions make use of different regions of the optical elements, which possibly have a different emissivity or differing amounts of surface contamination that cause additional self-radiation. This effect is not seen in visible-wavelength systems because system components at room temperature do not exhibit appreciable self-radiation at visible wavelengths. Scan noise can also be caused by vignetting in an image-side scanner, which reduces the amount of flux received by the detectors from off-axis portions of the scene. Vignetting is generally not an issue for object-side scanners.

Shading is a gradual falloff in scene brightness toward the edges of the display, and is caused by a cosine-to-the-fourth-power dependence of the effective brightness of a uniform source. We derive this radiometric effect in Chapter 2. Shading is controlled by optical design techniques that keep the chief ray angle small in image space.

Figure 1.27. Image showing narcissus effect.

Narcissus results from a reflected image of the cold detectors being formed back on the detectors. In narcissus, this reflected image is fairly well focused, resulting in a localized portion of the detected scene having less flux than do other portions of the scene. If the cold reflection is not well focused, there is an approximately uniform decrease in flux level over all portions of the scanned scene. The well-focused reflected image can arise at an appropriately curved surface in the optical train. This cold reflection generally occurs at the center of the scan, because the alignment needed to produce the reflection of the detector back onto itself usually occurs only at the center of the scan. For other angles the detector is more likely to receive reflected radiation from warmer structures. Narcissus is typically noticed as a dark region in the center of the video display, superimposed on the scene, as seen in Fig. 1.27. Narcissus is controlled by antireflection coatings on the optical elements, and by optical design techniques that ensure that any reflected images of the detector are out of focus at the detector plane.

1.7. IMAGE QUALITY

Thus far, we have made the assumption of perfect imaging, that is, that points in the object map to points in the image. However, because of the effects of diffraction and aberrations, points in the object are mapped to finite-sized blur spots in the image. Better image quality is obtained for smaller blur spots.

Diffraction is a fundamental consequence of the wave nature of electro-magnetic radiation. A well-corrected imaging system is capable of forming a minimum blur spot size that is related to the $F/\#$ of the marginal rays that converge to the image point. This minimum blur spot size is characteristic of diffraction, and can only get larger (yielding poorer image quality) from the effects of other image defects known as aberrations. The best performance obtainable from an optical system is called *diffraction-limited performance*. Assuming radiation with a wavelength λ and an imaging system with an $F/\#$ defined according to either Eqs. (1.14b) or (1.14c), we find the diameter of the blur caused by diffraction as

$$d_{\text{diffraction}} = 2.44\lambda F/\# \qquad (1.21)$$

The diffraction pattern of a point source has some ring structure in it, as seen in Fig. 1.28, and the diameter quoted in Eq. (1.21) is the diameter out to the first dark ring, which encloses 84% of the flux in the diffraction pattern.

For instance, a diffraction-limited $F/2$ system using infrared radiation of $\lambda = 10\ \mu$m will form a spot diameter of 48 μm. The same system operating at $F/8$ would form a spot diameter of 192 μm.

The effects of diffraction may also be expressed in angular terms, as seen in Fig. 1.29. The diffraction-limited angular blur, β, is the spot diameter (either

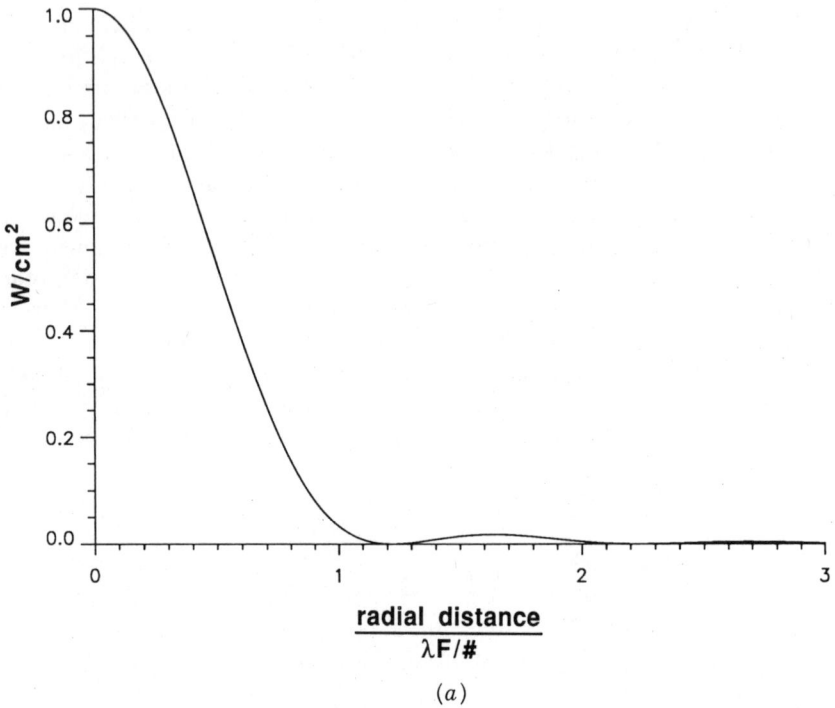

$$\frac{\text{radial distance}}{\lambda F/\#}$$

(a)

Figure 1.28. (a) Radial profile of diffraction-limited irradiance distribution.

in object space or image space) divided by the distance from the aperture stop to that blur spot, for a single-lens system with the aperture stop at the lens:

$$\beta = \frac{2.44\lambda(F/\#)}{\text{distance from aperture stop to blur spot}}$$

$$= \frac{2.44\lambda(F/D)}{F} = \frac{2.44\lambda(q/D_{lens})}{q} = \frac{2.44\lambda(p/D_{lens})}{p}$$

$$= \frac{2.44\lambda}{D_{lens}} \tag{1.22}$$

The spot diameter in object space $(2.44\lambda p/D_{lens} = \beta \times p)$ is interpreted as the projected size of the diffraction blur to the object plane. This represents the minimum resolution feature, referred to object space. For compound lenses, the angular blur β is different in object space and image space, and is the blur spot diameter divided by the distance from the exit pupil to the image-plane blur spot, or the distance from the entrance pupil to the object-plane blur spot:

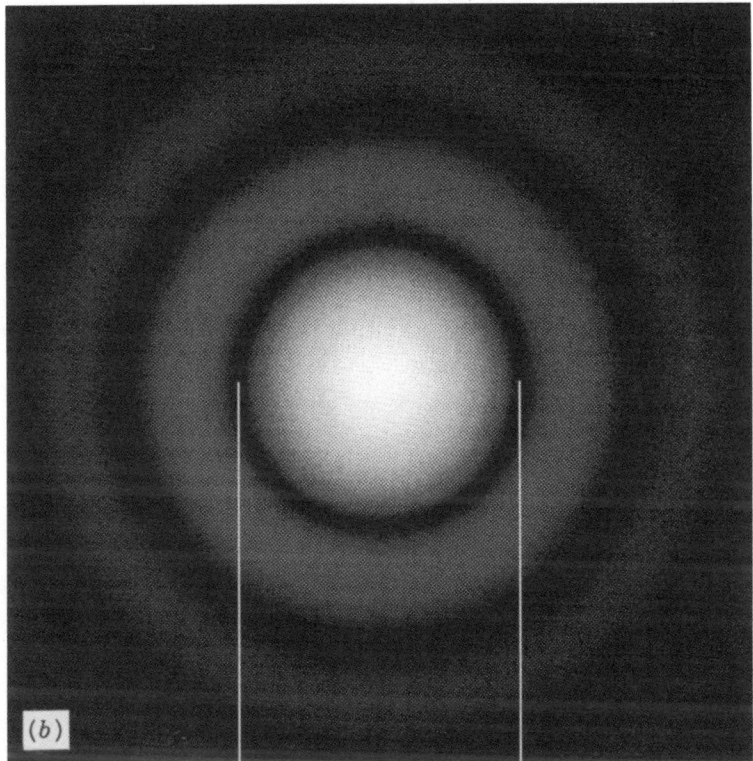

Figure 1.28. (*b*) Intensity image of diffraction-limited irradiance distribution. Lines indicate diameter to first zero.

$$\beta_{\text{obj}} = \frac{2.44\lambda}{D_{\text{enp}}} \tag{1.23a}$$

$$\beta_{\text{img}} = \frac{2.44\lambda}{D_{\text{exp}}} \tag{1.23b}$$

The corresponding expressions for the spot diameter $d_{\text{diffraction}}$ in object space and image space for a compound lens are

$$d_{\text{diffraction, obj}} = \frac{2.44\lambda}{D_{\text{enp}}} \times z_{o,\,\text{enp}} \tag{1.24a}$$

$$d_{\text{diffraction, img}} = \frac{2.44\lambda}{D_{\text{exp}}} \times z_{\text{exp},\,i} \tag{1.24b}$$

$$d_{\text{diffraction, img}} = \frac{2.44\lambda}{D_{\text{exp}}} \times z_{\text{exp},f_2} = \frac{2.44\lambda}{D_{\text{enp}}} \times F \qquad (1.24c)$$

which are consistent with Eqs. (1.14) and (1.21).

Aberrations arise because of the inability of optical systems with a finite number of spherical surfaces to produce geometrically perfect imaging for all possible object points. All aberrations worsen with decreasing $F/\#$ and/or increasing FOV. Aberration-control considerations generally dictate a minimum-allowable $F/\#$ and a maximum-allowable FOV. However, the system also has flux-transfer requirements in addition to field-coverage constraints. A basic tradeoff exists between the amount of flux in the image and the image quality. Two powerful free variables are available to the optical designer: the distribution of power between the surfaces of any element (lens bending) and the

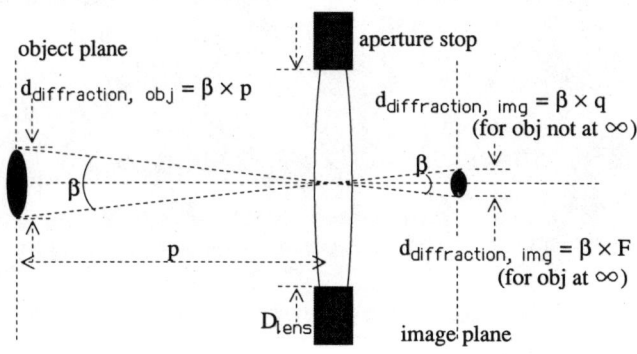

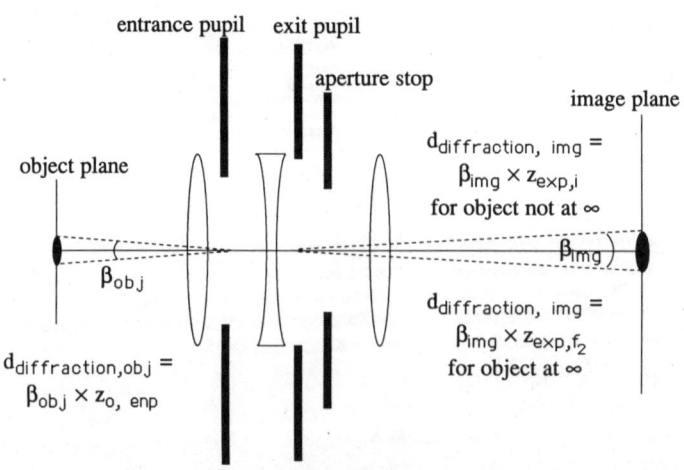

Figure 1.29. Diffraction-limited resolution angle.

location of the aperture stop. Neither of these affects the focal length of the lens, and both have large impact on image quality. After these variables are exploited, further optimization of image quality usually involves the creation of more degrees of freedom: for example, curvatures, thicknesses, and indices of refraction. A more complex system, with smaller aberrations, is the result.

In the optical-design process, the benefit gained by correcting to make the aberration spot size much smaller than the diffraction-limited spot size is negligible. The larger diffraction spot size at infrared wavelengths makes it possible for infrared-optical systems to be simpler in form than visible-wavelength optical systems of similar aperture and field requirements.

Systems operating at a low $F/\#$ have the smallest diffraction-limited spot sizes, and thus the best potential performance. However, the effects of aberrations generally become worse (producing larger spot sizes) as the $F/\#$ decreases. Thus, low $F/\#$ systems are harder to correct to a diffraction-limited level of performance.

1.8. MATERIALS CONSIDERATIONS

One basic material parameter is refractive index, n, defined as the ratio between the velocity of an electromagnetic wave in free space to its velocity in the material under consideration. Typically, refractive index varies with the wavelength of the radiation, a phenomenon called *material dispersion*. A plot of refractive index as a function of wavelength for common infrared-transmitting materials is shown in Fig. 1.30.

Whenever a ray crosses a boundary between two materials having a different refractive index, there will be some power transmitted and some power reflected. The direction of the transmitted ray is determined by Snell's law; the direction of the reflected ray is determined by the law of reflection. The distribution of power between the transmitted and reflected components is determined by the Fresnel equations:

$$r = \left(\frac{n_2 - n_1}{n_2 + n_1} \right)^2 \tag{1.25a}$$

$$t = \frac{4n_2 n_1}{(n_2 + n_1)^2} \tag{1.25b}$$

where n_1 and n_2 are the indices on each side of a boundary; r is the reflected power per surface (reflectance); and t is the transmitted power per surface (transmittance). These equations are used as follows:

$$\phi_{\text{reflected}} = \phi_{\text{incident}} \times r \tag{1.26a}$$

$$\phi_{\text{transmitted}} = \phi_{\text{incident}} \times t \tag{1.26b}$$

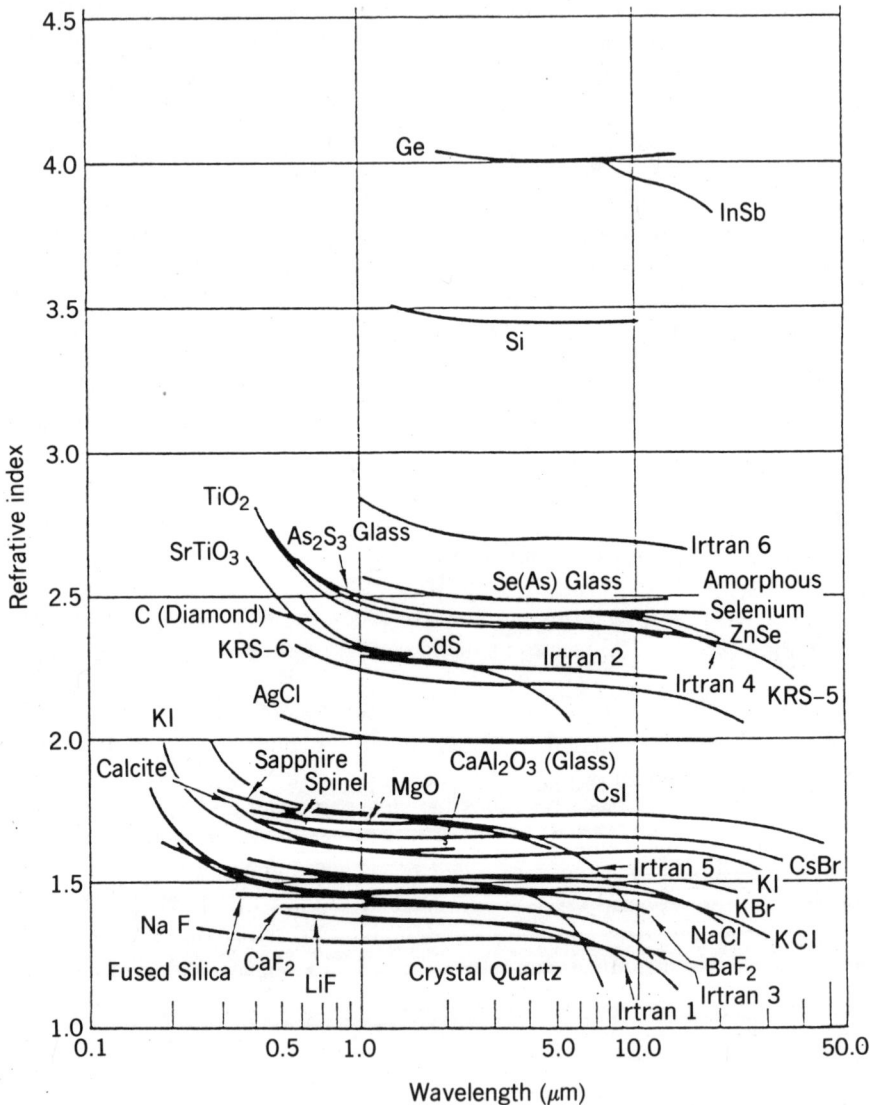

Figure 1.30. Index of refraction vs. wavelength. (Originally appeared in W. L. Wolfe and G. J. Zissis, *The Infrared Handbook*. Reprinted by permission of the Environmental Research Institute of Michigan.)

where the symbol ϕ represents flux (watts). If we add Eqs. (1.25a) and (1.25b), we find their sum is unity, which means that these equations assume that absorptance is zero. These equations are also approximations for small angle of incidence. At larger angles of incidence, the reflectance generally increases.

As an example of the use of the Fresnel equations, refer to Fig. 1.31. We

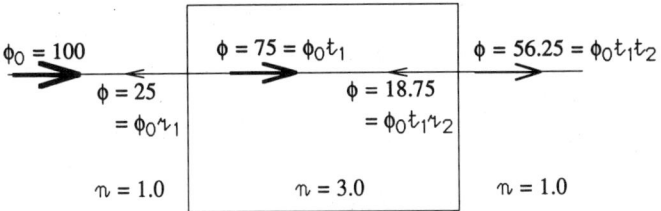

Figure 1.31. Fresnel equation example.

assume that the window material has a refractive index of 3.0, which, as we have seen in Fig. 1.30, is a typical index for infrared materials. From Eqs. (1.24a) and (1.24b), we find that $r = 0.25$ and $t = 0.75$. Using Eqs. (1.25a) and (1.25b) for both surfaces in the window, we find that the power transmitted through the window is $0.5625 \times \phi_{incident}$, the rest of the power having been lost to reflections at the first and second surface. For the high-index materials typically used in infrared applications, we find that thin-film antireflection coatings are often used to reduce such losses.

The choice of materials for transmissive elements in the optical train is primarily dictated by the spectral dependence of the transmittance of the material under consideration. Acceptable transmissive materials require low absorption over the wavelength band of interest. Absorption takes power out of a beam of radiation, and raises the temperature of the material by Joule heating. Absorption can be described by an attenuation coefficient, a (usually a function of wavelength), having units of inverse distance

$$\phi(z) = \phi_0 e^{-az} \qquad (1.27)$$

where z is the propagation distance from the origin of coordinates, ϕ_0 is the power in a beam of radiation at $z = 0$, and $\phi(z)$ is the power in the beam at a distance z.

We can define a quantity called internal transmittance, $t_{internal}$, as the transmittance through a distance in a medium, that does not include Fresnel reflection losses at the boundaries

$$t_{internal} \equiv \frac{\phi(z)}{\phi_0} = e^{-az} \qquad (1.28)$$

We can also define a quantity called *external transmittance*, $t_{external}$, as the transmittance through a block of material of length L, that includes Fresnel reflection losses at the boundaries:

$$t_{external} \equiv t^2 e^{-aL} = t^2 \times t_{internal} \qquad (1.29)$$

The distinction between internal and external transmittance is important when considering material specifications from a vendor. If the specification is in terms of external transmittance, there will be a baseline Fresnel transmittance, consistent with Eq. (1.24b).

Plots of external transmittance versus wavelength for a number of common infrared materials are shown in Fig. 1.32.

In Fig. 1.33 we plot internal transmittance of the atmosphere over a 1.8-

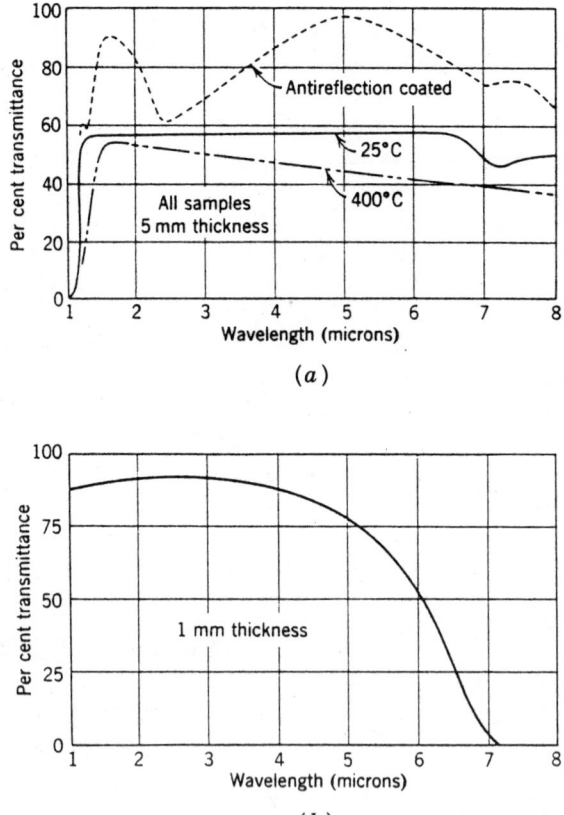

Figure 1.32. External transmittance vs. wavelength for various infrared materials. (a) Silicon. (Originally appeared in P. Kruse et al., *Elements of Infrared Technology*, 1962. Reprinted by permission of John Wiley & Sons, Inc. and Texas Instruments.) (b) Sapphire. (Courtesy of Linde Air Products, New York. Originally appeared in P. Kruse et al., *Elements of Infrared Technology*, 1962. Reprinted by permission of John Wiley & Sons, Inc.) (c) Germanium. (Courtesy of U.S. Air Force Wright Laboratory, Dayton, Ohio. Originally appeared in P. Kruse et al., *Elements of Infrared Technology*, 1962. Reprinted by permission of John Wiley & Sons, Inc.) (d) IRTRAN and arsenic trisulfide, for material thickness of 2 mm. (Data on IRTRAN courtesy of Eastman Kodak Company. Originally appeared in R. Hudson, *Infrared System Engineering*, 1969, John Wiley & Sons, Inc. Reprinted by permission of R. Hudson.)

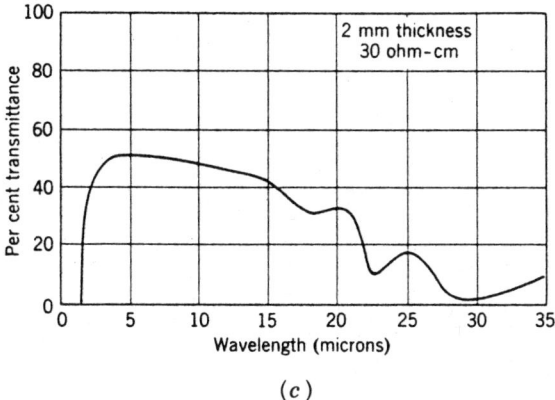

(c)

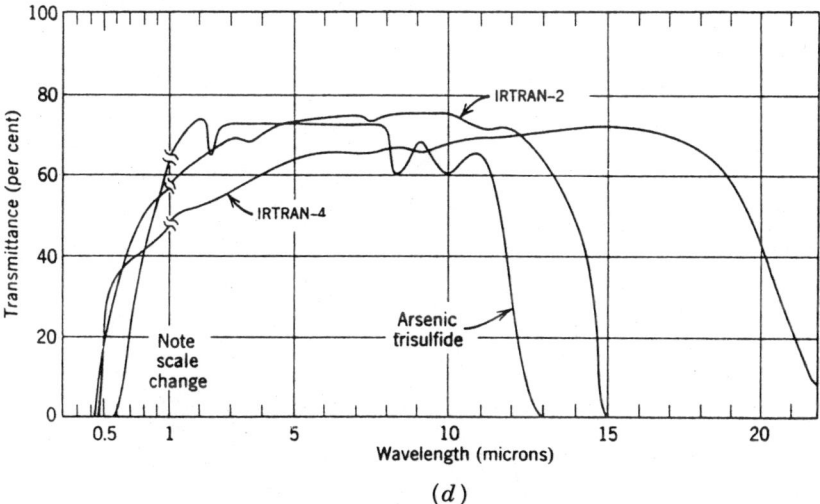

(d)

Figure 1.32. (*Continued*)

km path as a function of wavelength. Note that there are three main atmospheric windows: one that encompasses the visible portion of the spectrum, along with most of the shortwave infrared up to 2 μm; the second window encompasses most of the 3- to 5-μm band (the middle-wave infrared); and a third that ranges from 8 to 14 μm (the long-wave infrared). These windows are significant in that source, detector, material, and system technologies have evolved separately for each window, optimizing the properties of these components for operation in a particular spectral region.

Figure 1.34 shows several plots of reflectance versus wavelength for commonly used mirror materials. All of the metals exhibit a higher reflectance at long wavelengths.

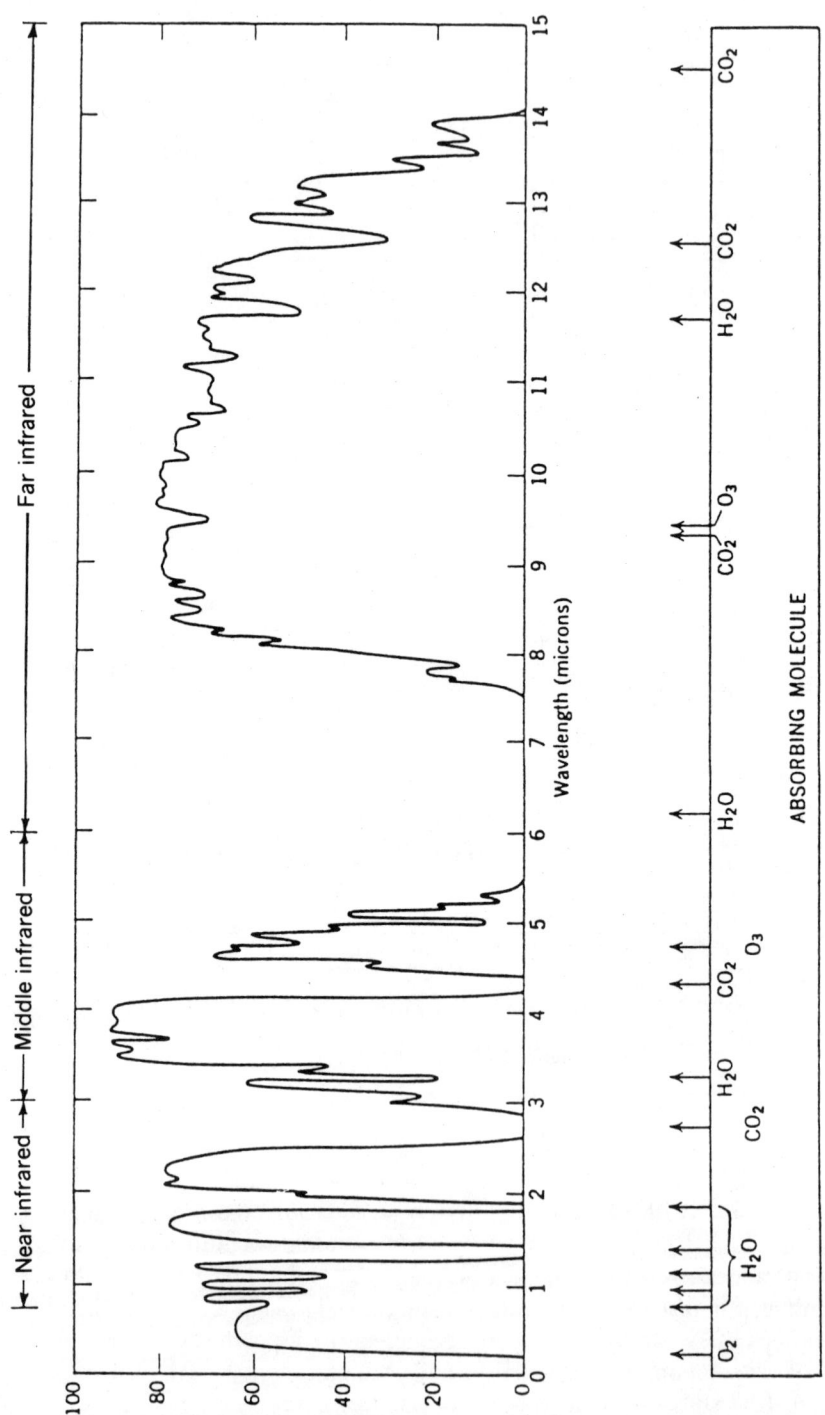

Figure 1.33. Transmittance vs. wavelength for the atmosphere, over a 6000 ft horizontal path at sea level, containing 17 mm of precipitable water. (Originally appeared in R. Hudson, *Infrared System Engineering*, 1969, John Wiley & Sons, Inc. Reprinted by permission of R. Hudson.)

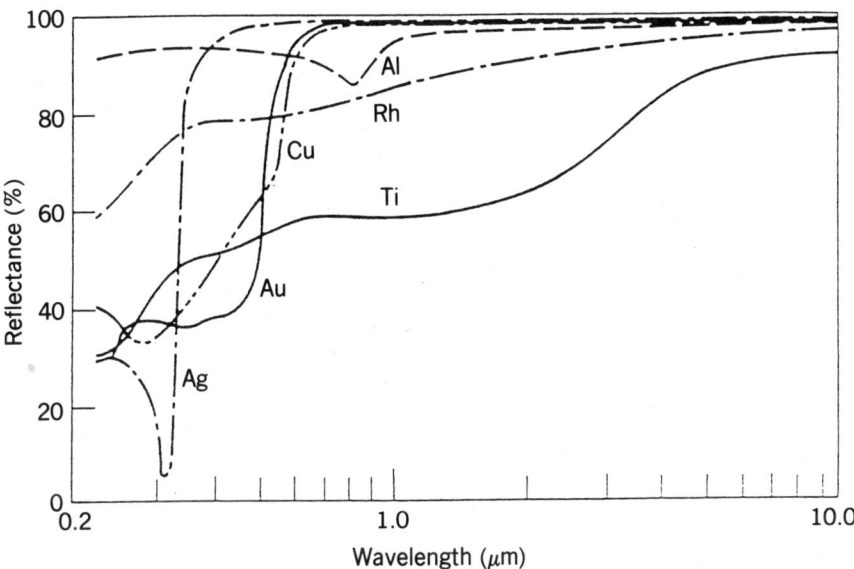

Figure 1.34. Mirror reflectance vs. wavelength. (Originally appeared in W. L. Wolfe and G. J. Zissis, *The Infrared Handbook*. Reprinted by permission of the Environmental Research Institute of Michigan.)

BIBLIOGRAPHY

Born, M., and E. Wolf, *Principles of Optics*, Pergamon, New York, 1975.

Hecht, E., and A. Zajac, *Optics*, Addison-Wesley, Reading, MA, 1974.

Hudson, Richard D., Jr., *Infrared System Engineering*, Wiley, New York, 1969.

Jenkins, F. A., and H. E. White, *Fundamentals of Optics*, McGraw-Hill, New York, 1976.

Kruse, Paul W., Laurence D. McGlauchlin, and Richmond B. McQuistan, *Elements of Infrared Technology: Generation, Transmission, and Detection*, Wiley, New York, 1962.

Longhurst, R. S., *Geometrical and Physical Optics*, Longman, New York, 1973.

Carter, D. R., *Burle Electro-Optics Handbook*, Burle Industries, Inc., Lancaster, PA, 1974.

Smith, W. J., *Modern Optical Engineering*, McGraw-Hill, New York, 1966.

Wolfe, W. L., and G. J. Zissis, *The Infrared Handbook*, Optical Engineering Press, Bellingham, WA, 1990.

PROBLEMS

1.1. What is the external transmittance of a block of material with index 2 immersed in air? The material has no absorption losses.

1.2.   A window made of material that has absorption is immersed in air. The external transmittance of the slab (at normal incidence) is 0.1. The attenuation coefficient of the material is 0.1/mm. The thickness of the window, in the direction of propagation, is 15 mm. What is the window's index of refraction?

1.3.   A thin lens has a focal length of 120 mm. The object is a transparency 35 mm (horizontal) by 24 mm (vertical), which is centered on the axis, at a distance of 30 cm to the left of the lens. Where will the image be formed and what will its dimensions be? What object distance would you use to obtain a magnification reciprocal to that previously obtained?

1.4.   Parallel rays from a point source at infinity are incident on a lens, making an angle of 5 degrees with the optic axis. If the lens has a focal length of 10 cm, how far from the axis will the rays come to focus?

1.5.   Using a 500-mm focal length lens that is focused on a target 400 m away, what are the dimensions of the region of object space that will be recorded on a piece of film having dimensions 35 mm by 24 mm? What is the image distance and the linear magnification? Repeat for a 50-mm focal length lens. Which situation has the larger FOV?

1.6.   If an $F/3$ system has its aperture stop made larger or smaller by 50% in diameter, what are the new $F/\#$s?

1.7.   A camera consisting of a 100-mm lens and detector array is focused on an object 1 km away. How much does the detector array need to be moved to focus on an object at 100 m? Is this motion of the detector array toward or away from the lens, compared to the situation where the lens was focused at 1 km?

1.8.   What is the FOV (full field; express in radians) of a system consisting of a 50-μm detector, a 10-power afocal telephoto, and a detector lens with $F = 20$ mm?

1.9.   A 2-cm-thick germanium window is measured to have an external transmittance of 38.5% at normal incidence. What external transmittance would you expect from a 10-cm-thick window of the same material?

1.10. If a laser beam of wavelength 0.9 μm is reflected successively off of ten aluminum mirrors, what fraction of the beam power would remain?

1.11. Given a detector array having 512-by-512 elements, where each detector is square, 15 μm on a side. Suppose you want to build a two-lens relay system to image a standard 35-mm slide transparency (dimensions of 35 mm by 24 mm) onto this detector array, without losing any of the image. If you already have a lens with a 25-mm focal length, what other lens would you need to do the job? Is there only one answer? Why?

1.12. Assume a lens of diameter 3 cm has a focal length of 10 cm. Given the location of the object plane (you will need to find this), the image plane of the lens is 12 cm to the right of the lens. A square detector is located in the image plane, of dimensions 20 μm by 20 μm. What are the dimensions of the footprint (at the object plane) of the detector? What is the diameter of the footprint (at the object plane) of the diffraction blur spot? Assume the usual criterion of 84% encircled energy for the diameter, and assume that the system operates in a narrow band around a wavelength of 5 μm.

1.13. A detector array consisting of 512 × 512 50-μm^2 pixels is looking at the earth from a height of 50,000 feet. The focal length of the optics is 2 m. What is the total size of the footprint of the array on the ground? What is the size of an individual pixel footprint on the ground?

1.14. What is the area of the detector footprint in object space for a system consisting of a 20-μm × 20-μm detector, a 5-power afocal telescope, and a detector lens with $F = 30$ mm. Express the area in square meters. Assume that the object plane is at a distance of 5 km from the system.

1.15. A particular tracking system has a focal length of 4 m. If the system is initially in focus for a point target at a distance of 10 km, how much will the image plane (a detector array) need to move along the axis to stay in focus if the target moves to a distance of 2 km?

1.16. What will be the diffraction-limited blur angle (full angle, express in radians) for a system with a 6-cm aperture diameter, which operates at the following wavelengths: 0.5 μm, 4 μm, 10 μm, 2 mm. In each case, what is the corresponding object-plane spot diameter, assuming that the object plane is 2 km away?

2

Radiometry

2.1. INTRODUCTION

This chapter covers the concepts of radiometry necessary for a quantitative understanding of flux transfer. The fundamental question to be answered is: Given the geometrical configuration of the source, detector, and any intervening optics, and given that the strength of the source is specified appropriately, how much energy from the source is collected and brought to the detector surface? This issue is critical to the overall signal-to-noise ratio achieved by the system.

Certain provisos and approximations allow us to simplify the calculations that apply to our discussion in this chapter. We specifically consider the radiometry of incoherent sources, such as thermal (blackbody) sources. Lasers and other sources that are partially or totally coherent are specifically excluded. Thus, the calculation of the energy distribution over some surface is the result of a scalar sum of energies, and not the magnitude of a vector sum of amplitudes, as would be necessary in the calculation of an interference pattern. Also, we will ignore the effects of diffraction, except in the consideration of unresolved point sources. Generally, we will make small-angle assumptions similar to those made for paraxial optics. The sine of an angle is approximated by the angle itself in radians, unless stated otherwise.

Many physical constants arise in this chapter. These are collected in Table 2.1 in an approximate form convenient for calculations.

2.2. SOLID ANGLE

In the work to follow, we will have occasion to use the concept of *solid angle*, measured in *steradians*. It is helpful to recall the definition of radians from plane geometry (see Fig. 2.1), in which the planar angular subtense θ is defined in radians (dimensionless), to be the ratio of arc length on a circle to the radius, s/r. Planar angular subtense can be thought of as the range of pointing angles from which one can go from the center of the circle to the specified arc. Any line in space, not necessarily the arc of the circle, may also be thought of as having an angular subtense. This is equivalent to the angular subtense of the

Table 2.1   Some Physical Constants

Speed of light in vacuum	$c \approx 3 \cdot 10^8$ meters/s
Boltzmann's constant	$k \approx 1.38 \cdot 10^{-23}$ joules/K
Planck's constant	$h \approx 6.6 \cdot 10^{-34}$ joule s
Stefan–Boltzmann constant	
Energy	$\sigma_e \approx 5.67 \cdot 10^{-12}$ watt/(cm^2 K^4)
Photon	$\sigma_q \approx 1.52 \ (10^{11})$ photon/(s cm^2 K^3)

Table 2.2   Rule-of-Thumb Approximation for Angular Measure

1 milliradian ≈ vehicle 15 feet long at distance of 2.84 miles
1 microradian ≈ vehicle 15 feet long at distance of 2841 miles
Diameter of sun's disc ≈ 0.5°
Diameter of moon's disc ≈ 0.5°
1 arc sec ≈ 4.85 μrad
1 arc min ≈ 291 μrad
1 degree ≈ 17.5 mrad

arc that just occludes the line of interest, as seen in Fig. 2.1. Table 2.2 lists some useful orders of magnitude and conversions for angular subtense.

Similarly, in three-dimensional geometry, one can define steradians of solid angular subtense by the ratio of an area on the surface of a sphere to the square of the radius, as shown in Fig. 2.2:

$$\Omega = \frac{A}{r^2} \tag{2.1}$$

A sphere contains $4\pi r^2/r^2 = 4\pi$ steradians of solid angle. As with the example for planar angular subtense, the solid angle subtended by a surface that is not on the sphere can be calculated by the subtense of that portion of the surface

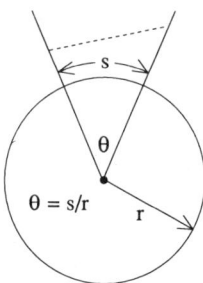

Figure 2.1.   Angular subtense θ measured in radians; s is arc length; dashed line subtends same angle as s.

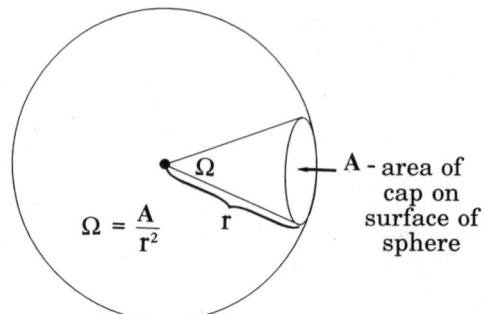

Figure 2.2. Solid angle subtense measured in steradians.

of the sphere that just occludes the surface of interest. Solid-angle subtense defines the range of pointing angles in two dimensions from which one can go from a point to a surface. The solid angle may be expressed in differential form as

$$d\Omega = \frac{dA}{r^2} \tag{2.2}$$

and is used to find the solid angular subtense of any surface of interest. If we use the spherical coordinate system seen in Fig. 2.3, and use $dA = r^2 \sin \theta \, d\theta \, d\varphi$, we can write an expression for the solid-angle subtense of a flat disc of planar half-angle θ_{max} as

$$\Omega = \int d\Omega = \int_0^{2\pi} d\varphi \int_0^{\theta_{max}} \sin \theta \, d\theta = 2\pi(1 - \cos \theta_{max}) \tag{2.3}$$

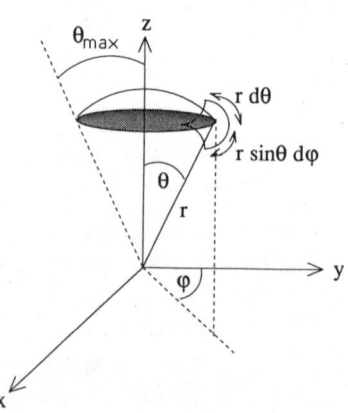

Figure 2.3. Relationship of solid angle to planar angle.

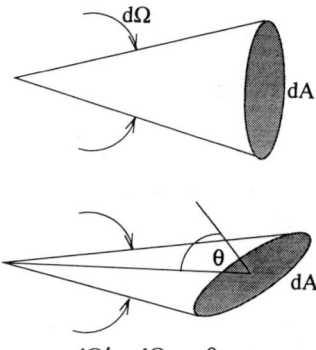

$d\Omega' = d\Omega \cos\theta$ **Figure 2.4.**   Solid angle of a tilted area.

If the disc is sufficiently small with respect to the line-of-centers distance (θ_{max} small in radians), then Eq. (2.3), for the solid-angle subtense of the flat disc, becomes the area of the disc divided by the square of the distance. In this limit, the difference between the disc and the portion of the surface of the sphere of equal area is negligible.

It is also helpful to note that if this small disc is tilted at some angle θ between the normal line of the disc and the line of sight (Fig. 2.4), then its solid angular subtense is reduced by a factor of cos θ.

$$\Omega_{\text{tilted surface}} = \left(\frac{A}{r^2}\right) \cos\theta \tag{2.4}$$

2.3.   RADIOMETRIC QUANTITIES

Quantitative characterization of radiant sources requires a precise system of units. Table 2.3 summarizes the pertinent quantities, symbols, and units. Examples of typical calculations with each quantity will follow. We will eventually characterize any source of interest in these terms, but for the moment we will assume that the strength of the source has already been specified.

A note is also in order regarding units. We will exclusively use units based on the SI system, though occasionally common usage dictates that these are not always MKS units for certain quantities. SI is the preferred system currently in use for radiometry, but the reader needs to be aware that a large body of literature uses other systems of units. A very complete conversion table is found in Wolfe and Zissis (1990).

The units listed in Table 2.3 are based on joule as the fundamental quantity, and are thus termed *energy derived*. An analogous set of quantities, based on number of photons, is termed *photon derived*. The photon-derived quantities are always written with a subscript q. The energy-derived quantities are written

Table 2.3   Radiometric Quantities

Symbol	Quantity	Units
Q_e	Energy	joule
ϕ_e	Flux	watt
I_e	Intensity	watt/sr
E_e	Irradiance	watt/cm^2
M_e	Exitance	watt/cm^2
L_e	Radiance	watt/(cm^2 sr)

with a subscript e if it is necessary to distinguish, or without a subscript if it is obvious from context.

For example, ϕ_q is the photon flux and is measured in photons/s. Similarly, I_q is termed photon-flux intensity, and is measured in photons/s-sr. A conversion between the two sets of units is easily accomplished using the relationship for the amount of energy carried per photon: $\mathcal{E} = hc/\lambda$. For example,

$$\phi_e(\text{joule/s}) = \phi_q(\text{photon/s}) \cdot \mathcal{E}(\text{joule/photon}) \tag{2.5}$$

This alternate set of quantities will be useful when we consider detectors that respond directly to photon events (so-called *photon detector*), rather than thermal energy (so-called *thermal detector*). Because the energy per photon $\mathcal{E} = hc/\lambda$, a long-wavelength photon carries less energy than a short-wavelength photon. Thus, the conversion between the two sets of units depends on λ. This affects the shape of functions that are plotted with respect to wavelength, depending on whether the vertical axis of the plot is in terms of energy-based or photon-based units.

Irradiance and exitance have the same units, but have different interpretations. Irradiance is the amount of power with respect to unit area that falls on a surface, while exitance is the amount of power per unit area that leaves a surface. Existance thus characterizes a self-luminous source that is producing energy, while irradiance characterizes a passive receiver surface. Exitance can be written in differential form as $M = \partial\phi/\partial A$, and can be approximated for small areas as $M = \phi/A$.

For example, a 1-mm^2 radiating surface has an exitance of 3 W/cm^2. What is the total radiated power?

$$\phi = A_s \times M = (0.1)^2 \times 3 = 30 \text{ mW} \tag{2.6}$$

This power is radiated into the entire hemisphere in front of the source (Fig. 2.5). It is not radiated uniformly in angle, however, as we will see in our discussion of radiance. The surface of the sphere is not uniformly irradiated.

For example, to calculate irradiance, assume that the irradiance falling on a (1-cm)^2 surface is 4 W/cm^2, and assume it is uniform over the surface. How

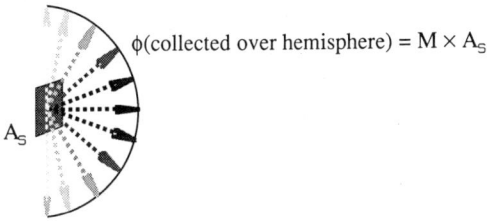

ϕ(collected over hemisphere) $= M \times A_s$

Figure 2.5. Exitance of a source.

much power is received by a 1-mm by 1-mm portion of the receiving surface?

$$\phi = A \times E = (0.1)^2 \times 4 = 40 \text{ mW} \tag{2.7}$$

Figure 2.6 shows that this represents the calculation of how much energy a small detector would receive in the image plane, if the image irradiance were given.

We now consider the concept of intensity, given in watts per solid angle, $I = \partial\phi/\partial\Omega$.

Intensity is the only quantity that can be used to specify the radiation from a point source. Because a point source has negligible area (compared to the square of the viewing distance), the quantities with area in the denominator cannot be used. Intensity can also be used to specify the watts per unit solid angle radiated from an extended source. As we will see in our discussion of radiance, intensity from an extended source will vary with the view angle.

Consider a point source with a uniform intensity of 3 W/sr. What is the total radiated power? If the total solid angle into which a point source radiates equals 4π, we find (as seen in Fig. 2.7)

$$\phi = I \times \Omega = 3 \times 4\pi = 12\pi \tag{2.8}$$

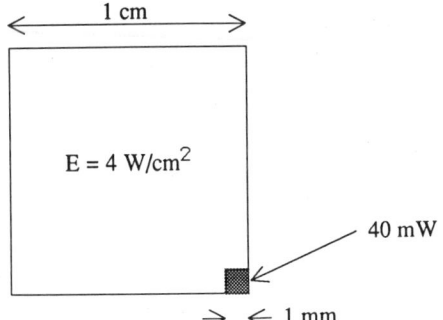

1 cm

$E = 4 \text{ W/cm}^2$

40 mW

1 mm

Figure 2.6. Uniform irradiance (W/cm^2) incident on surface.

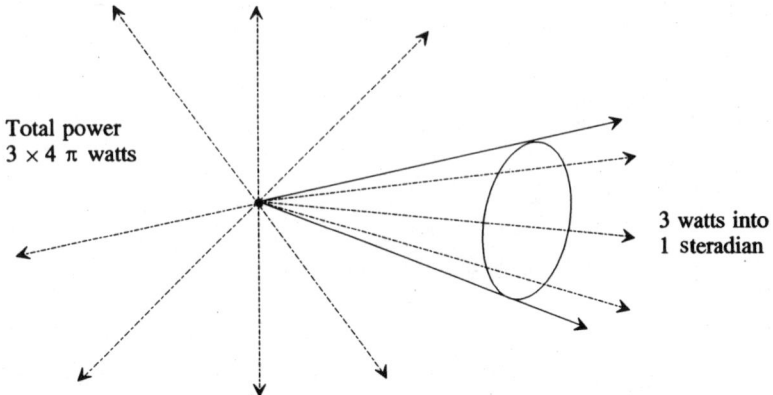

Figure 2.7. Intensity of a point source.

The total power radiated from a source is usually not a concern, because an optical system or a detector can collect only the power within a limited solid angle. We are usually concerned with flux collection over a limited solid angle; for example, how much power is radiated through a circular hole in a screen of radius 1 mm, which is 3 m from the point source? As Fig. 2.8 illustrates, we first calculate the solid angle of the hole, as seen by the point source:

$$\Omega = A/r^2 = \pi(1 \times 10^{-3} \text{ m})^2/(3m)^2 = 3.5 \times 10^{-7} \text{ sr} \qquad (2.9)$$

We then proceed to calculate power:

$$\phi = I \times \Omega = 3 \times (3.5 \times 10^{-7}) = 1.05 \times 10^{-6} \text{ W} \qquad (2.10)$$

A common rule of thumb is the "one-over-r-squared falloff," and this applies strictly to point sources. As seen in Fig. 2.9, we consider a receiver of

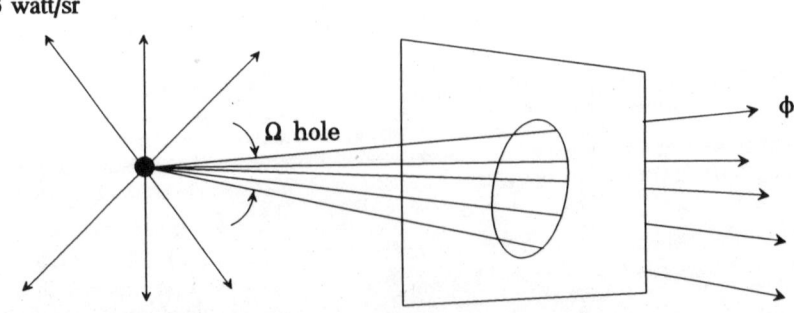

Figure 2.8. Power collected through an aperture.

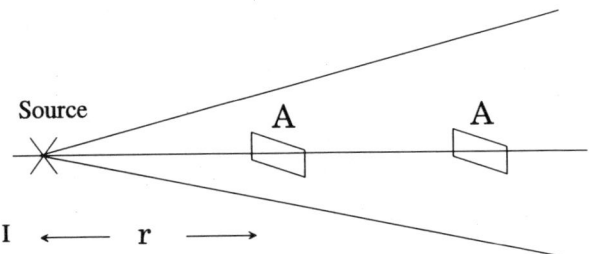

Figure 2.9. Irradiance falloff as a function of r^2 from source.

area A, which is placed at various distances from a point source having a uniform intensity I.

$$\phi = I \times A/r^2 \tag{2.11a}$$

$$E = \phi/A = I/r^2 \tag{2.11b}$$

Because the solid angle subtended by the detector falls off as $1/r^2$, the collected flux and the irradiance also decrease proportionally. This is an accurate and simple means of verifying the linearity of a detector, and is important for laboratory measurements.

2.4. RADIANCE

We now consider the concept of radiance. It is used to characterize an extended source, that is, one that has appreciable area compared to the square of the viewing distance. It is the amount of power radiated per unit projected source area per unit solid angle. In differential form:

$$L = \frac{\partial^2 \phi}{\partial A_s \cos \theta_s \, \partial \Omega_d} \tag{2.12}$$

Some additional explanation of this definition is necessary. Referring to Fig. 2.10, the notation $\partial^2 \phi$ indicates that the power received by the detector is differential with respect to both incremental projected area of the source and incremental solid angle of the detector. We can integrate separately with respect to either projected source area or detector solid angle, leaving a single differential in either case.

For example, rearranging the defining equation for radiance,

$$\partial^2 \phi = L \, \partial A_s \cos \theta_s \, \partial \Omega_d \tag{2.13}$$

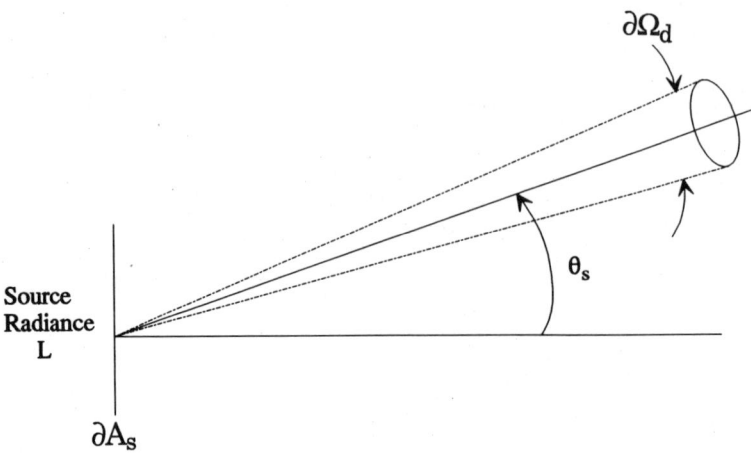

Figure 2.10.   Radiance of an extended source.

we integrate once with respect to source area to obtain intensity

$$I = \frac{\partial \phi}{\partial \Omega_d} = \int_{A_s} L \, \partial A_s \cos \theta_s \qquad (2.14)$$

Similarly, we integrate once with respect to detector solid angle using Eq. (2.13) to obtain an expression for exitance

$$M = \frac{\partial \phi}{\partial A_s} = \int_{\Omega_d} L \cos \theta_s \, \partial \Omega_d \qquad (2.15)$$

A Lambertian radiator is one whose radiance is independent of view angle θ_s. Radiance for a Lambertian radiator is the same in all directions from the emitting surface. Radiance is analogous to the visual sensation of perceived brightness of a surface. A non-Lambertian source will appear to change brightness as the angle θ_s is varied. (Examples of calculations involving non-Lambertian sources can be found in the problems at the end of this chapter.) We show later in this chapter that an ideal thermal source (blackbody) is perfectly Lambertian, while certain special diffusers also closely approximate the condition. An actual source is typically approximately Lambertian within a range of view angles θ_s that is less than 20 degrees.

However, the reader should note that even for a Lambertian source, the amount of flux per solid angle (intensity) will depend on θ_s. Referring to Eq. (2.14), the factor $L \cos \theta_s$ will come outside of the source-area integral whether or not L is a function of θ_s. Making the assumption of L independent of source position, we can complete the integration to yield:

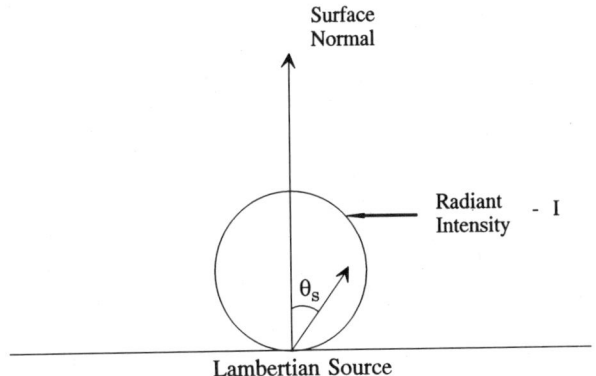

Figure 2.11. Radiant intensity as a function of θ_s for a Lambertian source.

$$I = \frac{\partial \phi}{\partial \Omega_d} = \int_{A_s} L \, \partial A_s \cos \theta_s = L A_s \cos \theta_s \qquad (2.16)$$

Note that intensity has a cosine dependence on θ_s for the case of radiance independent of angle, as seen in Fig. 2.11. This is because of the definition of radiance in terms of flux per unit projected area per unit solid angle. For non-Lambertian sources, the radiance L will be a function of angle itself, and the falloff of I with θ_s will be faster than $\cos \theta_s$.

For a planar Lambertian source, the following relationship exists between radiance and exitance:

$$M[\text{watt/cm}^2] = L[\text{watt/(cm}^2 \text{ sr)}] \pi[\text{sr}] \qquad (2.17)$$

The proportionality factor is π instead of 2π because of the integration of the projected area factor for various portions of the hemisphere into which the source radiates (Fig. 2.5). The hemisphere will be illuminated nonuniformly. To derive this proportionality factor, we can return to Eq. (2.15) and integrate

$$M = \frac{\partial \phi}{\partial A_s} = \int_{\Omega_d} L \cos \theta_s \, \partial \Omega_d = \int_0^{2\pi} d\varphi \int_0^{\pi/2} L \cos \theta_s \sin \theta \, d\theta$$

$$= 2\pi L \frac{1}{2} = \pi L \qquad (2.18)$$

where the Lambertian-source assumption has been used to pull L outside of the angular integrals. For a non-Lambertian source, the integration would yield a proportionality constant different from π.

For the simplest calculational example of radiance, we let θ_s be zero. Consider the geometrical configuration shown in Fig. 2.12, given an extended

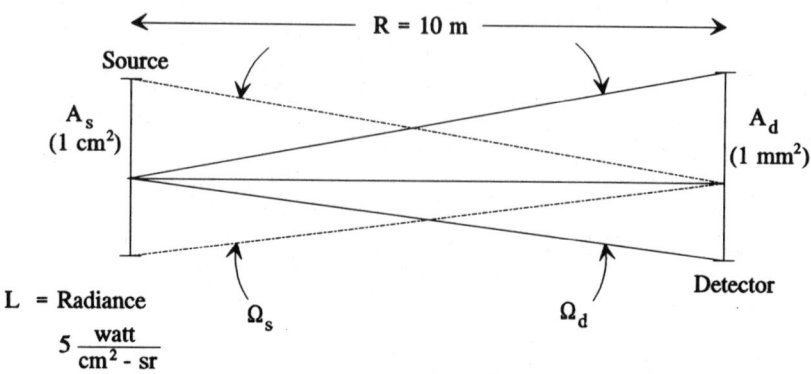

Figure 2.12.   Radiant power transfer from source to detector.

source of $A_s = (1 \text{ cm})^2$ that is specified by a radiance (L_e) of 5 W/cm²-sr. A detector of $A_d = (1 \text{ mm})^2$ is located at an axial distance of 10 m. How much power falls on the detector?

We first calculate the solid angle of the detector:

$$\Omega_d = (1 \times 10^{-3}m)^2/(10m)^2 = 10^{-8} \text{ sr} \qquad (2.19)$$

We multiply this solid angle by the area of the source and the radiance of the source to obtain the radiant power on the detector:

$$\phi_d = LA_s\Omega_d \qquad (2.20a)$$

$$\phi_d = 5(\text{W/cm}^2\text{-sr})(1 \text{ cm})^2 \times 10^{-8} \text{ sr} = 5 \times 10^{-8}\text{W} \qquad (2.20b)$$

We can note from Eq. (2.20a) that

$$\phi_d = LA_s\Omega_d = \frac{LA_sA_d}{r^2} = L\Omega_sA_d \qquad (2.21)$$

We are free to group the r with whichever area we choose, and thus the flux on the detector is expressed as the radiance of the source multiplied by an area-times-solid-angle ($A\Omega$) product. Two provisos are in order for this observation. First, a small-angle assumption was inherent for the approximation of solid angle of a flat surface by A/r^2. Second, the flux transfer is assumed to be unaffected by absorption losses in the system.

We can use our expression for the solid angle of a tilted receiver to write the expression for the flux transfer to an on-axis tilted detector (Fig. 2.13). The source normal is along the line of centers, so $\theta_s = 0$ in this case. We define the angle θ_d, which is the angle between the line of centers and the

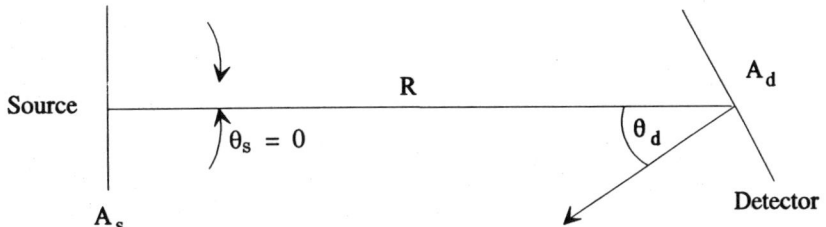

Figure 2.13. Radiant power transfer to a tilted detector.

normal of the detector surface. θ_d is nonzero in this example. We can write the following expression:

$$\phi_d = LA_s\Omega_d \tag{2.22}$$

Using Eq. (2.4), we get a solid angle of $\Omega_d = A_d \cos \theta_d/r^2$ for the tilted detector, and find

$$\phi_d = LA_s(A_d \cos \theta_d/r^2) \tag{2.23}$$

Thus the flux collected and the irradiance (ϕ/A_d) have both been decreased by a factor of $\cos \theta_d$.

Under the assumption of a flat Lambertian source, we now proceed to calculate the flux on the detector when both θ_s and θ_d are nonzero, leading to the so-called "cosine to the fourth law" (see Fig. 2.14). A cosine falloff factor arises at both the source and the receiver, and two cosine factors arise from the "stretch" of the line-of-centers distance, referenced to r, the original perpendicular distance to an on-axis receiver:

$$\phi_d = LA_s \cos \theta_s \frac{A_d \cos \theta_d}{(r/\cos \theta_s)^2} \tag{2.24}$$

We now consider the case of flux transfer in image-forming systems. Within the limitations of paraxial optics (small angles), flux transfer is described sim-

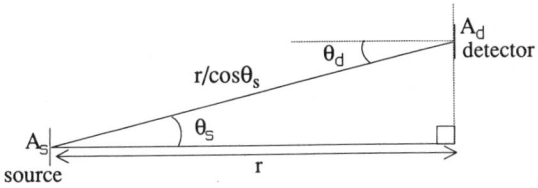

Figure 2.14. Cosine-to-the-fourth ($\cos^4$) law.

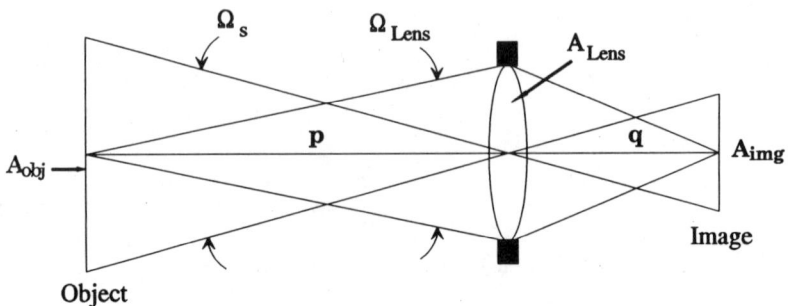

Figure 2.15.   Radiant power collected by an optical system.

ply in terms of the $A\Omega$ product. Referring to Fig. 2.15, a certain amount of flux ϕ is collected by the optical system. This flux may be calculated by a standard flux-transfer calculation, letting the lens aperture (A_{lens}) act as an intermediate receiver (in a more complex optical system, the entrance pupil is this intermediate receiver)

$$\phi = LA_{obj}\Omega_{lens} = LA_{lens}\Omega_{obj} \qquad (2.25)$$

Note: The product $A_{lens}\Omega_{obj}$ is the area-solid-angle product of the optical system. The entrance pupil area and the field of view can be used to get this $A\Omega$ product from lens specifications.

This flux collected will be reformatted because of the lens, and will form an image of the original object, at an appropriate magnification, the size of which can be found by the methods of Chapter 1. The calculation of image irradiance is important from the point of view of exposure levels for a detector or a film receiver. The irradiance may be found simply by dividing the flux collected in Eq. (2.25) by the image area:

$$\phi = LA_{lens}\Omega_{lens} = LA_{lens}\Omega_{obj} = L\frac{A_{lens}A_{obj}}{p^2} = L\frac{A_{lens}A_{img}}{q^2} \qquad (2.26)$$

where the last equality was obtained using $A_{img} = A_{obj}(q/p)^2$. We can now calculate the image irradiance E_{img}:

$$E_{img} = \frac{\phi}{A_{img}} = L\frac{A_{lens}}{q^2} \qquad (2.27)$$

When the chief ray of the optical system makes an appreciable angle with the optic axis (Fig. 2.16), a cosine-to-the-fourth falloff in image irradiance is seen, relative to the on-axis irradiance calculated in Eq. (2.27).

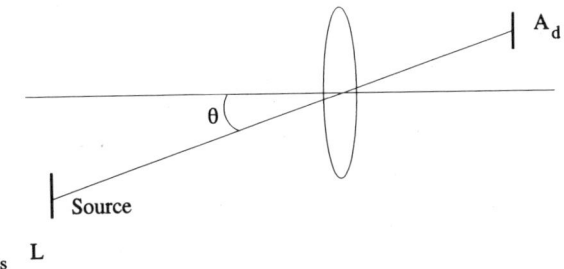

Figure 2.16. Cosine-to-the-fourth falloff in an imaging system.

2.5. SOURCE CONFIGURATIONS

2.5.1. Point Sources

Often the source object of interest for a detection system is effectively a point source (an unresolved object). Sources of this type are specified by their intensity (flux per solid angle). Referring to Fig. 2.17, we may simply calculate the amount of flux collected as

$$\phi = I\Omega_{\text{lens}} \tag{2.28}$$

Because of diffraction, at best only 84% of this flux is concentrated into a spot of diameter $2.44 \, \lambda q / D_{\text{lens}}$, which can be written as $2.44 \, \lambda F/\#$, using the definition of image space $F/\#$. The distribution of flux in this spot is not uniform, but follows the functional form seen in Fig. 2.18 for a diffraction-limited lens. An operational definition of "point source" is a source that can be made just a bit larger and not appreciably affect the size of the blur spot in the image.

If the point source is sufficiently far from the lens, the image will be formed approximately at the focus, at a focal length (F) distance behind the lens. In this case the image space $F/\#$ is F/D_{lens}. We obtain the following expression

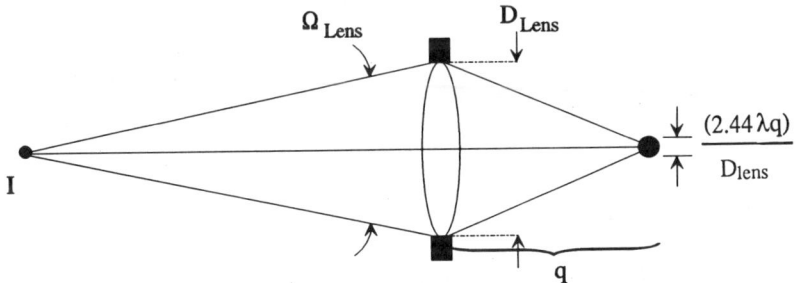

Figure 2.17. Point-source radiant power transfer.

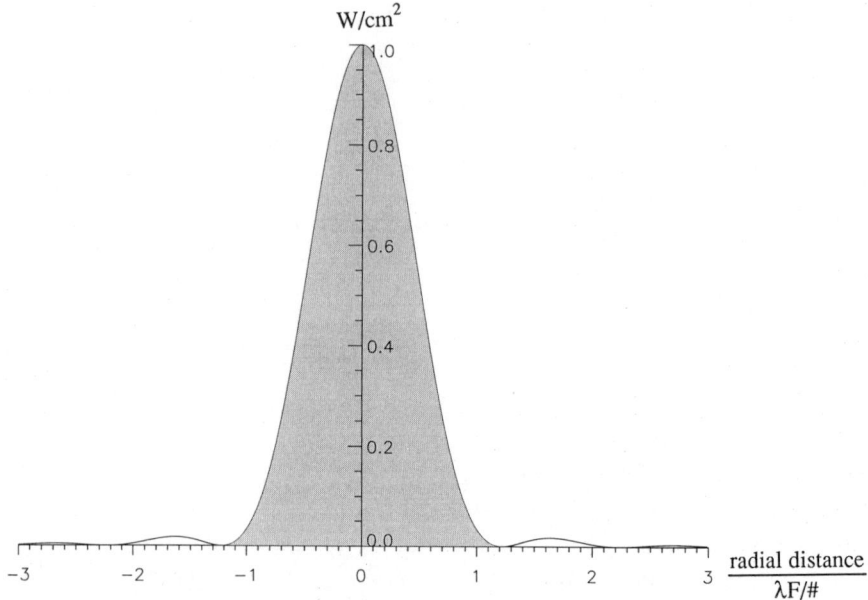

Figure 2.18. Airy-disc distribution of irradiance in the image plane for a point source. Shaded area corresponds to 84% of the flux in the image.

for the average irradiance in the diffraction-limited spot:

$$E = \frac{IA_{\text{lens}}}{p^2} \times \frac{0.84}{\frac{\pi}{4}\,[2.44\ \lambda F/\#]^2} \tag{2.29}$$

If the area of the collecting optic is held constant, the amount of flux collected is a constant. If the $F/\#$ is decreased by decreasing the focal length, then (provided that the system is still diffraction limited at the new $F/\#$) the irradiance will increase, because the same flux is concentrated into a smaller diffraction-limited spot.

2.5.2. Extended Sources

Another application of these techniques is to find the irradiance in front of a large Lambertian disc, which can be either an actual source or an intermediate source, such as a lens, as seen in Fig. 2.19. We integrate over the solid angle over which the detector views the extended source. This solid angle is bounded by the system's marginal rays, limited by the aperture stop.

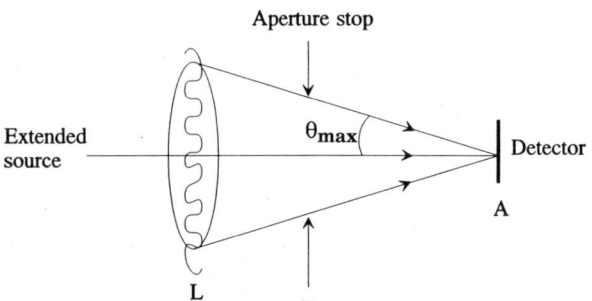

Figure 2.19. Extended source filling field of view.

We begin with the fundamental equation for radiometry:

$$\partial^2\phi_d = L\ \partial A_s \cos\theta_s\ \partial\Omega_d = L\ \partial A_s \cos\theta_s \frac{\cos\theta_d\ \partial A_d}{r^2} \tag{2.30}$$

Referring to Fig. 2.20, we integrate over source area and detector area, with the variable r as the line-of-centers distance between the incremental source and detector areas.

$$\phi_d = \int_{As}\int_{Ad} L\ \partial A_s \cos\theta_s \frac{\cos\theta_d\ \partial A_d}{r^2} \tag{2.31}$$

We identify the following relationships: $\theta_s = \theta_d \equiv \theta$; $dA_s = 2\pi r\ dr$; $r = z\tan\theta$; $r_{12} = z/\cos\theta$; $dr = d(z\tan\theta) = z(d\theta/\cos^2\theta)$; $dA_s = 2\pi z \tan\theta z(d\theta/\cos^2\theta)$. With these substitutions:

$$\phi_d = L\int_\theta\int_{Ad} 2\pi z \tan\theta z \frac{d\theta}{\cos^2\theta} \cos^2\theta \left(\frac{z}{\cos\theta}\right)^{-2} dA_d \tag{2.32a}$$

$$\phi_d = 2\pi L A_d \int_0^{\theta_{max}} \cos\theta \sin\theta\ d\theta = 2\pi L A_d \frac{1}{2}\sin^2\theta_{max} \tag{2.32b}$$

$$E = \phi_d/A_d = \pi L \sin^2\theta_{max} \tag{2.32c}$$

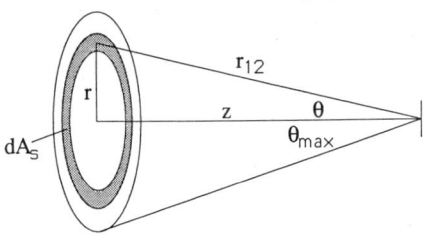

Figure 2.20. Calculation for extended source.

Equation (2.32c) is exact, and does not assume small angles. Thus it reduces to the small-angle approximation (using $\phi = LA\Omega$) of $E = 4\pi L \sin^2(\theta_{max}/2)$ only in the limit of small angles. The factor of $\sin^2\theta_{max}$ can be expressed in terms of the $F/\#$, which is often more convenient for discussion of optical systems:

$$\sin^2 \theta = \frac{1}{4(F/\#)^2 + 1} \tag{2.33}$$

so the irradiance on a detector from an extended source is

$$E = \frac{\pi L}{4(F/\#)^2 + 1} \tag{2.34}$$

Recall that camera lenses are stopped in irradiance steps of 2, 4, 8, and 16, which corresponds to $F/\#$s of 1.1, 2, 2.8, and 4, so one stop change produces a factor of 2 irradiance.

Consider two further examples of the concept of extended sources. Shown in Fig. 2.21 is a detector in an enclosure at absolute zero (0 K), which looks out at a room-temperature source of radiance ($L_q = 1.3 \times 10^{18}$ photon/s-cm^2) through a circular hole with a half-angle of 30°. What is the irradiance (E_q) on the detector?

$$E_q = \pi(1.3 \times 10^{18}) \sin^2 30° \approx 10^{18}(\text{photon/s-cm}^2) \tag{2.35}$$

Then if a (lossless) lens is placed in front of the detector, as shown in Fig. 2.22, with the same configuration as shown in Fig. 2.21, the irradiance on the detector is not changed, as long as the detector receives radiation over the same half-angle. The various sources we have discussed and the corresponding applicable equations for irradiance on the focal plane for a lens system shown in

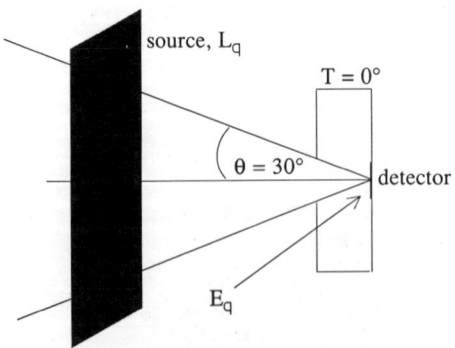

Figure 2.21.   Extended source example.

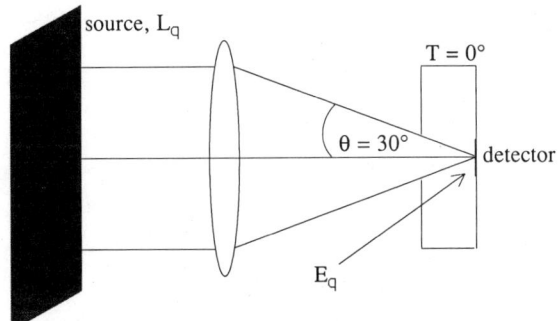

Figure 2.22. Extended source in front of lens system.

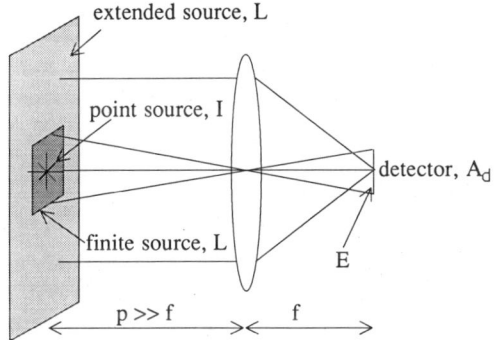

Figure 2.23. Radiant power transfer for various sources.

Fig. 2.23 is summarized:

$$\text{Point source:} \quad E = \frac{0.84ID_{\text{lens}}^2}{p^2\lambda^2(F/\#)^2 2.44^2} \quad (2.36)$$

$$\text{Finite source:} \quad E = \frac{LA_{\text{lens}}}{q_2} \quad (2.37)$$

$$\text{Extended source:} \quad E = \pi L \sin^2\theta = \frac{\pi L}{4(F/\#)^2 + 1} \quad (2.38)$$

2.6. BLACKBODY RADIATION

A blackbody is a perfect radiator. A blackbody radiates the maximum number of photons per unit time from a surface area in a wavelength interval that any body can radiate at a given kinetic temperature. No surface that is in thermodynamic equilibrium can radiate more photons, unless it contains fluorescent or radioactive materials.

Note that the usual definition of temperature, in terms of the ratio of numbers of electrons at various energy levels, does not apply to lasers, which have population inversions in their gain media. Lasers can easily have spectral radiances exceeding any other source, blackbody or otherwise, because of their extremely narrow emission wavelength interval and their small beam angle.

2.6.1. Blackbody Radiation Theory

In this section we derive the analytical expression for blackbody emission. This expression was first derived formally by Planck in 1900. His significant contribution was the assumption that the radiation field was not a continuously varying quantity with energy but was quantized into steps, each having a value of $h\nu$. This revolutionary idea led to the photoemission experiment and theory of Einstein in 1905 of the photoelectric effect.

The derivation of the blackbody radiation law (often called Planck's radiation law) will follow a semiclassical approach. An alternative quantum-mechanical approach can be found in Leighton (1959) or Loudon (1973). We will develop the normal modes for a cavity, calculate the density of modes in the cavity using the wave equation, and then apply the average occupancy per mode from statistical mechanics to obtain an expression for the number of photons per second and per wavelength interval emitted by a blackbody. Table 2.4 illustrates blackbody radiance as a function of different variables.

Consider a cube cavity of lengths ℓ_x, ℓ_y, ℓ_z, as shown in Fig. 2.24. Using the wave equation, we can find all the modes (solutions to the wave equation) in the cavity. We will include the effect of a nondispersive refractive index n. We will also assume that the cavity walls are perfect conductors, so no tan-

Table 2.4 Blackbody Radiance as a Function of Different Variables

Variable	L_q	L_e
k Wave vector	$\dfrac{ck^2 dk}{4\pi^3(e^{hck/2\pi kT} - 1)}$	$\dfrac{k^3 hc^2 dk}{8\pi^4(e^{hck/2\pi kT} - 1)}$
λ Wavelength	$\dfrac{2cd\lambda}{\lambda^4(e^{hc/k\lambda T} - 1)}$	$\dfrac{2hc^2 d\lambda}{\lambda^5(e^{hc/k\lambda T} - 1)}$
ν Frequency	$\dfrac{2\nu^2 d\nu}{c^2(e^{h\nu/kT} - 1)}$	$\dfrac{2h\nu^3 d\nu}{c^2(e^{h\nu/kT} - 1)}$
x Parameter	$2c\left(\dfrac{kT}{hc}\right)^3 \dfrac{x^2 dx}{(e^x - 1)}$	$\dfrac{2(kT)^4}{h^3 c^2} \dfrac{x^3 dx}{(e^x - 1)}$
Units	$\dfrac{\text{photon}}{\text{s-cm}^2\text{-sr}}$	$\dfrac{\text{watt}}{\text{cm}^2\text{-sr}}$

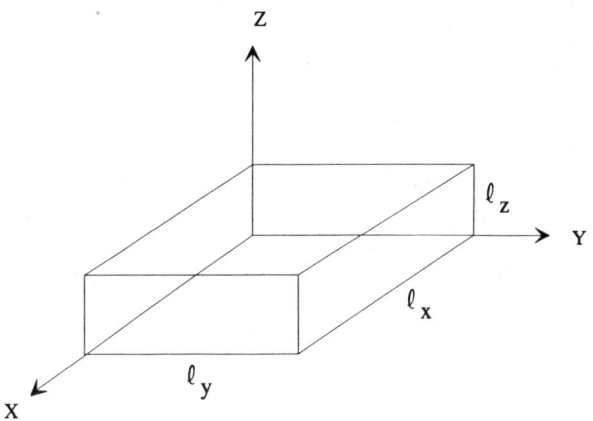

Figure 2.24. Rectangular cavity full of modes.

gential electric field can exist in these surfaces. The wave equation is

$$\nabla^2 \mathbf{E} = \frac{1}{c_n^2} \frac{\partial^2 \mathbf{E}}{\partial t^2} \qquad (2.39)$$

where

$$\mathbf{E} = \mathbf{E}(r, t) = \mathbf{E}(x, y, z, t) = [\mathbf{E}_{ox}\hat{x} + \mathbf{E}_{oy}\hat{y} + \mathbf{E}_{oz}\hat{z}]e^{j(\nu t + k \cdot r)} \qquad (2.40)$$

is the space and time-varying electric field and c_n is the speed of light in the medium of index n: $c_n = c/n$. The wave vector k has magnitude $\nu n/c$.

We will solve the wave equation for $\mathbf{E}$ in rectangular coordinates using separation of variables by assuming a solution of the form

$$\mathbf{E}(x, y, z, t) = \mathbf{E}_o X(x) Y(y) Z(z) T(t) \qquad (2.41)$$

The wave equation for the x-component of the electric field is

$$\frac{\partial^2 \mathbf{E}_x}{\partial x^2} + \frac{\partial^2 \mathbf{E}_y}{\partial y^2} + \frac{\partial^2 \mathbf{E}_z}{\partial z^2} - \frac{1}{c_n^2} \frac{\partial^2 \mathbf{E}_x}{\partial t^2} = 0 \qquad (2.42)$$

yielding the auxiliary equation, on substitution,

$$\frac{X''}{X} + \frac{Y''}{Y} + \frac{Z''}{Z} - \frac{1}{c_n^2} \frac{T''}{T} = 0 \qquad (2.43)$$

where

$$X = X(x) = X_o e^{jk_x x} \qquad Y = Y(y) = Y_o e^{jk_y y}$$

$$Z = Z(z) = Z_o e^{jk_z z} \qquad T = T(t) = T_o e^{-jvt}$$

$$X'' = \frac{\partial^2 X(x)}{\partial x^2} \qquad Y'' = \frac{\partial^2 Y(y)}{\partial y^2} \qquad Z'' = \frac{\partial^2 Z(z)}{\partial z^2} \qquad T'' = \frac{\partial^2 T(t)}{\partial t^2} \qquad (2.44)$$

Solving for the x-component of the field

$$E_x = X_o Y_o Z_o T_o e^{j(k_x x + k_y y + k_z z - vt)} = E_{ox} e^{j(k_x x + k_y y + k_z z - vt)} \qquad (2.45)$$

In terms of sines and cosines, we can write

$$E_x = E_{ox} e^{-j(vt)} \begin{pmatrix} \sin k_x x \\ \cos k_x x \end{pmatrix} \begin{pmatrix} \sin k_y y \\ \cos k_y y \end{pmatrix} \begin{pmatrix} \sin k_z z \\ \cos k_z z \end{pmatrix} \qquad (2.46)$$

where the choice of sine or cosine is determined by the boundary conditions at the perfectly conducting walls of the cavity. Restricting k_x, k_y, and k_z to be nonnegative quantities does not restrict the possible solutions.

The boundary conditions require that the tangential electric field is zero in the perfectly conducting walls of the box. This forces a choice in the sines and cosines in the preceding equation. One implication is that $E_x(y = 0)$ and $E_x(z = 0)$ are both zero, yielding

$$E_x = E_{ox} e^{-j(vt)} \begin{pmatrix} \sin k_x x \\ \cos k_x x \end{pmatrix} \sin k_y y \sin k_z z \qquad (2.47a)$$

Similarly

$$E_y = E_{oy} e^{-j(vt)} \sin k_x x \begin{pmatrix} \sin k_y y \\ \cos k_y y \end{pmatrix} \sin k_z z \qquad (2.47b)$$

and

$$E_z = E_{oz} e^{-j(vt)} \sin k_x x \sin k_y y \begin{pmatrix} \sin k_z z \\ \cos k_z z \end{pmatrix} \qquad (2.47c)$$

Maxwell's equations also require that inside the box $(0 < x < \ell_x, 0 < y < \ell_y,$ and $0 < z < \ell_z)$,

$$\nabla \cdot E = \frac{\partial E_x}{\partial x} + \frac{\partial E_y}{\partial y} + \frac{\partial E_z}{\partial z} = 0 \qquad (2.48)$$

which must be satisfied everywhere inside the box. Because

$$\frac{\partial E_y}{\partial y} + \frac{\partial E_z}{\partial z} \propto \sin k_x x \qquad (2.49)$$

the condition

$$\frac{\partial E_x}{\partial x} \propto \sin k_x x \qquad (2.50)$$

is required for Eq. (2.48) to stand for all x values in the range $0 < x < \ell_x$. Therefore, we get

$$E_x = E_{ox} e^{-j\nu t} \cos k_x x \sin k_y y \sin k_z z \qquad (2.51a)$$

likewise,

$$E_y = E_{oy} e^{-j\nu t} \sin k_x x \cos k_y y \sin k_z z \qquad (2.51b)$$

and

$$E_z = E_{oz} e^{-j\nu t} \sin k_x x \sin k_y y \cos k_z z \qquad (2.51c)$$

Furthermore, Eq. (2.48) requires that

$$E_{oz} k_x + E_{oy} k_y + E_{oz} k_z = 0 \qquad (2.52a)$$

or equivalently,

$$E_o \cdot k = 0 \qquad (2.52b)$$

Another boundary condition also requires that

$$E_y(x = \ell_x) = 0 \qquad (2.53)$$

implying

$$k_x \ell_x = m\pi \qquad (2.54a)$$

or

$$k_x = m\pi/\ell_x \qquad (2.54b)$$

where m is an integer. The boundary condition constrains the x-component of the wave vector to take on only discrete values, spaced by π/ℓ_x. Each allowed value of k_x occupies a space of π/ℓ_x in one-dimensional k space. Because of the periodicity of the sine function, all solutions are included for $k_x \geq 0$. However, we want k_x to be both positive and negative, to represent light traveling in both directions. The total number of solutions to the boundary-value problem is the same. Hence, while we can say that k_x shares the half-k space of $k_x \geq 0$, with a spacing between allowed k_x values of π/ℓ_x, we prefer to say that k_x takes the range $-\infty \leq k_x \leq \infty$, with a spacing between allowed k_x values of $2\pi/\ell_x$.

The density of states (DOS) is defined as the number of states per unit cavity volume, per unit k-space volume. In one-dimensional terms, we can write

$$\text{DOS}_x = \frac{1}{2\pi/\ell_x} \times \frac{1}{\ell_x} = \frac{1}{2\pi} \tag{2.55}$$

Because the refractive index affects the magnitude of k, we find that the DOS is higher for a cavity filled with a medium other than vacuum. The wave crests of the electromagnetic wave are closer together. Thus more modes can be fit into a given size box. This is the major result of carrying the refractive index in the derivation. From this point on in the development, we shall assume that $n = 1$ for notational convenience.

Now, if we include similar conditions for boundary conditions at $y = \ell_y$ and $z = \ell_z$, we get our three-dimensional DOS

$$\text{DOS} = \left(\frac{1}{2\pi/\ell_x}\right)\left(\frac{1}{2\pi/\ell_y}\right)\left(\frac{1}{2\pi/\ell_z}\right) \times \frac{1}{(\ell_x\ell_y\ell_z)} = \frac{1}{(2\pi)^3} \tag{2.56}$$

Now, for each mode, represented by a distinct k value, Eq. (2.52b) only provides one constraint. Consequently, there are two sets of independent solutions allowed by Eq. (2.52b). We allow a free multiplicative constant depicting the field amplitude. These two independent mathematical solutions correspond to the two orthogonal polarization modes, as shown in Fig. 2.25.

Therefore the DOS of Eq. (2.56) must be multiplied by 2 because two po-

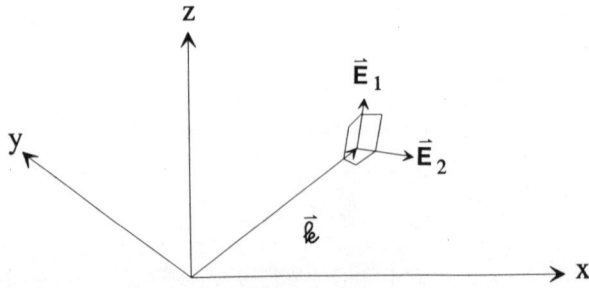

Figure 2.25.   Two orthogonal polarization modes exist for any k vector.

larizations can exist, and the DOS is

$$\text{DOS} = \frac{1}{4\pi^3} \tag{2.57}$$

The number of photons per unit cavity volume N is the DOS multiplied by the average number of photons $(\bar{n})$ per k state, integrated over all k space:

$$N = \int \text{DOS } \bar{n} \, dk \tag{2.58}$$

$$\left[\frac{\text{number of photons}}{\text{volume}} \right] = \left[\frac{\text{number of } k\text{-states}}{\text{volume} \times k\text{-space volume}} \right]$$

$$\times \left[\frac{\text{number of photons}}{k\text{-state}} \right] \times [k\text{-space volume}] \tag{2.59}$$

The DOS is from Eq. (2.57), and dk denotes the differential volume in k space. Considering a spherical coordinate system, $dk = 4\pi k^2 \, dk$, *which is the volume of a spherical shell of k space between k and $k + dk$.* This counts the number of k states with wave-vector magnitude lying in the range $k < |k| < k + dk$.

To calculate average number of photons $(\bar{n})$ that occupy a given k state requires further analysis. The major breakthrough Planck set forth was that the permitted energies for each mode are quantized in integer multiples of $h\nu$, where these quanta of energy $(h\nu)$ are called *photons* with the frequency $\nu = kc/2\pi$.

The average number of photons per state $\bar{n}$ must be determined from a probability density function. For any temperature, T, the probability that n photons exist in a given energy state is given by the Boltzmann probability distribution

$$p(n) = \frac{e^{-nh\nu/kT}}{\sum\limits_{n=0}^{\infty} e^{-nh\nu/kT}} \tag{2.60}$$

where

$k =$ Boltzmann's constant
$T =$ absolute temperature

Figure 2.26 shows a plot of $p(n)$ versus n.

Let $x = h\nu/kT$, referred to as the x parameter. The summation over infinity of the geometric series can be written as

$$\sum\limits_{n=0}^{\infty} e^{-nx} = \frac{1}{1 - e^{-x}} \tag{2.61}$$

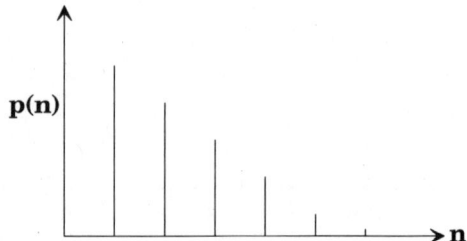

Figure 2.26.   Boltzmann probability distribution function.

Substituting, we can write for $p(n)$

$$p(n) = (1 - e^{-h\nu/kT})e^{-nh\nu/kT} \tag{2.62}$$

The average number of photons per state $\bar{n}$ corresponding to the expected number of photons in any mode with energy $\mathcal{E} = h\nu$ is the expected value of n

$$\bar{n} = \sum_{n=0}^{\infty} np(n) = \sum_{n=0}^{\infty} n(1 - e^{-h\nu/kT})e^{-nh\nu/kT}$$

$$= (1 - e^{-h\nu/kT}) \sum_{N=0}^{\infty} n(e^{-nh\nu/kT}) \tag{2.63}$$

Recall the expression for the derivative of an exponential

$$\frac{d}{dx} e^{-nx} = -ne^{-nx} \tag{2.64}$$

We can now write

$$\bar{n} = (1 - e^{-x}) \frac{d}{dx} \sum_{n=0}^{\infty} (-1)(e^{-nx})$$

$$= (1 - e^{-x})(-1) \frac{d}{dx} \left(\frac{1}{1 - e^{-x}} \right) \tag{2.65a}$$

$$\bar{n} = \frac{(1 - e^{-x})e^{-x}}{(1 - e^{-x})^2} = \frac{e^{-x}}{(1 - e^{-x})} = \frac{1}{(e^x - 1)} = \frac{1}{(e^{h\nu/kT} - 1)} \tag{2.65b}$$

We can now substitute Eqs. (2.65b) and (2.57) into Eq. (2.58) to find N, the number of photons per unit cavity volume.

$$N = \int \frac{1}{4\pi^3} \frac{1}{(e^{h\nu/kT} - 1)} 4\pi k^2 \, dk = \int \frac{k^2}{\pi^2} \frac{1}{(e^{h\nu/kT} - 1)} \, dk \qquad (2.66)$$

We identify the integrand of Eq. (2.66) as the number of photons per unit cavity volume per unit wave vector. This quantity, denoted N_k, is the spectral density of N:

$$N_k \, dk = \frac{k^2}{\pi^2} \frac{1}{(e^{h\nu/kT} - 1)} \, dk \qquad (2.67)$$

(Wolfe and Zissis, 1990). We will also develop alternative expressions, based on wave-vector magnitude (k), wavelength (λ), frequency (ν), and x-parameter ($x = h\nu/kT$).

Now that we have an expression for the photon density inside of our cavity, we can develop an expression for photon exitance by letting some photons escape. Consider Fig. 2.27, which shows a small hole of area dA in the wall of an infinite conducting cavity full of photons. Photons within the cavity move at the speed of light, c, in all directions (4π sr). The hole will act as a source of blackbody radiation.

Consider a small volume with base area dA, and length equal to the average x-component of photon velocity, $\bar{v}_x$, multiplied by a time interval dt. The number of photons, Q_q, that will escape in a time, dt, is equal to half of the photons in the volume, $dA \, \bar{v}_x \, dt$, because only those photons with a positive x-component of velocity v_x (one-half of the photons) have a chance of getting out of the hole. The other half do not have a chance to escape:

$$Q_q = \tfrac{1}{2} N \, dA \, \bar{v}_x \, dt \qquad (2.68)$$

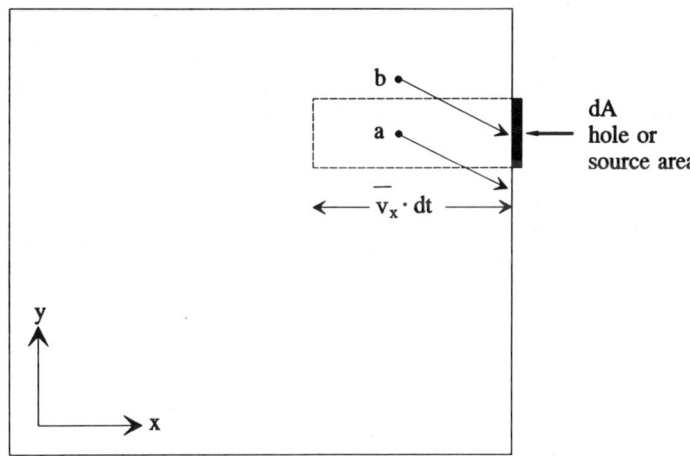

Figure 2.27.   Source area from an infinite conducting box full of photons.

Of those photons with $v_x > 0$ contained in the volume at any instant in time, not all will emerge through the hole. Conceptually, for each photon that escapes through the side of the volume (photon "a" in Fig. 2.27), there is a corresponding one ("b") that enters. We can perform this calculation explicitly by finding the average x-component of the velocity $\bar{v}_x$, the average velocity of those photons going toward the area dA. The x-component of velocity is

$$v_x = c \cos \theta \tag{2.69}$$

Integrating over the field of view of the hole (2π sr) for the photons traveling in the positive velocity direction, we can determine the average velocity $\bar{v}_x$.

$$\bar{v}_x = \frac{\int_{\text{hemisphere}} c \cos \theta \, d\Omega}{\int_{\text{hemisphere}} d\Omega} = \frac{2\pi c \int_0^{\pi/2} \cos \theta \sin \theta \, d\theta}{2\pi \int_0^{\pi/2} \sin \theta \, d\theta} = \frac{c}{2} \tag{2.70}$$

Combining Eqs. (2.66), (2.68), and (2.70), we find for M_q the photon exitance, or the number of photons per area per unit time emitted by the blackbody

$$M_q = \frac{Q_q}{dA \, dt} = \frac{1}{2} N \bar{v}_x = \frac{c \, N}{4} \tag{2.71a}$$

$$M_q = \int \frac{c}{4} \frac{k^2}{\pi^2} \frac{1}{(e^{h\nu/kT} - 1)} \, dk \tag{2.71b}$$

Finally we can write M_q in its spectral density form

$$M_q = M_{q,k} \, dk = \frac{c}{4} \frac{k^2}{\pi^2} \frac{1}{(e^{h\nu/kT} - 1)} \, dk \tag{2.71c}$$

Reviewing Fig. 2.27 further from a source-emission consideration, we can determine that the source is Lambertian because of the angular dependence of the photon flux, so the relationship between exitance (M) and radiance (L) can be determined using $L = M/\pi$. The photon radiance (photon/s-cm^2-sr) is

$$L_q = L_{q,k} \, dk = \frac{c}{4\pi} \frac{k^2}{\pi^2} \frac{1}{(e^{h\nu/kT} - 1)} \, dk \tag{2.72}$$

This is the spectral radiance of blackbody radiation that can now be expressed in wavelength (λ) or frequency (ν):

$$k = \frac{2\pi}{\lambda} \tag{2.73a}$$

$$dk = -\frac{2\pi}{\lambda^2} d\lambda \tag{2.73b}$$

Substituting for k and dk into Eq. (2.72), we find an expression for spectral photon radiance. Spectral quantities with respect to wavelength are written with a subscript λ:

$$L_{q,\lambda}(\lambda, T)\, d\lambda = \frac{(2\pi)^2}{\lambda^2\pi^2(e^{hc/\lambda kT} - 1)} \frac{c}{4\pi} \frac{2\pi}{\lambda^2}\, d\lambda \tag{2.74}$$

where the negative sign is taken up by the change in limits of integration

$$L_{q,\lambda}(\lambda, T)\, d\lambda = \frac{2\,c}{\lambda^4(e^{hc/\lambda kT} - 1)}\, d\lambda \left[\frac{\text{photons}}{\text{s-cm}^2\text{-sr}}\right] \tag{2.75}$$

To obtain the spectral radiance in terms of power, the photon radiance can be multiplied by hc/λ (energy per photon)

$$L_{e,\lambda}(\lambda, T)\, d\lambda = \frac{2\,h\,c^2}{\lambda^5(e^{hc/\lambda kT} - 1)}\, d\lambda \left[\frac{\text{W}}{\text{cm}^2\text{-sr}}\right] \tag{2.76}$$

What has been developed is that heated bodies emit radiation that is concentrated in the infrared region for most earthly temperatures and that such heated bodies emit their radiation in a continuum of wavelength. We have derived the spectral radiance and spectral exitance of a blackbody. To get the total radiance or exitance from a source over a finite spectral region, an integration is required:

$$L_q = \int_{\lambda_1}^{\lambda_2} L_{q,\lambda}(\lambda, T)\, d\lambda \tag{2.77}$$

$$M_q = \int_{\lambda_1}^{\lambda_2} M_{q,\lambda}(\lambda, T)\, d\lambda = \pi L_q \tag{2.78}$$

This integration cannot be carried out analytically for any limits except zero and infinity. An approximate trapezoidal integration usually suffices to calculate in-bands value over small intervals. Appendix A contains accurate numerical results.

The radiant properties of blackbodies are described by Planck's equations derived earlier, Eqs. (2.75) and (2.76). Given the source temperature, we can obtain the source spectral radiance by application of these formulas. Rewriting these equations without the differentials, thus looking at their spectral characteristics,

$$L_{e,\lambda}(\lambda, T) = \frac{2hc^2}{\lambda^5[e^{hc/\lambda kT} - 1]} \left[\frac{\text{watt}}{\text{cm}^2\text{-sr-}\mu\text{m}}\right] \quad (2.79)$$

$$L_{q,\lambda}(\lambda, T) = \frac{2c}{\lambda^4[e^{hc/\lambda kT} - 1]} \left[\frac{\text{photon}}{\text{s-cm}^2\text{-sr-}\mu\text{m}}\right] \quad (2.80)$$

The only variables that occur in the preceding equations are λ and T. Temperature must be specified in kelvins. Appendix B has a computer program written in BASIC to calculate these functions. Because the equations are complicated, we prefer to work with plots of the functions involved when possible.

Figure 2.28 and 2.29 are plots of spectral photon radiance, Eq. (2.80), for blackbody temperatures from 300 K to 6000 K. Figures 2.30 and 2.31 are plots of spectral radiance, Eq. (2.79), over the same blackbody source temperature. The corresponding equations for spectral radiant exitance [$M_{e,\lambda}(\lambda, T)$] and spectral photon flux exitance [$M_{q,\lambda}(\lambda, T)$] are related by $M = \pi L$:

$$M_{e,\lambda}(\lambda, T) = \frac{2\pi hc^2}{\lambda^5[e^{hc/\lambda kT} - 1]} \left[\frac{\text{watt}}{\text{cm}^2\text{-}\mu\text{m}}\right] \quad (2.81)$$

$$M_{q,\lambda}(\lambda, T) = \frac{2\pi c}{\lambda^4[e^{hc/\lambda kT} - 1]} \left[\frac{\text{photon}}{\text{s-cm}^2\text{-}\mu\text{m}}\right] \quad (2.82)$$

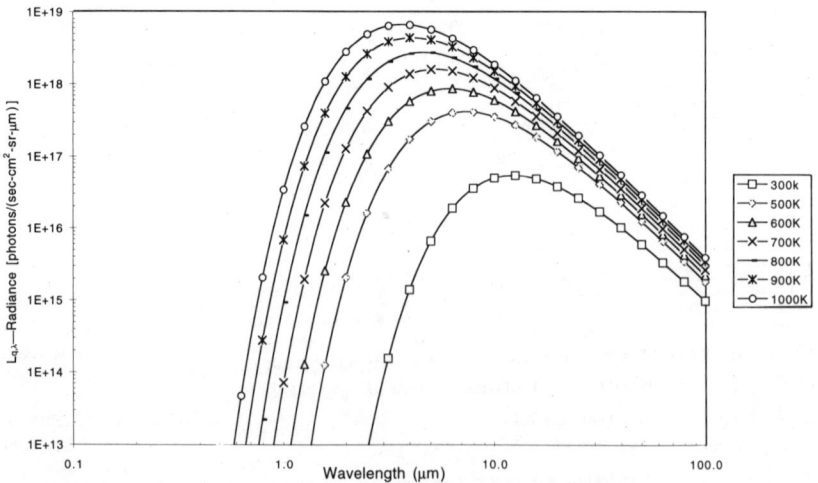

Figure 2.28. Spectral photon radiance vs. wavelength for blackbody temperatures from 300 K to 1000 K.

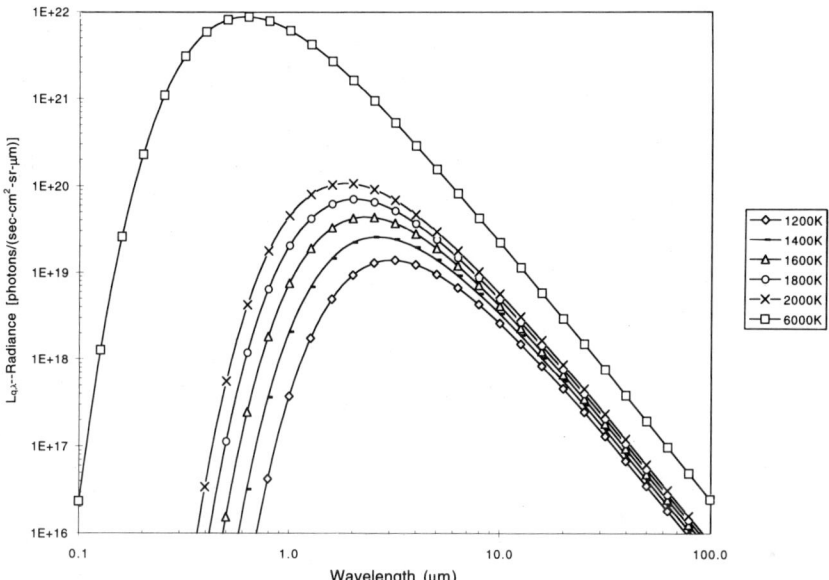

Figure 2.29. Spectral photon radiance vs. wavelength for blackbody temperatures from 1200 K to 6000 K.

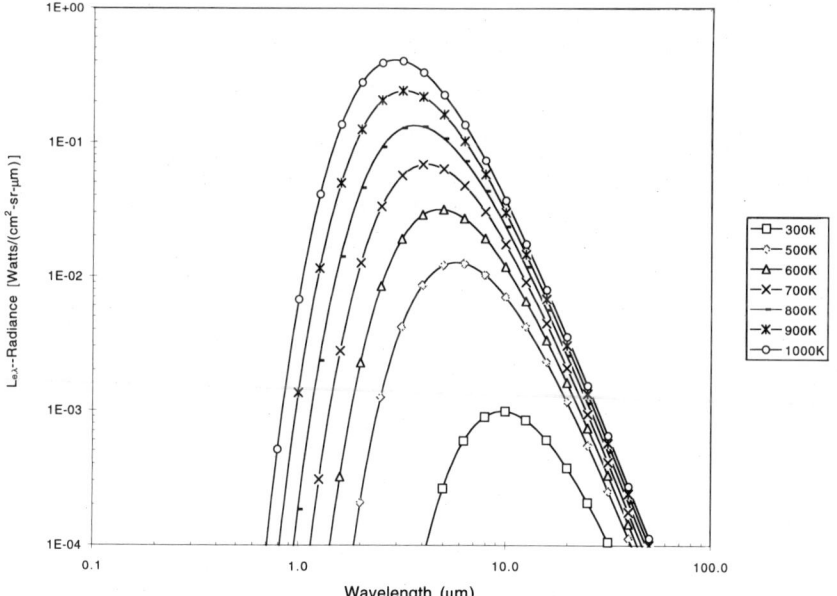

Figure 2.30. Spectral radiance vs. wavelength for blackbody temperatures from 300 K to 1000 K.

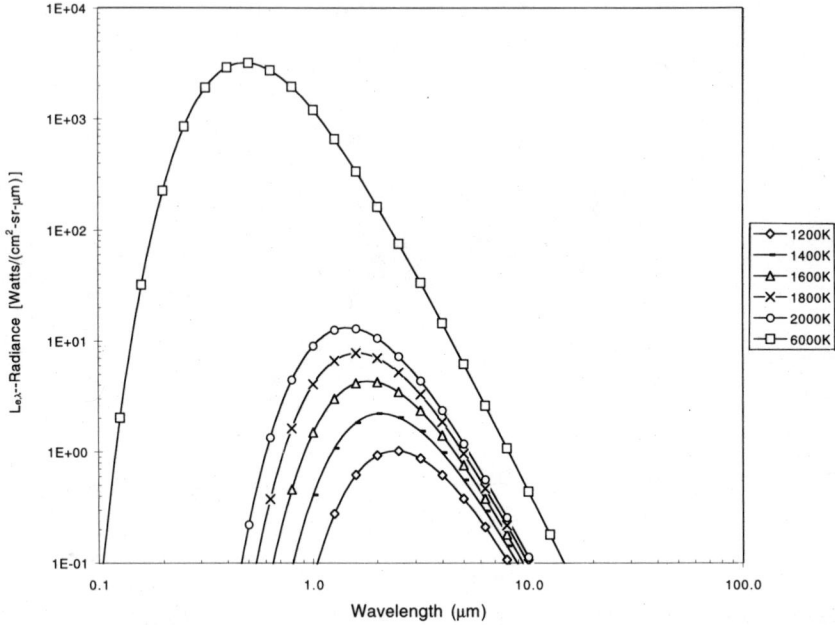

Figure 2.31. Spectral radiance vs. wavelength for blackbody temperatures from 1200 K to 6000 K.

2.6.2. Stefan–Boltzmann Law

Total radiant exitance from a blackbody at temperature T is the integral over all wavelengths

$$M_e(T) = \int_0^\infty M_{e,\lambda}(\lambda, T)\, d\lambda$$

$$= \int_0^\infty \frac{2\pi hc^2}{\lambda^5 [e^{hc/\lambda kT} - 1]}\, d\lambda \tag{2.83}$$

This can be interpreted as the area under the spectral entrance curves for a given temperature, as shown in Fig. 2.32.

Carrying out this integral over all wavelengths (Bramson, 1966):

$$M_e(T) = \frac{2\pi^5 k^4}{15\, c^2 h^3}\, T^4 \tag{2.84}$$

$$M_e(T) = \sigma_e T^4 \tag{2.85}$$

where σ_e is called the Stefan–Boltzmann constant and has an approximate value of

$$\sigma_e \approx 5.67 \times 10^{-12}\ \text{watt/(cm}^2\ \text{K}^4) \tag{2.86}$$

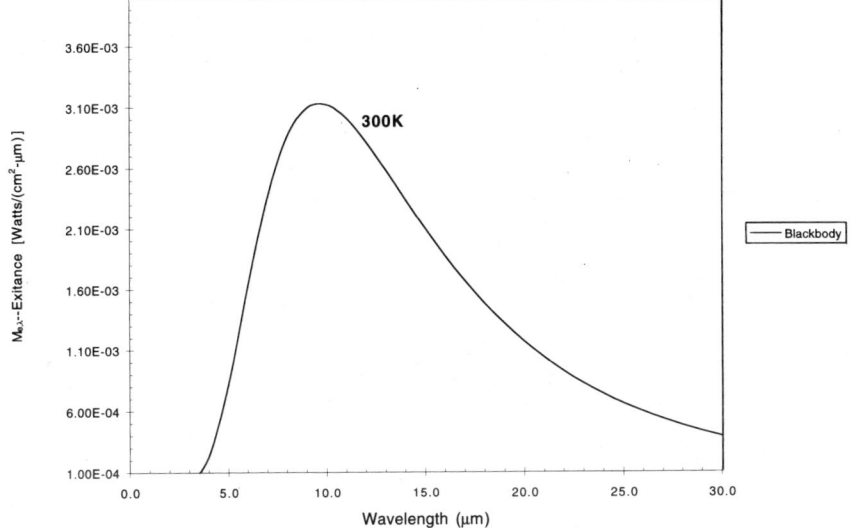

Figure 2.32. Total radiant exitance from a blackbody at 300 K is the area under the spectral exitance curve.

The relationship between the total exitance of a blackbody and its temperature (Eq. 2.87) is called the Stefan–Boltzmann law.

Similarly, the total photon exitance over all wavelengths is

$$M_q(T) = \int_0^\infty M_{q,\lambda}(\lambda,\ T)\ d\lambda \tag{2.87}$$

$$M_q(T) = \sigma_q T^3 \ \text{photon/s-cm}^2 \tag{2.88}$$

where σ_q is evaluated and is approximately

$$\sigma_q \approx 1.52 \times 10^{11} \ \text{photon/(s-cm}^2 \ \text{K}^3) \tag{2.89}$$

For example, a 300-K blackbody total photon exitance is

$$M_q(300) = 1.52\ (10^{11})\ 300^3 = 4 \times 10^{18} \ \text{photon/(s-cm}^2). \tag{2.90}$$

2.6.3. Wien Displacement Laws

The decrease in the wavelength of peak exitance as the temperature increases is quantified in the Wien Displacement Law. The analytic relationship can be derived from the condition for the peak of the exitance function by setting the partial derivative equal to zero and solving for wavelength at maximum exitance:

$$\frac{\partial M_{e,\lambda}(\lambda, T)}{\partial \lambda} = 0 \tag{2.91}$$

This produces a constraint on the wavelength of maximum exitance, λ_{max}

$$\lambda_{max} = \frac{2898 \ \mu m \ K}{T} \quad \text{or} \quad \lambda_{max} T = 2898 \ \mu m \ K \tag{2.92}$$

This is called the *Wien Displacement Law*. The plot of λ_{max} versus source temperature is a hyperbola. For example, a blackbody source at $T = 300$ K (room temperature) has its maximum exitance at approximately 9.7 μm. If the source in question were at 1000 K, the value of λ_{max} would be at 2.9 μm. It is interesting to note that the λ_{max} for the sun is near 0.5 μm, very close to the peak of sensitivity of the human eye.

Similarly, the corresponding Wien Displacement Law for maximum wavelength for photon flux exitance can be found by taking the partial derivative of the photon exitance with respect to wavelength and setting it equal to zero.

$$\frac{\partial M_{q,\lambda}(\lambda, T)}{\partial \lambda} = 0 \tag{2.93}$$

This produces a maximum wavelength temperature product of

$$\lambda_{max} T = 3662 \ (\mu m \ K) \tag{2.94}$$

2.6.4. Finite Spectral Regions

The radiant exitance over a finite spectral region of interest is

$$\int_{\lambda_1}^{\lambda_2} M_{e,\lambda}(\lambda, T) \ d\lambda \tag{2.95}$$

which can be written as a difference of two integrals starting from zero wavelength:

$$\int_{\lambda_1}^{\lambda_2} M_{e,\lambda}(\lambda, T) \ d\lambda = \int_{0}^{\lambda_2} M_{e,\lambda}(\lambda, T) \ d\lambda - \int_{0}^{\lambda_1} M_{e,\lambda}(\lambda, T) \ d\lambda \tag{2.96}$$

Now one can calculate the integral using an approximation (Williams, 1988):

$$\int_{0}^{\lambda} M_{e,\lambda}(\lambda, T) \ d\lambda = f_e(\lambda, T) \ \sigma_e T^4 \tag{2.97}$$

where $\sigma_e T^4$ = Stefan–Boltzmann law for total exitance. Using the x parameter previously defined, we can express $f_e(\lambda, T)$:

$$f_e(\lambda, T) = \frac{90}{\pi^4} e^{-x} \left(1 + x + \frac{x^2}{2} + \frac{x^3}{6} \right) \left(1 + \frac{\lambda T - 2500}{150{,}000} \right)$$

$$\text{for} \quad \lambda T < 2500 \qquad (2.98a)$$

$$f_e(\lambda, T) = \frac{90}{\pi^4} e^{-x} \left(1 + x + \frac{x^2}{2} + \frac{x^3}{6} \right) \quad \text{for} \quad \lambda T > 2500$$

$$(2.98b)$$

where λ is in microns and T is in kelvins.

Similarly to integrate the photon exitance over a spectral region,

$$\int_{\lambda_1}^{\lambda_2} M_{q,\lambda}(\lambda, T)\, d\lambda = \int_0^{\lambda_2} M_{q,\lambda}(\lambda, T)\, d\lambda - \int_0^{\lambda_1} M_{q,\lambda}(\lambda, T)\, d\lambda \quad (2.99)$$

one can use the difference-of-integral expression of Eq. (2.96) and the Williams approximation:

$$\int_0^{\lambda} M_{q,\lambda}(\lambda, T)\, d\lambda = f_q(\lambda, T)\, \sigma_q T^3 \qquad (2.100)$$

where $\sigma_q T^3$ = Stefan–Boltzmann law for total photon exitance

$$f_q(\lambda, T) = \frac{81}{\pi^4} e^{-x} \left(1 + x + \frac{x^2}{2} \right) \left(1 + \frac{\lambda T - 2600}{100{,}000} \right) \quad \text{for} \quad \lambda T < 2600$$

$$(2.101a)$$

$$f_q(\lambda, T) = \frac{81}{\pi^4} e^{-x} \left(1 + x + \frac{x^2}{2} \right) \quad \text{for} \quad \lambda T > 2600 \qquad (2.101b)$$

2.6.5. Exitance Contrast

A consideration of how much the exitance changes with temperature is also important to the sensitivity of an infrared (IR) system. For a system operating within a finite passband ($\Delta\lambda$), the question becomes: For a given source temperature, what is the wavelength where the source exitance changes the most with temperature?

This consideration of the exitance contrast involves the following second partial derivative:

$$\frac{\partial}{\partial \lambda} \left[\frac{\partial M_{e,\lambda}(\lambda, T)}{\partial T} \right] = 0 \qquad (2.102)$$

which produces a constraint of similar form to the Wien Displacement Law, and yields

$$\lambda_{\text{max contrast}} = \frac{2410 \ (\mu m \ K)}{T \ (K)} \tag{2.103}$$

At a source temperature of 300 K, for example, we have a maximum contrast $\partial M_{\lambda, e}/\partial T$ occurring at a wavelength of about 8 μm, which is not the wavelength for maximum exitance (recall Eq. (2.94)).

2.7. EMISSIVITY

The question that naturally arises is, how does one model real sources; that is, sources that do not have the strict ideal behavior of blackbody sources? For example, a vehicle that one desires to detect can be thought of as having an effective temperature, but its spectral dependence is only approximately that of a blackbody at that temperature. Indeed, the blackbody curve provides only an *upper limit* of the overall spectral exitance of a source, for any specific temperature. The spectral exitance curve for any actual thermal source will be bounded by the corresponding blackbody curve at the source temperature.

How closely the radiation spectrum of a real heated body corresponds to that of a blackbody is related to its emissivity. Emissivity (ε) is a dimensionless number ≤ 1. The ratio between the exitance of the actual source and the exitance of a blackbody at the same temperature is defined as emissivity. In general, ε depends on λ and T:

$$\varepsilon(\lambda, \ T) = \frac{M_{e, \lambda}(\lambda, \ T)_{\text{source}}}{M_{e, \lambda}(\lambda, \ T)_{\text{blackbody}}} \tag{2.104}$$

A blackbody source will have $\varepsilon = 1$ for all wavelengths. The emissivity of a graybody is independent of λ. A selective radiator has an emissivity that depends on wavelength. The spectral exitance of any real source at a given temperature is bounded by the spectral exitance of a blackbody at the same kinetic temperature; hence ε is constrained to be ≤ 1.

If emissivity is independent of wavelength, then the source in question is called a *graybody*. The exitance of the graybody at any wavelength is a constant fraction (ε) of what the corresponding blackbody would produce. The graybody and the blackbody have the same shape spectrally as shown in Fig. 2.33. Thus, a graybody's emissivity multiplies the corresponding blackbody curve; both as a function of wavelength and, because ε is constant, integrated over all λ's as well. The total exitance for graybody at all λ's is

$$M_e^{gb}(T) = \varepsilon \sigma_e T^4 \tag{2.105}$$

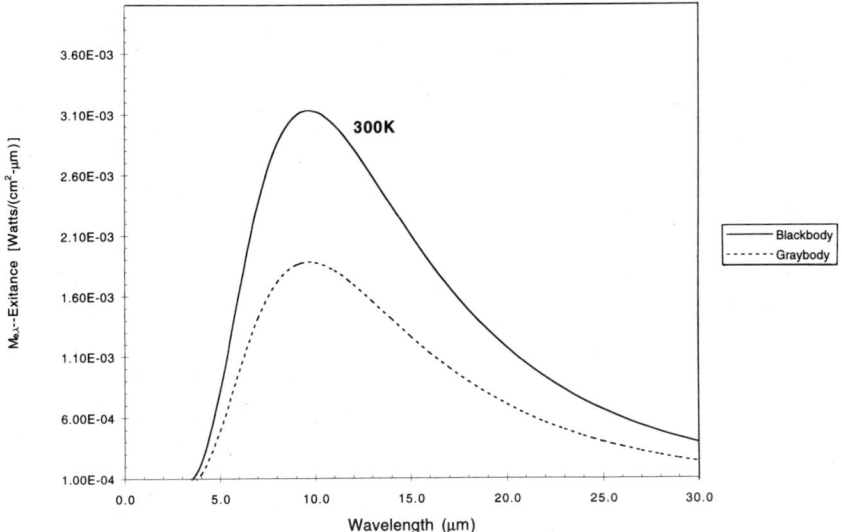

Figure 2.33. Spectral exitance of a blackbody and graybody at the same temperature.

Emissivity provides a convenient parameter for use in the modeling of real sources. The next step upward from the graybody in complexity and accuracy is the so-called "selective radiator," for which emissivity is a piecewise-constant function of wavelength, allowing realistic models of source behavior over a wider spectral range.

In Fig. 2.34, curves of emissivity are shown for a spectrally selective radiator and a blackbody at the same temperature. The selective radiator exitance

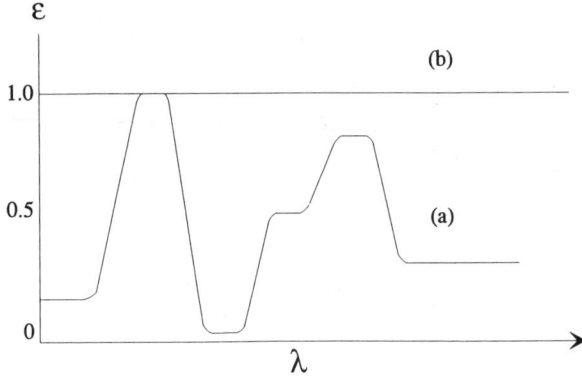

Figure 2.34. Spectral emissivity of a selective radiator.

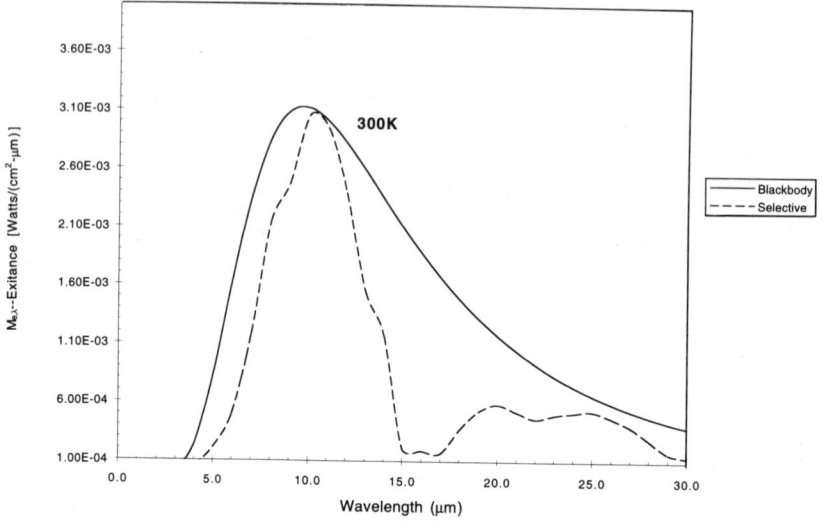

Figure 2.35.   Spectral exitance of a selective radiator.

is the point-by-point multiplication of emissivity and blackbody exitance on a spectral basis, shown in Fig. 2.35.

2.7.1. Kirchhoff's Law

If a body of mass is at thermal equilibrium with its surrounding environment, conservation of energy requires

$$\phi_{\text{incident}} = \phi_{\text{absorbed}} + \phi_{\text{reflected}} + \phi_{\text{transmitted}} \tag{2.106}$$

Dividing both sides of the equation by ϕ_{incident} yields

$$a + r + t = 1 \tag{2.107}$$

where a = absorptance, r = reflectance, t = transmittance. For an opaque body where there is no transmittance ($t = 0$), either the radiation is absorbed or reflected. We obtain

$$a + r = 1 \quad \text{or} \quad a = 1 - r \tag{2.108}$$

If the body does not absorb all of the radiation that is incident on it, it will emit less radiation to remain in thermal equilibrium. Therefore,

$$\text{watts absorbed} = a \times E \times \text{area}$$

$$= \varepsilon \times M \times \text{area} = \text{watts radiated}$$

According to Kirchhoff's law, the integrated absorptance equals integrated emissivity ($\alpha = \varepsilon$). Kirchhoff's law holds for spectral quantities also $a(\lambda, T) = \varepsilon(\lambda, T)$, where a and ε are functions of λ and T. A blackbody has $\varepsilon = 1$ and is also perfectly black, that is, it is a perfect absorber with $a = 1$. Good absorbers are good emitters. For example, matte black paint has a low reflectance, so absorptance and emissivity are high. Typically, good reflectors are poor emitters. Another example is polished aluminum, for which the reflectance is high, so absorptance and emissivity are low. A body can be absorbing radiation from a high-temperature source (such as the sun) and reradiating as a lower temperature source ($\approx$ 300 K). Recall that the peak exitance of the sun is at 0.5 μm and the peak exitance of a 300-K body is around 10 μm.

It must be remembered that on a spectral basis, the absorptance equals the emissivity: $a(\lambda) = \varepsilon(\lambda)$. However, it is not generally true that $a(\lambda = 0.5\ \mu m)$ equals $\varepsilon(\lambda = 10\ \mu m)$. Because the emissivity is a function of wavelength, we cannot estimate it in the infrared by visible appearance. An example of this is white paint (TiO_2). At 0.5 μm, its emissivity is 0.19, while at 10 μm, its emissivity is 0.94. Table 2.5 lists some materials and their solar absorptance and room temperature emissivity. For a more detailed list, see Wolfe and Zis-

Table 2.5   Select Materials Solar Absorptance and 300-K Emissivity

Material	Solar Absorptance α	Low-Temperature (300 K) Emissivity ε
Aluminum		
polished and degreased	0.387	0.027
foil, dull side, crinkled and smoothed	0.223	0.030
foil, shiny side	0.192	0.036
sandblasted	0.420	0.210
oxide, flame sprayed, 0.001 inch thick	0.422	0.765
anodized	0.150	0.770
Fiberglass	0.850	0.750
Gold		
plated on stainless steel and polished	0.301	0.028
Magnesium		
polished	0.300	0.070
Paints		
Aquadag, 4 coats on copper aluminum	0.782	0.490
aluminum	0.540	0.450
Microbond, 4 coats on magnesium	0.936	0.844
TiO_2, gray	0.870	0.870
TiO_2, white	0.190	0.940
Rokide A	0.150	0.770
Stainless steel		
type 18-8, sandblasted	0.780	0.440

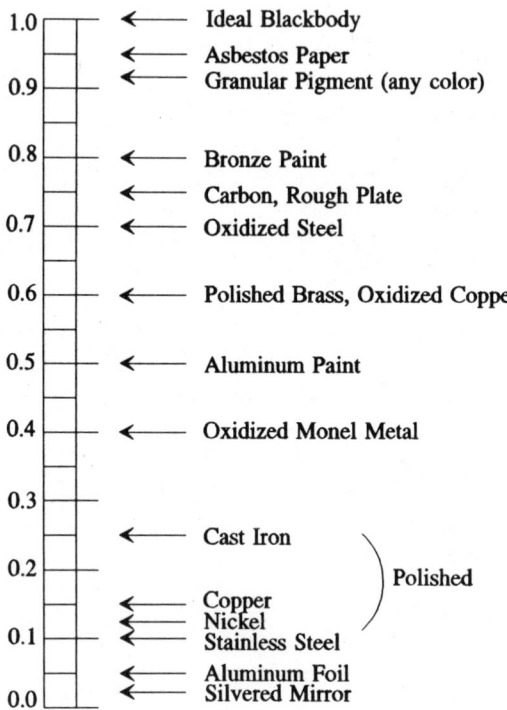

Figure 2.36. Emissivity values for common materials. (Originally appeared in R. W. Astheimer, *Handbook of Infrared Radiation Measurement*, 1983, Barnes Engineering Co. Reprinted by permission of R. W. Astheimer.)

sis (1990). Figure 2.36, adapted from Astheimer (1983), shows relative emissivity values for some common materials. The surface finish is very important to radiative properties.

Emissivity data for most materials are frequently a single number, and are seldom given as either a function of λ or T unless it is an especially well-characterized material. Usually we have limited information about the target; however, some general trends or observations about emissivity as a function of temperature follow. For nonmetallic substances, typically $\varepsilon > 0.8$ for room temperature and decreases with increasing temperature. For metallic substances, the emissivity is very low at room temperature, unless the surface is oxidized. Generally for metallic substances, the emissivity increases proportional to temperature. One obvious reason is the oxidation process that is taking place.

2.8. RADIOMETRIC MEASURES OF TEMPERATURE

Often we would like to know the actual kinetic temperature (T) of a remote object. However, we can only measure the flux that the object radiates, either over all λ or over some $\Delta\lambda$ interval. The radiant flux (ϕ), however, is a function of both temperature and emissivity. Therefore information on emissivity is necessary for an accurate estimate of temperature from flux measurements. If we know the emissivity $\varepsilon(\lambda)$ of the object, we can make a calculation to get back to T by an inversion of the Planck equation. If we have incomplete knowledge of emissivity, our measurement of temperature will have a built-in bias, and will not yield the correct temperature.

The three methods of measuring temperature yield three different estimates of temperature: radiation temperature (T_R); brightness temperature (T_B); and color temperature (T_C).

2.8.1. Radiation Temperature

The total exitance over all wavelengths from a body at a kinetic temperature follows the Stefan–Boltzmann law. The blackbody temperature that gives the same exitance as that measured (M_{meas}) is the radiation temperature (T_R). Mathematically

$$M_{\text{meas}} = \sigma T_R^4 \qquad (2.109)$$

If the source is a blackbody, then $T_R = T$; however, if the source in question is a graybody of known emissivity, then we can calculate the true temperature (T) from the radiation temperature:

$$M_{\text{meas}} = \varepsilon\sigma_\ell T^4 = \sigma_\ell T_R^4 \qquad (2.110)$$

$$T = \varepsilon^{-1/4}T_R \qquad (2.111)$$

Conceptually, radiation temperature is the temperature of the blackbody that would give the same area under the spectral exitance curve as the source being measured. Consider an example of a graybody with $\varepsilon = 0.8$ and true temperature 900 K. What is the radiation temperature? It will surely be less than T because there is less area under the spectral exitance curve:

$$900 = (0.8)^{-1/4}T_R \qquad (2.112a)$$

$$900 \times (0.8)^{1/4} = T_R = 851 \text{ K} \qquad (2.112b)$$

2.8.2. Brightness Temperature

Brightness temperature is the temperature of a blackbody that gives the same exitance in a narrow spectral region ($\Delta\lambda$) around a fixed wavelength (λ_o) as the remote object being measured. Knowing the object exitance at a given wavelength $M_{\text{obj}}(\lambda_o)$, the temperature can be determined. There is a unique solution to the Planck radiation expression at a given wavelength (λ_o) (see Fig. 2.37).

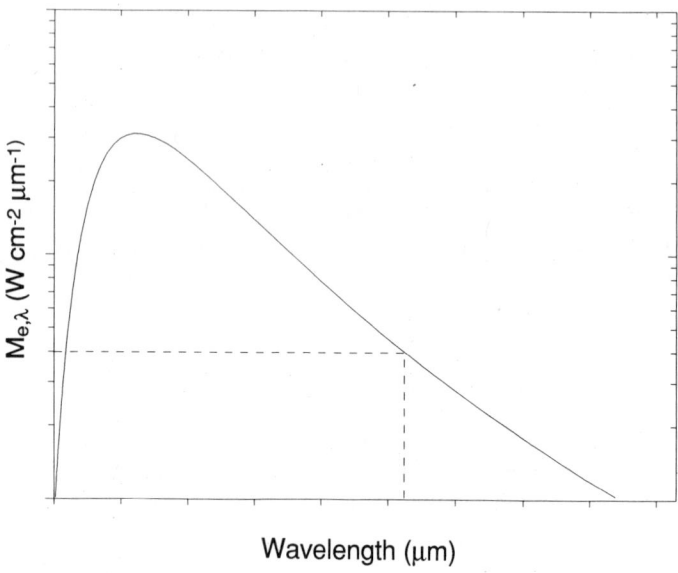

Figure 2.37. A particular point in $M_{e,\lambda}$ vs. λ space determines a unique blackbody temperature.

$$M(\lambda_o, T_B) = \frac{2\pi hc^2}{\lambda_o^5 (e^{hc/\lambda_o kT_B} - 1)} = M_{\text{obj}}(\lambda_o) \qquad (2.113)$$

One can solve for the brightness temperature (T_B) from the measured exitance. If the body in question is known to be a graybody, with emissivity ε, we can equate expressions for $M(\lambda_o)$ and solve for T:

$$\frac{2\pi hc^2}{\lambda_o^5 (e^{hc/\lambda_o kT_B} - 1)} = \varepsilon \times \frac{2\pi hc^2}{\lambda_o^5 (e^{hc/\lambda_o kT} - 1)} \qquad (2.114)$$

If the source in question is a blackbody, the brightness temperature and true kinetic temperature are equal.

2.8.3. Color Temperature

Color temperature is the temperature of a blackbody that best matches the spectral distribution of the source in question in at least two spectral regions. Usually, spectral distribution is defined as the ratio of the spectral exitance at two given wavelengths (λ_1 and λ_2). Given a measured ratio of exitance at two wavelengths $M(\lambda_1)/M(\lambda_2)$, the color temperature (T_C) is defined by the temperature that satisfies both Planck equations simultaneously:

$$\frac{M(\lambda_1)}{M(\lambda_2)} = \frac{\lambda_2^5 (e^{hc/\lambda_2 kT_C} - 1)}{\lambda_1^5 (e^{hc/\lambda_1 kT_C} - 1)} \qquad (2.115)$$

If the source in question is a graybody, the emissivity cancels in the numerator and denominator, and the color temperature is the true temperature. This property is useful in situations in which there may be an attenuation or loss of signal that is approximately independent of wavelength.

REFERENCES

Astheimer, R. W., *Handbook of Infrared Radiation Measurements*, Barnes Engineering Co., Stamford, CT, 1983.

Bramson, M. A., *Infrared: A Handbook for Applications*, Plenum Press, New York, 1966.

Leighton, R. B., *Principles of Modern Physics*, McGraw-Hill, New York, 1959.

Loudon, R., *The Quantum Theory of Light*, Oxford University Press, London, 1973.

Williams, O. M., "Infrared Photodetector Photon Formation: Extension and Application," *Infrared Physics* **27**(3), 167–179 (1987).

Wolfe, W. L., and G. J. Zissis, *The Infrared Handbook*, Optical Engineering Press, Bellingham, WA, 1990.

BIBLIOGRAPHY

Boyd, Robert W., *Radiometry and the Detection of Optical Radiation*, Wiley, New York, 1983.

Budde, W., *Optical Radiation Measurements: Vol. IV. Physical Detectors of Optical Radiation*, Academic Press, New York, 1983.

Dereniak, Eustace L., and Devon G. Crowe, *Optical Radiation Detectors*, Wiley, New York, 1984.

Grum, Franc, and Richard J. Becherer, *Optical Radiation Measurements: Vol. I. Radiometry*, Academic Press, New York, 1979.

Hudson, R. D., *Infrared System Engineering*, Wiley, New York, 1969.

Johnson, R. Barry, and William L. Wolfe (eds.), *Selected Papers on Infrared Design*, SPIE Milestone Series, Vol. 513 (parts 1 and 2), Optical Engineering Press, Bellingham, WA, 1985.

Nicodemus, F. E., et al., *Self-Study Manual on Optical Radiation Measurements*, NBS Technical Note, Number 910, National Bureau of Standards, Washington, DC, 1976.

Planck, M., *The Theory of Heat Radiation*, Dover, New York, 1991.

Spiro, Irving J. (ed.), *Selected Papers on Radiometry*, SPIE Milestone Series, Vol. 14, Optical Engineering Press, Bellingham, WA, 1990.

Spiro, Irving J., and Monroe Schlessinger, *Infrared Technology Fundamentals*, Dekker, New York, 1989.

Stimson, Allen, *Photometry and Radiometry for Engineers*, Wiley, New York, 1974.

Taylor, Jack H., *Radiation Exchange: An Introduction*, Academic Press, San Diego, 1990.

Wyatt, Clair L., *Radiometric Calibration: Theory and Methods*, Academic Press, 1978.

Wyatt, Clair L., *Radiometric Systems Design*, Macmillan, New York, 1987.

PROBLEMS

2.1 A flat receiving screen is located at a distance R from a point source with uniform intensity I. The line joining the point source with the center of the screen is perpendicular to the screen (see figure). Express the irradiance for a general point on the screen $E(x)$, in terms of the on-axis irradiance, $E(x = 0)$.

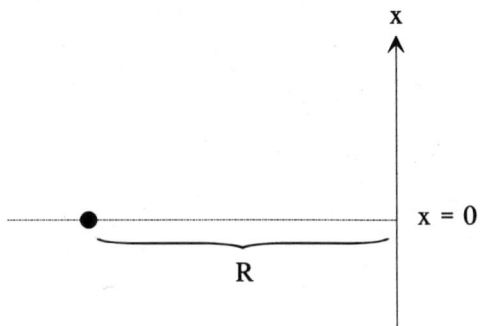

Figure P2.1

2.2. In the configuration shown in the figure, how much flux in watts falls on a 1-mm × 1-mm detector that is placed on-axis in the image plane? The source is 2 cm × 2 cm, Lambertian, with radiance 5 W/cm²-sr. The object distance, p, is 50 cm; the image distance, q, is 25 cm. The lens diameter is 5 cm. Neglect diffraction effects and Fresnel losses.

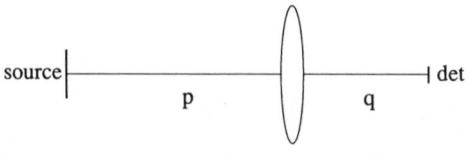

Figure P2.2

2.3. What is the solid angle of an $F/3$ cone of radiation?

2.4. Calculate the flux transferred and the on-axis image irradiance for the single-thin-lens imaging system described below:

$$L = 5 \text{ W/cm}^2\text{-sr}; \quad A_s = (3 \text{ mm})^2; \quad D_{\text{lens}} = 2 \text{ cm};$$

$$F = 20 \text{ cm}; \quad p = 50 \text{ cm}$$

Assume no transmission loss in the lens and that the source normal is parallel to the optical axis.

2.5. Given a Lambertian source with $L = 5 \text{ W/cm}^2$-sr, and area 1 cm^2 (see figure). A sphere of radius 1 meter is the receiver, and its center is located 3 m from the source. What is the irradiance on the sphere at a point that is 0.5 meter from the axis?

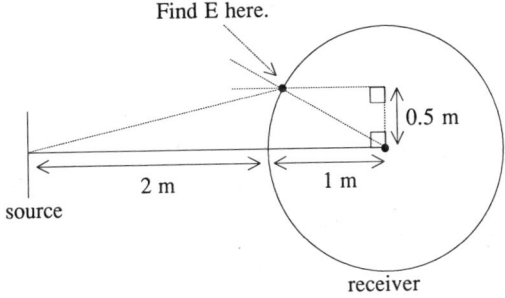

Figure P2.5

2.6. Consider an optical system with an exit pupil that is a uniform radiance circular disc (3 W/cm^2-sr) (see figure). Plot the normalized irradiance $E(\theta)/E(0)$ in the focal plane as a function of field angle θ for the two cases of an $F/1$ and $F/10$ system.

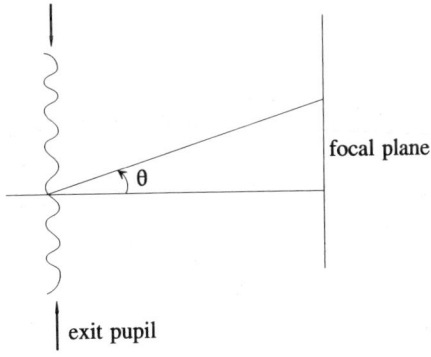

Figure P2.6

2.7.   Calculate the radiant power collected on a detector at 10-m distance from a tungsten halogen source as shown in the figure (assume the source is a blackbody at 2856 K).

$$A_s = 1 \text{ cm}^2; \qquad A_d = 1 \text{ mm}^2$$

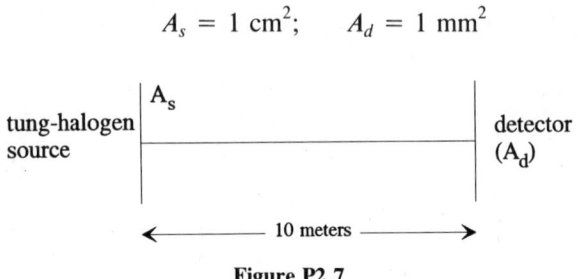

Figure P2.7

2.8.   Given a Lambertian source with radiance L and a non-Lambertian source with radiance $L_o \cos \theta_s$, where L and L_o are not equal, if the two sources have the same area, what must the relationship be between L and L_o such that the two sources emit the same amount of total power?

2.9.   Calculate the irradiance (E) on a detector from a circular source of uniform radiance L_e as illustrated in the schematic. Plot on log-log paper

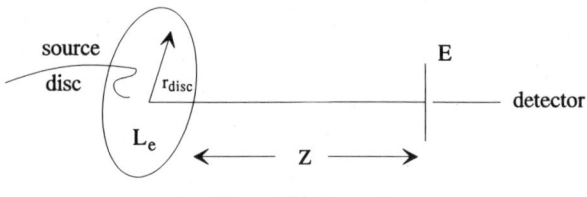

Figure P2.9

the irradiance (E) versus z/r (ratio of range-to-source disc radius) for values of 0 to 100. At what ratio can the disc be considered a point source (irradiance falls off as inverse square law) with less than 1% error?

2.10. In the configuration shown in the figure, what is the flux in watts that falls on the detector at all wavelengths? The source is a 1-mm × 1-mm

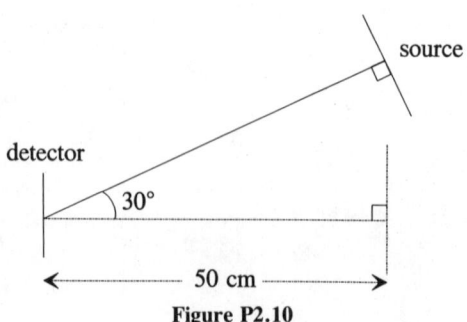

Figure P2.10

blackbody, at a temperature of 900 K. The detector is 1 mm × 1 mm, and the surrounding environment is deep space at 2.7 K, which is negligible.

2.11. Plot a curve of spectral radiant exitance $M_e(\lambda, T)$ for a blackbody of temperature 300 K, with at least 10 points on the curve between the λ's at which the value of M_λ has fallen to $\approx 10\%$ of its peak value. What wavelengths are the 50% points?

2.12. Consider a blackbody source at 1650 K, which is a planar disc of 1-cm diameter. Calculate, using the Planck radiation expression (do not use plots), how much flux (in watts) this source will emit between 3.0 μm and 3.2 μm.

2.13. The distance (r) between a graybody ($\varepsilon = 0.75$) at 900 K and a detector is 0.25 m. Source normal and detector normal are coincident with the optical axis:

$$\text{Given:} \quad A_s = 1 \text{ cm}^2; \quad A_d = 1 \text{ mm}^2$$

(a) What is the total radiant power (flux) emitted by the source?
(b) How much total radiant power (flux) falls on the detector?
(c) How much radiant power (flux) falls on the detector from 8 to 14 μm?

2.14. For the spectral regions shown, calculate the photon radiance from a 500-K blackbody using Eq. (2.100) and compare it to the values found from tables or exact integration: (a) 0–10 μm; (b) 8–13 μm; (c) 0–5 μm; (d) 3–5 μm.

2.15. Calculate the radiant exitance (in watts/cm^2) for a 500-K blackbody over the spectral region (0–10 μm). What is the mean energy per photon over this spectral region? How does it compare to a 5-μm photon energy?

2.16. On the same graph (use polar graph paper), plot intensity as a polar function (I versus θ) for the two different extended sources below:
(a) Source "A" is Lambertian, with radiance $L = 10$ W/cm^2-sr.
(b) Source "B" is non-Lambertian, with $L = [10 \text{ W/cm}^2\text{-sr}] \times \cos(\theta)$.

2.17. Find the value of flux on the detector for the following two situations (see figure):
(a) The source is Lambertian, with radiance 2 W/cm^2-sr; source size is 2 cm × 2 cm; $\theta_s = 30°$. Line-of-centers distance from source to detector is 0.75 m. Detector is 1 mm × 1 mm, with $\theta_d = 20°$.
(b) Repeat the calculation for an identical situation, except that the source is non-Lambertian, with radiance 2 W/cm^2-sr × $\cos^2\theta$.

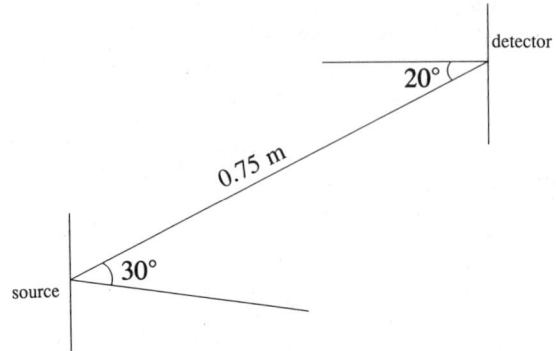

Figure P2.17

2.18. The local camera store salesperson suggests that you replace the 300-W light bulb in your slide projector with a 500-W bulb to get more light on the screen. A slide projector bulb is (for both cases) a tungsten lamp whose temperature is 2856 K. By simply replacing the lamp, how much increased screen irradiance do you get?

2.19. Derive the Wien Displacement Law for:
   (a) $M_{e,\lambda}(\lambda, T)$ W/cm^2-μm.
   (b) $M_{q,\lambda}(\lambda, T)$ photon/s-cm^2-μm.

2.20. Beyond what wavelength does the earth's spectral exitance exceed the solar spectral irradiance received on the earth? The sun (5500-K blackbody) is a disc that subtends 30 arc minutes from the earth. The earth's temperature is 300-K blackbody. Neglect atmospheric effects.

2.21. A detector/spectral filter whose temperature is 4 K (liquid helium) has a cold stop defining a 30° full field of view. The detector views an extended source at temperature 1000 K. The spectral filter has $a = 5\%$, $\iota = 0\%$, $t = 95\%$ for 3 μm $< \lambda < 5$ μm. Outside of this band (for $\lambda > 5$ μm and $\lambda < 3$ μm), the spectral filter has $a = 5\%$, $\iota = 95\%$, $t = 0\%$. Calculate the photon irradiance on the detector. If the spectral filter is not cooled, and is at room temperature of 300 K, recalculate the photon irradiance on the detector.

2.22. Find the value of the partial derivative of radiant exitance $(\partial M_{e,\lambda})/(\partial T)$, at $\lambda = 10$ μm, for a blackbody source of temperature 300 K. Find the value of $(\partial M_{e,\lambda})/(\partial T)$ at $\lambda = 8$ μm for the same source.

2.23. Plot the radiant exitance $M_{e,\lambda}(\lambda, T)$ versus the product λT for blackbody curves at 500 K and 1000 K. At what λT value do these curves peak?

2.24. Given a graybody source, with $\varepsilon = 0.6$, and a true temperature of 800 K:
   (a) The brightness temperature of the source in the region of 7 μm is

closest to:

$$700 \text{ K} \qquad 650 \text{ K} \qquad 675 \text{ K?}$$

(b) What is the color temperature? (Use $\lambda_1 = 2 \ \mu\text{m}$ and $\lambda_2 = 6 \ \mu\text{m}$.)

2.25. What is the brightness temperature of a graybody ($\varepsilon = 0.9$) with true temperature of 900 K, at $\lambda_o = 9 \ \mu\text{m}$? What is the radiation temperature?

2.26. A point source has uniform monochromatic intensity of 100 W/sr at 3.39 μm. It is 2 km from the collecting optic, which has a 0.25-m diameter and focal length (F):

(a) Case a: $F = 250$ cm   (b) Case b: $F = 1000$ cm

For each case, what is the average irradiance in the central lobe of the on-axis image, if the system is diffraction limited?

2.27. Derive the relationship between exitance (M) and radiance (L) for:

(a) Lambertian source: radiance $L = L_o$

(b) Non-Lambertian source: radiance $L = L_o \cos \theta$

(c) Non-Lambertian source: radiance $L = L_o \cos^2 \theta$

2.28. You are a soldier in an open field and have the choice of sleeping in a tent (40 ft $\times$ 40 ft) or outside near a fire to stay warm. The tent is a 310-K source and the fire is a 1650-K source, 1-ft^2 area. Either location is sufficient for sleeping. Which source produces a larger signal (assume that the signal is proportional to the total in-band flux radiated) to an enemy remote sensor operating in the 4–5-μm spectral region? Where would you sleep?

2.29. How many photons/s make 1 W of radiation at $\lambda_1 = 0.5 \ \mu\text{m}$? at $\lambda_2 = 10 \ \mu\text{m}$? If we have a detector that responds equally to any photon event (between λ_1 and λ_2), and we plot its output per watt of incident radiation, what will the curve look like, as a function of λ?

2.30. Prove that the irradiance on the wall of an integrating sphere (see figure) is a constant if it is coated with a Lambertian diffusing material. An integrating sphere has a highly reflective diffuse coating inside. (*Hint:* Assume that a small Lambertian source is on the interior surface of the sphere.)

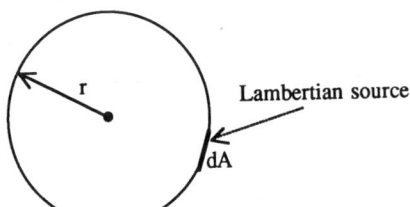

Lambertian source

dA

Figure P2.30

3

Optical-Detection Processes

3.1. OPTICAL-DETECTOR CLASSIFICATION

In this book, a sensor is assumed to detect optical radiation in the spectrum of 0.2 μm to 1000 μm (50,000 to 10 wave numbers). The detectors discussed are analogous to the rods and cones of the eyes; however, the physics or chemistry of the detection process is somewhat different. Although optical-sensor systems traditionally have been modeled after the human eye, recent research indicates that modalities other than image-forming systems can be used to detect, identify, and monitor targets of interest. The sensors that are used to detect optical radiation are confined to two classes: photodetectors and thermal detectors.

Photon detectors are devices or materials that detect light by a direct interaction of the radiation with the atomic lattice of the material. This interaction of light and matter produces parameter changes that are detected by associated circuitry or interfaces. Physical parameters that can change in a photon detector because of optical radiation are resistance, inductance, voltage, and current (external emission of electrons). Table 3.1 lists examples of these photon detectors.

Thermal detectors respond to the heating effects of the absorbed energy of the optical radiation by changing the temperature of the sensor. This process requires two steps: the radiation must change the temperature of the detector, and then this temperature change causes or induces some measurable parameter change. The physical parameter change is then detected by associated instrumentation. The most typical example would be the resistance change in a semiconductor caused by a temperature change. Another example is that of a superconductor switched between zero resistance and some finite resistance because of the temperature-induced heating of the radiation. Examples of thermal detectors and their mechanisms include:

Bolometers	Resistance
Pyroelectrics	Capacitor
Thermocouples	Voltage
Golay Cells	Mechanical displacement
Superconductors	Resistance

Table 3.1 Examples of Optical Photon Detectors

Si	PC or PV
PbS	PC
PbSe	PC
Ge:Hg	PC
Ge:Cu	PC
CdS	PC
Ge	PV
Si:As	PC
Si:Ga	PC
HgCdTe	PC or PV
GaAs	PV
InSb	PV
AgO:Cs	PE

Notes: PC indicates photoconductive, PV indicates photovoltaic, and PE indicates photoemissive.

Because thermal-detection processes require the absorbed radiation to change the temperature of the detector element, the time response is slower than that in the corresponding photodetectors, for which absorption of a photon generates a hole–electron pair faster.

First we examine the photon-detection process. Thermal detectors are discussed in detail in Chapter 9, where their response to optical radiation is developed analytically. However, some elementary detection-process concepts must be understood to fully appreciate the limitation of sensitivity imposed by noise processes within these devices. (For additional readings, see Boyd, 1983; Dennis, 1986; Kruse et al., 1962; Vincent, 1990.)

3.2. PHOTON-DETECTION MECHANISM

To understand the photon-detection process, the concept of converting a photon of light into an electron must be reviewed. The efficiency of converting a photon to an electron is defined as quantum efficiency (η). It is a normalized value always less than one and relates the number of photons incident on the detector's active area to the number of independent electrons generated. The restriction to count only the independent electron events excludes any post-detection gain mechanisms internal to the device, which are not counted as part of the quantum efficiency.

The quantum efficiency is the number of independent electrons produced per photon, normally expressed in percent. Quantum efficiency takes into account reflectance, absorptance, scattering, and electron recombination.

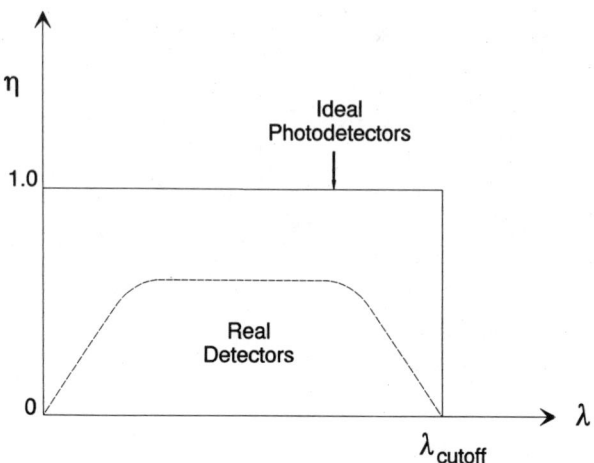

Figure 3.1.   Ideal quantum efficiency vs. wavelength.

The ideal quantum efficiency is a binary function of wavelength. Either the photon has sufficient energy to produce a photogenerated electron or it does not. As monochromatic optical radiation moves further into the infrared, the photon energy decreases and does not have sufficient energy to create a free electron. This limit defines the upper cutoff wavelength (λ_c) of response.

A plot of quantum efficiency versus wavelength is shown in Fig. 3.1 for the ideal case, where λ_c is the cutoff wavelength, or the longest wavelength at which the detector can detect optical radiation. Again, not all the photons in the spectral region from 0 to λ_c produce conduction electrons. Table 3.2 lists different detectors and their quantum efficiencies, which is the percentage of photons that produce conduction electrons. Ideal detectors would have 100% quantum efficiency.

The electron–hole generation rate (number/volume-s) as a function of distance into the detector is (Sze, 1969)

$$g(x) = (1 - \imath)E_q a e^{-ax} \tag{3.1}$$

Table 3.2   Quantum Efficiency Values for Given Photon Detectors

Detector Type	η (%)
Photoconductor (intrinsic)	~60
Photoconductor (extrinsic)	~30
Photovoltaic	~60
Photoemissive	~10
Photographic film	~1

where

t = Fresnel reflectance
E_q = photon irradiance
a = absorption coefficient

and the photon irradiance is incident on a detector as shown in Fig. 3.2. The magnitude of the current density (amp/cm^2) is then

$$J = q \int_0^{\ell_x} g(x)\, dx$$

$$= (1 - t)\, q\, E_q \int_0^{\ell_x} a e^{-ax}\, dx \tag{3.2}$$

where t is the Fresnel reflectance and q is the magnitude of the charge on an electron. Performing the integration yields

$$J = (1 - t)\, q\, E_q (1 - e^{-a\ell_x}) \tag{3.3}$$

where ℓ_x is the detector dimension in the direction of propagation as shown in Fig. 3.2. Rearranging terms, the quantum efficiency can be expressed as

$$\frac{\text{number electrons}}{\text{number photons}} = \frac{J}{q\, E_q} \equiv \eta = (1 - t)(1 - e^{-a\ell_x}) \tag{3.4}$$

The quantum efficiency tends to be high when the Fresnel reflectance is low because the radiation can get into the detector material. The absorption path length (ℓ_x) needs to be large to ensure absorption of the optical radiation. The

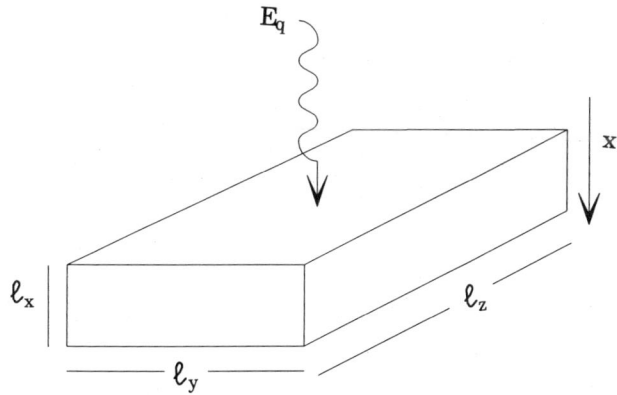

Figure 3.2. Irradiance on an optical detector.

absorption coefficient is a strong function of wavelength and, in particular, whether the photon has more or less energy than the energy gap of the detector material. For high-energy photons, the absorptivity is (Pankove, 1971)

$$a = a_0(h\nu - \mathcal{E}_g)^{1/2} + a_0' \quad \text{for} \quad h\nu > \mathcal{E}_g \quad (3.5)$$

For photon with energy less than the energy gap of the material (see Fig. 3.3):

$$a = a_0' \exp\left\{\frac{h\nu - \mathcal{E}_g}{kT}\right\} \quad \text{for} \quad h\nu < \mathcal{E}_g \quad (3.6)$$

Typically for InSb (Schoolar and Tenescu, 1986), $a_0 \approx 1.9 \, (10^4) \, \text{cm}^{-1}$ and $a_0' = 800 \, \text{cm}^{-1}$. For HgCdTe material, these values are similar within 25%.

Quantum efficiency is a function of wavelength to the extent that the Fresnel reflectance and absorption coefficient are related to the wavelength of the radiation. Fresnel reflectance, which varies because the index of refraction is a function of wavelength, usually affects short-wavelength response. Long-wavelength photons tend to be transmitted by the detector material and are not absorbed (a_0 is a function of wavelength as seen in Fig. 3.3), causing a loss in quantum efficiency (Bratt, 1977).

The cutoff wavelength is a generic detector parameter; however, the photon energy ($h\nu$) must be sufficient to cause an electron in the valence band to be released to the conduction band and be detected.

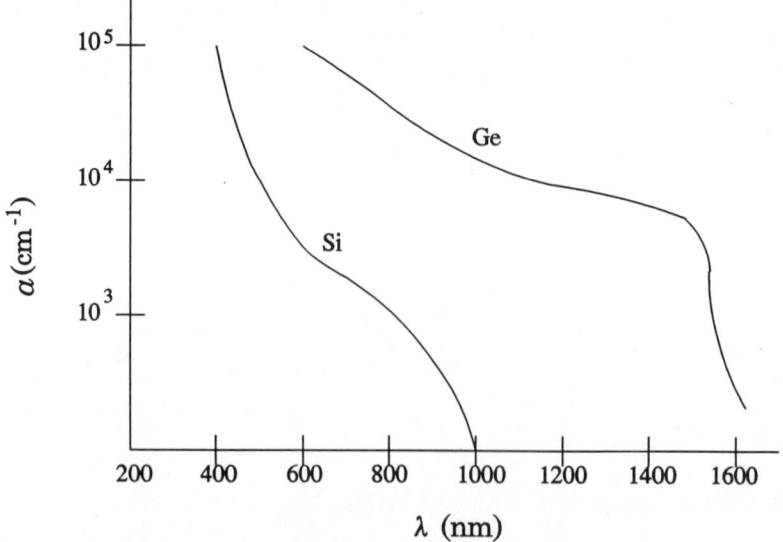

Figure 3.3. Absorption coefficient a of silicon and germanium at 300 K.

The energy of the photon must be equal to or greater than the energy gap of the detector material for the external physical parameter to change. Mathematically,

$$hv \geq \mathcal{E}_g \tag{3.7}$$

where

h = Planck constant (6.6×10^{-34} J s)
v = optical frequency of photon (cycles/s)
$\mathcal{E}_g$ = energy gap of detector material (electron volts).

Because the optical frequency can be related to the speed of light (c) and wavelength of radiation, as illustrated by Eq. (1.1), and the lowest energy photon that can be detected is at wavelength λ_c is one whose energy just equals the energy gap:

$$\mathcal{E}_g = \frac{hc}{\lambda_c} \tag{3.8}$$

where the equal sign assumes the longest wavelength (lowest energy) photon. Rearranging terms, substituting for the constants, and correcting for units such that wavelength is in microns and energy gap in electron volts yields

$$\lambda_c \ (\mu m) = \frac{hc}{q \ \mathcal{E}_g} = \frac{1.238}{\mathcal{E}_g(eV)} \tag{3.9}$$

Therefore, the energy gap of the detector material determines the cutoff wavelength. Small energy-gap materials are capable of detecting radiation far into the infrared, that is, $\mathcal{E}_g \approx 0.01$ eV corresponds to $\lambda_c \approx 123.8 \ \mu m$.

Particularly for detector materials with small energy gaps, electrons can be thermally excited into the conduction band (dark current). However, photogenerated electrons are the desired signal, not thermally generated electrons. The Boltzmann distribution tells us that the population of energy levels follows an exponent, $\exp\{-\mathcal{E}_g/kT\}$; thus, as $\mathcal{E}_g$ becomes smaller, keeping temperature T constant, the number of thermally generated electrons increases. These dark-current-generated electrons are prevented from outnumbering the photogenerated electrons by cooling the detectors to a cryogenic temperature. This cooling prevents the thermally generated (kT) electrons from dominating the electron-generation process. By cryogenic cooling, essentially no "phonon"-generated electrons are present, which also eliminates the shot noise associated with these dark-current-generated electrons. Obviously, detectors with a longer cutoff wavelength must be cooled to a lower temperature than detectors with a shorter cutoff wavelength (Beyen and Pagel, 1967). We now consider some generic photon detectors and their corresponding quantum efficiencies.

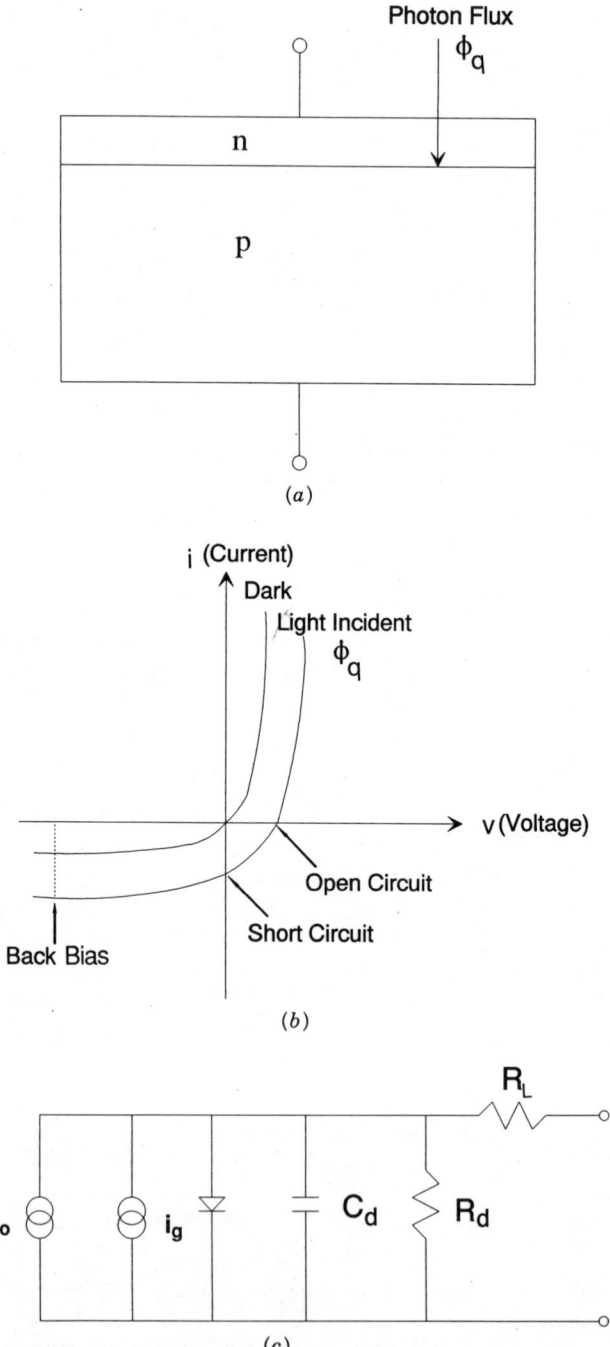

Figure 3.4. (a) Photovoltaic detector and electrical symbol. (b) Characteristic curve for voltage and current for photodiode. (c) Photovoltaic detector equivalent circuit.

3.2.1. Photovoltaic Detectors

The photovoltaic detector produces a voltage or current from incident optical radiation. It is often called a photodiode because it is a semiconductor diode (p-n junction), which is light sensitive. Figure 3.4a is a cross section of the p-n junction as a two-port electrical device, and Fig. 3.4b shows the corresponding characteristic diode plot relating voltage and current by the Schottky equation (Streetman, 1990).

The photogenerated current (i_g) produced by photons is

$$i_g = \eta \phi_q q \tag{3.10}$$

The diode equation for the characteristic curve, Fig. 3.4b, is modified by the addition of this photogenerated current

$$i = i_0(e^{qv/bkT} - 1) - i_g \tag{3.11}$$

where

i_o = reverse saturation current
v = voltage across diode
T = temperature
k = Boltzmann's constant
b = nonideality factor

The electrical equivalent circuit is shown in Fig. 3.4c for a photovoltaic.

The expected quantum efficiency is around 65% for this type of detector. Some examples of photovoltaic detectors and corresponding cutoff wavelengths are shown in Table 3.3. The corresponding electrical parameters (detector resistance R_d, detector capacitance C_d, load resistance R_L, and cross-sectional area A_d) are shown in Fig. 3.4c. Note particularly the temperature dependence of the resistance-area product for HgCdTe-type detectors. Typically, the resistance is described in ohm/square, which means that detectors with the same length-to-width ratio have equal resistances. The exact dimensions are not important.

3.2.2. Photoconductors

Photoconductors respond to light by changing resistance or conductance. The resistance (R_d) is inversely proportional to the photon flux (Bube, 1978):

$$R_d \propto \frac{1}{\phi_q} \tag{3.12}$$

Table 3.3 Examples of Photovoltaic Detectors

	η (%)	R_d (Ω/square)	λ_c (μm)	C_d/A_d (pF/mm^2)	T Nominal (K) Temperature	R_d or C_d Dependence
InSb	45	10^9	5.5	450	77	$R_d \propto \dfrac{1}{T^2}\, e^{1.24/k\lambda T}$
HgCdTe						
$x = 0.3$	65	10^7	5		77	$R_d A_d = 7.6(10^{10})\, e^{-0.124704T}$
$x = 0.2$	65	$2(10^3)$	11		40	$R_d A_d = 6988 - 163.6T + 0.954T^2$
InAs	40	10^8	3.3	450	77	$R_d \propto \dfrac{1}{T^2}\, e^{1.24/k\lambda T}$
InGaAs	86	10^8	1.7	30 to 75	Room	
PIN (Si)	80		1.1	2 to 8	Room	
Aval (Si)	70		1.1	5 to 66	Room	
GaAsP	60	$13(10^8) \ll 43(10^8)$	0.7	180 to 400	Room	
Si	65	$10^8 \ll 35(10^8)$	1.1	35 to 100	Room	
Ge	64	10^4 to 10^6	1.8	250 to 333	Room	$C_d(v) = C_o[0.996 - 0.541v + 0.189v^2 - 0.035v^3 + 0.0031v^4 - 0.0001v^5]$

Table 3.4 Intrinsic Photoconductor Detectors

Material	Typical η	λ_c (μm)	Resistance R_d (Ω/square)	Time Constant τ (μs)	Nominal Temp (K) T	$R_d \propto fn\,(T)$	$\tau\,(\mu s) \propto fn\,(T)$
PbS	50	3	1.8 (10^5)	200	300	$R_d = 1.7(10^9)e^{-0.030405T}$ [190 < T < 300]	$\tau = 8(10^5)\,e^{-0.028375T}$
PbSe	50	5	2.5 (10^5)	1.5	300	$R = 7(10^9)\,e^{-0.0341312T}$ [190 < T < 300]	$\tau = 2.5(10^4)e^{-0.0315896T}$
HgCdTe	60	25	20–150	1.0	77		

R function of initial fabrication techniques. However, the temperature coefficient remains constant.

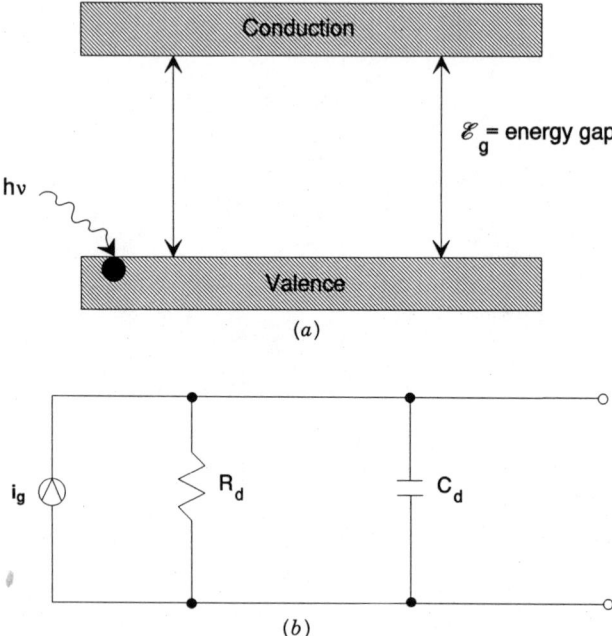

Figure 3.5. (*a*) Intrinsic semiconductor energy-band diagram. (*b*) Photoconductor detector equivalent circuit.

Examples of intrinsic-photoconductor optical detectors are shown in Table 3.4. The energy band of an intrinsic semiconductor is shown in Fig. 3.5*a*, where the photon must have sufficient energy to cause an electron to move from the valence band to the conduction band. The equivalent circuit of a photoconductor is shown in Fig. 3.5*b*.

The corresponding quantum efficiency is around 60% for the intrinsic materials. To get a longer wavelength response toward the infrared region (i.e., smaller energy gap) extrinsic photoconductors can be used. Extrinsic photoconductors, however, have two less-desirable characteristics. Because of their smaller bandgap, they require a lower operating temperature than do intrinsics. Also, their quantum efficiency is substantially lower than intrinsic detectors, around 30%, because the extrinsic dopant species is necessarily more sparse than the host material, resulting in a smaller optical-absorption cross section. A schematic of the mechanism for extrinsic photoconductive detectors is shown in Fig. 3.6. Typical properties, including cutoff wavelengths, are shown in Table 3.5 (Sclar, 1984).

The change in resistance in the case of the extrinsic photoconductors is always attributable to the majority carrier concentration for the doped semiconductors. To use a photoconductive detector, one must utilize a biasing circuit to detect the change in resistance. A standard biasing circuit is shown in

(a) P-Type Extrinsic

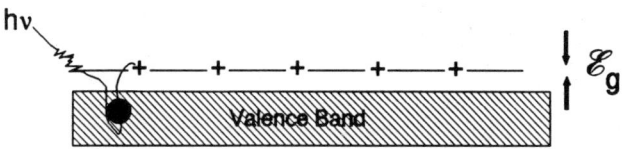

(b) N-Type Extrinsic

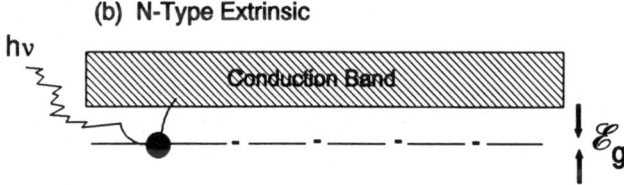

Figure 3.6. Extrinsic semiconductor energy-band diagram.

Table 3.5 Extrinsic Photoconductors

Material	Typical η (%)	λ_c (μm)	Time Constant τ (μs)	Nominal Temperature (K) T
Ge : Hg	30	14	0.1	4
Ge : Cu	30	27	0.1	4
Ge : In	30	120	0.1	4
Si : As	40	24	—	4
Si : In	40	8	—	45
Si : Ga	40	19	—	18
Si : P	40	29	—	12

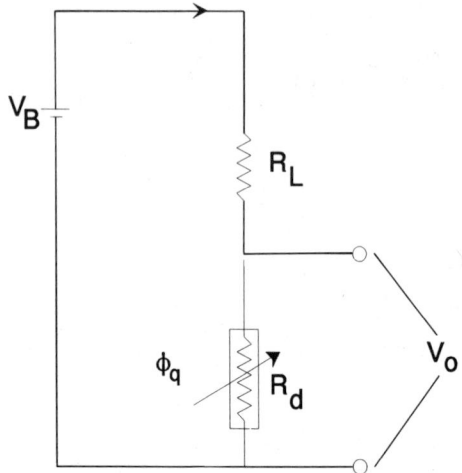

Figure 3.7. Photoconductor biasing circuit.

Fig. 3.7 with a load resistor (R_L), a dc voltage bias source (v_B), and the output voltage (v_o) that varies in response to the photon flux incident on the detector (ϕ_q).

From the circuit, the output voltage, v_o, is

$$v_o = \frac{v_B R_d}{R_d + R_L} \tag{3.13}$$

The signal voltage attributable to a change in detector resistance is

$$dv_0 = \frac{v_B R_L}{(R_d + R_L)^2} \, dR_d \tag{3.14}$$

where small-signal approximations are made. By differentiating Eq. (3.12) to obtain the change in resistance and substituting into Eq. (3.13),

$$dv_0 \propto \frac{v_B R_L}{(R_d + R_L)^2} \frac{d\phi_q}{\phi_q^2} \tag{3.15}$$

Therefore, a change in photon flux incident on the detector produces a change in signal voltage.

3.2.3. Photoemissive Detectors

Photovoltaic and photoconductive detectors operate with internal photoelectron emission. This section describes those detectors with *external* photoelectron emission. The excited electron physically leaves the detecting material

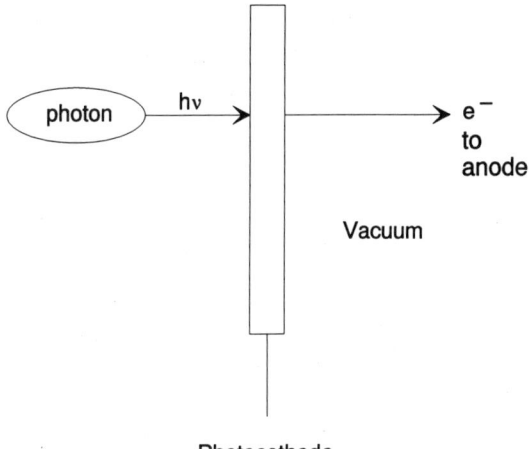

Figure 3.8. Detection process for photoemissive detector.

and moves in free space. The sequence of events is illustrated in Fig. 3.8, where the photocathode is the detecting material (Engstrom, 1980).

A photon is absorbed in the photocathode material that has deposited on an optically transparent substrate. An energetic electron is formed by absorption of the energy of the photon, which travels to the vacuum-side surface of the photocathode. The electron has sufficient kinetic energy to overcome the work function, which tends to contain it in the photocathode material, and thus is externally emitted into the vacuum as a free electron. Using an electric field, the electron can be pulled to an anode and be counted as current. The photocathode must have strong optical absorption, a long mean-free path for the electron to travel, and a low electron affinity ($\sim$ work function). This is the fundamental event that makes a photoemissive detector operate.

Photocathode materials are listed in Table 3.6. Their quantum efficiency is low and the spectral response of these devices is limited to the visible and near infrared. The photo-generated current (i_g) for a photon flux (ϕ_q) is $i_g = \eta \, \phi_q \, q$.

The quantum efficiency of photoemissive materials in this wavelength region is higher than that of photographic film, which has a quantum efficiency around 1%.

Table 3.6 Photocathode Characteristics

Material	$\lambda_c(\mu m)$	η (%)
Ag-O-Cs (S1)	1.1	1
Cs-Nag-K-Sb (S20)	0.9	20
GaAsP (NEA)	0.9	30

3.3. THERMAL-DETECTION MECHANISMS

Optical radiation incident on a thermal detector causes the detector temperature to increase because of absorbed radiation. This temperature rise causes some physical parameter to change (such as resistance or voltage). Thermal detectors are energy devices where the energy absorbed depends on the absorption properties of the detector. The spectral response of thermal detectors is determined by the spectral dependence of the surface absorptance. As previously discussed, absorptance is related to emissivity so the thermal detector's spectral dependence is directly related to the emissivity spectral dependence $[\varepsilon = fn(\lambda)]$. This is in contrast to the photon detectors in which the radiation interacts directly with the lattice sites, giving rise to excess carriers, which in turn produce a resistance or current change in the photon detector material.

Values of thermal mass (specific heat × mass) and thermal conductance are associated with the thermal detector, and determine the response time. Several methods or technologies convert (transduce) the temperature change within the detector material into a usable physical parameter. Some examples of thermal detectors that use heat to cause a mechanical displacement are shown below:

EXAMPLE	THERMAL ABSORBER
Thermometer	Hg in a cylindrical glass lens
Crooke's radiometer	Four-panel windmill absorber in a vacuum
Golay cell	Gas absorber in a chamber with membrane
Thermocouple strip	Bimetal spring in furnace temperature control

Some examples of thermal detectors that respond to optical radiation and produce an electrical parameter change are shown below:

EXAMPLE	PARAMETER CHANGE DUE TO TEMPERATURE CHANGE
Bolometer	Resistance
Thermocouple	Voltage
Pyroelectric	Capacitance/ferroelectric material
Superconductor	Resistance

To examine the differences between thermal detectors and photon detectors, some generalities are appropriate in terms of spectral, temporal, and operation temperature. Thermal detectors are typically slower responding devices because their thermal mass must experience a rise and fall in temperature, which is a slow process. Obviously, we want the largest temperature change per unit radiant power, which means a small thermal mass or well-isolated detector. In this case, the heat will not dissipate very fast and therefore the decay time is longer. Thus, response speed and responsivity are a tradeoff. In addition, the spectral response is much broader for the thermal detector than for the photo-detector. Typically, for a thermal detector, the wavelength response may extend from the visible to 40 μm in the infrared. The operating temperature for thermal detectors used in the infrared is typically room (ambient) temperature, not cryogenic as is the case for photon detectors (a notable exception is the low-temperature bolometer). The thermal detector can be thought of as two essential parts: the absorber and the temperature sensor. Some examples of thermal detectors and their properties are shown in Table 3.7.

3.3.1. Bolometers

Bolometer operation is most commonly based on the change in the electrical resistance of semiconductor materials as a function of temperature. These semiconductor bolometers are often called *thermistor bolometers*, or *thermally sensitive resistors* (Astheimer, 1983). Bolometer technology takes advantage of semiconductor materials, which have a high temperature coefficient of resistance (about 3.5% change per degree Celsius).

Bolometer–semiconductor materials include synthetic diamond, germanium, sintered oxides of manganese, cobalt, and/or nickel. These chips are mounted to heat-dissipating substrates or thermal sinks. When the radiation strikes the detector, the detector temperature increases. When the radiation is removed, the detector returns to the temperature of the thermal sink. Practical limitations dictate the size of the detector and mounting. Recall that smaller detector chips produce larger temperature changes but longer response times. The apparent optically active area is often increased by using a hemispherical or hyperhemispherical lens secured to the detector to increase the radiant power collected by the detector. These immersion lenses use optical gain to cause a larger apparent detector area. An immersion lens is typically fabricated from a high-index-of-refraction material such as silicon or germanium. Performance improvements from 3.5 to 16 are typically attained. Figure 3.9 shows a cross section of a hemispherical immersion lens mounted on a bolometer.

The bolometer is similar to the photoconductor in end results, that is, optical radiation produces a resistance change of the chip. A photoconductive electrical interface could be used, as shown in Fig. 3.7; however, a more typical and more accurate method is to use a bridge circuit as shown in Fig. 3.10. The resistance of a semiconductor bolometer can be expressed exponentially as

$$R_d = R_0 e^{\mathscr{b}/T} \tag{3.16}$$

where R_0 = ambient resistance at nominal temperature, $\mathscr{b}$ = material constant.

The resistance change that results from an optically induced temperature change can be expressed as the differential of Eq. (3.16):

$$dR_d = R_0 e^{\mathscr{b}/T} \left(-\frac{\mathscr{b}}{T^2} \right) \tag{3.17}$$

Table 3.7 Thermal Detector Examples

Types	Detectivity D^* cmv/Hz W^{-1}	Capacitance (pF)	Resistance (Ω)	Area (mm²)	Operational Temperature (K)	Temperature Coefficient α	Spectral Range	Time Constant (ms)	Comments
Pyroelectric	10^9	15–250	10^{13}	0.78–63.6	—	218–358	1 nm–1 mm	10^{-4}	$C(d) \approx -34.2 + 29.5\, d$, high detectivity at high frequencies
Thermistor	$3(10^8)$	—	Type 1: 2500 Ω-cm Type 2: 250 Ω-cm	0.5×0.5 to 5×5	300	$\alpha_1 = -4.2\%/K$ $\alpha_2 = -3.8\%/K$	Flat response over entire infrared spectrum	5	Thermal runaway, immersion lenses increase detectivity
Thermopile	$7(10^8)$	—	$(6–15)\,10^3$	2×2	208–343	—	200 nm–30 μm	5	Ideal for dc applications, Johnson noise, preamp, performance, long time constants
Bolometer (Ge)	BLIP	—	$(3–9.6)10^6$	0.25×0.25 1×1 4×4 (composite)	0.3–2.0	-2 to -45 K^{-1}	1.6 μm–millimeter region	0.5	Higher values of α in Ge-diamond composite used for long wavelengths

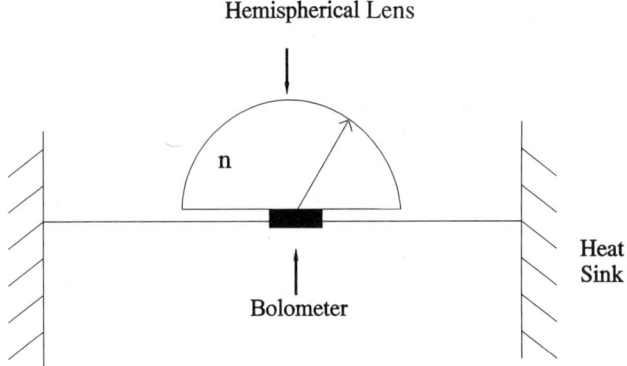

Figure 3.9.   Use of an immersion lens with a bolometer.

The relative resistance change can be expressed as a ratio of Eqs. (3.16) and (3.17) or the temperature coefficient (α) as

$$\frac{1}{R_d} \frac{dR_d}{dT} = \alpha \tag{3.18}$$

The bridge circuit is constructed with two identical bolometers (R_d in Fig. 3.10) mounted close together with one covered or encapsulated to prevent optical radiation impingement. The second bolometer is used as a detector. When there is no infrared radiation on the bolometer, the bridge is balanced and no current flows through resistor R_2, if $R_1 = R_3$. Incident optical radiation causes the resistance to decrease in the variable bolometer (see Eq. (3.16)). This unbalances the bridge circuit and a voltage crosses R_2. The signal voltage output

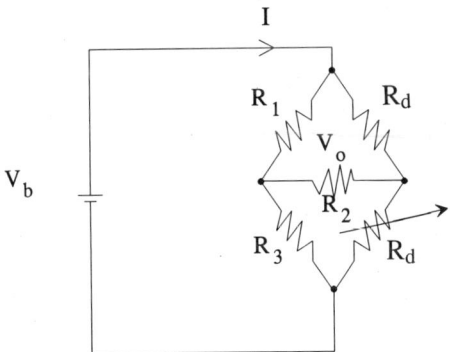

Figure 3.10.   Bridge circuit for bolometer electrical interface.

across R_2 for the circuit shown is

$$dv = \frac{v_B}{4R_d} dR_d \qquad (3.19)$$

3.3.2. Thermopile Detectors

The thermopile is the most direct means of producing an electrical voltage from optical radiation. It is a series connection of thermocouples used to increase the voltage output and is a junction of two dissimilar materials. Alternate junctions are defined as hot and cold junctions. *Hot junctions* receive radiant power by application of a high-emissivity surface and thermal isolation from the thermal heat sink. The *cold junctions* or reference junctions are connected to the thermal heat sink and remain relatively stable in temperature. When optical radiation is incident on the thermopile, the temperature of the hot junction changes and a voltage is generated proportional to the temperature difference between the hot and cold junctions. The optical radiation must be absorbed on the thermopile to be detected, so the surface is typically coated with a black substance.

Three thermoelectric effects describe the physical processes of thermopotential development. These effects have coefficients that are used to evaluate various limited junctions: (1) Seebeck coefficient, (2) Peltier coefficient, and (3) Thompson coefficient (Stevens, 1970).

The series connections between junctions raise the voltage to a level that is sufficiently high to be handled by electronic amplification. Thermopiles are relatively slow-responding devices and are typically used with chopped radiation at frequencies less than 25 Hz. Because the thermocouple is a metal-to-dissimilar-metal junction, the resistance is very low (a few ohms), and the preamplifier design must be modified relative to the conventional detector circuit. The classic approach is to transform couple the thermopiles; however, operational amplifier circuits are currently more commonly used. Figure 3.11 shows both the transform-coupled thermopile and the amplifier-circuit preamplifier design.

3.3.3. Pyroelectric Detectors

Pyroelectric detectors are the most common thermal detector. They are the most sensitive of the room-temperature thermal detectors and are fairly inexpensive. The pyroelectric effect was discovered over 2000 years ago by the Greeks. They observed that tourmaline, placed in hot ashes, would produce a change in surface charge. When hot, the material would attract ashes, and upon cooling would repel them (Dietrich, 1985).

In a pyroelectric device, a change in temperature creates a change in polar-

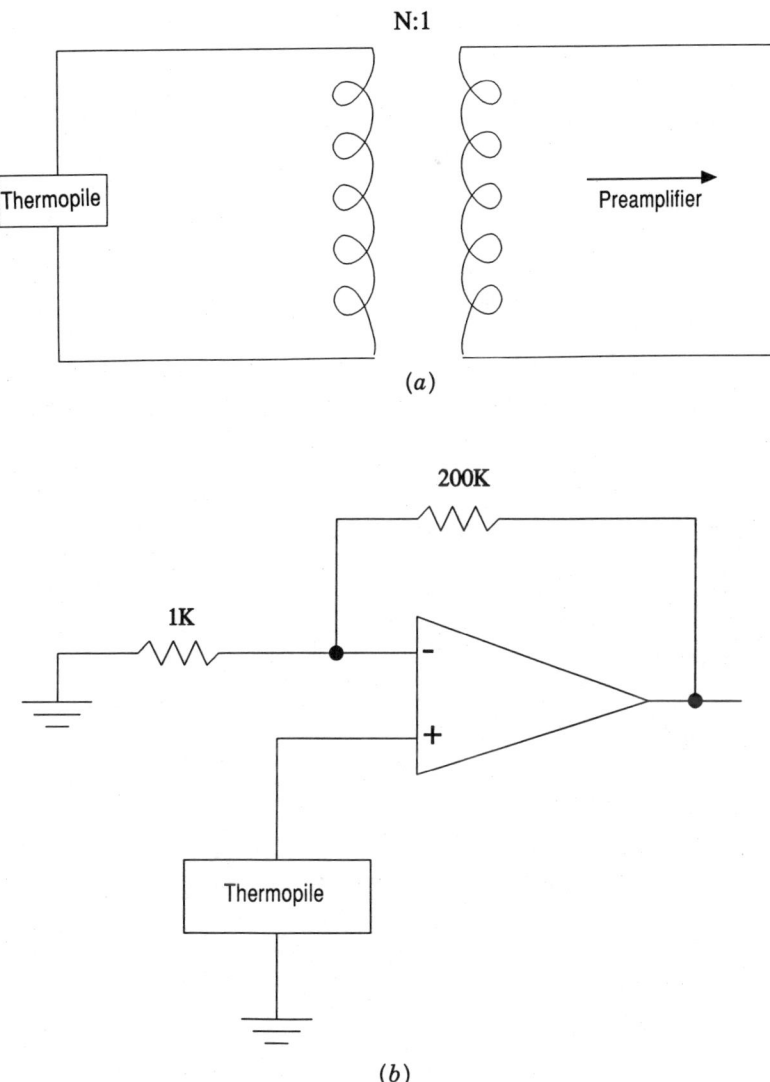

Figure 3.11. (a) Thermopile transformer coupling. (b) Operational-amplifier interface circuit for the thermopile.

ization. Electrical polarization change is related to a surface-charge change with respect to time (an electrical current). Thus a pyroelectric detector produces current only as it experiences a temperature rise or fall. If the temperature is constant, no current is produced. A pyroelectric effect is unlike the thermoelectric effect used in thermopiles, where a constant voltage is produced from two junctions of two dissimilar metals when they are held at different fixed temperatures. A pyroelectric detector can be operated at very high chopping frequencies, provided the loss in sensitivity can be tolerated (Marquardt et al., 1989). Pyroelectric detectors are also becoming very important in two-dimensional staring arrays (Hanson, 1993; Goss, 1987).

Pyroelectric devices rely on materials (such as lithium tantalate) that exhibit a polarization change when heated. In optical detectors, this heating is provided by incident photons. Because pyroelectric detectors are thermal detectors, their spectral response is very broad and thus of interest for infrared applications. Typical coatings used to enhance absorption are gold black, carbon black, and organic black. Paints typically are not used because they increase the mass of the sensor.

The electrical property sensed is the change in the spontaneous polarization of a dielectric material. The polarization change itself is very fast. However, the thermal properties of the detector limit the high-frequency response. In spite of this, and because of the relatively large magnitude of the pyroelectric effect, higher detectivities at higher frequencies have been obtained with pyroelectrics than with any other thermal detectors.

A pyroelectric material has a high degree of crystalline asymmetry, and thus possesses an electric dipole moment (without an applied electric field). An observable change in the surface charge density can be produced by changes in the internal dipole moment when the material undergoes a temperature change (Putley, 1970). Pyroelectric devices differ from piezoelectric devices only in that piezoelectrics use strain instead of temperature as a means to produce polarization change. Figure 3.12 shows a generic pyroelectric element with electrodes on top and bottom (Marshall, 1978).

The pyroelectric coefficient p measures the rate of change of electric polarization with respect to temperature. In conventional pyroelectrics, the magnitude of the internal polarization, P, is equivalent to the magnitude of the elec-

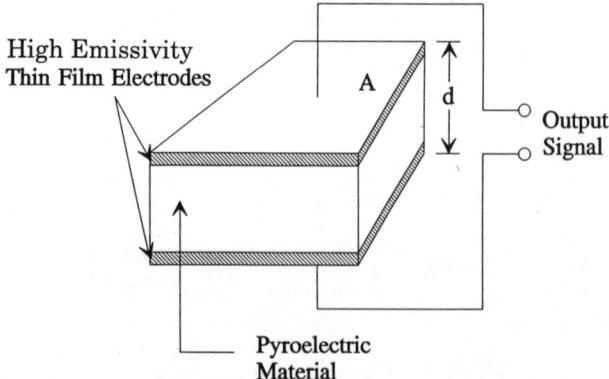

Figure 3.12.   Standard pyroelectric detector geometry.

tric displacement, D, which is stable in the absence of a bias electric field. The pyroelectric coefficient is defined as (Shorrocks, 1990)

$$p = \frac{dP}{dT} \approx \left(\frac{dD}{dT}\right)_{E=0}$$ (3.20)

A broader range of detector materials requires a bias field to induce the displacement, resulting in a more general coefficient

$$p(E) = \left(\frac{dD}{dT}\right)_E$$ (3.21)

This produces a stable response in ferroelectric materials operated below and above their temperature of phase transition (the Curie temperature T_c). This is known as the *dielectric bolometer mode of operation* and can result in large pyroelectric coefficients. The spontaneous polarization has a temperature dependence as shown in Fig. 3.13. The derivative of the polarization is the pyroelectric characteristic versus temperature. The Curie temperature shown in Fig. 3.13 is the temperature above which ferromagnetic materials become paramagnetic (Halliday and Resnick, 1981).

Loss of polarization in the ferromagnetic material can occur from temperature, age, or mechanical shock. The material can be returned to its ferromagnetic state by application of an external field while the material is cooled from a temperature above the Curie temperature. This process is known as repoling. The basic pyroelectric detector current output is

$$i = A_d p \frac{dT}{dt}$$ (3.22)

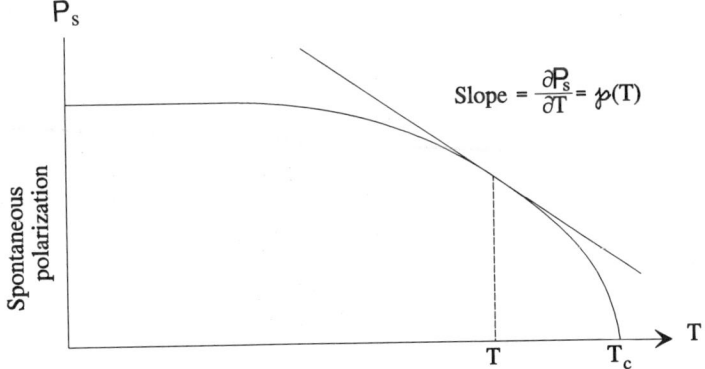

Figure 3.13. Spontaneous polarization vs. temperature.

where

    p = pyroelectric coefficient at operating temperature
    A_d = detector area
    dT = temperature change
    dt = time change

The principal materials used for pyroelectric devices are triglycerine sulphate (TGS), lithium tantalate ($LiTaO_3$), strontium barium niobate (SBN), ceramic materials based on lead zirconate titanate (PZT), and polymer films such as polyvinyl fluoride (PVF) and polyvinylidene fluoride (PVF_2). These are listed with their relevant properties in Table 3.8. The literature has a fairly wide range of values for these parameters.

TGS is the most sensitive material but it is water soluble, relatively fragile, and has a low Curie temperature (49°C). Doping TGS with L-alanine, an amino acid, causes the material to retain a preferred poling direction even after heating above the Curie temperature. Also, the doped material has a higher pyroelectric coefficient, lower dielectric constant, and lower dielectric loss, all of which improve detector performance (Putley, 1970, 1977). Ideally the material should possess a high pyroelectric coefficient, low dielectric constant, high thermal conductivity, low specific heat, and low density.

A low-noise preamplifier is imperative in pyroelectric detection circuits. Typically, a junction-field-effect-transistor (JFET) circuit such as shown in Fig. 3.14 is used. Figure 3.14 also shows a high-impedance pyroelectric detector, effectively, a capacitor, and a common preamplifier that uses an operational amplifier. Because of the high impedance of this capacitor-type detector, a field-effect transistor circuit is often built into the detector capsule to reduce the effective output impedance and eliminate stray capacitance.

3.3.4. Golay Cells

The Golay cell as an infrared detector was first developed in 1947 by Golay (1947, 1949). Today, its sole use is in the very far infrared ($\lambda > 300 \ \mu$m). It is the most sensitive room-temperature far-infrared detector available. The detector assembly consists of three parts: the gas chamber, optical readout, and electrical readout. An infrared-radiation absorbing element is placed in a chamber filled with pressurized gas. An infrared transmitting window allows radiation into the chamber; when absorbed, it heats up and by convection heats the gas in the chamber. The increased temperature of the gas results in increased pressure in the chamber. By the ideal gas law

$$\mathfrak{P}V \propto T \tag{3.23}$$

If the volume is constant, the pressure and temperature are directly related.

The chamber volume is controlled by a solid structure completely sur-

Table 3.8 Properties of Pyroelectric Materials

Material	Curie Temperature °C	Pyroelectric Coefficient $C\,cm^{-2}\,K^{-1}$	Dielectric Constant	Thermal Conductivity $W\,cm^{-1}\,K^{-1}$	Specific Heat $J\,g^{-1}\,K^{-1}$	Density $g\,cm^{-3}$
TGS	49	3×10^{-8}	30	6.8×10^{-3}	0.96	1.69
LiTaO$_3$	618	6×10^{-9}	58		0.42	7.45
BaTiO$_3$	126	2×10^{-8}	160	9×10^{-3}	0.5	6.0
LiNbO$_3$	1190	4×10^{-9}	30			4.64
SBN	115	6×10^{-8}	380		0.4	5.2
PVF$_2$	120	3×10^{-9}	10	4×10^{-5}		
PZT	200	3.5×10^{-8}	250			

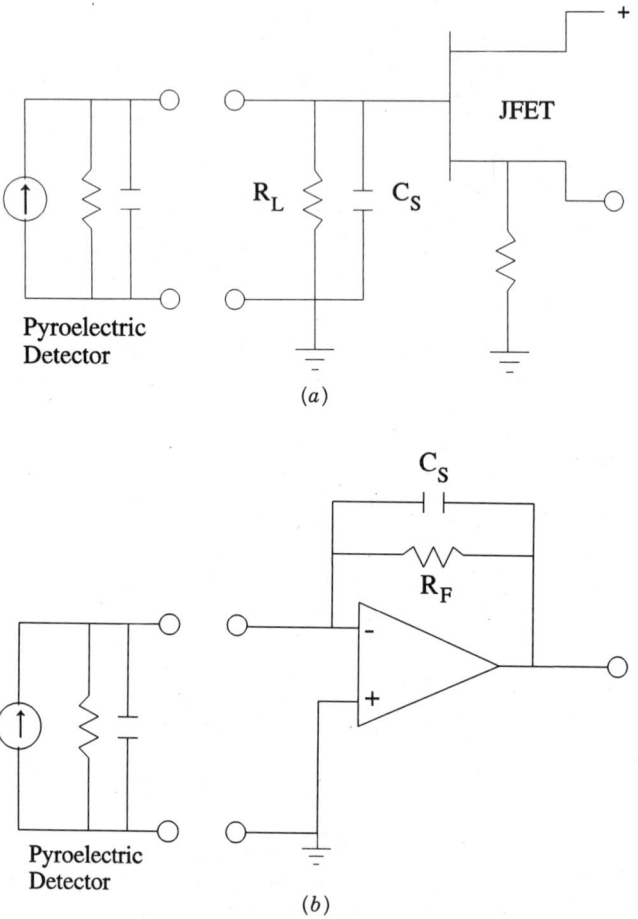

Figure 3.14. Pyroelectric detector electronic interfaces: (*a*) detector connected with source follower; (*b*) detector connected with high-impedance amplifier.

rounded except for a small flexible membrane with a reflective coating on the exterior of the membrane, which acts as a mirror. As the pressure changes (because of changes in temperature), the mirror flexure changes. The amount of distortion is sensed by a separate optical system consisting of a light source and photodetector. The entire system is shown in Fig. 3.15.

The use of light-emitting diodes (LEDs), modern solid-state detectors, low-noise electronics, and basic advances in plastic-membrane technology have improved the sensitivity of Golay cells. For low-frequencies (5 to 25 Hz) and wide bandwidths (1 to 1000 μm), this may be the best room-temperature detector available.

A chamber containing a gas is sealed at one end with an infrared transparent window through which radiation reaches a thin absorbing film as shown in

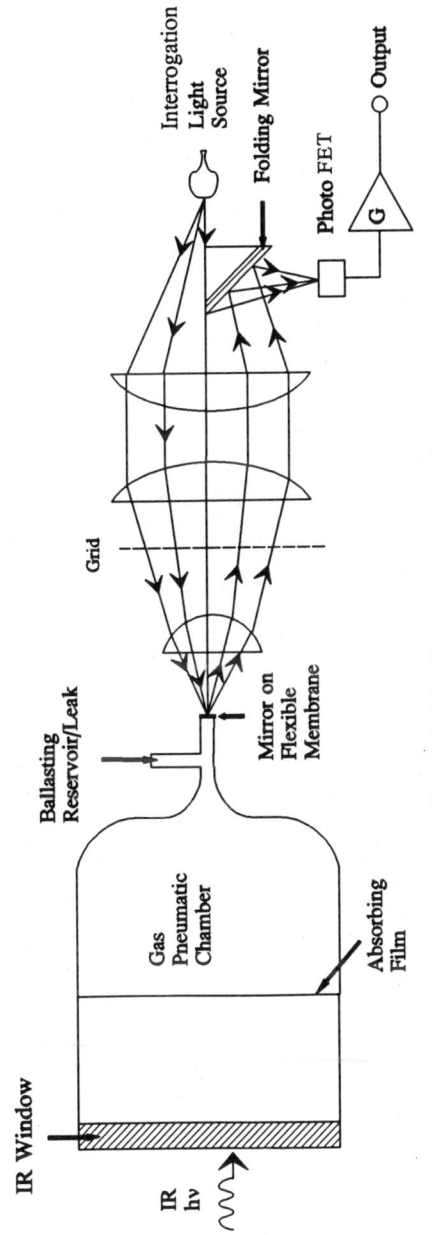

Figure 3.15. Golay detector assembly.

Fig. 3.15. The absorbing film in turn warms the gas with which it is in contact. A rise in temperature of the gas in the chamber produces a rise in pressure, and therefore a distortion of the mirror membrane. To prevent changes in ambient temperature from affecting the detector, a fine leak is provided that connects the detector chamber with a ballasting reservoir of gas on the other side of the mirror membrane. In the absence of a changing radiation signal, pressures are the same on both sides of the membrane, which remains flat. An alternating radiation signal of about 5 to 25 Hz will produce a corresponding deformation of the mirror membrane at 5 to 25 Hz which, by a suitable optical system and an amplifier, can be converted into an alternating voltage.

Light from the source passes through a condenser lens to the line grid and is concentrated onto the mirror membrane. A lens between the line grid and the mirror membrane focuses the beam so that, in the absence of any deformation, an image of one part of the line grid is superimposed on another part of the same grid. If the image of a gap between lines coincides with a gap in the grid, light will be transmitted and may be detected by a detector housing by the mirror.

Deformation on the mirror membrane will produce a corresponding change in the relations of line image and grid, and hence a change in the intensity of light reaching the sensor. The light sensor is a photo field-effect transistor (FET) also acting as the input stage for the solid-state amplifier circuit incorporated in the same box. The light source is a stabilized LED. This servo system is in effect a photoelectric amplifier, which has sufficiently high signal level to be photon-noise limited and the noise in the following amplifier to be negligible by comparison.

3.3.5. Superconductor Detectors

The recent discovery and development of high-temperature superconductivity (HTSC) in ceramic-oxide materials have created a renewed interest in the use of the superconductor for infrared (IR) detector applications (Lueng et al., 1987, 1988; Ryan, 1990). High temperature in this context is about 77 K, the temperature of liquid nitrogen (Sheng et al., 1988). Previously, superconducting detectors had been constructed with metallic materials and operated in the bolometric mode, but their resistance-versus-temperature curve was so steep in the superconducting transition region that operation required complex temperature-control systems that relegated such detectors to the research laboratory. Temperature control on the order of degrees milli-kelvin was required, and Keyes (1980) reported as recently as 1980 that interest in such detectors had largely vanished. More modern materials such as niobium nitride (NbN) and yttrium barium copper oxygen (YBCO) have more gradual superconducting transitions and are currently being used to fabricate HTSC IR detectors. In addition, the transition temperature of HTSC materials is typically in the 80- to 120-K range, and thus the required cryogenic systems are much less complex than those for metallic superconductors.

Three possible modes of operation of an HTSC detector are described by Kruse (1990). The first is the conventional bolometric mode, which relies on the sharp resistance change with temperature in the superconducting transition region as shown in Fig. 3.16. The second is the "nonequilibrium" or "non-bolometric" response that involves the breaking of Cooper electron pairs by the incident photons, thus destroying the superconductivity. The third is termed "photon-assisted tunneling" and is associated with a tunnel junction such as a Josephson junction device. In this mode the incident photons again break Cooper pairs, which results in a change of the i-v characteristic of the junction.

Because HTSC detectors are a fairly recent technology, no complete and conclusive theory is found in the literature for the second and third modes described earlier. In fact, there often is confusion as to which mode is being discussed. Experimental work to date has been directed toward the first two modes just described, and some researchers have endeavored to differentiate which mode is being observed (Thiede and Dereniak, 1990).

The bolometric mode exploits the large temperature coefficient of resistance, α, defined in Eq. (3.18), exhibited by HTSC materials in the superconducting transition region in the vicinity of the critical temperature, T_s, as seen in Fig. 3.16. A typical value of α for an HTSC is on the order of 0.2/K in the superconducting transition region, as compared to 0.002/K for a metal or 0.01/K for a semiconductor. The bolometer is operated by stabilizing its temperature in the middle of the superconducting transition region where α is large. The bolometer is usually biased from a constant current source, and the voltage drop across the device is measured. Incident optical radiation will heat the bolometer, causing a dramatic increase in R, and a subsequent dramatic increase in the measured voltage if used in a circuit such as was shown for a

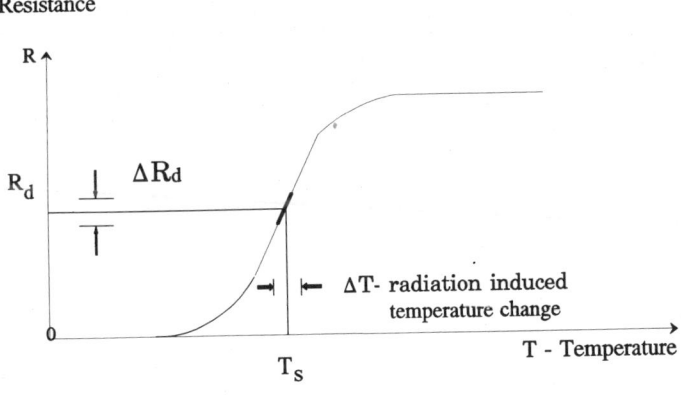

$$\Delta\phi_e \rightarrow \Delta T \, @T_s \rightarrow \Delta R_d \, @ \, R_d$$

Figure 3.16. Superconducting bolometer detector.

114

Table 3.9 Infrared Detector Superconductor Performance

Detector Type	E_g (eV)	λ_c (μm)	Temperature (K)	R	τ	NEP (W)	Spectral Range (μm)	D^* (cm Hz$^{1/2}$/W)
BaPb$_{0.7}$Bi$_{0.3}$O$_3$	$\cong 0.001$	1200	13	10^4 V/W	$\cong 1$ ns	7×10^{-14}	To FIR	3×10^{10}
NbN/BN	$\cong 0.002$	600	1.6–15	0.7 V/W	< 1 ns		To 500 + μm	$\cong 10^{10}$
YBa$_2$Cu$_3$O$_7$ Bolometer	—	—	$\cong 50$	4×10^3 V/W	$\cong 1$ μs			$\cong 10^8$
YBa$_2$Cu$_3$O$_7$ Nonequilibrium	0.25	50	30–90	$\cong 10^4$ V/W	0.1 ns	10^{-6}	To 500 + μm	$\cong 10^6$

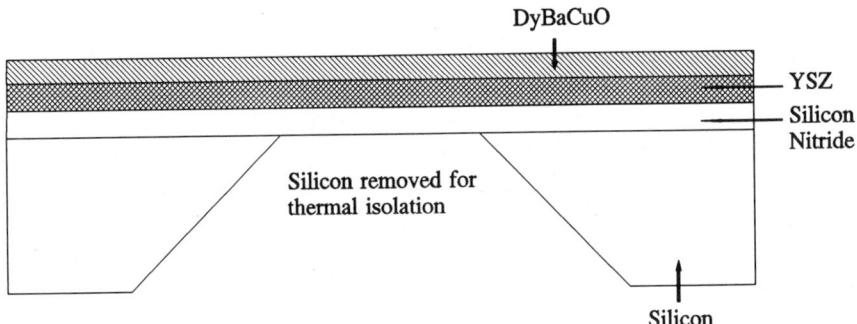

Figure 3.17. Microbolometer using high-temperature superconductor material.

photoconductor in Fig. 3.7. The general theory developed for bolometers is applicable to HTSC bolometers as well.

An HTSC bolometer is a thermal detector, and thus its spectral sensitivity is determined by the spectral dependence of the optical absorptance. The absorptance is determined by the optical properties of the HTSC material, any coatings applied to the HTSC, and the optical properties of the substrate. HTSC bolometers are expected to find application in the wavelength range from approximately 40 μm to 1 mm. Table 3.9 lists some examples of superconductors as optical detectors.

The fabrication of "microbolometers" is also of current interest. These extremely small structures can ultimately be extended to entire focal-plane arrays consisting of many such microbolometers. Here, an antenna-coupling technique has been employed to efficiently couple the collected radiation into the microbolometer because the dimensions of the bolometer are smaller than the wavelength of interest. A second example of a microbolometer is shown in Fig. 3.17. In this case, the HTSC material is DyBaCuO, which is closely related to YBCO. The HTSC has been prepared as part of a multilayer "microbridge" to achieve thermal isolation from the silicon substrate. This technique requires chemical etching of silicon in the correction crystal orientation to remove silicon in pyramid structure. The total thickness of the multilayer is about 1 μm, and the sensitive surface is 75 μm square. This structure constitutes one pixel in what will ultimately be a microbolometer array. Superconducting two-dimensional arrays are also of considerable interest (Quelle, 1990).

3.4. THE PHOTOELECTROMAGNETIC DETECTOR

The photoelectromagnetic (PEM) detector is a unique semiconductor photodetector that relies on the application of a steady-state magnetic field to produce separation and detection of photo-generated charge. This device produces either an open-circuit voltage or a short-circuit current that is directly proportional to the magnitude of the applied optical flux.

The detector is named for the photoelectromagnetic effect, which was first observed by Kikoin and Noskov (1934) in Cu_2O. Frenkel (1935) is credited with first explaining the effect in terms of diffusion of electron–hole pairs. The PEM effect has enjoyed wide use in the characterization of semiconductor materials, especially in the determination of carrier lifetimes.

The use of the PEM effect in detection has been hindered somewhat by the necessity of packaging a strong permanent magnet with the detector crystal. In spite of this packaging problem, however, it exhibits several attractive features that make it an excellent choice for certain applications. The ability to operate at very high speeds with reasonable detectivity at room temperature has motivated much of the current commercial interest in PEM detectors.

A conceptual representation for a PEM device is shown in Fig. 3.18, where ℓ_x is the detector width in the direction of photon arrival, ℓ_y is the length of the detector in the direction of current flow, and A is the cross-sectional area, normal to the current flow. Incoming photons are absorbed close to the front surface of the detector and form free electron–hole pairs. Under carrier diffusion and recombination effects, an excess carrier gradient appears in the y-direction. As the carriers diffuse from the front surface, in the y-direction, they are deflected by the applied magnetic field by the Lorentz force, thus giving rise to a separation of photogenerated holes and electrons. This is the basis for the PEM detection mechanism.

To gain a mathematical appreciation of how a PEM photodetector operates, a simple derivation by Moss et al. (1973) yields the output current and voltage. These expressions enable us to derive the responsivity and detectivity. In these derivations, it shall be assumed that the detector is under uniform optical illumination, and that the illumination has occurred sufficiently long that the detector has reached steady-state operation. All material parameters will be taken to be independent of the magnetic field. For simplicity, it will also be assumed that all of the incoming photons are absorbed at the front surface of the detector. For the interested reader, Lile (1973) has developed a more thorough treatment in which he accounts for bulk absorption effects.

Referring again to Fig. 3.18, note that if the detector is sufficiently large

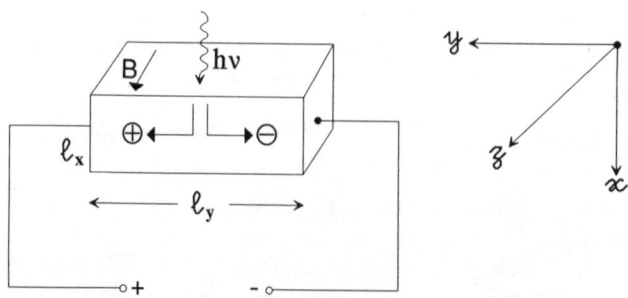

Figure 3.18.   Conceptual representation of a PEM detector.

that we can ignore edge effects, we may analyze the device on a one-dimensional basis. The relevant current components through the crystal are then four in number, namely the hole and electron currents in the y-direction, and the hole and electron currents in the x-direction. As an example, note that Eq. (3.24) is the hole-current density, per unit sample width (measured in the z-direction), in the downward, or positive y-direction, taking into account the processes of drift, diffusion, and the Lorentz force:

$$J_y^+ = -q\, D_h \frac{d}{dy} \Delta p - B\mu_h J_x^+ + q p \mu_h E_y \qquad (3.24)$$

where D_h is the diffusion constant for holes, μ_h is the hole mobility, and p is the carrier concentration for holes. Equation (3.25) is similar to Eq. (3.24), but representing the electron constituent of current density in the y-direction. Note all relevant subscript and sign changes.

$$J_y^- = q\, D_e \frac{d}{dy} \Delta n + B\mu_e J_x^- + q n \mu_e E_y \qquad (3.25)$$

where D_e is the diffusion constant for electrons, μ_e is the electron mobility, and n is the carrier concentration for electrons. In Eqs. (3.24) and (3.25), Δp and Δn refer to the excess (photogenerated) carrier concentration at a particular depth (y-direction) in the crystal. The same type of equations apply to the hole and electron current densities for the x-direction, with the understanding that no diffusion effect occurs under the assumption of uniform illumination:

$$J_x^+ = 0 + B\mu_h J_y^+ + q p \mu_h E_x \qquad (3.26)$$

$$J_x^- = 0 - B\mu_e J_y^- + q n \mu_e E_x \qquad (3.27)$$

We can combine these expressions to obtain an equivalent version of Eq. (3.24), written only in terms of the detector's material parameters, by applying the current continuity equation for the y-direction. Under our assumption of steady-state conditions, and further assuming that the photogenerated carrier population is small compared to thermal equilibrium values (i.e. $p \sim p_0$ and $n \sim n_0$), the continuity equation is

$$\frac{d}{dy} J_y^+ = -\frac{q \Delta p}{\tau} \qquad (3.28)$$

The symbol τ in Eq. (3.28) represents the carrier lifetime. By considering the short-circuit mode of operation, one may set the electric field in the x-direction (E_x) to zero in Eqs. (3.26) and (3.27). Because the net current out of the device is zero in the y-direction, the hole and electron currents of Eqs. (3.24) and

(3.25) are equal and opposite at all points in the crystal. To further simplify the analysis, note additionally that under an assumption of direct recombination with no trapping mechanisms, the excess electron concentration, Δn, may be replaced by Δp in Eq. (3.25). Furthermore, the familiar intermediate variable $\mathfrak{b}$ ($\mathfrak{b} = \mu_e/\mu_h = D_e/D_h$) may be used to eliminate μ_e and D_e from Eqs. (3.25) and (3.27). Applying these simplifications and assumptions, and combining Eqs. (3.24) to (3.28), yields the following desired expression for the hole current in the y-direction:

$$\frac{d^2 J_y^+}{dy^2} = \frac{(p + \mathfrak{b}n + p\mu_e^2 B^2 + n\mu_e\mu_h B^2) J_y^+}{\mathfrak{b}D_h\tau (p + n)} \tag{3.29}$$

This second-order linear differential equation has a solution of the following form:

$$J_y^+ = \mathfrak{D} \cosh\left(\frac{y}{\mathfrak{f}}\right) + \mathfrak{B} \sinh\left(\frac{y}{\mathfrak{f}}\right) \tag{3.30}$$

where

$$\mathfrak{f} = \left[\frac{\mathfrak{b}D_h\tau (p + n)}{(p + \mathfrak{b}n + (\mathfrak{b}p + n)\mu_e\mu_h B^2)}\right]^{1/2} \tag{3.31}$$

The intermediate variable "$\mathfrak{f}$," is referred to as the effective ambipolar magnetic diffusion length. It reduces to the usual expression for the ambipolar diffusion length if the magnetic field is zero.

To apply the general solution of Eq. (3.30), we must now select and apply appropriate boundary conditions at the upper and lower surface of the detector ($y = 0$ and $y = \ell_x$) and then solve for the constants $\mathfrak{D}$ and $\mathfrak{B}$. To represent the hole current at the surfaces, we must use the concept of surface recombination velocity, to account for the fact that surface rather than bulk recombination dominates at the boundaries of the device. For simplicity, we will let the surface recombination velocities (S) at the upper and lower surfaces be equal. Under these conditions, the current densities immediately below and above the crystal surfaces are as given by Eqs. (3.32) and (3.33). At $y = 0$:

$$J_y^+ = (E_q - S \Delta p) \, q \tag{3.32}$$

while at $y = \ell_x$:

$$J_y^+ = S \Delta p \, q \tag{3.33}$$

The term E_q represents the photon irradiance at the $y = 0$ surface. Applying these boundary conditions yields the following solution for the hole current flowing in the positive y-direction.

$$J_y^+ = qE_q \left[\frac{\sinh\left(\frac{\ell_x - y}{f}\right) + \left(\frac{\tau S}{f}\right)\cosh\left(\frac{\ell_x - y}{f}\right)}{\left(1 + \left(\frac{\tau S}{f}\right)^2\right)\sinh\left(\frac{\ell_x}{f}\right) + 2\left(\frac{\tau S}{f}\right)\cosh\left(\frac{\ell_x}{f}\right)} \right] \tag{3.34}$$

This result may now be used to solve for the short-circuit current flowing in the detector. This is done by integrating the sum of the electron and hole current densities flowing in the x-direction, as shown in Eq. (3.35).

$$i_{sc} = \int_0^{\ell_x} (J_x^+ + J_x^-)\, dy \tag{3.35}$$

We may now combine Eqs. (3.26) and (3.27), (again setting E_x to 0, to modify Eq. (3.35) to a form that utilizes our result of Eq. (3.34):

$$i_{sc} = B\,(\mu_e + \mu_h) \int_0^{\ell_x} J_y^+\, dy \tag{3.36}$$

The result of this integration over the thickness of the crystal yields the final result for the short-circuit output current

$$i_{sc} = \frac{B(\mu_e + \mu_h)\, q\, E_q f}{\left(\frac{\tau S}{f}\right) + \coth\left(\frac{\ell_x}{2f}\right)} \tag{3.37}$$

The result is easiest to interpret if we assume that the crystal is thick in relation to the effective diffusion length f ($\coth(\ell_x/2f) \approx 1$), the magnetic field is low ($f \approx L$, the ambipolar diffusion length), and that the surfaces have been well etched (S is small, and $\tau S/f \ll 1$). Under these assumptions, the short-circuit current is as shown in Eq. (3.38).

$$i_{sc} = B\,(\mu_e + \mu_h)\, q\, E_q L \tag{3.38}$$

One may conclude that for high PEM photocurrents, a large diffusion length L is desirable, as are large electron and hole mobilities. The photocurrent increases linearly with the strength of the applied magnetic field, at least within modest limits.

By applying Ohm's law, the open-circuit voltage for the PEM detector may easily be obtained

$$v_{oc} = \frac{i_{sc}\, \ell_y}{\sigma A_n} \tag{3.39}$$

Using the usual definition for voltage responsivity, and the results of Eqs. (3.37) and (3.39), we now have the following:

$$\mathcal{R}_v = \frac{\eta\, r\, (\mathfrak{b} + 1)\, \mathsf{B}\, \mathfrak{f}\, \lambda}{(\mathfrak{b}n + \mathfrak{p})\, A_n hc \left[\dfrac{\tau\, S}{\mathfrak{f}} + \coth\left(\dfrac{\ell_x}{2\mathfrak{f}}\right)\right]} \tag{3.40}$$

The term η has been introduced to account for nonideal quantum efficiency. Additionally, the Hull scattering coefficient r has been added, to improve upon earlier, idealized representations of the Lorentz-force currents appearing in Eqs. (3.24) to (3.27).

Careful examination of Eq. (3.40) reveals that in low-magnetic fields, the responsivity is optimized by minimizing the intrinsic carrier concentration, and by using a p-type sample (with $\mathfrak{p} \approx (2\mathfrak{b})^{1/2}\, n_i$) with a large mobility ratio ($\mathfrak{b}$ large). In high magnetic field cases, the responsivity becomes independent of the field, and no general conclusions may be made regarding optimal material parameters.

Under open-circuit operation, troublesome $1/f$ noise may be eliminated, giving rise to Johnson-noise-limited detectivities.

$$D^*(\lambda) = \frac{\eta\, r\, (\mu_e + \mu_h)\, \mathsf{B}\, \mathfrak{f}\, \lambda\, q^{1/2}}{(\mu_e n + \mu_h \mathfrak{p})^{1/2}\, w^{1/2} hc \left[\dfrac{\tau\, S}{\mathfrak{f}} + \coth\left(\dfrac{\ell_x}{2\mathfrak{f}}\right)\right] (4kT)^{1/2}} \tag{3.41}$$

Analysis of this expression indicates that in weak-field cases, a low intrinsic carrier concentration is again desirable, and that both the hole mobility and the mobility ratio "$\mathfrak{b}$" should be as large as possible. The p concentration should again be $\sim (2\mathfrak{b})^{1/2} n_i$. In high-field cases, the detectivity becomes independent of the field, and the only general conclusion is that a low electron mobility becomes desirable.

Table 3.10 shows a number of observed detectivity values obtained from a survey paper by Nowak (1987).

As reported by Piotrowski (1991), the ultimate detectivity (uncooled) achievable in an optimized 10-μm PEM detector is about 3.4×10^7, with a responsivity of 0.6 V/W. This requires a substantial magnetic field of around $2T$. This performance is about an order of magnitude worse than the best values obtainable from an uncooled photoconductor. One may obtain about an order of magnitude improvement in the detectivity of a PEM detector by cooling the device to 200 K or so, but even then it is still inferior to the performance offered by competing photoconductors, and this approach suffers from cooling difficulties associated with the rather bulky permanent magnet.

In practice, the PEM detector is most often used at or near room temperature. Detectivities of recently reported uncooled HgCdTe and ZnCdTe 10.6-μm detectors are now within a factor of 3 of the theoretical limit ($D^* \approx 1 \times$

Table 3.10 Representative Measured Parameters for PEM Detectors

Material	D^* (λ_{max}) (cm Hz$^{1/2}$/W)	$\mathfrak{R}_v$ (V/W)	τ (s)
InSb	3×10^6 $(6.6\ \mu m)$		$<2 \times 10^{-7}$
	1.9×10^6 $(6.2\ \mu m)$		$<10^{-6}$
HgCdTe	5×10^4 $(10.6\ \mu m)$	0.02 $(10.6\ \mu m)$	
	4×10^4 $(10.6\ \mu m)$	0.02 $(10.6\ \mu m)$	
	4×10^4 $(10.6\ \mu m)$		5×10^{-9}
	10^5 $(10.6\ \mu m)$	0.01 $(8\text{-}12\ \mu m)$	2×10^{-10}
ZnCdTe	10^5 $(12\ \mu m)$	0.04	

10^7). The primary advantage of these room-temperature devices is their speed of operation. The frequency response of a PEM detector extends from dc to a high-frequency cutoff that is determined by an "effective recombination time" that is faster than the usual bulk recombination time. Response times as short as 50 ps have been reported, with little compromise in responsivity. This is faster than most other room-temperature detectors. This speed of operation is perhaps the most important factor that leads one to use a PEM detector.

Additional advantages of the PEM device derive from its open-circuit mode of operation. As noted earlier, low-frequency noise ($1/f$ noise) is not present, and the construction of electrical contacts is less critical than in the case of photoconductors. The PEM detector requires no electrical bias, and offers an inherently low electrical resistance that is particularly well-suited for direct connection to transmission lines and wide-band amplifiers. A PEM detector can handle high optical-power levels, making it especially useful in laser-power measurements. Finally, it is relatively insensitive to temperature-related drift effects and is rugged and easy to reproduce.

PEM detectors will continue to improve. New concepts under development include the use of optical cavities to enhance responsivity and detectivity at specific user-defined wavelengths, the use of immersion lenses, and exploitation of new materials and growth techniques.

REFERENCES

Astheimer, R. W., "Thermistor Infrared Detectors," *SPIE Proc.* **443** (1983).

Beyen, W. J., and B. R. Pagel, "Cooling Requirements for Intrinsic Photoconductive Infrared Detectors," *Infrared Physics* **6**(4), 161–166 (1967).

Boyd, Robert W., *Radiometry and the Detection of Optical Radiation*, Wiley, New York, 1983.

Bratt, P. R., "Impurity Germanium and Silicon Infrared Detectors," in *Semiconductors and Semimetals* (R. K. Williardson and A. C. Beer, eds.), Academic Press, New York, pp. 39–141. 1977.

Bube, Richard H., *Photoconductivity of Solids*, R. F. Krieger Publ., Huntington, NY, 1978.

Dennis, P. N. J., *Photodetectors: An Introduction to Current Technology*, Plenum, New York, 1986.

Dietrich, R. V. *The Tourmaline Group*, Van Nostrand Reinhold, New York, 1985.

Engstrom, Ralph W., *Photomultiplier Handbook*, RCA, 1980.

Frenkel, J., *Phys. Z. Sowjetunion* **8**, 185 (1935).

Golay, M. J. E., "A Pneumatic Infrared Detector," *Rev. Sci. Instr.* **18**(5), 347 (1947).

Golay, M. J. E., "The Theoretical and Practical Sensitivity of the Pneumatic Infrared Detector," *Rev. Sci. Instr.* **20**(7), 816 (1949).

Goss, A. J., "The Pyroelectric Vidicon—A Review," *SPIE Proc.* **807,** 25–32 (1987).

Halliday, D., and R. Resnick, *Fundamentals of Physics*, p. 617, Wiley, New York, 1981.

Hanson, C. M., *Uncooled Thermal Imaging* at Texas Instruments *SPIE Proc.* **2020,** 330–339 (1993).

Keyes, Robert J., *Optical and Infrared Detectors*, Springer-Verlag, Berlin/New York, 1980.

Kikoin, I. K., and M. M. Noskov, *Phys. Z. Sowjetunion* **5**, 586 (1934).

Kruse, P. W., "High T_c Superconducting IR Detectors," *SPIE Proc.* **1292,** 108–177 (1990).

Kruse, Paul W., Laurence D. McGlauchlin, and Richmond B. McQuistan, *Elements of Infrared Technology: Generation, Transmission, and Detection*, Wiley, New York, 1962.

Leung, M., U. Strom, et al., "NbN/BN Granular Films—A Sensitive, High-Speed Detector for Pulsed Far-Infrared Radiation and Other Activity," *Appl. Phys. Lett.* **50**(23) (1987).

Leung, M., et al., "Sensing, Discrimination, and Signal Processing and Superconducting Materials and Instrumentation," *SPIE Proc.* **879** (1988).

Lile, D. L., "Generalized Photoelectromagnetic Effect in Semiconductors," *Phys. Rev. B* **8**(10), 4708–4722 (1973).

Marquardt, P., Nimtz, G., and Galeczki, G., "On Speeding-Up Pyroelectric Detectors," *SPIE Proc.* **1106,** 134–141 (1989).

Marshall, D. E., "A Review of Pyroelectric Detector Technology," *SPIE Proc.* **132,** 100–117 (1978).

Moss, T. S., G. J. Burrell, and B. Ellis, *Semiconductor Opto-Electronics*, Wiley, New York, 1973.

Nowak, M., "Photoelectromagnetic Effect in Semiconductors and Its Applications," *Prog. Quant. Electron.* **11**(3,4), 319–331 (1987).

Pankove, J. I. *Optical Properties of Semi Conductors*, Dover, NY, 1971.

Piotrowski, J., W. Galus, and M. Grudzien, "Near-Room Temperature IR Photodetectors," *Infrared Physics* **31** (1), 29–35 (1991).

Putley, E. H., "The Pyroelectric Detector," in *Semiconductors and Semimetals* (R. K. Willardson and A. C. Berr, eds., pp. 259–285. Academic Press, New York, 1970.

Putley, E. H., "The Pyroelectric Detector—An Update," in *Semiconductors and Semimetals* (R. K. Willardson and A. C. Beer, eds. Academic Press, New York, 1977, pp. 441–449.

Quelle, F. W., "Superconducting IR Focal Plane Arrays," *SPIE Proc.* **1243,** 206–213 (1990).

Ryan, P. A. "High-Temperature Superconductivity for EW and Microwave Systems," *J. Electronic Defense*, May 1990.

Schoolar, R., and E. Tenescu, *SPIE Proc.* **686,** 2 (1986).

Sclar, N., "Properties of Doped Silicon and Germanium Infrared Detectors," *Prog. Quant. Electron.* **9**(3), 145–257 (1984).

Sheng, E. Z., et al., "Superconductivity at 90K in the TLBa-Co-O System," *Phys. Rev. Lett.* **60**(10), 937 (1988).

Shorrocks, N. M., et al., "Uncooled Infrared Thermal Detector Arrays," *SPIE Proc.* **1320,** 88–94 (1990).

Stevens, N. B., "Infrared Detectors," in *Semiconductors and Semimetals*, vol. 5, R. K. Willardson and A. C. Berr (eds), Academic Press, New York, 1970.

Streetman, Ben G., *Solid State Electronic Devices*, Prentice Hall, Englewood Cliffs, NJ, 1990.

Sze, S. M., *Physics of Semiconductor Devices*, Wiley-Interscience, New York, 1969.

Tebo, A. R., "Superconducting Detectors Show Promise," *Laser Focus World*, Penwell Publishing, Tulsa, OK, 1991, pp. A37–A39.

Thiede, D. A., and E. L. Dereniak, "High T_c Superconducting Thin Films as Optical Radiation Detectors," *SPIE Proc.* **1287,** (1990).

Vincent, John David, *Fundamentals of Infrared Detector Operation and Testing*, Wiley, New York, 1990.

PROBLEMS

3.1. In which of the three groups of materials—metals, semiconductors, and insulators—are photoconductive effects most easily observed? Why are they more difficult to observe in the other two types of materials?

3.2. Describe the factors that determine the spectral response of photoemissive, photoconductive, photovoltaic, and thermal detectors.

3.3. How would you design a detection system that approaches 100% quantum efficiency for some limited spectral region? (*Hint:* More than one detector can be used.)

3.4. A photodiode has just been fabricated from the newly discovered Dereniakium, which has a direct bandgap of 0.85 eV. Sketch its absolute spectral responsivity (A/W) as a function of wavelength and point out relevant features of the curve. Use a fixed value of 60% for the in-band quantum efficiency. Three different lasers can be directed toward the detector, which has an active area of 5 mm^2. The detector receives 1 mW at 632.8 nm (He-Ne), 1 mW at 1060 nm (Nd-YAG), and 2 mW at 1532 nm (Ir-Ne). What is the output current of the detector for each laser?

3.5. Discuss the factors that influence the quantum efficiency versus wavelength. Why isn't it flat and equal to one (1)?

3.6. If a 1 mm $\times$ 1 mm $\times$ 10 μm (active area is 1 mm $\times$ 1 mm) detector is made of germanium, index = 4 for all wavelengths, and absorptivity

$$a_0 = 10^4 \text{ cm}^{-1} \text{ eV}^{1/2}$$

$$a_0' = 10^3 \text{ cm}^{-1}$$

what is the maximum quantum efficiency (η)? What limits the quantum efficiency at the short wavelength?

3.7. Why is the cutoff wavelength of a photovoltaic detector independent of the energy gap of the impurity levels (unlike an extrinsic photoconductor)?

3.8. An arsenic-doped silicon (Si : As) photoconductor has an energy gap from donor to conductor band of 0.052 eV operating at 4 K. What is the cutoff wavelength? At what wavelength has the quantum efficiency dropped to 50% of original value.

3.9. A silicon detector with a polished surface ($n = 3.5$) is used to measure radiant incidence (irradiance). Describe and plot its responsivity as a function of angle of incidence of the incoming radiation. How could you use this detector to measure the radiance of an extended source?

3.10. The spectral response of a silicon p-n junction photocell shows the following profile (see figure):

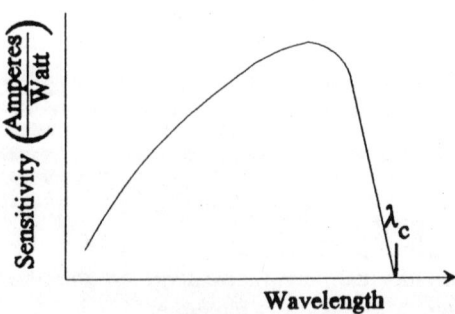

Figure 3.10

(a) Describe the reason for the steep drop on the long-wavelength side of the maximum.

(b) Give at least two reasons for the gradual dropoff toward shorter wavelengths.

3.11. A cesium–antimony photocathode has an energy gap of 1.6 eV and an electron affinity of 0.45 eV.

(a) Make a plot of the velocity (maximum value) of ejected photoelectrons versus illumination wavelength in microns.

(b) Draw an energy-band diagram that illustrates the photocathode's behavior for illumination wavelengths above and below the threshold wavelength.

3.12. A new superconducting microbolometer has the following experimentally determined resistance versus temperature characteristic (for 79 K $\le T \le$ 94 K).

$$R(T) = -1,213,460 + 57,468.1(T) - 1,015.11(T^2)$$
$$+ 7.919720(T^3) - 0.0229932(T^4)$$

(a) Plot the $R(T)$ relationship.

(b) Derive an expression for the temperature coefficient α, and plot it on the same graph used for part (a).

(c) At what temperature should the detector be operated.

(d) Over what temperature range must the detector be stabilized to have a response that is always within 95% of the peak value? How does this compare to metallic superconducting bolometers?

3.13. Sketch the ideal (quantum efficiency equals one) spectral responsivity versus wavelength curve for a photodetector and explain the following:

(a) What determines the cutoff wavelength?

(b) How does a nonunity quantum efficiency affect the curve?

(c) How does the responsivity of a real detector vary versus wavelength? (Sketch on earlier curve and label.)

(d) Explain why the wavelength response is degraded over the spectral range of a photoconductive detector's response.

(e) Explain why the wavelength response is degraded over the spectral range of a photovoltaic detector's response.

4

Probability and Statistics for Optical Detection

4.1. INTRODUCTION

This chapter explains probability theory necessary for us to understand and apply probability theory to the problem of noise in optical detection. The chapter bibliography lists additional texts for further investigation of probability theory.

4.2. DEFINITION OF PROBABILITY

The three distinct methods for obtaining an estimate for the probability of a certain event are as follows.

1. *A Priori (Classic):* If an event E can occur n different ways out of a total number of m equally likely means, then the probability of the event is n/m. Mathematically,

$$P(E) = \frac{n}{m} \tag{4.1}$$

   Note that the estimate is made beforehand, or *a priori*, based on a hypothetical experiment.

2. *A Posteriori (Relative Frequency of Occurrence):* If an event E is observed to have occurred in n trials of m repetitions of an experiment, m being very large, then the probability of the event is n/m. Mathematically,

$$P(E) = \lim_{n \to \infty} \frac{n}{m} \tag{4.2}$$

   This is also termed the *empirical probability*, because it is an estimate based on the analyzed results of collected data.

These estimate approaches are insufficient because of vague wording such as "equally likely" and "very large," and, referring to the *a posteriori* definition, because we will never be able to observe an infinite number of trials.

The axiomatic definition of probability, which comes from set theory, is more abstract than the estimate approaches, but it avoids the mathematical pitfalls of both the *a priori* and *a posteriori* definitions.

3. *Axiomatic: Probability* is a mapping or rule of assignment between an event E in a set of events and the probability of that event $P(E)$ with respect to some or all other possible events.

A *set* is any group of events, and any event that belongs to a set is a member of that set. The certain event is the set comprising all possible events. The certain-event set is often denoted with the letter Ω. The union of two (or more) sets is the set that includes all members of either (or all) set(s). The union operator is denoted with the symbol $\cup$ and is functionally equivalent to a logical OR. The intersection of two (or more) sets includes only those events that are members of both (or all) set(s). The intersection operator is denoted with the symbol $\cap$, and is functionally equivalent to a logical AND. These concepts may be better understood by referring to Fig. 4.1.

$P(E)$ obeys the following three axioms (Kolmogoroff, 1933):

$$P(E) \geq 0 \qquad (4.3)$$

the probability is never negative;

$$P(\Omega) = 1 \qquad (4.4)$$

the "certain" event has unit probability; and

$$P(E_1 \cup E_2 \cdots) = P(E_1) + P(E_2) + \cdots \qquad (4.5)$$

for mutually exclusive or disjoint events $E_1, E_2, \ldots$.

These three axioms, Eqs. (4.3) to (4.5), imply all the rules of probability.

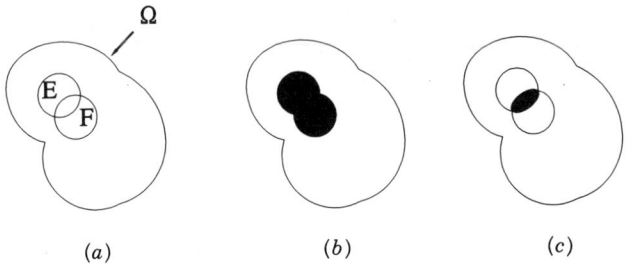

Figure 4.1. Illustrations of set theory: (*a*) the sets E, F, and Ω; (*b*) union of E and $F(G = E \cup F)$; (*c*) intersection of E and $F(G = E \cap F)$.

A function P that obeys all three of these axioms is called a *probability function* and $P(E)$ is called the *probability of the event E*.

From the axioms, Eqs. (4.3) to (4.5), we can establish the following additional results:

$$P(E) + P(E^c) = 1 \qquad (4.6)$$

where E^c is the complement of E;

$$P(\varnothing) = 0 \qquad (4.7)$$

where $\varnothing$ is the complement of Ω, often called the empty set;

$$0 \geq P(E) \geq 1 \qquad (4.8)$$

from Eqs. (4.3) and (4.6);

$$P(EF^c) = P(E) - P(E \cap F) \qquad (4.9)$$

and

$$P(E \cup F) = P(E) + P(F) - P(E \cap F) \qquad (4.10)$$

The last two results, Eqs. (4.9) and (4.10), are left as exercises for the reader.

These are not the only axioms from which a valid probability theory could be developed. The ultimate test for any theory is its usefulness in solving real problems in the physical world. Any theory must be consistent with the relative frequency of occurrence definition, that is, it must model the empirical probability.

4.3. JOINT AND CONDITIONAL PROBABILITY

Given two events/subsets E and F of Ω that may or may not be separate and distinct, the joint event G is the event/subset of Ω that includes all of the members of Ω that are members of both E and F. Mathematically this is written

$$G \equiv E \cap F \qquad (4.11)$$

which is read "G is the intersection of E and F." The joint probability is the probability that a joint event will occur. It is often abbreviated

$$P(G) = P(E \cap F) = P(EF) \qquad (4.12)$$

The joint event is never more probable than either of its constitutive events.

For example, suppose that on any particular day E is the event that it will be snowing in Rochester, New York, and F is the event that it will be raining in Tucson, Arizona. As we just discussed, the joint event G is the event that it would be both snowing in Rochester and raining in Tucson on the same day. Suppose now that we went down to our local weather bureau and found that during the last 10,000 days event E occurred 3650 times, event F occurred 1100 times, and event G occurred 750 times. By the *a posteriori* definition, the probabilities of these events occurring would be approximated by

$$P(E) \cong \frac{n_E}{m} = 0.365$$

$$P(F) \cong \frac{n_F}{m} = 0.110$$

$$P(G) = P(EF) \cong \frac{n_G}{m} = 0.075 \tag{4.13}$$

Now we wish to know the likelihood that it snowed in Rochester on a particular day, given that it rained in Tucson on that same day as well. This answer is simply the ratio $P(G)/P(F)$. A subset of F, G, is the likelihood of the event in which we are interested, and F replaces Ω as the new certain event.

This example illustrates the *a posteriori* definition of conditional probability. The conditional probability is denoted and defined as

$$P(E|F) \equiv \frac{P(EF)}{P(F)} \tag{4.14}$$

It is read "the probability of E given F," and for our example it is given by

$$P(E|F) \equiv \frac{P(EF)}{P(F)} = \frac{n_{EF}/m}{n_F/m} = \frac{n_G}{n_F} = \frac{750}{1100} = 0.681 \tag{4.15}$$

The probability of F given E for the same example is

$$P(F|E) \equiv \frac{P(EF)}{P(E)} = \frac{n_{EF}/m}{n_E/m} = \frac{n_G}{n_E} = \frac{750}{3650} = 0.205 \tag{4.16}$$

In other words, if the optics students in Tucson woke up to a miserable rainy day, those students would enjoy the slight satisfaction that there was a 68% chance that their colleagues in Rochester were being snowed upon. By contrast, the Rochester students, upon waking up to a glorious blizzard, could predict that there was an almost 80% chance that their Arizona colleagues were basking in the sun's rays.

4.4. CONTINUOUS RANDOM VARIABLES AND PROBABILITY DENSITY

Until now, we have dealt with purely discrete events; however, many random phenomena are continuous in nature. We therefore define a continuous random variable (rv) x that can have a continuum of possible values $-\infty \leq x \leq \infty$. To define a probability in this case, it is necessary to introduce the concept of a probability density and then consider the event E associated with an interval along the rv axis, such as $a \leq x \leq b$. The probability of the event E is then defined as

$$P(E) \equiv \int_a^b p(x) \, dx \qquad (4.17)$$

Note that the continuous probability density, the so-called probability density function (pdf) is labeled with a lower case p. All of the previously defined axioms are obeyed by the pdf, with minor modifications. A partial list follows:

$$p(x) \, dx \geq 0 \qquad (4.18a)$$

$$P(\Omega) \equiv \int_{-\infty}^{\infty} p(x) \, dx = 1 \qquad (4.18b)$$

$$p(x|y) = \frac{p(x, y)}{p(y)} \qquad (4.18c)$$

where $p(x, y)$ is read "p of x and y."

The continuous situation can be regarded as the general case, and we take the discrete limit because when modeling a discrete law it is considered a "sampled" version of a corresponding continuous law:

$$p(x) = \sum_{n=0}^{n_{max}} P_n \, \delta(x - x_n) \qquad (4.19)$$

Because of the sifting property of the dirac delta function, the right-hand side of Eq. (4.19) is zero, except where $x = x_n$.

This correspondence allows any relation involving a continuous probability density to be rewritten for a discrete probability. For example, the normalization property for a density function becomes, after substituting Eq. (4.19),

$$\int_{-\infty}^{\infty} dx \left[\sum_{n=0}^{\infty} P_n \, \delta(x - x_n) \right] = \sum_{n=0}^{\infty} P_n \int_{-\infty}^{\infty} \delta(x - x_n) \, dx = \sum_{n=0}^{\infty} P_n = 1$$

$$(4.20)$$

4.5. MEANS AND MOMENTS

The mean, or average, of a probability function is a very useful piece of information because it allows us to predict quantitatively what value the random variable will come to have. Remember that, at least in this text, these probability functions will eventually be describing parameters such as photon flux and signal current. Another useful piece of information about the probability function is the variation of the function about the mean. This spread about the mean, the variance, allows us to hedge our bets regarding the function's actual tendency to be near its mean value. Other higher moments yield even more information about the probability function. It will be useful for us to define these quantities.

The mean, or expected value, of a probability function is defined by

$$E\{x\} \equiv \bar{x} \equiv \int_{-\infty}^{\infty} xp(x)\, dx \tag{4.21}$$

The second moment of a probability function is defined by

$$\overline{x^2} \equiv \int_{-\infty}^{\infty} x^2 p(x)\, dx \tag{4.22}$$

Continuing in a straightforward fashion, the higher-order moments of a probability function are defined by

$$\overline{x^i} \equiv \int_{-\infty}^{\infty} x^i p(x)\, dx \tag{4.23}$$

where x^i refers to the ith moment.

The central moments of a pdf are defined by

$$\mu_i \equiv \overline{(x - \bar{x})^i} = \int_{-\infty}^{\infty} (x - \bar{x})^i p(x)\, dx \tag{4.24}$$

Central moments are shifted or biased so that the moment is calculated with respect to the mean.

Variance, σ^2, is defined as the second central moment of the pdf and is given by

$$\sigma^2 = \overline{(x - \bar{x})^2} \tag{4.25}$$

Expanding and reducing,

$$\sigma^2 = \overline{x^2 - 2x\bar{x} + \bar{x}^2} = \overline{x^2} - \overline{2x\bar{x}} + \overline{\bar{x}^2}$$
$$\sigma^2 = \overline{x^2} - 2\bar{x}^2 + \bar{x}^2 = \overline{x^2} - \bar{x}^2 \tag{4.26}$$

Variance provides an indication of the spread of the pdf about the mean. However, the measure of spread is usually normalized to the original scale of the rv by taking the square root of the variance, the standard deviation. It is given by

$$\sigma = \sqrt{\overline{x^2} - \overline{x}^2} \tag{4.27}$$

All of the preceding means and moments can also be defined for a discrete probability function. They are

$$\overline{n} = \sum_{n=0}^{n_{max}} nP(n) \tag{4.28}$$

$$\overline{n^2} = \sum_{n=0}^{n_{max}} n^2 P(n) \tag{4.29}$$

$$\overline{n^i} = \sum_{n=0}^{n_{max}} n^i P(n) \tag{4.30}$$

$$\mu_i = \sum_{n=0}^{n_{max}} (n - \overline{n})^i P(n) \tag{4.31}$$

4.6. MOMENT-GENERATING AND CHARACTERISTIC FUNCTIONS

The moment-generating function provides a straightforward way of computing all of the moments of a random variable. It is defined as the expectation value of e^{sx}, where s is a complex variable. Therefore, it is given by

$$M(s) = E\{e^{sx}\} = \int_{-\infty}^{\infty} e^{sx} p(x) \, dx \tag{4.32}$$

Except for the sign in the exponent, the moment-generating function is equivalent to the double-sided Laplace transform of $p(x)$.

The moment-generating function is used to compute all of the moments of $p(x)$. It can be shown that

$$\overline{x^i} = \frac{d^i}{d_s^i} [M(s)]\big|_{s=0} \tag{4.33}$$

If the reader wishes to verify this property of the moment-generating function, it is necessary only to substitute Eq. (4.32) into Eq. (4.33), and expand e^{sx} in its infinite series.

For discrete random variables, the moment-generating function is given by

$$M(s) = \sum_{n=0}^{n_{max}} P(n) \, e^{sn} \tag{4.34}$$

The characteristic function is just the moment-generating function with s replaced by $j\omega$. It is given by

$$\Phi(\omega) = \int_{-\infty}^{\infty} e^{j\omega x} p(x) \, dx \tag{4.35}$$

Except for the sign in the exponent, the characteristic function is equivalent to the double-sided Fourier transform of $p(x)$.

The characteristic function is defined for discrete probability functions by

$$\Phi(\omega) = \sum_{n=0}^{n_{max}} P(n) \, e^{j\omega n} \tag{4.36}$$

4.7. RELEVANT PROBABILITY DISTRIBUTIONS

Part of our objective in optical detection is to predict the noise on the output signal. In this text, noise is quantified in terms of the statistical properties of the photon emission and detection processes. Because these processes can be modeled using electromagnetic theory, statistical thermodynamics, and quantum mechanics, it is important to discuss some of the probability distributions associated with these subjects. Such a discussion will lead to the remarkable insight that these probability distributions are all closely related.

The results of the preceding sections are summarized in Table 4.1, to provide a convenient reference for the reader.

4.7.1. Binomial Distribution

The binomial distribution is one of the most widely applicable probability distributions. It is given by

$$P_m(n) = \binom{m}{n} p^n \, q^{m-n} \tag{4.37}$$

where $q = 1 - p$ and

$$\binom{m}{n} = C_{mn} = \frac{m!}{n!(m-n)!} \tag{4.38}$$

Table 4.1 Summary of Probability Definitions

Term Name	Continuous	Discrete
Probability density function	$p(x)\, dx$	$P(n)$
Mean	$\bar{x} \equiv \displaystyle\int_{-\infty}^{\infty} xp(x)\, dx$	$\bar{n} = \displaystyle\sum_{n=0}^{n_{max}} nP(n)$
ith moment	$\overline{x^i} \equiv \displaystyle\int_{-\infty}^{\infty} x^i p(x)\, dx$	$\overline{n^i} = \displaystyle\sum_{n=0}^{n_{max}} n^i P(n)$
Variance	$\sigma^2 = \overline{x^2} - \bar{x}^2$	$\sigma^2 = \overline{n^2} - \bar{n}^2$
Standard deviation	$\sigma = \sqrt{\overline{x^2} - \bar{x}^2}$	$\sigma = \sqrt{\overline{n^2} - \bar{n}^2}$
Moment-generating function	$M(s) = \displaystyle\int_{-\infty}^{\infty} e^{sx} p(x)\, dx$	$M(s) = \displaystyle\sum_{n=0}^{n_{max}} P(n)\, e^{sn}$
Characteristic function	$\Phi(\omega) = \displaystyle\int_{-\infty}^{\infty} e^{j\omega x}\, p(x)\, dx$	$\Phi(\omega) = \displaystyle\sum_{n=0}^{n_{max}} P(n)\, e^{j\omega n}$

C_{mn} is normally referred to as the binomial coefficient, with m being the number of trials and n being the number of successes.

The binomial distribution describes the probable outcome of a discrete and finite set of repeated trials, often referred to as *Bernoulli trials*. Flipping a coin and tossing a die are both examples of Bernoulli trials.

The form of the binomial distribution is straightforward. Suppose there existed an experiment with two possible outcomes, such as flipping an unfair coin. If the probability of a head (success) were labeled p and the probability of a tail (failure) were labeled q, then the second axiom would necessitate $p + q = 1$. If this experiment is repeated m times, the probability of getting at least n successes in m flips ($n \le m$) is p^n, leaving the other $m - n$ flips as "don't cares." To calculate the probability of getting exactly n successes in m flips, one needs to count as a definitive failure every one of the previous $m - n$ "don't cares," so that this probability becomes $p^n q^{m-n}$.

What we just calculated is the probability of getting exactly n successes in m trails. However, there are many ways to achieve this result. For example, we could have successes on the first n flips, and failures on the last $m - n$, or the first flip could be a failure, the next n flips successes, and then the final $m - n - 1$ flips failures. Many different combinations result in getting exactly n successes in m flips. To find the total probability of getting exactly n successes in m tries in any order, we must multiply our previous expression by a degeneracy factor, which is a function of m and n. This degeneracy factor is the binomial coefficient, C_{mn}, which tells us how many different ways we can get exactly n successes in m trials. Combining all these expressions together, we get the binomial distribution, Eq. (4.37).

Let's confirm that the distribution is normalized, and then calculate the distribution's mean and variance. Testing for normalization,

$$\sum_{n=0}^{n_{max}} P_m(n) = \sum_{n=0}^{m} \binom{m}{n} p^n q^{m-n}$$

$$= (p + q)^m \tag{4.39}$$

$$= (p + (1 - p))^m$$

$$= 1$$

where we used the binomial theorem in the second step and the fact that $p + q = 1$ in the third.

The mean, defined by Eq. (4.28), is given by

$$\bar{n} = \sum_{n=0}^{m} n P_m(n)$$

$$\bar{n} = \sum_{n=1}^{m} n P_m(n) \tag{4.40}$$

where the last step is possible because the $n = 0$ term of the series does not contribute to the mean. Proceeding

$$\bar{n} = \sum_{n=1}^{m} n \binom{m}{n} p^n q^{m-n}$$

$$= \sum_{n=1}^{m} n \frac{m!}{n! \, (m - n)!} p^n q^{m-n} \tag{4.41}$$

$$= \sum_{n=1}^{m} \frac{m!}{(n - 1)! \, (m - n)!} p^n q^{m-n}$$

Using the variable substitution, $k = n - 1$ so that $n = k + 1$, we get

$$\bar{n} = \sum_{k=0}^{m-1} \frac{m!}{(k)! \, (m - (k + 1))!} p^{k+1} q^{m-(k+1)}$$

$$= mp \sum_{k=0}^{m-1} \frac{(m - 1)!}{(k)! \, (m - (k + 1))!} p^k q^{m-(k+1)}$$

$$= mp \sum_{k=0}^{m-1} \frac{(m - 1)!}{(k)! \, (m - (k + 1))!} p^k q^{(m-1)-k} \tag{4.42}$$

$$= mp \, (p + q)^{m-1}$$

$$= mp$$

The last two steps were made using the binomial theorem and the fact that $q = 1 - p$.

We can also indirectly compute the mean and other moments of the binomial distribution using the moment-generating function and its associate derivatives. Substituting Eq. (4.37) into Eq. (4.34),

$$M(s) = \sum_{n=0}^{m} P_m(n) \, e^{sn}$$

$$= \sum_{n=0}^{m} \left[\binom{m}{n} p^n \, q^{m-n} \right] e^{sn} \tag{4.43}$$

$$= \sum_{n=0}^{m} \binom{m}{n} (pe^s)^n \, q^{m-n}$$

$$= (pe^s + q)^m$$

Remembering that the mean of $P_m(n)$ is the first derivative of $M(s)$ with respect to s, evaluated at $s = 0$,

$$\bar{n} = \frac{d}{ds} M(s) \big|_{s=0}$$

$$= \frac{d}{ds} [(pe^s + q)^m] \big|_{s=0} \tag{4.44}$$

$$= [m(pe^s + q)^{m-1} \times pe^s]_{s=0}$$

$$= mp(p + q)^{m-1}$$

$$= mp$$

We can continue in this fashion and compute all of the higher-order statistical properties of the binomial distribution. The first- and second-order properties are summed up in Table 4.2.

4.7.2. Poisson Distribution

The Poisson probability distribution can be derived both independently and as in the limiting case of Bernoulli trials. In this section, we first show how the Poisson distribution can follow from the binomial distribution and then we derive the Poisson distribution from first principles.

Recalling the binomial distribution, we have

$$P_m(n) = \binom{m}{n} p^n \, q^{m-n}$$

$$= \frac{m!}{n!(m-n)!} p^n (1 - p)^{m-n} \tag{4.45}$$

Let us now allow the total number of trials m to increase without bound

Table 4.2 Binomial Distribution Properties

$P_m(n)$	$\binom{m}{n} p^n q^{m-n}$
$\bar{n}$	mp
$\overline{n^2}$	$mpq + m^2 q^2$
σ^2	mpq
σ	$\sqrt{mpq}$
$\Phi(\omega)$	$(pe^{j\omega} + q)^m$

while keeping constant the distribution mean, that is,

$$\text{let } m \rightarrow \infty \text{ while } \bar{n} = mp = \text{constant} \qquad (4.46)$$

which suggests that as

$$m \rightarrow \infty, p \rightarrow 0 \qquad (4.47)$$

This limiting case of the binomial distribution is sometimes referred to as the "rare-event" limit.

Continuing from Eq. (4.45) and substituting $n = mp$

$$P_m(n) = \frac{m!}{n!(m-n)!} \left(\frac{\bar{n}}{m}\right)^n \left(1 - \frac{\bar{n}}{m}\right)^{m-n} \qquad (4.48)$$

We now investigate the limiting behavior (as m goes to infinity) of the first and third expression in the product just given. Working on the first expression,

$$\lim_{m \rightarrow \infty} \frac{m!}{n!(m-n)!} = \frac{1}{n!} \lim_{m \rightarrow \infty} \frac{m!}{(m-n)!}$$

$$= \frac{1}{n!} \lim_{m \rightarrow \infty} [m(m-1)(m-2) \cdots (m-n+1)]$$

$$= \frac{1}{n!} m^n \qquad (4.49)$$

Skipping to the third expression of Eq. (4.48) and proceeding to find its limiting behavior,

$$\lim_{n \rightarrow \infty} \left(1 - \frac{\bar{n}}{m}\right)^m \left(1 - \frac{\bar{n}}{m}\right)^{-n}$$

$$= \lim_{m \rightarrow \infty} \left(1 - \frac{\bar{n}}{m}\right)^m \times \lim_{m \rightarrow \infty} \left(1 - \frac{\bar{n}}{m}\right)^{-n} \qquad (4.50)$$

$$= e^{-\bar{n}} \times 1$$

$$= e^{-\bar{n}}$$

where we used the definition

$$\lim_{m \to \infty} \left(1 - \frac{u}{m} \right)^m = e^u \tag{4.51}$$

Combining the preceding results back into Eq. (4.28), we get

$$\begin{aligned} P(n) &= \left(\frac{m^n}{n!} \right) \left(\frac{\bar{n}}{m} \right)^n e^{-\bar{n}} \\ &= \frac{\bar{n}^n e^{-\bar{n}}}{n!} \end{aligned} \tag{4.52}$$

This is the Poisson probability distribution.

We would also expect the respective characteristic functions to show the same limiting behavior. Starting from Table 4.2

$$\begin{aligned} \Phi_{\text{binomial}}(\omega) &= [p \, e^{j\omega} + q]^m \\ &= [p \, e^{j\omega} + (1 - p)]^m \\ &= [1 + p(e^{j\omega} - 1)]^m \\ &= \left[1 + \frac{\bar{n}(e^{j\omega} - 1)}{m} \right]^m \end{aligned} \tag{4.53}$$

and using Eq. (4.51) again

$$\lim_{m \to \infty} \Phi_{\text{binomial}}(\omega) = e^{\bar{n}(e^{j\omega} - 1)} = \Phi_{\text{poisson}}(\omega) \tag{4.54}$$

which is the characteristic function of the Poisson distribution.

As stated earlier, the Poisson distribution can also be derived from more fundamental principles. Although the results of the derivation are consistent with Eq. (4.52), this derivation will offer some additional insight into some of the basic assumptions implicit in Poisson statistics. However, before we state the hypotheses, let's begin by defining some terms and nomenclature:

ϕ_q = event emission or arrival rate; for our purposes, ϕ_q = a constant

τ = time interval for observation; typically, $\tau = 1$ s

Δt = an infinitesimally small interval; only one event can either occur or not occur in it

t = time coordinate of observation

m = number of small intervals in observation time: $m = \tau/\Delta t$; $m \to \infty$ as $\Delta t \to 0$

$\bar{n}$ = mean number of events counted in time τ; $\bar{n} = \phi_q \tau$

(t_1, t_2) = time interval starting at t_1 and ending at t_2

$P(n; t, t + \tau + \Delta t)$ = probability of "n" events occurring in time interval $(t, t + \tau + \Delta t)$.

For now, we assume that $\phi_q(t)$ is constant with time. Therefore, all time intervals will be referenced to $t = 0$. For example, the time intervals $(t, t + \tau + \Delta t)$ and $(t, t + \Delta t)$ will be represented as $(0, \tau + \Delta t)$ and $(0, \Delta t)$. We can then simplify our nomenclature, representing $P(n; t, t + \tau + \Delta t)$ by $P(n; 0, \tau + \Delta t)$ or even more simply as $P(n$ in $\tau + \Delta t)$. Later, we will consider what effect the assumption of a constant event rate has on our final result.

The fundamental hypotheses needed to derive the Poisson probability distribution are as follows.

1. *Statistical Independence of Time Intervals:* The number of events occurring in nonoverlapping time intervals is statistically independent. For this hypothesis to be valid, the total number of events must be allowed to vary. This is very different from the binomial distribution where the total number of events is finite and known.

2. *Single-Event Probability Limit:* As Δt becomes sufficiently small, the probability for a single event occurring in a time interval of duration Δt is the product of Δt and a real nonnegative function, $\phi_q(t)$, which is the event rate. Mathematically,

$$\lim_{\Delta t \to 0} P(1 \text{ in } \Delta t) = \phi_q \Delta t \qquad (4.55)$$

3. *Normalization:* As Δt becomes sufficiently small, the probability of multiple events occurring in a time interval of duration Δt tends toward zero. Therefore,

$$\lim_{\Delta t \to 0} [P(0 \text{ in } \Delta t) + P(1 \text{ in } \Delta t)] = 1 \qquad (4.56a)$$

or

$$\lim_{\Delta t \to 0} P(0 \text{ in } \Delta t) = 1 - \phi_q \Delta t \qquad (4.56b)$$

The event occurrence rate is thus modeled as a stream of randomly spaced delta functions, each occurring—or not occurring—in a time slot of width Δt. This is illustrated in Fig. 4.2 for an observation time of $\tau = m\Delta t$.

One final note on the relationship between the binomial and Poisson distributions before we derive the Poisson distribution independently: The binomial distribution describes the probabilities associated with a set of events whose total number is finite and does not vary. The Poisson distribution, however,

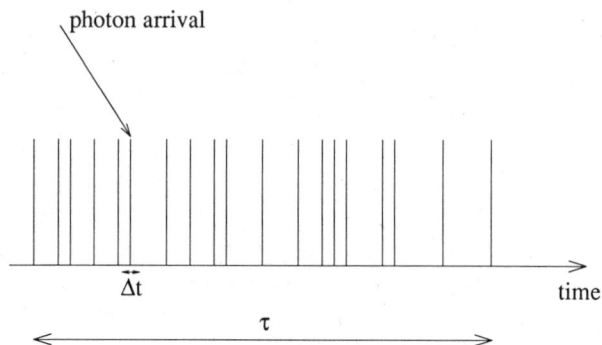

Figure 4.2.   Rate of event occurrences from an event source.

describes a set of events whose total number most certainly varies. In fact, the total event number must be a random variable for hypothesis (1) above to hold true.

How then does the binomial distribution follow the Poisson distribution in the rare-event limit? Remember that the binomial distribution is a conditional probability, conditioned on the total number of events m. The Poisson distribution is a simple probability. Mathematically,

$$P_{\text{binomial}}: P_m(n)$$
$$P_{\text{Poisson}}: P(n) \tag{4.57}$$

As m increases, the conditional probability gradually becomes conditioned on the infinite set, making it a simple probability by default:

$$P_{m \to \infty}(n) \to P(n) \tag{4.58}$$

Still, one should take heed that the binomial distribution is always exact and the Poisson is always an approximation to it (albeit, sometimes a very good one).

With the three preceding assumptions, we are now equipped to calculate $P(n$ in $\tau + \Delta t)$, that is, the probability that n events will occur in any time interval of duration $\tau + \Delta t$. Let's begin by calculating the zero event probability $P(0$ in $\tau + \Delta t)$.

The only way for no events to occur in $(0, \tau + \Delta t)$ is if no events to occur during $(0, \tau)$ and no events to occur during $(\tau, \tau + \Delta t)$. Mathematically,

$$P(0 \text{ in } \tau + \Delta t) = P(0 \text{ in } \tau) \cap P(0 \text{ in } \Delta t)$$
$$= P(0 \text{ in } \tau) \, P(0 \text{ in } \Delta t) \tag{4.59}$$
$$= P(0 \text{ in } \tau) \, [1 - \phi_q \Delta t]$$

where hypothesis (1) is implicit in the first two steps and hypotheses (2) and (3) are implicit in the last. Rearranging and dividing through by Δt,

$$\frac{P(0 \text{ in } \tau + \Delta t) - P(0 \text{ in } \tau)}{\Delta t} = -\phi_q \, P(0 \text{ in } \tau) \tag{4.60}$$

Taking the limit as $\Delta t \to 0$,

$$\frac{d}{dt} P(0 \text{ in } \tau) = -\phi_q \, P(0 \text{ in } \tau) \tag{4.61}$$

The solution of this differential equation is readily seen to be

$$P(0 \text{ in } \tau) = P_0 \, e^{-\phi} q^\tau \tag{4.62}$$

where P_0 is determined by the initial conditions. In this case, $P(0 \text{ in } 0) = 1$, which forces $P_0 = 1$. Therefore,

$$P(0 \text{ in } \tau) = e^{-\phi} q^\tau \tag{4.63}$$

We can calculate $P(n \text{ in } \tau + \Delta t)$ under two possible conditions:

  i. n events occur in $(0, \tau)$ and 0 events occur in $(\tau, \tau + \Delta t)$.
  ii. $n - 1$ events occur in $(0, \tau)$ and 1 event occurs in $(\tau, \tau + \Delta t)$.

Mathematically,

$$P(n \text{ in } \tau + \Delta t) = [P(n \text{ in } \tau) \cap P(0 \text{ in } \Delta t)] \cup [P(n - 1 \text{ in } \tau) \cap P(1 \text{ in } \Delta t)] \tag{4.64}$$

Assuming statistical independence of time intervals and that the events i and ii are disjoint, the intersections are direct multiplications and the union is a sum. Hence,

$$P(n \text{ in } \tau + \Delta t) = [P(n \text{ in } \tau) \times P(0 \text{ in } \Delta t)] + [P(n - 1 \text{ in } \tau) \times P(1 \text{ in } \Delta t)] \tag{4.65}$$

Substituting the expressions of hypotheses (2) and (3) into Eq. (4.65),

$$P(n \text{ in } \tau + \Delta t) = [P(n \text{ in } \tau) \times (1 - \phi_q \Delta t)] + [P(n - 1 \text{ in } \tau) \times (\phi_q \Delta t)]$$

$$= P(n \text{ in } \tau) - (\phi_q \Delta t \times P(n \text{ in } \tau)) + (\phi_q \Delta t \times P(n - 1 \text{ in } \tau)) \tag{4.66}$$

Rearranging terms and dividing through by Δt,

$$\frac{P(n \text{ in } \tau + \Delta t) - P(n \text{ in } \tau)}{\Delta t} + \phi_q P(n \text{ in } \tau) = (\phi_q P(n - 1 \text{ in } \tau)) \quad (4.67)$$

Taking the limit as $\Delta t \to 0$,

$$\frac{d}{d\tau} P(n \text{ in } \tau) + \phi_q P(n \text{ in } \tau) = \phi_q P(n - 1 \text{ in } \tau) \quad (4.68)$$

Define

$$f(\tau) \equiv P(n \text{ in } \tau) \quad (4.69a)$$

$$g(\tau) \equiv P(n - 1 \text{ in } \tau) \quad (4.69b)$$

Then Eq. (4.68) becomes

$$\frac{d}{d\tau} f(\tau) + \phi_q f(\tau) = \phi_q g(\tau) \quad (4.70)$$

Multiplying through by an integrating factor,

$$e^{\phi_q \tau} \frac{d}{d\tau} f(\tau) + \phi_q \, e^{\phi_q \tau} \, f(\tau) = \phi_q e^{\phi_q \tau} \, g(\tau) \quad (4.71)$$

The left-hand side should be recognized as a perfect differential. Proceeding,

$$\frac{d}{d\tau} (f(\tau) \, e^{\phi_q \tau}) = \phi_q \, e^{\phi_q \tau} \, g(\tau) \quad (4.72)$$

Multiplying through by $d\tau$ and integrating both sides,

$$f(\tau) \, e^{\phi_q \tau} = \int_0^\tau \phi_q \, e^{\phi_q \tau'} \, g(\tau') \, d\tau' \quad (4.73)$$

Finally,

$$f(\tau) = \phi_q \, e^{-\phi_q \tau} \int_0^\tau e^{\phi_q \tau'} \, g(\tau') \, d\tau'; \quad f(0) = 0 \quad (4.74)$$

Returning to our previous notation,

$$P(n \text{ in } \tau) = \phi_q \, e^{-\phi_q \tau} \int_0^\tau e^{\phi_q \tau'} \, P(n - 1 \text{ in } \tau') \, d\tau' \quad (4.75)$$

Equation (4.75) implies a recursion relation. Knowing $P(0 \text{ in } \tau)$, we can calculate $P(1 \text{ in } \tau)$. We can then use $P(1 \text{ in } \tau)$ to determine $P(2 \text{ in } \tau)$ and continue until we ultimately find a relation for $P(n \text{ in } \tau)$ by induction. Substituting Eq. (4.63) into Eq. (4.75),

$$
\begin{aligned}
P(1 \text{ in } \tau) &= \phi_q \, e^{-\phi_q \tau} \int_0^\tau e^{\phi_q \tau'} P(0 \text{ in } \tau') \, d\tau' \\
&= \phi_q \, e^{-\phi_q \tau} \int_0^\tau e^{\phi_q \tau'} \, e^{-\phi_q \tau'} \, d\tau' \\
&= \phi_q \, e^{-\phi_q \tau} \int_0^\tau d\tau' \\
&= \phi_q \, \tau \, e^{-\phi_q \tau}
\end{aligned}
\tag{4.76}
$$

Substituting this result back into Eq. (4.75),

$$
\begin{aligned}
P(2 \text{ in } \tau) &= \phi_q \, e^{-\phi_q \tau} \int_0^\tau e^{\phi_q \tau'} P(1 \text{ in } \tau') \, d\tau' \\
&= \phi_q \, e^{\phi_q \tau} \int_0^\tau e^{\phi_q \tau'} \phi_q \tau' \, e^{\phi_q \tau'} \, d\tau' \\
&= \phi_q^2 \, e^{-\phi_q \tau} \int_0^\tau \tau' \, d\tau' \\
&= \phi_q^2 \, \frac{\tau^2}{2} \, e^{-\phi_q \tau}
\end{aligned}
\tag{4.77}
$$

By induction,

$$
P(n \text{ in } \tau) = \phi_q^n \, \frac{\tau^n}{n!} \, e^{-\phi_q \tau}
$$

$$
P(n \text{ in } \tau) = \frac{\bar{n}^n e^{-\bar{n}}}{n!}; \qquad \text{for } \bar{n} = \phi_q \tau
\tag{4.78}
$$

which the reader will recognize as Poisson's law.

For an event rate that is not constant with respect to time, the form of the probability function remains exactly the same. The only generalization that must be made is that $\bar{n} = \phi_q \tau$ must now be replaced by the expression

$$
\bar{n} = \phi_q \tau \rightarrow \int_0^{t+\tau} \phi_q(t) \, dt
\tag{4.79}
$$

Therefore, the most general expression for the Poisson discrete probability law is

$$P(n) = \frac{\left[\int_0^{t+\tau} \phi_q(t) \, dt \right]^n \exp\left[- \int_0^{t+\tau} \phi_q(t) \, dt \right]}{n!} \qquad (4.80)$$

We now determine that the distribution is normalized, and then calculate its mean and variance. The normalization condition is

$$\sum_{n=0}^{n_{max}} P(n) = \sum_{n=0}^{\infty} \frac{\bar{n}^n e^{-\bar{n}}}{n!}$$

$$= e^{-\bar{n}} \sum_{n=0}^{\infty} \frac{\bar{n}^n}{n!} \qquad (4.81)$$

$$= e^{-\bar{n}} e^{\bar{n}} = 1$$

The expression in indeed normalized. Calculating the mean,

$$\bar{n} = \sum_{n=0}^{\infty} nP(n)$$

$$= \sum_{n=1}^{\infty} nP(n) \qquad (4.82)$$

where, again, the first term does not contribute to the mean. Substituting the Poisson probability law for $P(n)$,

$$\bar{n} = \sum_{n=1}^{\infty} n \left[\frac{\bar{n}^n e^{-\bar{n}}}{n!} \right]$$

$$= \sum_{n=1}^{\infty} \frac{\bar{n}^n \, e^{-\bar{n}}}{(n-1)!} \qquad (4.83)$$

$$= \bar{n} \sum_{n=1}^{\infty} \frac{\bar{n}^{n-1} \, e^{-\bar{n}}}{(n-1)!}$$

Using the variable substitution, $k = n - 1$,

$$\bar{n} = \bar{n} \sum_{k=0}^{\infty} \frac{\bar{n}^k \, e^{-\bar{n}}}{(k)!}$$

$$= \bar{n} \qquad (4.84)$$

In the last step we used the normalization condition previously established.

Finally, we calculate the variance by first calculating the second moment and then substituting the result into Eq. (4.26). By definition

$$\overline{n^2} = \sum_{n=0}^{\infty} n^2 P(n) \tag{4.85}$$

We rewrite n^2 as

$$n^2 = n(n - 1) + n \tag{4.86}$$

Substituting this expression back into Eq. (4.26),

$$\overline{n^2} = \sum_{n=0}^{\infty} [n(n - 1) + n] P(n)$$

$$= \sum_{n=0}^{\infty} [n(n - 1)P(n)] + \sum_{n=0}^{\infty} [nP(n)] \tag{4.87}$$

$$= \sum_{n=2}^{\infty} [n(n - 1)P(n)] + \overline{n}$$

Substituting Poisson's law for $P(n)$,

$$\overline{n^2} = \left(\sum_{n=2}^{\infty} n(n - 1) \frac{\overline{n}^n e^{-\overline{n}}}{n!} \right) + \overline{n}$$

$$= \left(\sum_{n=2}^{\infty} \left[\frac{\overline{n}^n e^{-\overline{n}}}{(n - 2)!} \right] \right) + \overline{n} \tag{4.88}$$

$$= \overline{n}^2 \left(\sum_{n=2}^{\infty} \left[\frac{\overline{n}^{n-2} e^{-\overline{n}}}{(n - 2)!} \right] \right) + \overline{n}$$

Again, using a variable substitution, $k = n - 2$,

$$\overline{n^2} = \left(\overline{n}^2 \sum_{k=0}^{\infty} \frac{\overline{n}^k e^{-\overline{n}}}{k!} \right) + \overline{n} \tag{4.89}$$

$$= \overline{n}^2 + \overline{n}$$

Finally, substituting our result for the second moment of n back into our expression for the variance,

$$\sigma^2 = \overline{n^2} - \overline{n}^2$$

$$= \overline{n}^2 + \overline{n} - \overline{n}^2 \tag{4.90}$$

$$= \overline{n}$$

These results are fundamental to the Poisson pdf; the variance equals the mean.

Table 4.3 Poisson Distribution Properties

$P(n)$	$\dfrac{\bar{n}^n e^{-\bar{n}}}{n!}$
$\bar{n}$	$\phi_q \tau$
$\overline{n^2}$	$\bar{n} + \bar{n}^2$
σ^2	$\bar{n}$
σ	$\sqrt{\bar{n}}$
$\Phi(\omega)$	$e^{\bar{n}(e^{j\omega} - 1)}$

First- and second-order statistical properties of the Poisson distribution are listed in Table 4.3.

4.7.3. Normal (Gaussian) Density Function

The most frequently used continuous pdf is the normal distribution, chiefly because of its relationship to the central limit theorem. It is given by

$$p(x) = \frac{1}{\sqrt{2\pi\sigma^2}} \exp\left\{\frac{-(x - \bar{x})^2}{2\sigma^2}\right\} \tag{4.91}$$

The normal law, because of its frequent use, has developed its own notation, which is $N(\bar{x}, \sigma^2)$.

For the special case of $\bar{x} = 0$, the probability law is called a Gaussian distribution. A conversion between a normal distribution and a Gaussian distribution is often made by making a transformation of variables from x to a standardized random variable z, given by

$$z = \frac{(x - \bar{x})}{\sigma} \tag{4.92}$$

The related Gaussian distribution is then

$$p(x) = \frac{1}{\sqrt{2\pi}} \exp\left\{\frac{-z^2}{2}\right\} \tag{4.93}$$

First- and second-order statistical properties are listed in Table 4.4. Derivation of the quantities in the table are left to the reader.

In the limit of $m \to \infty$, the binomial distribution can be approximated by a Gaussian distribution provided that neither p nor q is near zero, that is, by staying away from the extreme regions of success and failure. This condition can also be stated by requiring n not to be far (on the scale of the variance) from its mean value. We can state the constraint as follows.

Table 4.4 Normal Distribution

$p(x)$	$\dfrac{1}{\sqrt{2\pi\sigma^2}} \exp\left\{\dfrac{-(x - \bar{x})^2}{2\sigma^2}\right\}$
$\bar{x}$	$\bar{x}$
x^2	$\bar{x}^2 + \sigma^2$
σ^2	σ^2
σ	σ
$\Phi(\omega)$	$\exp\left\{j\omega\bar{x} - \dfrac{\sigma^2\omega^2}{2}\right\}$

1. *No extremes*

$$|n - \bar{n}| \le \sigma \qquad (4.94a)$$

which is equivalent to

$$|n - mp| \le \sqrt{mpq} \qquad (4.94b)$$

An additional constraint is the smoothness of the curve $P(n)$. If we limit the amount by which $P(n)$ can change from one value of n to the next to be much less than the standard deviation, this produces a curve that is locally smooth. For the discrete case being considered, $\Delta n = 1$. We can now state our constraint mathematically.

2. *Smoothness*

$$\sigma \gg \Delta n \qquad (4.95a)$$

or

$$mpq \gg 1 \qquad (4.95b)$$

Because there is a relation between the binomial and normal distributions and a relation between the binomial and Poisson distributions, we can expect a relation between the Poisson and normal distributions. This is indeed the case. It can be proved that if n is the Poisson random variable of Table 4.3 and

$$z = \frac{n - n'}{\sqrt{n'}} \qquad (4.96)$$

is the corresponding standardized random variable, then as $n' \to \infty$, the Poisson distribution approaches the normal distribution, that is,

$$\lim_{n' \to \infty} z \to \text{normal } rv \qquad (4.97)$$

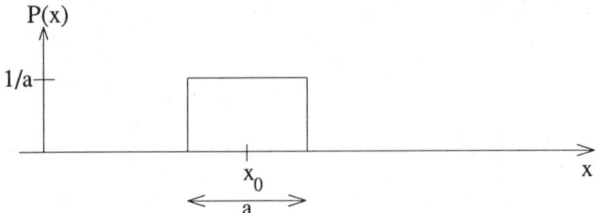

Figure 4.3.   Uniform distribution function.

4.7.4.   Uniform Distribution

Figure 4.3 illustrates a uniform distribution function and Table 4.5 summarizes uniform distribution properties:

$$p(x) = \frac{1}{a} \operatorname{rect}\left(\frac{x - x_0}{a}\right) \tag{4.98}$$

4.7.5.   Thermal Distributions of Ideal Gases

4.7.5.1.   Bose–Einstein Distribution. Bose–Einstein statistics describe the statistical behavior of a system of particles with symetric wavefunctions (bosons) in thermal equilibrium at temperature T. Photons can be thought of as bosons whose total number is unrestricted. As we will see in Chapter 5, the Bose–Einstein distribution can be used to derive the Planck radiation law for photons leaving a thermal source. Table 4.6 is an illustration of the Bose–Einstein distribution properties, where k is Boltzmann's constant.

4.7.5.2.   Fermi–Dirac Distribution. The Fermi–Dirac distribution describes the probability associated with fermions including electrons, holes,

Table 4.5   Uniform Distribution

$p(x)$	$\dfrac{1}{a} \operatorname{rect}\left(\dfrac{x - x_0}{a}\right)$
$\bar{x}$	x_0
$\overline{x^2}$	$x_0 + \dfrac{a^2}{12}$
σ^2	$\dfrac{a^2}{12}$
σ	$\dfrac{a}{\sqrt{12}}$
$\Phi(\omega)$	$e^{j\omega x_0} \operatorname{sinc}\left(\dfrac{a\omega}{2\pi}\right)$

Table 4.6 Bose–Einstein Distribution Properties

$P(n)$	$\dfrac{\bar{n}^n}{(1 + \bar{n})^{n+1}}$
$\bar{n}$	$(\exp\{hc/\lambda kT\} - 1)^{-1}$
$\overline{n^2}$	$\bar{n} + 2\bar{n}^2$
σ^2	$\bar{n} + \bar{n}^2$
σ	$\sqrt{\bar{n}(1 + \bar{n})}$
$\Pi(\omega)$	$(1 + \bar{n}(1 - e^{j\omega}))^{-1}$

Table 4.7 Fermi–Dirac Distribution Properties

$P(n)$	Problem 2.9
$\bar{n}$	$(\exp\{hc/\lambda kT\} + 1)^{-1}$
$\overline{n^2}$	Problem 2.9
σ^2	$\bar{n} - \bar{n}^2$
σ	$\sqrt{\bar{n}(1 - \bar{n})}$
$\Phi(\omega)$	Problem 2.9

protons, and many other particles of interest. The derivation of the properties is similar to that of the Bose–Einstein distribution, and the results are listed in Table 4.7. Some of the quantities are left for the reader to calculate.

4.7.5.3. Maxwell–Boltzmann Distribution. The Maxwell–Boltzmann distribution is often called as the *classical distribution* because in the limit of low occupancy (also called the *high-energy limit*, for obvious reasons), it is the limiting distribution of both the Bose–Einstein and Fermi–Dirac distributions. For the Maxwell–Boltzmann distribution, we get

$$p(x) = 1/\bar{x}e^{-x/\bar{x}} \tag{4.99}$$

where x is the energy level and $\bar{x} = kT$. Lower-order properties are listed in Table 4.8.

Table 4.8 Maxwell–Boltzmann Distribution

$p(x)$	$\dfrac{1}{\bar{x}} \exp\left\{ -\dfrac{x}{\bar{x}} \right\}$
$\bar{x}$	kT
$\overline{x^2}$	$2\bar{x}^2$
σ^2	$\bar{x}^2$
σ	$\bar{x}$
$\Phi(\omega)$	$\dfrac{1}{1 - j\omega\bar{x}}$

REFERENCE

Kolmogoroff, A., "Grundbegriff der Wahrscheinlichkeitsrechnung," *Ergeb. Mat. und ihrer Grenzg.* **2**(3) (1933).

BIBLIOGRAPHY

Frieden, B. R., *Probability, Statistical Optics, and Data Testing*, Springer-Verlag, Berlin, New York, 1991.

Gagliardi, R. M., and S. Karp, *Optical Communications*, Wiley, New York, 1976.

Loudon, R., *The Quantum Theory of Light*, Oxford University Press, Oxford, 1979.

Papoulis, A., *Probability, Random Variables and Stochastic Processes*, McGraw-Hill, New York, 1965.

Parzen, E., *Stochastic Processes*, Holden-Day, San Francisco, 1967.

Planck, M., *Treatise on Thermodynamics*, Dover, New York, 1945.

Saleh, B., *Photoelectron Statistics*, Springer-Verlag, Berlin, 1978.

PROBLEMS

4.1. Prove that the variance equals the mean $\overline{n^2} - \bar{n}^2 = \bar{n}$ for the Poisson distribution,

$$P(n) = \frac{\bar{n}^n e^{-\bar{n}}}{n!}$$

Note that

$$\bar{n} = \sum_{n=0}^{\infty} nP(n)$$

$$\overline{n^2} = \sum_{n=0}^{\infty} n^2 P(n)$$

and

$$e^{\bar{n}} = \sum_{n=0}^{\infty} \frac{\bar{n}^n}{n!}$$

(*Hint:* $(-1)! = \infty$ and $(-2)! = \infty$.)

4.2. Blackbody photons follow Bose–Einstein statistics: Under what wavelength and blackbody temperatures can they be approximated by Poisson statistics?

4.3. Starting with the binomial distribution

$$P_m(n) = \binom{m}{n} p^n q^{m-n}$$

where

m = number of trials

n = number of events in m trials

p = probability of an event occur

$q = 1 - p$ (not occurring)

$$\binom{m}{n} = \frac{m!}{n!\,(m-n)!}$$

obtain the Poisson distribution function when m increases to infinity (∞) and p decreases in such a manner that $mp = \Lambda$ (Λ is a constant). (*Hint:* $\lim_{m \to \infty} (1 - (\Delta/m))^m = e^{-\Lambda}$.)

4.4. Prove $P(EF^c) = P(E) - P(EF)$.

4.5. Prove $P(E \cup F) = P(E) + P(F) - P(EF)$.

4.6. Prove that the ith derivative of the moment-generating functions is equal to the ith moment.

4.7. Find the expression for the variance of the binomial distribution.

4.8. Find the expression for the mean and variance, from the characteristic function of the normal distribution.

4.9. Fill in the table values for $P(n)$, $\overline{n^2}$, and $\Phi(\omega)$ for the Fermi–Dirac distribution (Table 4.7).

5

Noise in Optical Detection

5.1 INTRODUCTION

Thus far, we developed the concept of optical-detector response to optical radiation or photon flux; however, we have not determined how fast a change in or how low a radiant power level can be detected. We refer to optical radiation in terms of watts because of the existing figures of merit. For the detectors we discussed, an analytical expression has been found that relates a measurable parameter influenced by the presence of optical radiation.

Now let's do a Gedanken experiment. Let us reduce the optical radiant power on the detector as low as is possible and yet maintain a signal (i.e., Δv or Δi) proportional to the radiation. As we decrease the radiant power on the detector, we eventually reach a radiant power level where the photo-generated current is equal to the noise current.

The term "noise" comes from audio radio static. This minimum amount of detectable radiant power is the minimum detectable radiant signal or noise-equivalent power (NEP). The NEP is the rms incident radiant power necessary to produce a signal-to-noise ratio (SNR) of one. Figure 5.1 is an example of oscilloscope traces with an SNR of one, where the sine and square waves are the signal.

Noise is the random fluctuation in electrical output from a detector. The noise limits set the lower level of sensitivity. The noise determines this lower limit. A detector that can detect to a picowatt (10^{-12}) is more sensitive than one that detects to a microwatt (10^{-6}). (*Note:* Later we will discuss spatial variations across a two-dimensional array and also call it noise.)

The sources of noise can be man-made (e.g., transformers, motors, television stations), natural (e.g., lightning, earthquakes, sunspots), or intrinsic to the optical detector. The most interesting noise sources are the optical-detector-associated noises, assuming that the other noise sources can be solved by good engineering practice and geometrical constraints.

Optical-detector noise can be classified as either internal or external (Fig. 5.2). The further to the left side of Fig. 5.2 that the limiting noise appears, the better the system performance. External sources include photon flux, the interface electronics (preamplifier), and microphonic noise. Seven possible noise sources are internal to the detector:

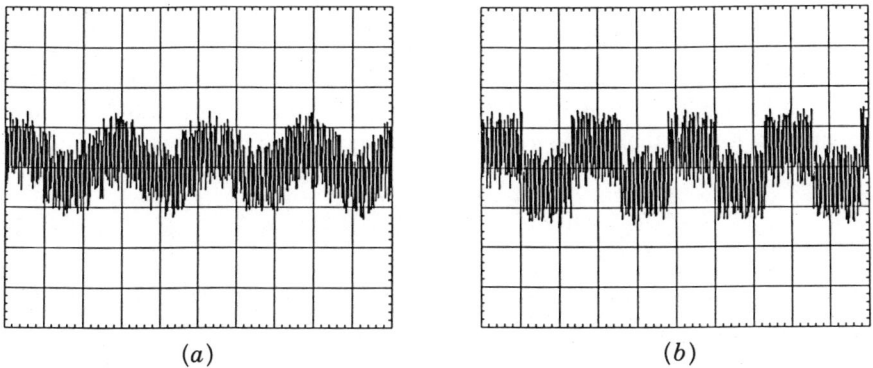

Figure 5.1. Scope photographs for SNR of one: (a) sine wave; (b) square wave.

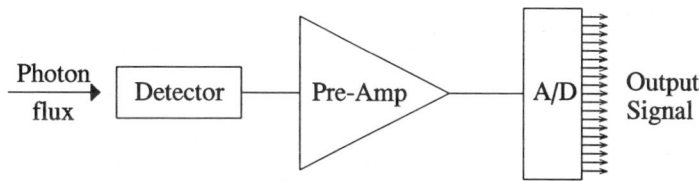

Figure 5.2. Noise sources related to optical-detector interface.

Johnson

Shot

Generation recombination

One over frequency ($1/f$)

Temperature fluctuation

Microphonics

Popcorn (Barkhuesen)

In most detector cases, we are concerned primarily with three or four noise sources for a particular detector. For the case of optical detectors, noise can be expressed in terms of voltage, current, or electrical power. If we consider one of these parameters a random variable and assign to it a probability-density function, we can find its statistics. The standard deviation is defined as the noise of the random variable.

For a voltage source with an average value of $\bar{v}$, the variance (standard deviation squared) is

$$\overline{(\Delta v)^2} = \overline{(v - \bar{v})^2} = \frac{1}{\Delta t} \int_0^{\Delta t} (v - \bar{v})^2 \, dt \tag{5.1}$$

where Δt = time interval (assuming an ergodic function, where the time average equals the ensemble average). Noise fluctuates in time because the generation rates fluctuate. To characterize a nonperiodic, fluctuating quantity, we will use the concept of a noise-power spectrum. Assume we have a voltage as a function of time as shown in Fig. 5.3a. To simplify the algebra, we will assume the average voltage is zero (ac-coupled through a capacitor). The autocorrelation $\Re$ of that voltage waveform is shown in Fig. 5.3b:

$$\Re(\Delta t) = \int_{-\infty}^{\infty} v(t)\, v(t + \Delta t)\, dt \tag{5.2}$$

The Fourier transform of the autocorrelation $\Re(\Delta t)$ is the power spectral density $\mathfrak{S}(f)$ shown in Fig. 5.3c (from Gaskill, 1978). The central ordinate theorem for Fourier transforms yields

$$\Re(\Delta t = 0) = \int_{-\infty}^{\infty} \mathfrak{S}(f)\, df \tag{5.3}$$

or the total power for no time shift ($\Delta t = 0$). Also

$$\Re(\Delta t = 0) = \int_{-\infty}^{\infty} v(t)\, v(t)\, dt = \int_{-\infty}^{\infty} v^2(t)\, dt \tag{5.4}$$

where this is just the mean-square voltage fluctuation.

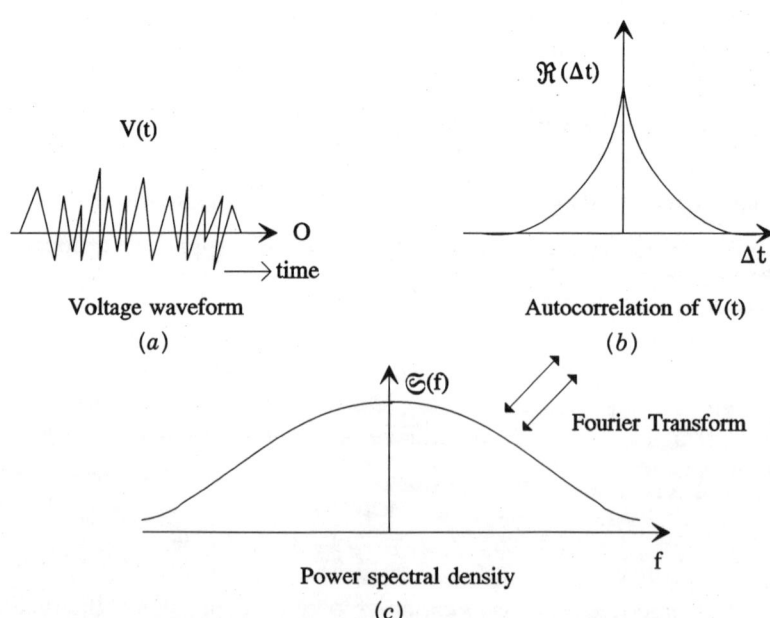

Figure 5.3. Noise processes related to autocorrelation and power spectral density.

Because the noise power of each frequency component adds, voltage noises add in quadrature. By adding or integrating over the entire power spectral density (adding noise powers of various frequencies), we get the total noise power. If the noise power is constant (independent of frequency (f)), it is called *white noise*. The noise power or variance is proportional to the electrical bandwidth (Δf):

$$\overline{(\Delta v)^2} \propto \Delta f \tag{5.5}$$

and the root-mean-square (rms) noise voltage is proportional to the root of the bandwidth

$$\sqrt{\overline{(\Delta v)^2}} \propto \sqrt{\Delta f} \tag{5.6}$$

Similarly, independent random-variable noise-power spectra add (conceptually we know that two sinusoidal voltages do not add, their power outputs add, or their equivalent heating effects add), so the variances of the independent random variables also add. Because variances add to obtain total noise, one adds variances (power), not standard deviations (voltage noise). That is, for independent voltage-noise sources, the total noise power is

$$\overline{(\Delta v)^2} = \overline{(\Delta v)_1^2} + \overline{(\Delta v)_2^2} + \overline{(\Delta v)_3^2} \tag{5.7}$$

where the mean-square voltage fluctuation is proportional to power. The square root is the rms

$$\Delta v_{\text{rms}} = \sqrt{\overline{(\Delta v)^2}} = \sqrt{\overline{(\Delta v)_1^2} + \overline{(\Delta v)_2^2} + \overline{(\Delta v)_3^2}} \tag{5.8}$$

Noise powers add, and thus variances add in a summation fashion just as the frequency-component powers add in the power-spectral-density expression Eq. (5.8). For example, if three noise sources are present, the total noise is

$$(\text{Total noise})^2 = \text{Johnson}^2 + \text{shot}^2 + \text{photon}^2 \tag{5.9}$$

For example, assume Johnson noise and shot noise are present with the following values:

$$V_j = 6 \ \mu V \tag{5.10a}$$

$$V_s = 10 \ \mu V \tag{5.10b}$$

The total noise voltage is

$$v_{\text{rms}} = \sqrt{6^2 + 10^2} = 11.7 \ \mu V \tag{5.11}$$

The two important messages here are that (1) noise powers add (i.e., noise voltages are summed in quadrature), and (2) the noise voltage is proportional to square-root bandwidth for white-noise sources.

5.2. PHOTON NOISE: EMISSION AND INTERACTION NOISE

As stated in Chapter 2, photons follow Poisson statistics for all practical detector applications. That is, for visible and near-infrared radiation, low-temperature blackbodies, and short wavelength ($h\upsilon \gg kT$), the photon noise follows Poisson statistics. In the photon-noise-limited situation, the photon noise is the lowest possible noise. Here we are talking about the noise of the photons incident on the detector, and the detector has no influence on it and cannot increase it. In no application can we have lower noise than in the photon-noise-limited case. Note that the two sources of photons are signal photons and background photons.

From Chapter 2, the number of photons from a blackbody source on a detector can be calculated, but inherent in that radiation measurement is an uncertainty in that number of photons. If one calculates the photon exitance from a 300-K source (i.e., $\sigma_q T^3$), the average value is 10^{18} photon/s-cm^2. The variance in this number is noise. Figure 5.4 illustrates the point graphically.

In Fig. 5.4, σ is a measure of the spread or noise on the average value $M_q(T)$. Recall that this has nothing to do with the detection mechanism. Thus, we can talk about the ultimate sensitivity of a detector without knowing the detector.

The instantaneous rate at which photons are emitted by a source is a result of several processes that vary in time. The number of photons emitted by a

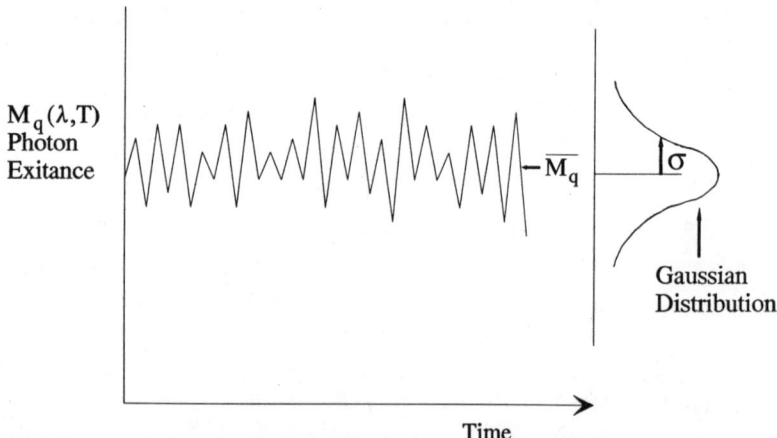

Figure 5.4. Noise of photon exitance vs. time.

source fluctuates about an average value because of the random nature of the photon emission. This variation, or uncertainty, gives rise to photon noise. This approach to photon-noise analysis will derive the variance for the photon flux. We assume that the photons are indistinguishable particles of the same energy. A laser source produces such a photon flux or stream. We will find that the noise is proportional to the square root of the average number of photons in time interval Δt, and that the photons obey Poisson statistics.

Recall that, strictly speaking, blackbody photons have different energies and follow Bose–Einstein statistics. Because the Poisson assumption applies for laser sources, but not for blackbodies, a correction factor for bosons must be added. We now set up the restrictions or constraints for these conditions.

Consider a photon-noise analogy: We count vehicles moving on a freeway as they pass an exit. We count the cars per minute and then plot the result. The plot, shown in Fig. 5.5, follows Poisson distribution around, for example, 100 vehicles per minute. If the vehicles are spaced at exactly the same distance, for example 66 ft apart, we would count exactly 100 per one minute, if there were no variance and no noise. The analogy to our photon problem is that, if photons are emitted at exactly 100/s, the flux would have no noise.

Now we add some additional information to our freeway example. We know more information about the vehicles (more statistical information). For example, semitrailer trucks run in convoys (McCall, 1976) and motorcycle gangs ride in groups. Previously we did not distinguish by energy. We assumed only cars on the freeway and all cars were the same.

We know that the vehicles have different energy distributions, in fact, the vehicle energies follow a Bose–Einstein distribution. Two separate points are made here: (1) particles travel with different momenta, and (2) particles with less energy group or clump. A semitruck has more energy than a motorcycle.

We have additional information about the particles we are counting in time. Their energy distribution is known. We now do the counting experiment again.

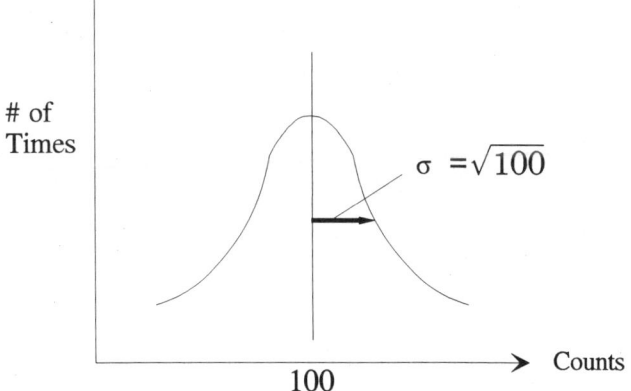

Figure 5.5. Distribution of counts around mean.

The variance or standard deviation differs from the Poisson distribution:

$$\sigma^2 = \bar{n} \left[\frac{e^x}{e^x - 1} \right] = \bar{n} + \overline{n^2} \qquad (5.12)$$

This variation arises because photons clump, that is, they are not necessarily emitted as individual, distinguishable particles.

5.2.1. Noise Processes in Laser Sources (Poisson)

Consider a laser source of radiation and the noise inherent within it. Let us find the statistical properties of a laser beam of photons incident on a photon detector (counter). This stream of particles, at a fixed wavelength or frequency, is a monochromatic source. Assume the photodetector has a perfect detection process, so that each photon is detected. Terms are defined as

ϕ_q = rate of arrival of photons
Δt = time interval over which the photons will be counted, typically 1 s
$\bar{n}$ = average number of photons counted ($\bar{n} = \phi_q \Delta t$).

The photon emission rate can be modeled as a stream of delta functions, and is shown pictorially in Fig. 5.6. Only if the photons were exactly spaced in time would there be no noise.

Conceptually (Fig. 5.6), we will divide the time interval (Δt) into a smaller interval (δt), where

$$\Delta t = m \, \delta t \qquad (5.13)$$

but this smaller interval is of such short duration that only one event or photon occurs during this interval. In any given interval, δt, a photon is either present or not. The probability of either finding no photon in δt or finding one photon

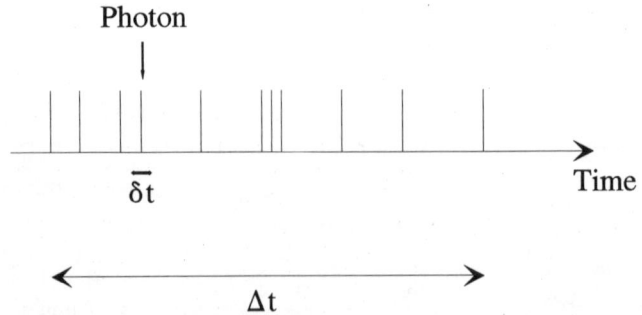

Figure 5.6. Emission rate of photons from a source. (*Note:* If and only if the photons were exactly spaced in time would there be no noise.)

in δt is equal to one. We can model this as a binomial process:

$$P(0, \delta t) + P(1, \delta t) = 1 \tag{5.14}$$

The obvious assumptions one can draw are

1. $P(1, \delta t) = \phi_q \delta t$; (rate times time)
2. $P(n, \delta t) = 0$, for $n > 1$
3. statistical independence—for any time interval, δt; whether a photon occurs is not related to the previous time period (δt_{n-1})

From Frieden (1991), the binomial law is the probability of n events in the time interval Δt and is given by

$$P(n) = (\phi_q \delta t)^n \, (1 - \phi_q \delta t)^{m-n} \, \frac{m!}{n! \, (m-n)!} \tag{5.15}$$

where

$\phi_q \delta t = P(1, \delta t)$
  m = number of small time intervals δt under consideration

From the statistics for a binomial distribution:

$$\text{average:} \quad \bar{n} = m \, \phi_q \delta t \tag{5.16a}$$

$$\text{variance:} \quad \sigma^2 = m \, \phi_q \delta t (1 - \phi_q \delta t) \tag{5.16b}$$

As m approaches ∞,

$$\phi_q \delta t \rightarrow 0 \tag{5.17}$$

In the limit of large numbers, the binomial distribution becomes

$$P(n) = (1 - \phi_q \delta t)^m \, (1 - \phi_q \delta t)^{-n} \, \frac{m!}{n! \, (m-n)!} \, (\phi_q \delta t)^n \tag{5.18}$$

Looking at each part of Eq. (5.18),

$$(1 - \phi_q \delta t)^m = [(1 - \phi_q \delta t)^{1/(\phi_q \delta t)}]^{m(\phi_q \delta t)} = [(1 - \phi_q \delta t)^{1/(\phi_q \delta t)}]^{\bar{n}} = e^{-\bar{n}} \tag{5.19a}$$

and

$$\lim_{\phi_q \delta t \to 0} = (1 - \phi_q \delta t)^{-n} = 1 \tag{5.19b}$$

and

$$\lim_{m \to \infty, m\phi_q \delta t \to \bar{n}} \left[\frac{m!}{(m - n)!} (\phi_q \delta t)^n \right] = \lim_{m \to \infty} \left[\frac{m!}{(m - n)!} \frac{\bar{n}^n}{m^n} \right] \qquad (5.19c)$$

Substituting the preceding, along with the Stirling approximation to the factorial, gives the Poisson distribution

$$P(n) = \frac{\bar{n}^n}{n!} e^{-\bar{n}} \qquad (5.20)$$

which describes the photon fluctuation in time. $P(n)$ is the probability of n photons being detected in a given time when the average number of photons detected are $\bar{n}$.

The noise is the standard deviation about the mean. To calculate the mean value $\bar{n}$ we use

$$\bar{n} = \sum_{0}^{\infty} n P(n) \qquad (5.21)$$

To find variance, we write out the terms

$$\sigma^2 = \overline{(n - \bar{n})^2} = \overline{n^2} - \overline{2n\bar{n}} + \overline{\bar{n}^2} = \overline{n^2} - \bar{n}^2 \qquad (5.22)$$

The variance is the mean of the squared value minus the square of the mean value. Solving for the second moment, $\overline{n^2}$, in the variance expression Eq. (5.22):

$$\overline{n^2} = \sum_{n=0}^{\infty} n^2 P(n) \qquad (5.23a)$$

$$\overline{n^2} = \sum_{n=0}^{\infty} (n(n - 1) + n) P(n) \qquad (5.23b)$$

$$\overline{n^2} = \sum_{n=0}^{\infty} n(n - 1) P(n) + \sum_{n=0}^{\infty} n P(n) \qquad (5.23c)$$

and

$$\overline{n^2} = \sum_{n=2}^{\infty} n(n - 1) P(n) + \bar{n} \qquad (5.23d)$$

Substituting the Poisson density function:

$$\overline{n^2} = \sum_{n=2}^{\infty} n(n-1) \left(\frac{\overline{n}^n e^{-\overline{n}}}{n!} \right) + \overline{n} \tag{5.23e}$$

$$\overline{n^2} = \sum_{n=2}^{\infty} \left(\frac{\overline{n}^n e^{-\overline{n}}}{(n-2)!} \right) + \overline{n} \tag{5.23f}$$

$$\overline{n^2} = \overline{n}^2 \sum_{n=2}^{\infty} \left(\frac{\overline{n}^{n-2} e^{-\overline{n}}}{(n-2)!} \right) + \overline{n} \tag{5.23g}$$

Substituting $n' = n - 2$:

$$\overline{n^2} = \overline{n}^2 \sum_{n'=0}^{\infty} \left(\frac{\overline{n}^{n'} e^{-\overline{n}}}{(n')!} \right) + \overline{n} \tag{5.24}$$

Note that over all n' space, a Poisson distribution is produced in which the probability over all space must sum to 1:

$$\sum_{n'=0}^{\infty} \left(\frac{\overline{n}^{n'} e^{-\overline{n}}}{(n')!} \right) \equiv 1 \tag{5.25}$$

Substituting Eq. (5.25) into Eq. (5.24):

$$\overline{n^2} = \overline{n}^2 + \overline{n} \tag{5.26}$$

Now we can find our variance on the photon flux by substituting Eq. (5.26) into Eq. (5.22):

$$\sigma^2 = \overline{n}^2 + \overline{n} - \overline{n}^2 = \overline{n}. \tag{5.27}$$

For a Poisson distribution, the variance equals the mean. If we measured the rms current flow of the electrons, they would obey Poisson statistics. These results are fundamental to photon noise. Laser sources follow Poisson statistics because their output are particles (photons) of a single energy $\mathcal{E} = h\nu$. For laser sources, the variance equals the mean.

5.2.2. Noise Processes in Thermal Sources (Bose–Einstein)

In Chapter 2, we used Bose–Einstein statistics to derive Planck's radiation law. The approach was to define the available energy states of photons leaving a thermal source as $\mathcal{E}_1, \mathcal{E}_2, \mathcal{E}_3, \ldots, \mathcal{E}_m$. The corresponding number of photons in each of these energy states in $n_1, n_2, n_3, \ldots, n_m$, where n_1 is the photon number in energy state $\mathcal{E}_1$, and n_2 is the photon number in energy state $\mathcal{E}_2$.

The probability of a system at temperature T being in state m is

$$P(\mathcal{E}_m) = \frac{e^{-(\mathcal{E}_m/kT)}}{\sum\limits_m e^{-(\mathcal{E}_m/kT)}} \tag{5.28}$$

This development parallels that found in Chapter 4, and in addition lends relevance to noise on a thermal photon stream.

The system of energy states is distributed over the accessible states. The total energy of the volume of a source is

$$\mathcal{E}_{total} = \sum\limits_m n_m \mathcal{E}_m \tag{5.29}$$

where

$m =$ possible state of a photon
$\mathcal{E}_m =$ energy of a photon of a state m
$n_m =$ number of photons in state m

Rewriting the probability that n photons are emitted with energy $h\nu$ (at a single frequency or wavelength), the probability of n photons in the state with energy $h\nu$ is

$$P(n) = \frac{e^{-nh\nu/kT}}{\sum\limits_{m=0}^{\infty} e^{-mh\nu/kT}} \tag{5.30}$$

For normalization, the denominator is one. Rewriting using the identity

$$\sum\limits_{m=0}^{\infty} e^{-mh\nu/kT} = \frac{1}{1 - e^{-h\nu/kT}} \tag{5.31}$$

Eq. (5.30) becomes

$$P(n) = (1 - e^{-h\nu/kT}) e^{-nh\nu/kT} \tag{5.32}$$

The expected number of photons in a state is

$$\bar{n} = \sum\limits_{n=0}^{\infty} n\, P(n) \tag{5.33a}$$

$$\bar{n} = \frac{1}{e^{h\nu/kT} - 1} \tag{5.33b}$$

This is the average number of photons at this energy state from a blackbody source. To get Planck's radiation expression, one needs the density of energy states and the photon energy. However, to find the noise associated with these photons one must find the variance on these photons at energy $h\nu$. The variance expression for these photons at energy $(h\nu)$ is (similar to Eq. 5.22)

$$\sigma^2 = \overline{n^2} - \overline{n}^2 \tag{5.34}$$

Working on the algebra on the parts of Eq. (5.34) gives us

$$\overline{n^2} = \sum_{n=0}^{\infty} n^2 P(n)$$

$$\overline{n^2} = \sum_{n=0}^{\infty} n^2 (1 - e^{-x}) e^{-nx}$$

$$\overline{n^2} = (1 - e^{-x}) \frac{\partial^2}{\partial x^2} \left[\sum_{n=0}^{\infty} e^{-nx} \right]$$

$$\overline{n^2} = (1 - e^{-x}) \frac{\partial}{\partial x} \left[\frac{-e^{-x}}{(1 - e^{-x})^2} \right]$$

$$\overline{n^2} = (1 - e^{-x}) \left[\frac{(1 - e^{-x})^2 e^{-x} + e^{-x} \times 2(1 - e^{-x}) e^{-x}}{(1 - e^{-x})^4} \right]$$

$$\overline{n^2} = \left[\frac{e^{-x} - e^{-2x} + 2e^{-2x}}{(1 - e^{-x})^2} \right] = \left[\frac{e^{-x}(1 + e^{-x})}{(1 - e^{-x})^2} \right]$$

$$\overline{n^2} = \left[\frac{e^{-x}(1 + e^{-x})}{e^{-2x}(e^x - 1)^2} \right] = \left[\frac{e^x(1 + e^{-x})}{(e^x - 1)^2} \right] \tag{5.35a}$$

or

$$\overline{n^2} = \frac{(1 + e^x)}{(e^x - 1)^2} \tag{5.35b}$$

Substituting Eqs. (5.35b) and (5.33b) into the variance expression Eq. (5.34) for the number of photons emitted at energy $h\nu$ yields

$$\sigma^2 = \frac{(1 + e^x)}{(e^x - 1)^2} - \frac{1}{(e^x - 1)^2} = \frac{e^x}{(e^x - 1)^2} \tag{5.36}$$

$$\sigma^2 = \overline{n} \left[1 + \frac{1}{e^x - 1} \right] \tag{5.37}$$

This variance (σ^2) represents the number of photons around the mean (average) emission n, in that state of energy. This is the variance for each energy level of the continuum in the spectrum. Bose–Einstein statistics gives for photons of energy $\mathcal{E}_1$ (state 1)

$$\sigma_1^2 = \bar{n}_1 \left[1 + \frac{1}{e^{x_1} - 1} \right] = \bar{n}_1 + \bar{n}_1^2 \tag{5.38}$$

For photons of energy $\mathcal{E}_2$ (another state)

$$\sigma_2^2 = \bar{n}_2 \left[1 + \frac{1}{e^{x_2} - 1} \right] = \bar{n}_2 + \bar{n}_2^2 \tag{5.39a}$$

$$\bar{n}_2 = \frac{1}{e^{h\nu_2/kT} - 1} \tag{5.39b}$$

The variance over an entire spectral region is the sum of the variances for each frequency (ν)

$$\sigma_{\text{total}}^2 = \sum_m \sigma_m^2 \tag{5.40}$$

or

$$\sigma_{\text{total}}^2 = \bar{n} + \bar{n}^2 = \bar{n}[1 + \bar{n}] \tag{5.41}$$

The brackets in Eq. (5.41) and Eq. (5.37) identify the Bose factor, which gives excess noise above the noise that one obtains for the Poisson case (Eq. 5.27). This Bose factor makes the noise from thermal photons larger than that expected from laser-generated photons.

For optical radiation, if $e^x \gg 1$, from Eq. (5.37), $\sigma^2 = \bar{n}$. Let us evaluate this restriction:

$$h\nu > kT \tag{5.42}$$

For example, for a blackbody at 500 K this restriction causes the wavelength to be less than

$$\lambda \ll \frac{hc}{kT} = \frac{(6.6 \times 10^{-34})\,(3 \times 10^8)}{(1.38 \times 10^{-23})\,500} = 29\ \mu\text{m} \tag{5.43}$$

For values of the wavelength–temperature product (λT) $\ll$ 14,500 (μm-K), e^x will be larger than 1 and the variance or noise squared equals the mean numbers of photons. Then Bose–Einstein statistics can be approximated by the Poisson statistics.

The variance on photons integrated to $\lambda < 29~\mu$m is the sum of variances over states

$$\sigma^2_{total} = \sum_m \sigma^2_m = \sum_m \bar{n}_m \qquad (5.44a)$$

The average number of photons is

$$\bar{n} = \sum_m \bar{n}_m \qquad (5.44b)$$

Thus,

$$\sigma^2_{total} = \bar{n} \qquad (5.45)$$

The variance is similar to our Poisson distribution ($\sigma^2 = \bar{n}$), so photon noise in the practical spectral region and blackbody-temperature range is equal to the square root of the number of photons, the same as for Poisson statistics. Blackbody radiation follows Bose–Einstein statistics, which, in the restricted limit, is Poisson. Laser sources (monochromatic sources) follow Poisson statistics directly independent of wavelength.

The Poisson density function can be written as

$$P(m, \Delta t) = \frac{(E_q A_d \Delta t)^m e^{-E_q A_d \Delta t}}{m!} \qquad (5.46)$$

where

E_q = photon irradiance (photon/s-cm^2)
A_d = detector area (cm^2)
Δt = observation time (s)
m = number of times an event occurs

The average number of photons is

$$\bar{n}_q = E_q A_d \Delta t \qquad (5.47)$$

and the variance on the photon stream is

$$\sigma^2 = \bar{n}_q = E_q A_d \Delta t \qquad (5.48)$$

Detectors operate best when they are photon-noise limited, which means that the dominant noise process is external to the detector. However, as given in Eq. (5.27) and again in Eq. (5.45), the variance was equal to the mean number of photons (noise is equal to the square root of the mean number of photons), so the collection area of the detector is important in determining the mean number of photons. In other words, a larger detector collects more photons. If we consider a photon irradiance, the average number of photons collected is proportional to the area, therefore, the noise is proportional to the square root of the area. This result is very important, as will be shown later.

5.2.3. Photodetection as Bernoulli Trials

One can express the SNR of the photon flux input to the detector, assuming Poisson statistics, as

$$\left(\frac{S}{N}\right)_{input} = \frac{\bar{n}_q}{\sigma} = (E_q A_d \Delta t)^2 = \sqrt{\bar{n}_q} \tag{5.49}$$

If the SNR input to a detector is photon limited, then it is the best achievable value. However, the photon either produces an electron in the detector or it does not. The probability of an electron being produced by an absorbed photon is the quantum efficiency (η) and the probability of no photo-generated electron being produced is $1 - \eta$. The process follows a binomial distribution. The uncertainty in photo-generated electrons is also a noise. Therefore, the noise in the photo-generated electrons is attributable to both photon noise and detector noise in cascade.

The binomial distribution has an average number of photo-generated electrons ($\bar{n}_e$)

$$\bar{n}_e = \eta \, \bar{n}_q \tag{5.50}$$

and a variance

$$\sigma^2 = \eta \, (1 - \eta) \, \bar{n}_q \tag{5.51}$$

where the quantum efficiency is interpreted as

η = probability of emission of an electron for a given photon event
$1 - \eta$ = probability of no electron emitted for a given photon event

The total photoelectron noise can be expressed as the sum of two variances:

$$\sigma_{pe}^2 = \bar{n}_q \eta \, (1 - \eta) + \eta^2 \bar{n}_q \tag{5.52}$$

where

$\bar{n}_q \eta \, (1 - \eta)$ = variance due to the quantum efficiency of detector, considered a binomial distribution
$\eta^2 \bar{n}_q$ = variance caused by transfer of photon noise through detection process

The second term in Eq. (5.52) comes from

$$\sigma = \sqrt{\bar{n}_q} - \text{variance of photons} \tag{5.53a}$$

$$\sigma = \eta\sqrt{\bar{n}_q} - \text{variance of photoelectrons from photon noise} \tag{5.53b}$$

Now then

$$\sigma^2 = \eta^2 \bar{n}_q \tag{5.53c}$$

Canceling the terms in total variance of photoelectrons (σ_{pe}^2) Eq. (5.52)

$$\sigma_{pe}^2 = \eta\bar{n}_q \tag{5.54}$$

The SNR output from the detector, using the average number of photo-generated electrons, $\bar{n}_e$, is

$$\left(\frac{S}{N}\right)_{out} = \frac{\bar{n}_e}{\sqrt{\sigma_{pe}^2}} = \frac{\eta\bar{n}_q}{\sqrt{\eta\bar{n}_q}} = \sqrt{\eta\bar{n}_q} \tag{5.55}$$

and from Eq. (5.43)

$$\left(\frac{S}{N}\right)_{out} = \sqrt{\eta}\left(\frac{S}{N}\right)_{in} \tag{5.56}$$

The S/N input caused by photon noise is degraded by the square root of the quantum efficiency. Another figure of merit is detective quantum efficiency (DQE), which is defined as

$$\text{DQE} = \frac{(S/N)_{out}^2}{(S/N)_{input}^2} = \eta \tag{5.57}$$

DQE is a measured system parameter or figure of merit. If and only if a detector is signal photon-noise limited, the DQE equals the responsive quantum efficiency.

The output SNR that depends on quantum efficiency should be highlighted for two cases:

1. Photon-noise dominated: $(S/N)_{out} \propto \sqrt{\eta}$
2. Other-noise dominant: it will be shown that $(S/N)_{out} \propto \eta$

5.3. PRIMARY SOURCES OF DETECTOR NOISE

Noise sources in the detector body itself limit the detector's ability to measure optical flux, even if this flux is noise free. As listed in the first section of this chapter, these noise sources are Johnson, shot, generation/recombination, $1/f$, temperature, microphonics, and popcorn.

5.3.1. Johnson Noise

Johnson noise is the fluctuation caused by the thermal motion of the charge carriers in a resistive element. Local random thermal motion of carriers sets up fluctuating charge gradients, even though charge neutrality exists generally across a resistor. This noise is also called *Nyquist noise* (Johnson, 1928; Nyquist, 1928). The Johnson noise voltage is

$$v_j = \sqrt{4\,kTR\Delta f} \tag{5.58}$$

To derive this expression, consider the circuit shown in Fig. 5.7. In this circuit, the voltage across the capacitor–resistor combination is constantly fluctuating because of the random agitation of charge carriers in the resistor. From statistical mechanics, we know that for a classic system at equilibrium and with one degree of freedom (here, the voltage on the capacitor), the expected value for the energy of the system is $(1/2)\,kT$, where k is Boltzmann's constant and T is the absolute temperature of the system. For this system we have

$$\tfrac{1}{2}\,C\overline{v}^2 = \tfrac{1}{2}\,kT \tag{5.59}$$

and

$$\overline{v}^2 = \frac{kT}{C} \tag{5.60}$$

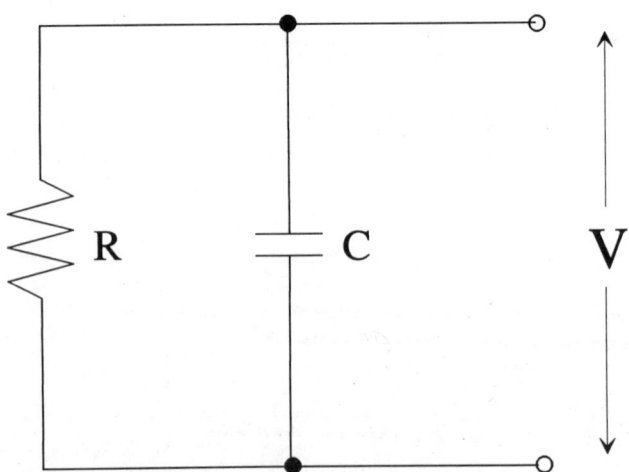

Figure 5.7.   Johnson noise equivalent circuit.

Here the mean-square value of the voltage across the capacitor is used in the familiar expression for the energy stored on a capacitor. Consider now the voltage that exists at a particular time t_0, and that which exists a short time τ later. From circuit theory, we have

$$v_i(t_0 + \tau) = v_i(t_0)e^{-\tau/RC} \tag{5.61}$$

Note that the i subscript represents a particular initial voltage (one value in a continuum of possible voltages) that one might observe at time t_0. If we multiply each side of this expression by $v_i(t_0)$ and take the ensemble average over all possible initial voltages, the statistical autocorrelation is obtained

$$\mathfrak{R}(\tau) = \overline{v_i(t_0)v_i(t_0 + \tau)} = \overline{v_i^2(t_0)}\ e^{-\tau/RC} \tag{5.62}$$

The Wiener–Kinchine formula allows us to solve for the power spectrum of the Johnson noise by integrating the autocorrelation function

$$\mathfrak{S}(f) = 4 \int_0^\infty \mathfrak{R}(\tau) \cos 2\pi f\tau\, d\tau \tag{5.63}$$

Substituting Eq. (5.62) into this formula yields the desired result of

$$\mathfrak{S}(f) = 4 \int_0^\infty \overline{v_i^2(t_0)}\ e^{-\tau/RC} \cos 2\pi f\tau\, d\tau \tag{5.64}$$

$$\mathfrak{S}(f) = 4\overline{v_i^2(t_0)} \left(\frac{RC}{1 + (2\pi fRC)^2} \right) \tag{5.65}$$

$$\mathfrak{S}(f) = 4\overline{v_i^2(t_0)}\ RC, \quad \text{for } f \ll \frac{1}{2\pi\ RC} \tag{5.66}$$

This power spectrum is plotted in Fig. 5.8. Substituting, using Eq. (5.60) gives us

$$\mathfrak{S}(f) = 4kTR \tag{5.67}$$

Notice that the mean-square voltage across the capacitor, as given by Eq. (5.60) was used in Eq. (5.66) under the assumption of equality between time and ensemble averages for this system. The rms Johnson noise is obtained in the usual way, integrating the noise power spectrum over the bandpass of interest, and taking the square root

$$(\overline{v^2})^{1/2} = v_j = \sqrt{4kTR\ \Delta f} \tag{5.68}$$

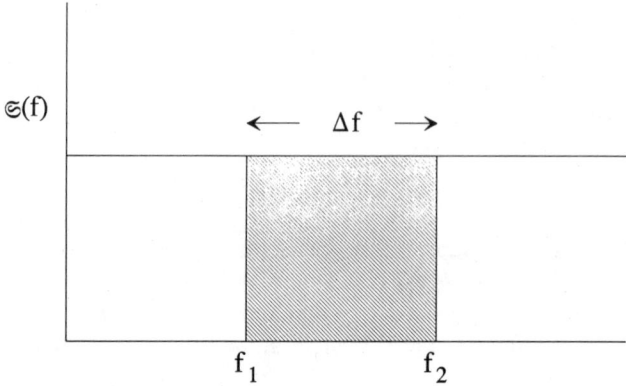

Figure 5.8. Calculation of the rms Johnson noise for a finite-frequency bandwidth Δf.

where

v_j = rms voltage
k = Boltzmann's constant
T = temperature (K)
R = resistance (ohm)
Δf = electrical bandwidth (Hz)

The simpler notation v_j has been adopted here for reference to the Johnson noise. The expression is the same as that presented in Eq. (5.58). The Thevenin equivalent circuit for a voltage source is shown in Fig. 5.9, where a true rms voltmeter would read Johnson noise voltage (v_j).

The Norton equivalent circuit is shown in Fig. 5.10. The current generator (i_j) is

$$i_j = \sqrt{\frac{4kT\Delta f}{R}} \tag{5.69}$$

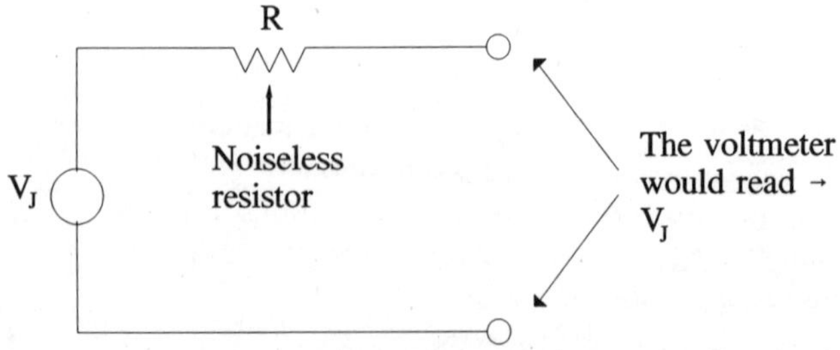

R

Noiseless
resistor

V_J

The voltmeter
would read →
V_J

Figure 5.9. Thevenin equivalent of resistor noise.

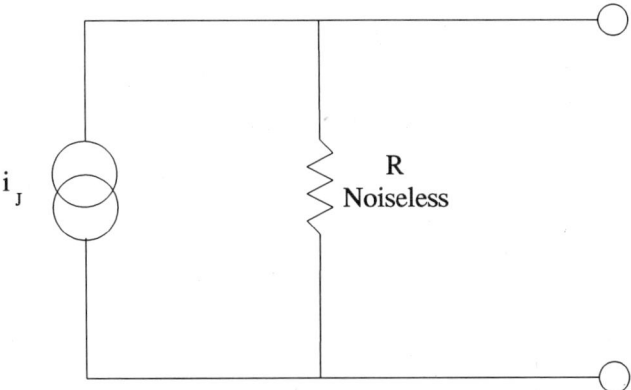

Figure 5.10. Norton equivalent of resistor noise.

For example, if we have two (10-kΩ) resistors in parallel, what is the Johnson noise voltage measured across them with a noise bandwidth (Δf) of 100 Hz? We can find out by substituting into Eq. (5.68)

$$v_j = \sqrt{4kTR\Delta f}$$

$$= [4 \times 1.38 \, (10^{-23}) \, 300° \times 5 \, (10^3) \, 100]^{1/2}$$

$$v_j = 90 \, (10^{-9}) \text{ volts (rms)} \tag{5.70}$$

The Johnson noise associated with the detector resistance is related to the square root of the optically active area (A_d). However, this is not always intuitively obvious. In the case of a photovoltaic (*PN* junction), the detector resistance is the inverse slope of the *i-v* curve. The resistance-detector-area product (often called *RA* product) can be shown to be a constant, depending on material parameters (Chap. 7).

If we substitute (RA_d) into the expression for Johnson noise current, Eq. (5.69), we get

$$i_j = \sqrt{\frac{4kT\Delta f}{(RA_d)}} \, A_d \tag{5.71}$$

indicating that this noise is proportional to the square root of the optically active area.

For a photoconductive detector, the situation is a little more ambiguous. However, the result is still consistent with the idea that the Johnson noise is proportional to the square root of the active area. The photoconductor can be irradiated in two orientations: (a) light propagating parallel to dc-bias current

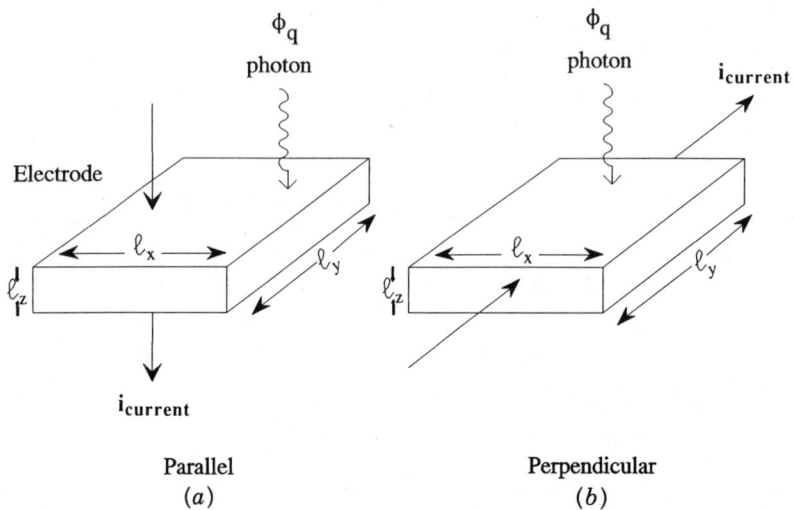

Figure 5.11. Photoconductor illumination relative to dc-bias current flow: (*a*) light propagating parallel to dc-bias current flow, and (*b*) light propagating perpendicular to dc-bias current flow.

flow, and (b) light propagating perpendicular to dc-bias current flow. These two cases are shown in Fig. 5.11*a* and *b*.

For the case of Fig. 5.11*a*, parallel illumination of a photoconductor such as a staring infrared focal-plane array, the detector resistance is

$$R_d = \frac{\rho \ell_z}{\ell_x \ell_y} \tag{5.72}$$

where

ρ = detector material resistivity
ℓ_z = absorption length
$\ell_x \ell_y$ = optical active area (A_d) for light

Substituting into the Johnson noise Eq. (5.69) shows that the noise is proportional to the square root of the detector area.

In the case of perpendicular illumination shown in Fig. 5.11*b*, detector resistance is

$$R_d = \frac{\rho \ell_y}{\ell_x \ell_z} \tag{5.73}$$

Note that when the detector active area is square ($\ell_x = \ell_y$), the resistance is fixed. For the case of perpendicular illumination, it makes no difference whether the detector is 1 cm × 1 cm or 1 mm × 1 mm. The resistance is simply the resistivity divided by thickness (assuming thickness does not change). Resistance is independent of optically active detector area. Multiplying by t/t; one can get the detector resistance in terms of detector area (A_d):

$$R_d = \frac{\rho \ell_y}{\ell_z \ell_x} \frac{\ell_x}{\ell_x} = \frac{\rho A_d}{\ell_z \ell_x^2} \tag{5.74}$$

The noise is not proportional to the square root of the detector's active area. However, the D^*s for the parallel and the perpendicular configurations are the same! (See problem 8.16.)

5.3.2. Shot Noise

Shot noise is associated with the dc current flowing across a potential barrier and is a series of independent events. The current through the diode possesses this noise because the charge carriers cross a potential barrier. (*Note:* The current through the resistor does not have shot noise.) Just as photons do not arrive or emit in discrete precise time intervals, neither do electrons. The noise on these electrons follow the Poisson equation, as will be shown. The current-noise expression for shot noise is

$$i_s = \sqrt{2q\overline{i}\Delta f} \tag{5.75}$$

If $\overline{i}$ is the result of photons, that is, a photocurrent, then the noise of the detector is photon-noise limited.

The shot-noise expression is derived as follows. Consider the current flow from the ideal photocathode:

$$i = \frac{n_e q}{\tau} \tag{5.76}$$

where

n_e = number of photogenerated electrons in time interval τ
q = electron charge—$1.6 \, (10^{-19})$ coulomb
τ = time interval of integration

If τ becomes very small, so that either only one electron is present in time τ or none are present, we can use the binomial statistics. This process in the limit of a large number of electrons also follows Poisson statistics (Sec. 5.2.1). The average current $\overline{i}$ can be related to the average number of electrons $(\overline{n}_e)$ created by

$$\overline{i} = \frac{\overline{n}_e q}{\tau} \tag{5.77}$$

The variance around this average photogenerated current is $\overline{(i - \bar{i})^2}$ (by definition of the variance of a random variable); more simply stated, the noise squared (standard deviation)^2:

$$\overline{(i - \bar{i})^2} = (\text{noise current})^2 = i_s^2 \tag{5.78}$$

Or, substituting Eqs. (5.76) and (5.77) into Eq. (5.78) gives us

$$i_s^2 = \frac{q^2}{\tau^2} \overline{(n_e - \bar{n}_e)^2} \tag{5.79}$$

The variance $\overline{(n_e - \bar{n}_e)^2}$ equals the mean number of electrons if Poisson statistics is assumed. Substituting in

$$i_s^2 = \frac{q^2}{\tau^2} \bar{n}_e \tag{5.80}$$

but from Eq. (5.77),

$$i_s^2 = \frac{q}{\tau} \bar{i} \tag{5.81}$$

Because the frequency bandwidth $\Delta f = 1/(2\tau)$ [see Eq. (5.153)], we have

$$i_s = \sqrt{2q\bar{i}\Delta f} \tag{5.82}$$

This noise is associated with a dc current traversing a potential barrier. Because the dc current is proportional to the cross-sectional area of the active detector (current density × area), this noise is also proportional to the square root of the detector area, if the potential barrier cross-sectional area is the optical active area, as is typical for a photodiode.

5.3.3. Generation–Recombination Noise

Generation and recombination noise results from statistical fluctuation in the rate of generation and rate of recombination of charged particles into an upper state within the detector material. These variations may be caused by variations in carrier lifetimes or the random-generation processes of the carriers. If the photons determine the fluctuation, we have a photon-noise case, which occurs in photoconductors. The statistical fluctuation in the concentration of carriers produces the white noise (constant), which has electrical frequencies less than the inverse of carrier lifetime. Also, the free electrons find empty donor levels where they randomly recombine in time because the carrier lifetime (τ) is short compared to the transit time (time to cross between electrodes). Therefore, the carriers are generated and recombined in the bulk before reaching electrodes. The current noise expression for generation–recombination noise is

$$i_{gr} = 2qG[\eta E_q A_d \Delta f + g_{th} A_d \Delta f \ell_x]^{1/2} \qquad (5.83)$$

where

G = photoconductive gain
q = charge of an electron
E_q = photon irradiance
A_d = detector area
Δf = noise equivalent bandwidth
η = quantum efficiency
g_{th} = thermal generation of carriers
ℓ_x = detector thickness in optical propagation direction

We will assume that the detector material is cooled so that the thermally generated carriers are insignificant. That is, the second term in Eq. (5.83) is dropped. We reinvestigate this requirement in Chapter 8.

The derivation of the generation–recombination current noise uses the power spectral density of current fluctuations

$$i_{gr}^2(t) = \int_0^\infty \mathfrak{S}(f)\, df \qquad (5.84)$$

where the power spectral density can be defined as (Van der Zeil, 1986)

$$\mathfrak{S}(f) = \left(\frac{\bar{i}}{\bar{n}}\right)^2 \mathfrak{D}(f) \qquad (5.85)$$

where

$\bar{i}$ = average current
$\bar{n}$ = total free electrons (average)

The variance on these electrons equals the mean by Poisson statistics

$$\overline{\Delta n^2} = \overline{(n - \bar{n})^2} = \bar{n} \qquad (5.86)$$

For the generation–recombination process, the variance for some time delay, Δt, following excitation or decay can be expressed as an autocorrelation

$$\overline{[n(t + \Delta t) - \bar{n}(t + \Delta t)]\,[n(t) - \bar{n}(t)]} = \overline{[n(t) - \bar{n}(t)]^2}\, e^{-\Delta t/\tau} \qquad (5.87)$$

This shows generation–recombination noise has correlation over some region of space (product of carrier mobility and carrier lifetime implies a distance when electric field is given). The spectrum $\mathcal{D}(f)$ is

$$\mathcal{D}(f) = 2 \int_0^\infty \overline{[n(t) - \bar{n}(t)]^2} \, e^{-\Delta t/\tau} \, e^{j2\pi f \Delta t} \, d\Delta t \tag{5.88}$$

Using the Fourier-transform relation (Gaskill, 1978):

$$\mathcal{D}(f) = \frac{4\tau \overline{[n(t) - \bar{n}(t)]^2}}{1 + (2\pi f \tau)^2} = \frac{4\tau \, \overline{(\Delta n)^2}}{1 + (2\pi f \tau)^2} \tag{5.89}$$

Substituting $\mathcal{D}(f)$ into the power spectral density:

$$\mathcal{S}(f) = \left(\frac{\bar{i}}{\bar{n}}\right)^2 \frac{4\tau \, \overline{(\Delta n)^2}}{1 + (2\pi f \tau)^2} = \frac{4 \, \bar{i}^2 \tau}{\bar{n}^2} \frac{\overline{(\Delta n)^2}}{1 + (2\pi f \tau)^2} \tag{5.90}$$

If $\overline{\Delta n^2}$ is photon generated (the detector is cooled, so thermal generation is negligible), then

$$\overline{(\Delta n)^2} = \eta E_q A_d \tau = \bar{n} \tag{5.91}$$

The current through the device can be expressed as the coulombs/s through the device, and is written as the product of the number of charge carriers moving and a velocity determined by the mobility, electric field, and electrode separation distance:

$$\bar{i} = \frac{q \, \bar{n} \mu \, \mathsf{E}}{\ell_x} \tag{5.92}$$

where

μ = mobility
E = electric field
ℓ_x = interelectrode spacing

Substituting $\bar{i}$ and $\overline{\Delta n^2}$ into $\mathcal{S}(f)$, Eq. (5.90), gives us

$$\mathcal{S}(f) = \left(\frac{q\bar{n}\mu \, \mathsf{E}}{\ell_x \bar{n}}\right)^2 \frac{4 \, \tau[\eta \, E_q A_d \, \tau]}{1 + (2\pi f \tau)^2}$$

$$= \frac{4q^2 \eta E_q A_d}{1 + (2\pi f \tau)^2} \frac{\tau^2}{\left(\dfrac{\ell_x}{\mu \mathsf{E}}\right)^2} \tag{5.93}$$

The interelectrode spacing divided by velocity (μE) is transit time, τ_t

$$\mathfrak{S}(f) = \frac{4q^2\eta E_q A_d}{1 + (2\pi f\tau)^2} \left(\frac{\tau}{\tau_t}\right)^2 \tag{5.94}$$

The photoconductive gain (G) is defined as

$$G = \frac{\tau}{\tau_t} \tag{5.95}$$

Rewriting

$$\mathfrak{S}(f) = \frac{4q^2 G^2 \eta E_q A_d}{1 + (2\pi f\tau)^2} \tag{5.96}$$

using Eq. (5.84) yields

$$i_{gr}^2(t) = \int_0^\infty \frac{4q^2 G^2 \eta E_q A_d}{1 + (2\pi f\tau)^2} \, df \tag{5.97}$$

If we remain in the white-noise region,

$$1 \gg (2\pi f\tau)^2 \tag{5.98a}$$

$$df \to \Delta f \tag{5.98b}$$

and

$$i_{gr} = \sqrt{i_{gr}^2(t)} \tag{5.98c}$$

$$i_{gr} = 2qG[\eta E_q A_d \Delta f]^{1/2} \tag{5.99}$$

Equation (5.99) is the rms generation–recombination current noise. Note that the rms generation–recombination noise is proportional to the square root of the detector area.

5.3.4. Temperature Noise

Temperature noise is the temperature fluctuation of the sensitive element. Causes include radiative exchange with the background and/or conductance with heat sink. The temperature of the detector has a mean value of T_0, but some fluctuation or noise is associated with it. In the case of thermal detectors, the response to a temperature change is a noise process. For thermal detectors, being temperature-fluctuation limited is the ultimate performance level. For a detector in thermodynamic equilibrium, a temperature (T) can be related to the equilibrium energy (u) of an ensemble of molecules by

$$u = kT \tag{5.100}$$

where u characterizes the equilibrium state of the molecules. Maxwell–Boltzmann told us that the mean energy of a "closed" system in thermodynamic equilibrium at a specific temperature is

$$\overline{\mathcal{E}} = \sum_{j} \mathcal{E}_j e^{(\mathcal{E}_s - \mathcal{E}_j)/u} \tag{5.101}$$

The Maxwell–Boltzmann expression can be used to investigate the fluctuations in the equilibrium distribution of radiation for excited modes in a blackbody hollow enclosure, as shown in Chapter 2.

The total state probability of the system must remain equal to one,

$$\sum_{j} e^{(\mathcal{E}_s - \mathcal{E}_j)u} = 1 \tag{5.102}$$

From the heat-transfer equation, the rate of heat flow is

$$\frac{d\Delta\mathcal{E}}{dt} = K\Delta T \tag{5.103}$$

Equaton (5.103) shows that the rate of flow of heat, $d\Delta\mathcal{E}/dt$, is proportional to ΔT with the proportionality constant being the thermal conductivity, K. The rate of heat flow can also be expressed as

$$\frac{d\Delta\mathcal{E}}{dt} = H\frac{d\Delta T}{dt} \tag{5.104}$$

where H is the heat capacity. If the material is at a higher temperature than the surroundings, $d\Delta\mathcal{E}/dt$ is negative. Heat flows both into the body from the radiating background and out of the body by conduction. This behavior is reflected in the heat-balance equation:

$$\phi_e = K\Delta T + H\frac{d\Delta T}{dt} \tag{5.105}$$

Assume a time-varying solution,

$$\phi_e e^{j\omega t} \tag{5.106}$$

and rearranging Eq. (5.105):

$$\frac{d\Delta T}{dt} + \frac{K}{H}\Delta T = \frac{\phi_e}{H} e^{j\omega t} \tag{5.107}$$

Let us multiply through by $\exp\{Kt/H\}$ to solve Eq. (5.106):

$$\frac{d\Delta T}{dt} e^{Kt/H} + \frac{K}{H}\Delta T e^{Kt/H} = \frac{e^{Kt/H}}{H} \phi_e e^{j\omega t} \tag{5.108a}$$

$$\frac{d}{dt}[\Delta T e^{Kt/H}] = \frac{e^{Kt/H}}{H} \phi_e e^{j\omega t} \tag{5.108b}$$

$$\Delta T e^{Kt/H} = \int \frac{e^{Kt/H}}{H} \phi_e e^{j\omega t}\, dt = \frac{\phi_e}{H} \int e^{(K/H + j\omega)t}\, dt \tag{5.108c}$$

Let us change variables so that $t' = (K/H + j\omega)t$ and $dt' = (K/H + j\omega)\, dt$. Substituting and solving

$$\Delta T e^{Kt/H} = \frac{\phi_e}{H} \int e^{t'} \frac{dt'}{(K/H + j\omega)} \tag{5.109a}$$

$$\Delta T e^{Kt/H} = \frac{\phi_e}{K + j\omega H} e^{(K/H + j\omega)t} \tag{5.109b}$$

Canceling common factors,

$$\Delta T = \frac{\phi_e e^{j\omega t}}{K + j\omega H} \tag{5.110}$$

The squared modulus of the solution is

$$\overline{\Delta T^2} = \frac{\overline{\phi_e^2}}{K^2 + \omega^2 H^2} \tag{5.111}$$

The term $\overline{\Delta T^2}$ in Eq. (5.111) is actually a function of frequency, so to find the total temperature variance, we must integrate over all frequencies:

$$\overline{\Delta T^2} = \int_0^\infty \frac{\overline{\phi_e^2}}{K^2 + \omega^2 H^2}\, df \tag{5.112}$$

Recalling $\omega = 2\pi f$ and the following integral identity:

$$\int_0^\infty \frac{dx}{a^2 + b^2 x^2} = \frac{1}{ab} \tan^{-1}\left[\frac{bx}{a}\right]\Bigg|_0^\infty \tag{5.113}$$

Equation (5.112) leads to

$$\overline{\Delta T^2} = \phi_e^2 \frac{1}{K2\pi H} [\tan^{-1} \infty - \tan^{-1} 0] \tag{5.114a}$$

$$\overline{\Delta T^2} = \phi_e^2 \frac{1}{K2\pi H} \left[\frac{\pi}{2}\right] = \frac{\phi_e^2}{4KH} \tag{5.114b}$$

Recalling from statistics, and from Eqs. (5.101) and Eq. (5.102), that

$$\overline{\mathcal{E}^2} = \frac{\sum\limits_{j} \mathcal{E}_j^2 \, e^{(\mathcal{E}_s - \mathcal{E}_j)/u}}{\sum\limits_{j} e^{(\mathcal{E}_s - \mathcal{E}_j)/u}} \tag{5.115}$$

To simplify, we substitute

$$z = \sum_{n} e^{a\mathcal{E}_n} \tag{5.116}$$

By using the characteristic function from Chapter 4 we solve for the variance in the energy, which is

$$\sigma^2 = \overline{\mathcal{E}^2} - \overline{\mathcal{E}}^2 \tag{5.117a}$$

$$\frac{dz}{da} = \sum_{n} e^{a\mathcal{E}_n} = \overline{\mathcal{E}_n} \sum_{n} e^{a\mathcal{E}_n} = \overline{\mathcal{E}_n} \, z \tag{5.117b}$$

$$\overline{\mathcal{E}_n} = \frac{1}{z} \frac{dz}{da} \tag{5.117c}$$

$$\frac{d^2z}{da^2} = \frac{d}{da}\left(\overline{\mathcal{E}_n} \sum_{n} e^{a\mathcal{E}_n}\right) \tag{5.118a}$$

$$\frac{d^2z}{da^2} = \overline{\mathcal{E}_n} \frac{dz}{da} = \left(\overline{\mathcal{E}_n} \times \overline{\mathcal{E}_n} \sum_{n} e^{a\mathcal{E}_n}\right) = (\overline{\mathcal{E}_n})^2 \, z \tag{5.118b}$$

$$\frac{1}{z} \frac{d^2z}{da^2} = (\overline{\mathcal{E}_n})^2 \tag{5.118c}$$

Squaring the expression for dz/da,

$$\left(\frac{dz}{da}\right)^2 = \left(\sum_{n} \mathcal{E}_n e^{a\mathcal{E}_n}\right)^2 \tag{5.119a}$$

$$\left(\frac{dz}{da}\right)^2 = \sum_{n} \mathcal{E}_n^2 e^{2a\mathcal{E}_n} = \overline{\mathcal{E}_n^2} \sum_{n} e^{2a\mathcal{E}_n} = \overline{\mathcal{E}_n^2} z^2 \tag{5.119b}$$

$$\overline{\mathcal{E}_n^2} = \frac{1}{z^2}\left(\frac{dz}{da}\right)^2 \tag{5.119c}$$

Differentiating Eq. (5.117c)

$$\frac{d}{da}(\overline{\mathcal{E}_n}) = \frac{d}{da}\left(\frac{1}{z}\frac{dz}{da}\right) \tag{5.120a}$$

$$\frac{d}{da}(\overline{\mathcal{E}_n}) = -\frac{1}{z^2}\left(\frac{dz}{da}\right)^2 + \frac{1}{z}\frac{d^2z}{da^2} \tag{5.120b}$$

$$\frac{d}{da}(\overline{\mathcal{E}_n}) = \overline{\mathcal{E}_n^2} - \overline{\mathcal{E}_n}^2 \tag{5.120c}$$

$$\frac{d\overline{\mathcal{E}_n}}{dz} = \frac{d\overline{\mathcal{E}_n}}{da}\frac{da}{dz} = \frac{d\overline{\mathcal{E}_n}}{da}\frac{1}{\left(\dfrac{dz}{da}\right)} \tag{5.121a}$$

$$\frac{d\overline{\mathcal{E}_n}}{dz} = \frac{(\overline{\mathcal{E}_n^2} - \overline{\mathcal{E}_n}^2)}{\overline{\mathcal{E}_n}z} \tag{5.121b}$$

$$\overline{\Delta\mathcal{E}^2} = (\overline{\mathcal{E}_n^2} - \overline{\mathcal{E}_n}^2) = \overline{\mathcal{E}_n}z\frac{d\overline{\mathcal{E}_n}}{dz} \tag{5.121c}$$

Because the heat capacity from Eq. (5.104) can be rewritten as

$$H = \frac{\partial\mathcal{E}}{\partial T} = \frac{\partial\mathcal{E}}{\partial z}\frac{\partial z}{\partial T} \tag{5.122}$$

with

$$\frac{\partial z}{\partial T} = \frac{\partial}{\partial T}\left(\sum_n e^{a\mathcal{E}_n}\right) = \frac{\partial}{\partial T}\left(\sum_n e^{-\mathcal{E}_n/kT}\right) \tag{5.123a}$$

where $a = -1/(kT)$, and

$$\frac{\partial z}{\partial T} = \frac{\overline{\mathcal{E}_n}}{kT^2}\left(\sum_n e^{-\mathcal{E}_n/kT}\right) = \frac{\overline{\mathcal{E}_n}z}{kT^2} \tag{5.123b}$$

then

$$H = \frac{\partial\mathcal{E}}{\partial T} = \frac{\partial\mathcal{E}}{\partial z}\frac{\partial z}{\partial T} \tag{5.124a}$$

$$H = \frac{(\overline{\mathcal{E}_n^2} - \overline{\mathcal{E}_n}^2)}{\overline{\mathcal{E}_n}z}\frac{\overline{\mathcal{E}_n}z}{kT^2} = \frac{(\overline{\mathcal{E}_n^2} - \overline{\mathcal{E}_n}^2)}{kT^2} \tag{5.124b}$$

This reduces Eq. (5.121) to

$$\overline{\Delta\mathcal{E}^2} = kT^2H \tag{5.125}$$

From Eq. (5.104), squaring and averaging:

$$\overline{\Delta\mathcal{E}^2} = H^2\overline{\Delta T^2} \tag{5.126}$$

Equating Eqs. (5.125) and (5.126)

$$kT^2 H = H^2\overline{\Delta T^2} \tag{5.127}$$

$$\overline{\Delta T^2} = \frac{kT^2}{H} \tag{5.128}$$

From Eq. (5.114) for the temperature variance:

$$\frac{kT^2}{H} = \frac{\overline{\phi_e^2}}{4KH} \tag{5.129}$$

or

$$\overline{\phi_e^2} = 4kKT^2 \tag{5.130}$$

Substituting into Eq. (5.111), the spectrum of the mean-square fluctuation in T is

$$\overline{\Delta T^2} = \frac{4\, kKT^2}{K^2 + (2\pi f)^2 H^2} \tag{5.131}$$

where

k = Boltzmann's constant
K = thermal conductance (W/K)
T = temperature (kinetic)
H = heat capacity (J/K).

This noise is especially important in cooled bolometers, which are low-temperature thermal detectors. For an alternative derivation, see Marasco and Dereniak (1993).

5.3.5. $1/f$ (One-over-f) Noise

This noise is a strong function of frequency, and most important at low audio frequencies. The noise power is approximately inversely proportional to the frequency:

$$i_f^2 \propto \frac{\overline{i}^{\sim 2}df}{f^{\sim 1}} \tag{5.132}$$

in terms of the current-noise power spectral density $\overline{i_f^2}$ where $\overline{i}$ is the dc bias current. The $1/f$ current noise expression is therefore inversely proportional to the square root of frequency:

$$i_f \propto \sqrt{\frac{\overline{i}^{-2}df}{f^{-1}}} \qquad (5.133)$$

Mathematically, the noise (i_f) expression blows up at zero frequency or dc. This is not a physical problem, because the finite length of any data record from which the spectrum is estimated will necessarily place a nonzero lower frequency limit on the measured one-over-f noise spectrum. Although the causes are not completely understood, the one-over-f noise magnitude is affected by nonohmic contacts at electrodes, and surface-state traps that cause current flow to be interrupted by variations in the trapping time constants. One-over-f noise is always present in both photoconductors and bolometers, because there is always a dc-bias current $\overline{i}$ present.

5.3.6. Microphonic Noise

Microphonic noise is caused by mechanical displacement. This noise results from changes in the interelectrode wire capacitance caused by displacement (d) from its position relative to ground. The basic capacitance equation is (see Fig. 5.12)

$$C = \frac{\epsilon A}{d} \qquad (5.134)$$

where ϵ is the dielectric constant of the intervening medium, and A is the effective capacitor area. If the displacement of the wire varies by an amount Δd, the capacitance changes. Thus, the voltage produced for a given amount of charge will change with mechanical vibration of the wire.

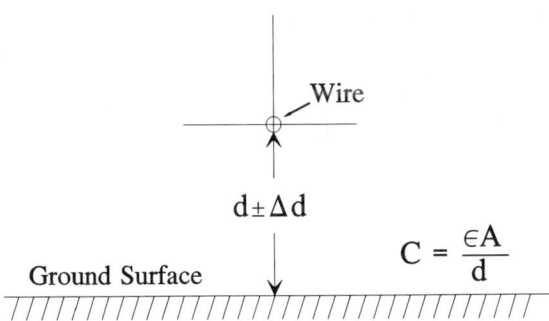

Figure 5.12. Microphonic noise model.

5.3.7. Popcorn Noise

Popcorn noise manifests itself as a spike or burst of voltage at the output of the detector. Popcorn noise sounds like corn popping in a speaker output. The spikes on the oscilloscope display could look like the wave shape shown in Fig. 5.13. Typically, the amplitude is an order of magnitude above the white noise.

The spiking is attributed to consequences of the impact ionization of neutral impurities by energetic electrons that gain the required energy to exceed the critical field value. The critical field is related to the balance between the energy gain in the field and loss to the lattice established by the electrons (Burstein et al., 1956). The critical field limits the voltage that may be effectively applied to a detector primarily because of the excess noise produced by the impact-ionization process. The bias for the optimum photodetector SNR, accordingly, is such that the electric field in the body of the detector lies below the critical field for impact ionization. This noise is caused by a manufacturing defect in the lattices, usually a metallic impurity, or results from the improper fabrication of electrical contacts (electrodes) on the detector. Improved manufacturing can minimize this problem. Popcorn noise is typically polarity sensitive.

The spiking occurs locally because of an increased electric field in the vicinity of the detector defects where the electric field has attained or exceeded the critical field value. Impact ionization and avalanche multiplication of injected or ionized electrons can only occur in the high electric-field region. As the electrons generated by impact ionization are swept into the body of the detector, they find the electric field too low to support avalanche multiplication and the process terminates. This produces a limited number of multiplied charges in a short period that represents a voltage spike whose magnitude is given by the capacitance-charge relation

$$v = \frac{Q}{C_d} \tag{5.135}$$

where

Q = charge-burst magnitude
C_d = detector capacitance

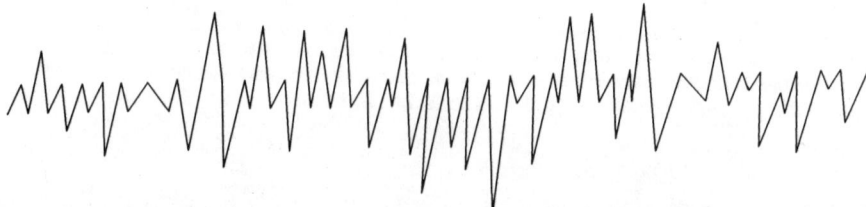

Figure 5.13.   Noise spikes above the white-noise spectrum.

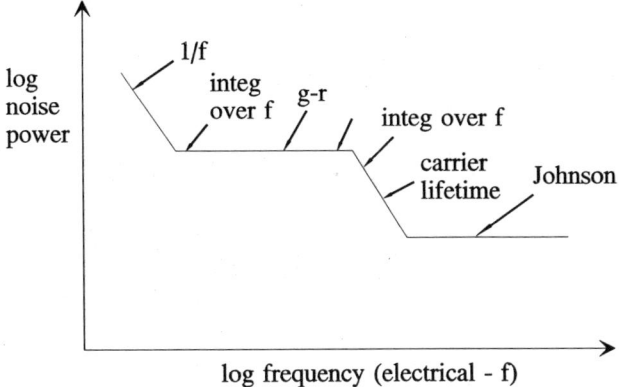

Figure 5.14. Typical noise power spectrum for a photoconductor vs. frequency.

The shortest duration of the voltage spike will be determined by the shortest transit time through the high-field region, while the decay time is controlled by the bandpass of the circuit or by the possibly longer time associated with the relaxation in the detector of a spatial charge produced by the generation and recombination of the charge carriers. The voltage spectrum of the spikes, according to this model, is determined by the number of events in the individual spikes, as related to the geometric location of the initiating events. The voltage dependence for charge injection at the contacts provides a vehicle for obtaining the voltage dependence of the spiking phenomena. Reverse biasing, if possible, sometimes minimizes this noise, especially in photoconductors.

5.3.8. Total Noise in Photoconductors

Noise spectrum of a photoconductor is shown in Fig. 5.14. The plot is in terms of power (voltage squared) versus electrical frequency.

Recall that noise voltages add in quadrature; therefore, the total noise is

$$v_{\text{total, rms}} = \sqrt{v_{1,\text{rms}}^2 + v_{2,\text{rms}}^2 + \cdots + v_{m,\text{rms}}^2} \tag{5.136}$$

5.4. ELECTRONIC-INTERFACE NOISE

Optical detectors must be connected to electronics to indicate the presence of a signal to auxiliary equipment. This electrical interface is very critical. In particular, one does not want to add noise or clutter to this signal. The immediate electrical interface is the preamplifier (see Fig. 5.2). Preamplifier and voltage/current noises are discussed in Section 7.3.

5.4.1. Quantization Noise

Quantization noise is associated with the conversion of analog signals to a digital representation whose resolution is directly proportional to the number of bits used. Ideally, one would like to represent the analog signal accurately, but this would require infinitely many bits. Realistically some sort of approximation must be made. The two options considered here are rounding and truncating.

The analog signal is assumed to have a range from 0 to v_{max} (v is not meant to constrain us to voltage; the following is generally applicable to any analog signal). Given m bits, this range can be separated into 2^m quantization levels, separated by

$$\Delta v = \frac{v_{max}}{2^m} \tag{5.137}$$

5.4.1.1. Rounding.

Rounding means representing an analog value by the nearest quantization level. Values that fall in the middle of the range require a special rule, such as always rounding up, down, or randomly up or down. In any case, the maximum error magnitude will equal $2^{-m}v_{max}/2$. Next, the assumption is made that the analog signals will be uniformly distributed within the range Δv. Any voltage level between $-\Delta v/2$ and $\Delta v/2$ sends the same digital signal from the analog-to-digital (A/D) converter. The probability density function is uniform between these two values (as shown in Fig. 5.15), where the uniform probability function can be integrated over all voltage within the voltage of a bit, and is equal to 1. Consequently, the error obeys the functional form

$$\int_{-\infty}^{\infty} p(v)\, dv = \int_{-\Delta v/2}^{\Delta v/2} \frac{1}{\Delta v}\, dv = 1 \tag{5.138}$$

where $\Delta v = 2^{-m}v_{max}$. To find the rms voltage-noise variance σ_v of this distribution we must calculate the mean and the second moment. The probability

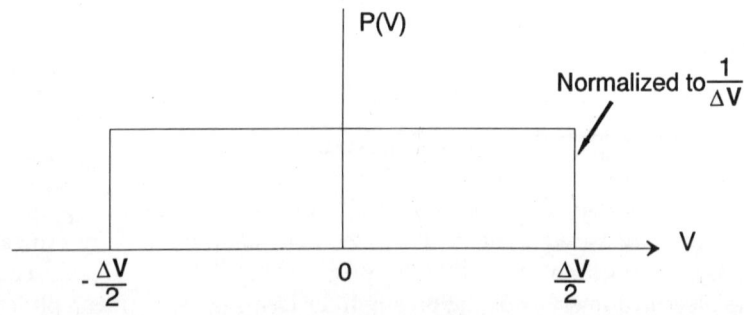

Figure 5.15.   Usual distribution for probability of a voltage within one bit.

density function $p(v)$ of this uniform distribution is

$$p(v) = \frac{1}{\Delta v} \text{rect} \left(\frac{v}{\Delta v} \right) \tag{5.139a}$$

$$\sigma_v = \sqrt{\overline{v^2} - \overline{v}^2} \tag{5.139b}$$

The mean is clearly zero:

$$\overline{v} = \int_{-\infty}^{\infty} vp(v) \, dv = \int_{-\Delta v/2}^{\Delta v/2} v \frac{1}{\Delta v} \, dv = \left. \frac{v^2}{2\Delta v} \right|_{-\Delta v/2}^{\Delta v/2} = 0 \tag{5.140}$$

The second moment is

$$\overline{v^2} = \int_{-\infty}^{\infty} v^2 p(v) \, dv = \int_{\Delta v/2}^{\Delta v/2} v^2 \frac{1}{\Delta v} \, dv = \left. \frac{v^3}{3\Delta v} \right|_{\Delta v/2}^{\Delta v/2} = \frac{\Delta v^2}{12} \tag{5.141}$$

Because the mean is zero, this result (Eq. 5.141) is equal to the voltage variance. Substituting for Δv from Eq. (5.137), we find that the noise variance and the rms noise are given by

$$\sigma_v^2 = \overline{v^2} = \frac{2^{-2m} v_{\text{max}}^2}{12} \tag{5.142}$$

and

$$\sigma_v = \sqrt{\overline{v^2}} = \frac{2^{-m} v_{\text{max}}}{\sqrt{12}} = v_{\text{rms}} \tag{5.143}$$

This noise is associated with reading an analog signal with an A/D converter having m bits and an input voltage range v_{max}.

5.4.1.2. Truncation. We follow much the same approach as used in deriving Eq. (5.143); however, in this case the error ranges from 0 to $-\Delta v$. Again, a uniform distribution of analog signals within Δv is assumed. The mean ($\overline{v}$) equals $-\Delta v/2$, and the second moment ($\overline{v^2}$) equals $\Delta v^2/3$. Consequently, the variance is given by

$$\sigma_v^2 = \overline{v^2} - \overline{v}^2 = \frac{\Delta v^2}{12} \tag{5.144}$$

We note that this is the same result as derived previously. Therefore, quantization noise is described by the same expression (5.144) for the cases of rounding and truncation. As expected, the higher the resolution Δv (more bits of digitization for a given voltage), the lower the rms noise.

5.5. NOISE BANDWIDTH

Thus far, we can recall that all the noise processes were a function of the electrical bandwidth (Δf). It is therefore necessary to define noise bandwidth. The noise bandwidth (Δf) is the frequency range that an ideal square passband in electrical frequency would have with the same power density as the real passband case. The noise-equivalent bandwidth can be considered an electrical filter with ideal uniform gain through the passband and zero gain outside: a rectangular function in frequency space. Real electrical-frequency responses cannot be made to be rectangular functions, so we must find an equivalent ideal bandwidth that would give an equivalent noise power. Referring to Fig. 5.16, $\mathcal{G}(f)$ is the voltage gain versus frequency, with $\mathcal{G}(f_0)$ as the maximum gain at the center frequency, f_0.

The total output noise power $v_{n,\text{out}}^2$ that one would measure using this filter with an input noise power spectrum $[v_{n,\text{in}}^2(f)]$ can be found by integrating over frequency and normalizing the gain to its maximum value at the center frequency $\mathcal{G}(f_0)$,

$$v_{n,\text{out}}^2 = \int_0^\infty \frac{v_{n,\text{in}}^2(f)\,\mathcal{G}^2(f)}{\mathcal{G}^2(f_0)}\,df \qquad (5.145)$$

If the noise power spectrum from the detector is flat, that is, white noise, we take $v_{n,\text{in}}^2(f)$ as a constant $v_{n,\text{in}}^2$, then

$$v_{n,\text{out}}^2 = \frac{v_{n,\text{in}}^2}{\mathcal{G}^2(f_0)} \int_0^\infty \mathcal{G}^2(f)\,df \qquad (5.146)$$

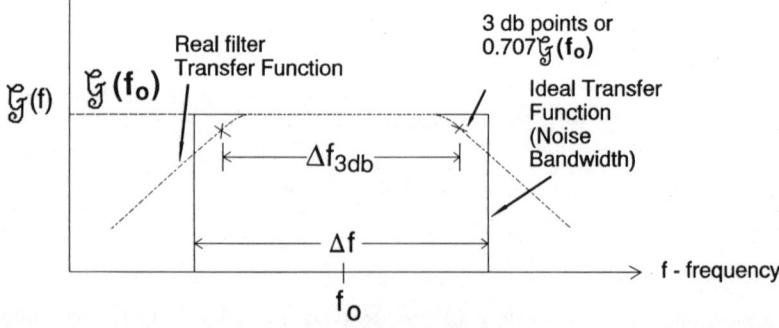

Figure 5.16.   Equivalent noise bandwidth related to a nominal voltage-gain transfer function.

where one can define equivalent noise bandwidth, Δf, as the second part of Eq. (5.146). This is the area under the filter transfer function over all frequencies:

$$\Delta f = \frac{1}{\mathcal{G}^2(f_0)} \int_0^\infty \mathcal{G}^2(f) \, df \qquad (5.147)$$

If we have a classical 1-pole filter rolloff (6 dB/octave) for $\mathcal{G}(f)$, then the relationship between the 3-dB bandwidth ($\Delta f_{3\text{dB}}$) as shown in Fig. 5.16 and Δf is $\pi/2$. Table 5.1 relates noise-equivalent bandwidth (Δf) and 3-dB bandwidth ($\Delta f_{3\text{dB}}$) for various orders of the filter.

If the noise power spectrum is not flat (i.e., $1/f$), then the we cannot discuss a noise-equivalent bandwidth, and the integration in Eq. (5.145) must be executed.

For example, what is the noise-equivalent bandwidth. Δf, required to reproduce a pulse of duration τ? The pulse can be expressed as a rectangular function (Gaskill, 1978):

$$g(t) = \text{rect}\left(\frac{t}{\tau}\right) \qquad (5.148)$$

The Fourier transform of a pulse of this duration τ is

$$\mathcal{G}(f) = \tau \left(\frac{\sin \pi\tau f}{\pi\tau f}\right) \qquad (5.149)$$

Table 5.1   Noise Bandwidth Related to 3-dB Bandwidth

Number of Poles in Filter	Δf
1	$\left(\dfrac{\pi}{2}\right) \Delta f_{3\text{dB}}$
2	$\left(\dfrac{\pi}{2}\right)^{1/4} \Delta f_{3\text{dB}}$
3	$\left(\dfrac{\pi}{2}\right)^{1/6} \Delta f_{3\text{dB}}$
4	$\left(\dfrac{\pi}{2}\right)^{1/8} \Delta f_{3\text{dB}}$
5	$\left(\dfrac{\pi}{2}\right)^{1/10} \Delta f_{3\text{dB}}$

The noise-equivalent bandwidth (Δf) from Eq. (5.147) is

$$\Delta f = \frac{1}{\tau^2} \int_0^\infty \tau^2 \frac{\sin^2(\pi\tau f)}{(\pi\tau f)^2} \, df \tag{5.150}$$

To integrate Eq. (5.150), let

$$f' = \pi\tau f \tag{5.151a}$$

and

$$df' = \pi\tau df \tag{5.151b}$$

Substituting,

$$\Delta f = \int_0^\infty \frac{\sin^2(f')}{(f')^2} \frac{df'}{\pi\tau} = \frac{1}{\pi\tau} \frac{\pi}{2} \tag{5.152}$$

$$\Delta f = \frac{1}{2\tau} \tag{5.153}$$

The relationship between observation time τ (also often called *integration time*) and noise-equivalent bandwidth is given by Eq. (5.153). As the integration time becomes shorter, the noise-equivalent bandwidth becomes wider.

5.6. DETECTOR RESPONSE SPEED

How fast a detector can respond to a pulse of optical radiation depends on its electrical response. To develop this concept, consider a detector's geometry and the current flowing through it. We will then use the continuity equation to obtain the time response. Figure 5.17 shows a cross section of a detector with

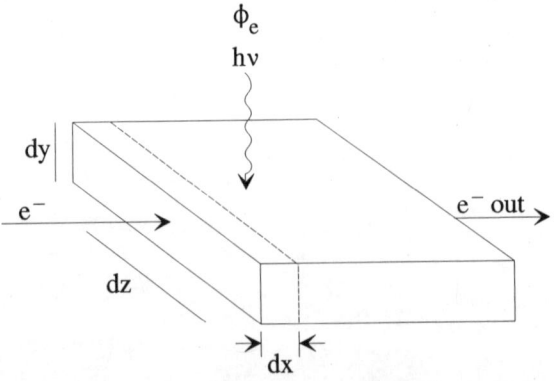

Figure 5.17. Detector cross section showing current flow and photon incidence.

current flowing in the x-direction, and optical radiation incident on the x-z plane.

The total number of electrons in this volume at an instant is $n(x, t) \, dx \, dy \, dz$ (Ambroziak, 1969). Using the change in the number of electrons in a time dt, we can set up the continuity equation

$$[n(x, t + dt) - n(x, t)] \, dx \, dy \, dz = \frac{\partial \Delta n}{\partial t} \, dx \, dy \, dz \, dt \qquad (5.154)$$

The change in charge is caused by photogeneration, recombination, and/or diffusion.

1. Photogeneration—(g is generation rate in number/cm³-s): $g \, dx \, dy \, dz \, dt$
2. Recombination—(τ is recombination time): $(\Delta n/\tau) \, dx \, dy \, dz \, dt$
3. Drift, Diffusion and Trapping:

$$-\frac{\partial \Delta n}{\partial x} \, v_x \, dx \, dy \, dz \, dt$$

where v_x = velocity of e- in the x-direction, and $q \Delta n \cdot v_x = J_x$ current density in the x-direction (amp/cm²)

This charge could be generated in the three dimensions

$$\frac{1}{q} \left[\frac{\partial J_x^-}{\partial x} \, dx \, dy \, dz \, dt + \frac{\partial J_y^-}{\partial y} \, dx \, dy \, dz \, dt + \frac{\partial J_z^-}{\partial z} \, dx \, dy \, dz \, dt \right]$$

$$= \frac{1}{q} [\nabla \cdot \mathbf{J}^- \times dx \, dy \, dz \, dt] \qquad (5.155)$$

Setting up the current-balance equation

$$\frac{\partial \Delta n}{\partial t} = g - \frac{\Delta n}{\tau} - \frac{1}{q} (\nabla \cdot \mathbf{J}^-) \qquad (5.156)$$

If we ignore diffusion, drift, and trapping (i.e., $\nabla \cdot \mathbf{J}^- = 0$), then the continuity equation becomes

$$\frac{\partial \Delta n}{\partial t} = g - \frac{\Delta n}{\tau} \qquad (5.157)$$

This differential equation has two solutions:

1. Steady state, $\partial \Delta n/\partial t = 0$, not of interest as an optical detector

2. Time-varying excess-charge conditions are more important for frequency-response considerations

$$\frac{\partial \Delta n}{\partial t} + \frac{\Delta n}{\tau} = g \qquad (5.158a)$$

(differential equation)

$$e^{t/\tau} \left\{ \frac{\partial \Delta n}{\partial t} + \frac{\Delta n}{\tau} \right\} = g e^{t/\tau} \qquad (5.158b)$$

$$\frac{\partial}{\partial t} (\Delta n\, e^{t/\tau}) = g e^{t/\tau} \qquad (5.158c)$$

$$\Delta n\, e^{t/\tau} = \int_0^t g e^{t'/\tau}\, dt' = \tau g e^{t/\tau} - \tau g + \text{const.} \qquad (5.158d)$$

By the boundary condition $\Delta n = 0$ at $t = 0$, we are able to set the constant equal to zero, yielding

$$\Delta n = \tau g (1 - e^{-t/\tau}) \qquad (5.159)$$

This is the excess charge concentration as a function of time shown in Fig. 5.18b, in response to a pulse of optical-radiation flux as shown in Fig. 5.18a. For a pulse of radiation on this detector, the output response would follow Eq. (5.159)

$$\Delta n = \eta E_q \propto \tau (1 - e^{-t/\tau}) \qquad (5.160)$$

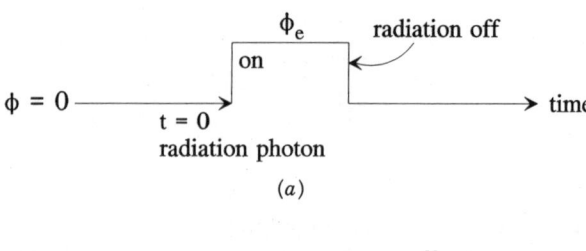

(a)

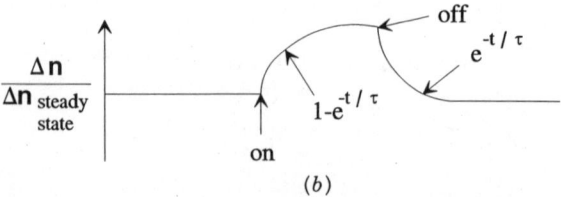

(b)

Figure 5.18.   (a) Pulse input of radiation onto a detector; (b) output waveform in time.

At the end of the radiation pulse, the radiation is turned off, and the differential equation Eq. (5.158a) has no generation term ($g = 0$):

$$\frac{\partial \Delta n}{\partial t} = g - \frac{\Delta n}{\tau} = -\frac{\Delta n}{\tau} \qquad (5.161)$$

Rearranging and solving:

$$\frac{d\Delta n}{\Delta n} = -\frac{dt}{\tau} \qquad (5.162a)$$

$$\ln \Delta n = -\frac{t}{\tau} + \text{const.} \qquad (5.162b)$$

$$\Delta n = \tau \, g e^{-t/\tau} \qquad (5.162c)$$

Thus, during the period of time when $g = 0$, as shown in Fig. 5.18b, the carrier charge drops off exponentially as $\exp\{-t/\tau\}$. Transforming this time-dependent expression into the frequency domain by a Fourier transform yields

$$\Delta n(f) = \frac{\tau g}{1 + j2\pi f\tau} \qquad (5.163)$$

which can be extended to responsivity versus frequency as

$$\mathcal{R}_i(f) = \frac{\mathcal{R}_i(0)}{1 + j2\pi f\tau} \qquad (5.164a)$$

$$|\mathcal{R}_i(f)| = \frac{\mathcal{R}_i(0)}{\sqrt{1 + (2\pi f\tau)^2}} \qquad (5.164b)$$

A typical responsivity-versus-frequency curve is plotted in Fig. 5.19.

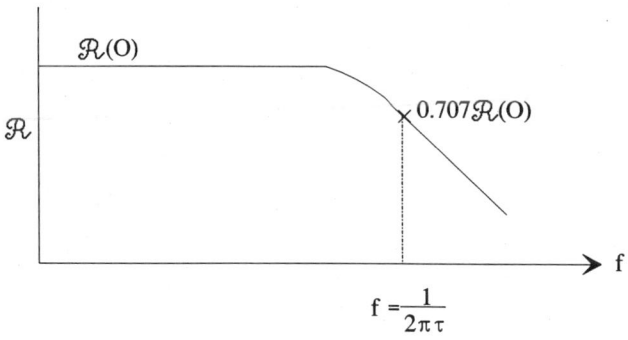

Figure 5.19.   General shape of responsivity vs. frequency.

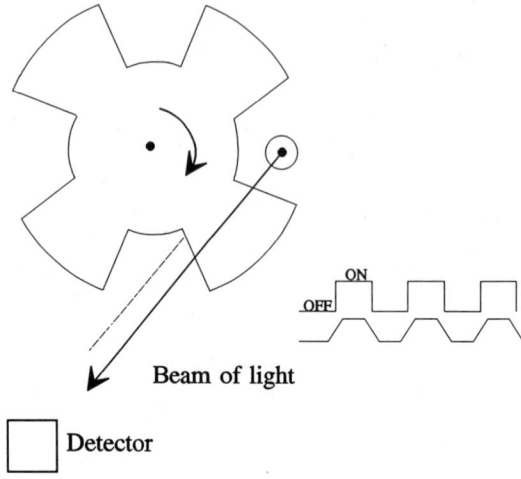

Beam of light

Detector

Figure 5.20.   Rotating chopper.

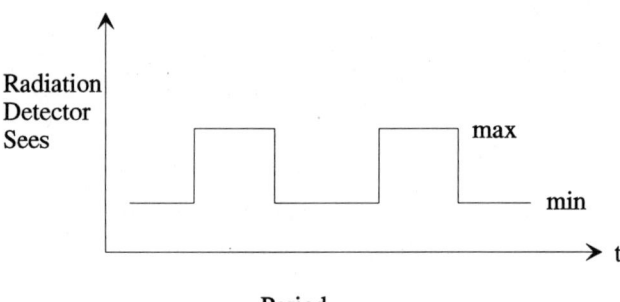

Radiation Detector Sees

Period

Figure 5.21.   Voltage waveform for chopper signal.

The carrier lifetime determines the high-frequency responsivity cutoff. The frequency response can be measured by modulating a source by a chopper, as shown in Fig. 5.20. The output voltage would vary from peak to valley as shown in Fig. 5.21. As the modulation frequency increases, eventually the peak-to-peak amplitude will decrease. By varying the chopper rotation rate, we plot the response as a function of modulation frequency (Fig. 5.19).

REFERENCES

Ambroziak, A., *Semiconductor Photoelectric Devices*, Gordon & Breach, New York, 1969.

Burstein, E., G. Pines, and N. Sclar, "Optical and Photoconductive Properties of Silicon and Germanium," in *Photoconductive Conference* (R. G. Breckenridge et al., eds.), Wiley, 1956, pp. 353–413.

Frieden, B. R., *Probability, Statistical Optics, and Data Testing*, Springer-Verlag, Berlin/New York, 1991.

Gaskill, J. D., *Linear Systems, Fourier Transforms and Optics*, Wiley, New York, 1978.

Johnson, J. B., "Thermal Agitation of Electricity in Conductors," *Phys. Rev.* **32,** 97 (1928).

Marasco, Peter L., and Eustace L. Dereniak, "Uncooled Infrared Sensor Performance," *SPIE Proc.* **2020,** 363–378 (1993).

McCall, C. W., "Convoy," Louis Davis Jr. and William Fries, Writer, American Gramaphone Publisher (1976).

Nyquist, H., "Thermal Agitation of Electric Charge in Conductors," *Phys. Rev.* **32,** 110 (1928).

Van der Ziel, Aldert, *Noise in Solid State Devices and Cricuits*, Wiley, New York, 1986.

BIBLIOGRAPHY

Keyes, Robert J., *Optical and Infrared Detectors Topics in Applied Physics*, vol. 19, Springer-Verlag, New York, 1980.

Kingston, Robert Hildreth, *Detection of Optical and Infrared Radiation*, Springer-Verlag, Berlin/New York, 1978.

Kruse, Paul W., L. D. McGlauchlin, and R. B. McQuistan, *Element of Infrared Technology: Generation, Transmission, and Detection*, Wiley, New York, 1962.

Reif, Frederick, *Fundamentals of Statistical and Thermal Physics*, McGraw-Hill, New York, 1965.

Smith, R. A., F. E. Jones, and R. P. Chasmar, *The Detection and Measurement of Infrared Radiation*, 2d ed., Oxford University Press, London, 1968.

Streetman, Ben G., *Solid State Electronic Devices*, Prentice Hall, Englewood Cliffs, NJ, 1990.

Sze, S. M., *Physics of Semiconductor Devices*, Wiley-Interscience, New York, 1969.

Van Vliet, K. M., "Noise Limitation in Solid State Photodetectors," *Appl. Opt.* **6,** (7), 1145 (1967).

Vincent, John David, *Fundamentals of Infrared Detector Operation and Testing*, Wiley, New York, 1990.

Wolfe, W. L., and G. J. Zissis, *The Infrared Handbook*, Optical Engineering Press, Bellingham, WA, 1990.

PROBLEMS

5.1.   Show that the noise-equivalent bandwidth (Δf) for a given observation time (τ) is related by $\Delta f = 1/(2\tau)$. Recall the Fourier-transform relation between time and frequency. Does this depend on the exact form of the impulse response? If so, provide an example.

5.2.   List and discuss noise sources associated with a detector system (equations are not necessary).

5.3.   How large is the excess radiation noise factor (Bose factor, b_{Bose}) for radiation at 118 μm ($\Delta\lambda = 1$ μm) from a 1000-K blackbody?

5.4.   Find the Johnson noise voltage v_J associated with two resistors (10^7 Ω @ 77 K and 10^6 Ω @ 4 K) that are in a parallel configuration as shown in the figure.

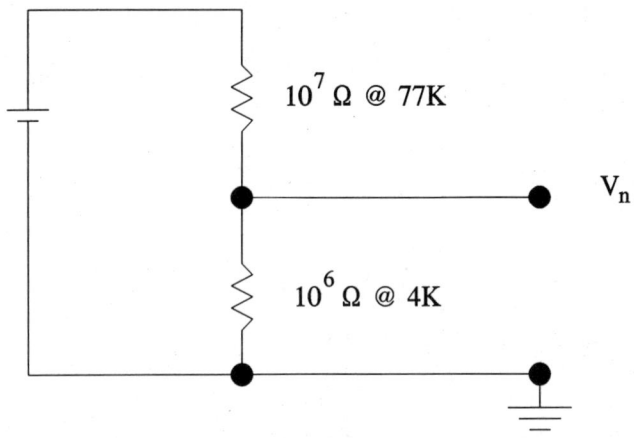

Figure 5.4

5.5.   Using Wien's approximation to the blackbody exitance expression, that is,

$$M_q(\lambda, T) = \frac{2\pi c}{\lambda^4(e^{hc/\lambda KT} - 1)} \cong \frac{2\pi c}{\lambda^4(e^{hc/\lambda KT})}$$

derive an expression for the SNR, interpreted as follows. Assume that the signal is the photon-exitance contrast, and the noise is some constant ($\mathfrak{B}$) times the blackbody exitance:

$$\frac{S}{N} = \frac{\partial M_q(\lambda, T)/\partial T}{\mathfrak{B}M_q(\lambda, T)}$$

Make observations regarding S/N versus spectral region for source temperatures of 300 K and 1000 K.

5.6.   Calculate the Johnson noise voltage (v_J) across the parallel combination of resistors shown ($\Delta f = 1000$ Hz).

(a) Refer to the figure.

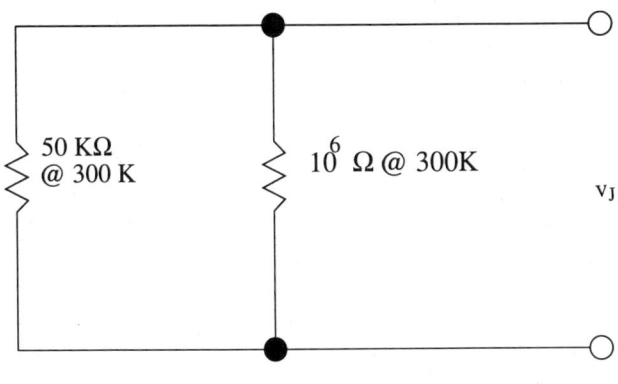

Figure 5.6(a)

(b) Cool the 50-kΩ resistor to liquid helium temperature, and recalculate v_J.

(c) Would cooling the 1-MΩ resistor have a larger effect? Why?

5.7. Calculate the noise-equivalent bandwidth Δf for a filter starting at dc ($f = 0$) and having voltage gain $\mathcal{G}(f)$, and a 6-dB/octave rolloff with 3-dB corner frequency $f_c =$ (a) 100 Hz and (b) 280 Hz.

5.8. Consider a "perfect" photon stream, where the photons arrive equally spaced in time on the detector with no variance ($\sigma^2 = 0$). The SNR on the photon stream is infinite $[(S/N)_{input} = \infty]$. What is the signal-to-noise output $[(S/N)_{output}]$ for a detector of quantum efficiency (η), assuming a binomial detection process inside the detector?

5.9. What is the total rms current noise (i_n) out of a photodiode with 1-mA bias and a resistance of 10^4 Ω? (Assume $\Delta f = 10$ Hz; $T = 300$ K; $f = 500$ Hz; $\alpha_{1/f} = 3$; $\beta_{1/f} = 1$; and $B_{1/f} = 1$.)

5.10. For a detector at 4 K shown, the heat capacity ($H = 2 \times 10^{-10}$ J/K) and thermal conductance ($K = 1.4 \times 10^{-7}$ W/K) for each wire. Find the temperature variance $\overline{\Delta T^2}$ at low frequency ($f = 10$ Hz) assuming a $\Delta f = 1$ (see the figure).

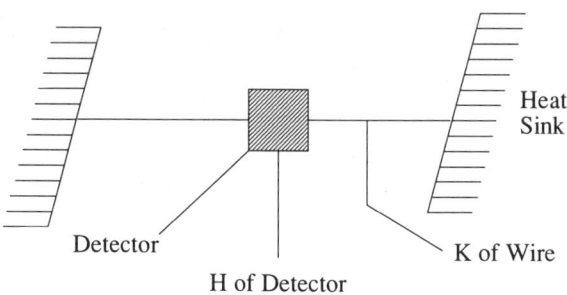

Detector

Heat Sink

K of Wire

H of Detector

Figure 5.10

5.11. Compute the equivalent noise bandwidth of an amplifier assuming a Gaussian voltage gain

$$\mathcal{G}(f) = \mathcal{G}(f_0) \exp \{-\pi(f - f_0)^2/\sigma^2\}$$

where $\sigma = 100$ Hz and $f_0 = 800$ Hz. Assume the noise is
(a) a white noise source.
(b) a $1/f$ noise source.

5.12. How could one test a cooled detector to determine if it is photon-noise limited?

5.13. Compute the rms Johnson, shot, and $1/f$ noise for a photovoltaic detector with the following parameters: $T = 300$ K; $R_d = 10$ kΩ; $\bar{i} = 10$ mA; $f = 500$ Hz; $\Delta f = 10$ Hz. What is the total detector noise? Which noise dominates?
Assume $\alpha_{1/f} = 2$, $\beta_{1/f} = 1$, $B_{1/f} = 1$.

5.14. The noise spectrum of a photoconductive detector is pictured in the figure. Describe the kinds of noise that are dominant in the four regions a–d and draw the corresponding relative D^* curve with a dashed line on the same figure.

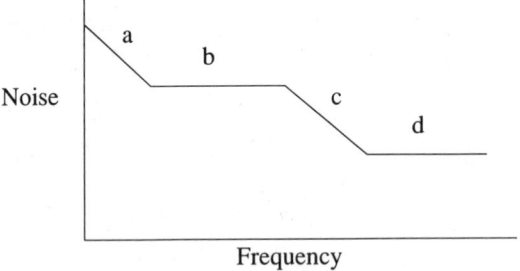

Figure 5.14

5.15. Why can the $1/f$ noise be eliminated in a photovoltaic detector but not in a photoconductor?

5.16. List the seven primary noises found in optical radiation detectors. Describe the salient characteristics of each, with equations where appropriate. Which noises are commonly found in photovoltaic, photoconductive, photoemissive, thermocouple, and pyroelectric detectors? Outline a procedure that can be used to distinguish the different forms of noise within a given experiment.

5.17. The mean-square shot noise in a highly reverse-biased photodiode with no radiation incident is given by

$$\overline{i_s^2} = 2qi_0\Delta f$$

where

i_0 = the reverse saturation current

q = the electronic charge

Δf = the noise-equivalent bandwidth

Using this expression and solving for resistance at zero bias ($v = 0$), show that the relationship is equivalent to a Johnson-noise expression.

5.18. Show that the equation for rms shot-noise current is within a factor of $2^{1/2}$ of the equation for rms generation–recombination noise, when the shot-noise current is entirely caused by photogenerated carriers, and the photoconductive gain $G = 1$.

6

Figures of Merit for Optical Detectors

6.1. INTRODUCTION

We use many different figures of merit to characterize or quantify the performance of optical detectors. These figures of merit have evolved over many years. Some were created and defined before photodetectors were discovered. Therefore, the figures of merit are often not a good representation of present-day photodetectors because of the units used that characterize them. In any case, the currently accepted figures of merit used in the electrooptic field are developed in this chapter. These parameters are general terms applicable to any type of detector, independent of the generic optical-detector processes.

Figures of merit enable the user to compare relative performance between detectors. However, many assumptions are hidden in the definition of these parameters. We will avoid photometric units and only deal with radiometric units, in terms of energy (subscript e) or quanta (subscript q). Table 6.1 defines the nomenclature that is used, consistent with Chapter 2.

We cannot overemphasize the need for standardization and calibration procedures. Like any other calibration or standardization, the true value can only be approached. We may never know the exact detectivity of a detector, but it will be close enough for all practical applications.

The first detectors discovered were thermal detectors, which responded to the heating effect of the radiation. The most common figures of merit used today still bear the imprint of the fact that they were developed to characterize thermal detectors. Although photon detectors would more normally be described in photon-based units, most figures of merit are developed in energy-based units for historical reasons.

One parameter we have already introduced is responsivity. However, it is difficult to directly compare responsivity of a photomultiplier to that of a bolometer. Thermal-detector responsivities are flat with respect to wavelength, but photodetector responsivity (in energy-based units) is a linear function of wavelength. The figures of merit that we develop first are for photon detectors.

6.2. RESPONSIVITY

The concept of responsivity has been developed that quantifies the amount of output seen per watt of radiant optical power input. We previously discussed

Table 6.1 Radiometric Nomenclature

Nomenclature for Quantity	Radiometric Symbol	Terms/Units
Radiant energy	Q_e	joule
Radiant power (flux)	ϕ_e	watt
Radiant intensity	I_e	watt/steradian
Irradiance (flux density)	E_e	watt/cm^2
Radiant exitance	M_e	watt/cm^2
Radiance	L_e	watt/cm^2-sr
Photon energy	Q_q	photon or quantum
Photon flux	ϕ_q	photon/s
Photon flux intensity	I_q	photon/s-sr
Incident photon flux density (irradiance)	E_q	photon/s-cm^2
Photon flux existance	M_q	photon/s-cm^2
Photon flux sterance	L_q	photon/s-cm^2-sr

responsivity as a function of temporal frequency. Now, we must discuss spectral-responsivity dependence as a function of wavelength. The nomenclature for spectral responsivity is $\mathfrak{R}(\lambda, f)$.

Our discussion has thus far avoided the spectral characteristic of the radiant power; a watt of radiant energy has been used only ambiguously. The spectral responsivity is defined as the output per watt of monochromatic radiation. The output can be either current or voltage output, per watt of monochromatic radiation. The nomenclature is $\mathfrak{R}_i(\lambda, f)$ for current spectral responsivity and $\mathfrak{R}_v(\lambda, f)$ for voltage spectral responsivity. Spectral responsivity is the output-signal (current or voltage) response to monochromatic radiation incident on the detector, modulated at a frequency (f).

The general equation for current responsivity is

$$\mathfrak{R}_i(\lambda, f) = \frac{\eta \lambda q}{hc \sqrt{1 + (2\pi f \tau)^2}} G \qquad (6.1)$$

where

η = quantum efficiency
λ = wavelength
q = electron charge
h = Planck's constant
c = speed of light
G = photoconductive gain
f = electrical chopping frequency
τ = time constant

Correspondingly, the blackbody responsivity nomenclature is $\Re_i(T, f)$ for current blackbody responsivity and $\Re_v(T, f)$ for voltage blackbody responsivity. Blackbody responsivity is interpreted as the output produced in response to a watt of input optical radiation from a blackbody at temperature T modulated at electrical frequency f. The detector community has standardized on two blackbody temperatures that are universally used as sources to evaluate detectors. The temperatures are 500 K for infrared measurements and 2856 K for tungsten–halogen, used for visible and near-IR measurements.

An item of note is that when measuring blackbody responsivity, the radiant power on the detector contains all wavelengths of radiation, independent of the spectral response curve of the detector. Some of the wavelengths produce an output and some do not, but all of the flux incident on the detector appears in the calculation of blackbody responsivity.

Although responsivity of a detector is useful to predict an expected signal level for a given radiant power on the detector, the responsivity is of limited usefulness from a sensitivity point of view. In addition to the signal level, the noise level must be considered also to quantify the signal-to-noise ratio (SNR). For example, even if the detector has a large responsivity, if the noise is large, the resulting SNR might be low.

6.3. NOISE EQUIVALENT POWER

The minimum radiant-flux level a detector can discern depends on the detector noise level. The signal current produced by the input power must be above the noise current to be easily detected. For example, the SNR can be expressed in terms of current responsivity as

$$\frac{S}{N} = \frac{\Re_i \phi_e}{i_n} \tag{6.2}$$

where

$\Re_i$ = current responsivity (amp/watt)
ϕ_e = radiant flux (watt)
i_n = noise current.

The radiant power, ϕ_e, incident on a detector (not necessarily the absorbed power) that yields SNR = 1 has a special name, the noise-equivalent power (NEP). Setting Eq. (6.2) equal to 1 and solving for the required power,

$$\phi_e = \frac{i_n}{\Re_i(\lambda, f)} \tag{6.3}$$

where ϕ_e in Eq. (6.3) is the NEP, the amount of radiant power collected on a detector that will produce an SNR of 1. Therefore, the NEP can be written as the root-mean-square (rms) noise divided by the responsivity:

$$\text{NEP} = \frac{i_n}{\mathfrak{R}_i} \tag{6.4}$$

In other words, the NEP times the responsivity must just equal the rms noise current. Similarly, in terms of voltage responsivity $\mathfrak{R}_v(\lambda, f)$

$$\text{NEP} = \frac{v_n}{\mathfrak{R}_v} \tag{6.5}$$

Another expression for NEP is to rewrite the responsivity as output signal current divided by input flux (i_{sig}/ϕ_e). Substituting into Eq. (6.4), the NEP is

$$\text{NEP} = \frac{\phi_e}{i_{\text{sig}}/i_n} \tag{6.6}$$

where i_{sig}/i_n is the current SNR. As the responsivity was a function of wavelength and frequency, so is NEP. Also, the NEP, in similar fashion to responsivity, can be either spectral or blackbody, depending on the type of incident radiation. From Eq. (6.4), if the responsivity in the denominator is spectral responsivity, then the NEP will be spectral NEP. If the responsivity is blackbody responsivity, then the NEP will be blackbody NEP. Thus the nomenclature for NEP is NEP(λ, f) for spectral NEP, which is the monochromatic radiant flux (watts) necessary to produce an rms SNR of 1 at the electrical frequency f, and NEP(T, f) for blackbody NEP, which is the blackbody radiant flux (watts) necessary to produce an rms SNR of 1 at frequency f.

An item of note is that when measuring blackbody NEP, the radiant power on the detector contains all wavelengths of radiation, independent of the spectral-response curve of the detector. Some of the wavelengths produce an output and others do not, but all of the incident flux appears in the calculation of NEP in Eq. (6.6), just as it did for the calculation of blackbody responsivity. Not all of the incident radiation produces an output. The unit of NEP is watts. A more sensitive detector will have a lower NEP. It is a defect function, rather than a figure of merit, in that smaller is better.

The NEP depends on many additional parameters that are often not taken into account by the equations. The parameters that must be specified along with NEP to compare detector performance of one detector to another are:

Optically active detector area—A_d

Noise-equivalent electrical bandwidth—Δf

Modulation or chopping frequency—f

Spectral operating region

Optimum bias

Detector operating temperature

The disadvantage to the use of NEP to describe detector performance is that NEP is a situation-specific descriptor. Through Eq. (6.6), NEP yields the SNR actually achieved in a given configuration. This information is useful for design purposes, but does not allow a direct comparison of the sensitivity of different detector mechanisms or materials. NEP (and the SNR actually achieved) is dominated by its dependence on the square root of the detector area (A_d), and the square root of the bandwidth (Δf) of the measurement. Only by specifying these two parameters can an intrinsic comparison of the sensitivities be made for the detector modality itself, independent of the sensor area and the bandwidth.

For instance, if two detectors have the same NEP, but detector 1 is 10 times the area of detector 2, which is the better detector material or sensor? Similarly, if we measured one detector with a Δf value 10 times the other, which detector is better? Both parameters affect the noise of the detector proportional to the square root of both area and bandwidth. To circumvent this problem with NEP, Jones (1953) defined a new term, D^*, normalizing the inverse of the NEP to a 1-cm^2 detector area and 1-Hz noise bandwidth.

6.4. DETECTIVITY

The figure of merit D^* (pronounced "Dee Star") called *normalized detectivity* is

$$D^* = \frac{\sqrt{A_d \, \Delta f}}{\text{NEP}} \qquad (6.7)$$

This figure of merit (bigger is better) is sensitivity normalized to a 1-cm^2 area and 1-Hz noise-equivalent bandwidth. It can be interpreted as an SNR out of a detector when 1 W of radiant power is incident on the detector, given an area equal to 1 cm^2 and noise-equivalent bandwidth of 1 Hz. The units of D^* are (Jones, 1953) defined as [cm $\sqrt{\text{Hz}}$/watt] = [Jones].

Just as spectral and blackbody NEP figures of merit were developed, spectral and blackbody D^* exist. From Eq. (6.7), depending on whether the NEP is spectral or blackbody, the D^* correspondingly can be either spectral or blackbody. Spectral D^*, or $D^*(\lambda, f)$, is the rms signal-to-noise output when 1 W of monochromatic radiant flux (modulated at electrical frequency (f)), is incident on a 1-cm^2 detector area, within a noise-equivalent bandwidth of 1 Hz.

Blackbody D^*, known as $D^*(T, f)$, is the rms signal-to-noise output when

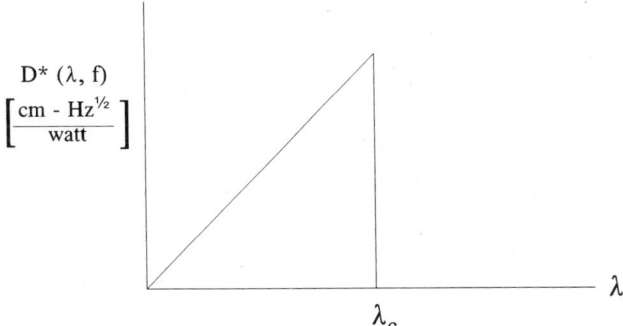

Figure 6.1. Spectral D^* vs. wavelength.

1 W of blackbody radiant power modulated at electrical frequency (f) is incident on a 1-cm^2 active area and noise-equivalent bandwidth of 1 Hz.

The assumptions made in arriving at the D^* expressions were:

1. noise is proportional to $\sqrt{\Delta f}$, a flat (white) noise power spectrum.

2. noise is proportional to $\sqrt{A_d}$

3. detector is at optimum bias

4. detector is at optimum operating temperature

5. frequency response is flat.

Figure 6.1 plots the spectral detectivity. The peak value of D^* is defined as peak spectral D^*. As stated in assumption (5), the D^* is in the flat region of the frequency response.

The maximum value of spectral D^* (called the peak spectral D^*) corresponds to the largest potential SNR. In addition, any optical radiation incident on the detector at a wavelength shorter than λ_c will have a D^* reduced from the peak spectral D^* in proportion to the ratio λ/λ_c. This relationship is linear, as seen in Fig. 6.1. The peak spectral D^* is always larger than a blackbody D^* for the same detector, because for a blackbody source, a wide range of wavelengths are incident, most of which will be well removed from the peak of the D^*-versus-λ curve.

This section has summarized the general figures of merit used to describe optical detectors. Any other performance parameters assume special operating conditions or circumstances.

6.5. BLACKBODY/SPECTRAL FIGURES OF MERIT RELATIONSHIPS

Typically, we measure blackbody responsivity and calculate the corresponding peak spectral responsivity. Therefore, we must determine the relationship between blackbody measured D^* and peak spectral values calculated from the blackbody measurements.

The expression for peak spectral voltage responsivity can be expressed as

$$\mathcal{R}_v(\lambda_c, f) = \frac{v_o}{\displaystyle\int_0^{\lambda_c} \frac{\mathcal{R}(\lambda)}{\mathcal{R}(\lambda_c)} L_e(\lambda, T) \, d\lambda \times \dfrac{A_s A_d}{r^2} \mathit{t} F_F} \qquad (6.8)$$

where

$$v_o = \text{output voltage produced by detector (volt)}$$
$$\frac{\mathcal{R}(\lambda)}{\mathcal{R}(\lambda_c)} = \text{normalized spectral response, which is linear}$$
$$L_e(\lambda, T) = \frac{2c^2 h}{\lambda^5 (\exp\{hc/k\lambda T\} - 1)} \left[\frac{\text{watt}}{\text{cm}^2\text{-sr } \mu m} \right]$$
$$A_s = \text{blackbody area}$$
$$A_d = \text{detector area}$$
$$\mathit{t} = \text{transmittance}$$
$$F_F = \text{form factor for conversion of peak-to-peak to rms}$$
$$r = \text{source-to-detector distance}$$

Because we can express the responsivity of a photodetector as a linear response to wavelength as Eq. (6.1), dropping the frequency dependence for convenience, we obtain

$$\mathcal{R}_i(\lambda) = \frac{\eta \lambda q}{hc} G \qquad (6.9)$$

The normalized spectral responsivity relative to maximum at λ_c is

$$\frac{\mathcal{R}_i(\lambda)}{\mathcal{R}_i(\lambda_c)} = \frac{\lambda}{\lambda_c} \qquad (6.10)$$

which, as we have seen in Fig. 6.1, shows a linear dependence with wavelength up to the cutoff wavelength. Substituting into Eq. (6.8),

$$\mathcal{R}_v(\lambda_c, f) = \frac{v_o}{\displaystyle\int_0^{\lambda_c} \frac{\lambda}{\lambda_c} \frac{2c^2 h}{\lambda^5 (\exp\{hc/k\lambda T\} - 1)} \, d\lambda \times \dfrac{A_s A_d}{r^2} \mathit{t} F_F} \qquad (6.11)$$

Figure 6.2 shows the layout of the radiation source (blackbody) relative to the detector for a typical measurement setup.

The terms in Eq. (6.11) can be rearranged as

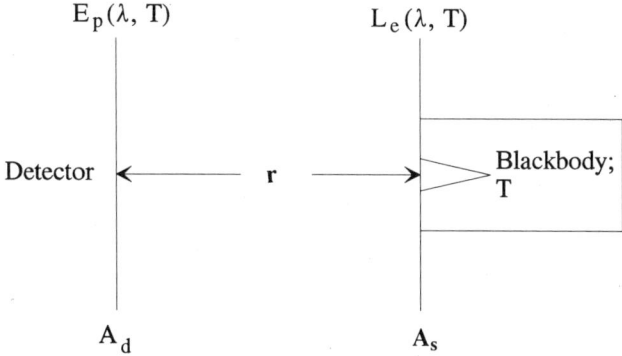

Figure 6.2.   Detector/blackbody source configuration for testing.

$$\mathcal{R}_v(\lambda_c, f) = \cfrac{v_o}{\left[\dfrac{hc}{\lambda_c}\right]\left[\displaystyle\int_0^{\lambda_c} \cfrac{2c\pi}{\lambda^4(\exp\{hc/h\lambda T\} - 1)} \, d\lambda\right]\left[\dfrac{A_s A_d}{\pi r^2} \ell F_F\right]} \tag{6.12}$$

photon $M_q(\lambda, T)$

energy   photon   geometrical

for λ_c exitance   configuration

           of blackbody   of blackbody/detector

The terms in the denominator can be grouped into three parts, photon energy at the peak wavelength, photon exitance (M_q) integrated over the spectral response of the detector, and a geometrical term relating source to detector geometry (see Fig. 6.2). The photon exitance accounts for the solid angle carried over from the radiance (L_q); thus, a π factor must be included to define the second term as exitance.

The interpretation of Fig. 6.2 is important in determining the irradiance incident on the detector. Where E_q is the photon irradiance at the detector surface from the blackbody at temperature, T, the denominator of Eq. (6.12) can be approximated as if all incident photons had an energy corresponding to the peak wavelength (λ_c). Recall the definition of blackbody responsivity, the output voltage that results from radiation from a blackbody source:

$$\mathcal{R}_v(T, f) = \frac{v_o}{\phi_e(T)} \tag{6.13}$$

Using the geometry shown in Fig. 6.2,

$$\mathcal{R}_v(T, f) = \cfrac{v_o}{\sigma T^4\left[\dfrac{A_s A_d}{\pi r^2} \ell F_F\right]} \tag{6.14}$$

We then substitute for v_o from Eq. (6.14) into Eq. (6.12) to find the factor relating the peak spectral responsivity to the blackbody responsivity:

$$\mathcal{R}_v(\lambda_c, f) = \frac{\sigma T^4}{\dfrac{hc}{\lambda_c} \displaystyle\int_0^{\lambda_c} M_q(\lambda, T)\, d\lambda} \, \mathcal{R}_v(T, f) \qquad (6.15)$$

This conversion factor is the ratio of total blackbody energy to the energy to which the photon detector responds (assuming all incident photons were at the peak wavelength). The conversion factor (u) is

$$u \equiv \frac{\mathcal{R}_v(\lambda_c, f)}{\mathcal{R}_v(T, f)} = \frac{\sigma T^4}{\dfrac{hc}{\lambda_c} \displaystyle\int_0^{\lambda_c} M_q(\lambda, T)\, d\lambda} \qquad (6.16)$$

which is also the conversion factor between blackbody and spectral D^*. Figure 6.3 shows a plot of the conversion ratio (u) between blackbody D^* to peak spectral D^*, as a function of the cutoff wavelength of the photodetector, for 300-K and 500-K blackbody temperatures.

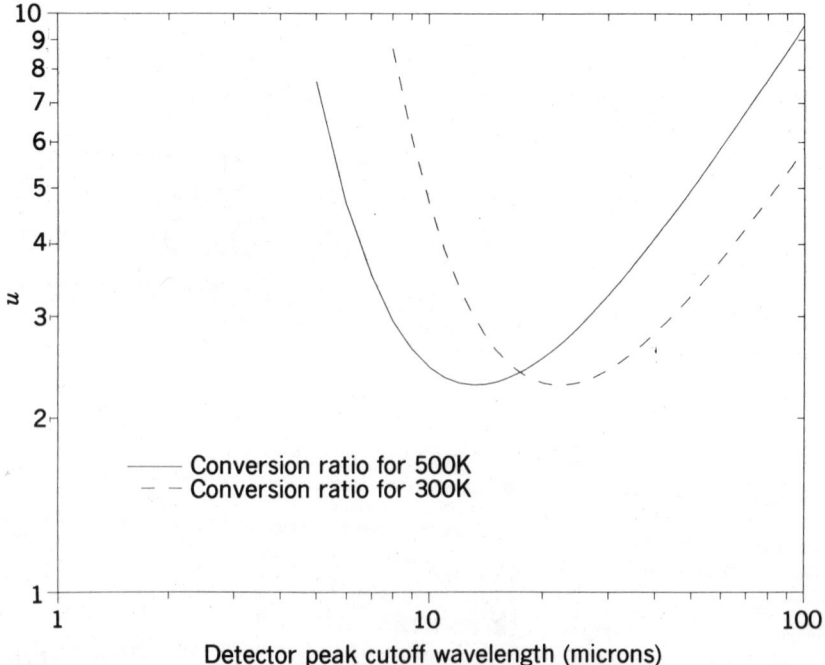

Figure 6.3. Conversion ratio between measured blackbody responsivity and peak spectral responsivity.

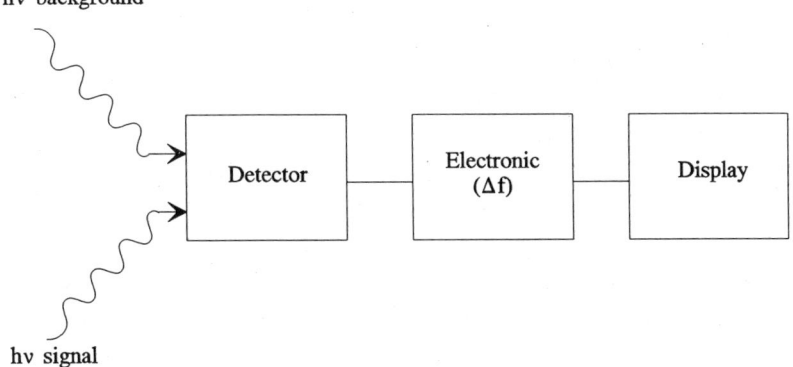

hv background

hv signal

Figure 6.4. Two photon sources incident on the detector.

6.6. PHOTON-NOISE-LIMITED PERFORMANCE

The photon-noise limit is approached when all other noise contributions are small compared to the noise associated with the incident photon flux. The photon-noise limit is thus the best possible condition. As shown in Fig. 6.4, both the signal and the background photons contribute photon noise.

6.6.1. Signal-Dependent Noise

The ideal situation is the case in which the signal photon noise is the dominant noise contribution (Seib and Aukerman, 1973). This condition is relatively rare. It can occur in the visible spectrum when using photomultipliers and in the infrared when using a cooled solid-state photomultiplier (Petroff et al., 1987). When the noise on the signal flux dominates we have a signal-dependent noise, and one can solve for the rms value of the fluctuation as

$$\phi_{q,\text{rms}} = \left[\frac{1}{\tau} \int_0^\tau \phi_q^2(t)\, dt \right]^{1/2} \tag{6.17}$$

Figure 6.5a shows the incident flux as a sinusoidal time signal. Substituting into Eq. (6.17), we find that

$$\phi_{q,\text{rms}} = \left[\frac{1}{\tau} \int_0^\tau \frac{\phi_q^2}{4} (1 + 2 \sin \omega t + \sin^2 \omega t)\, dt \right]^{1/2} \tag{6.18a}$$

$$\phi_{q,\text{rms}} = \frac{\phi_q}{2} \left[\frac{1}{\tau} \left(\tau - \frac{2}{\omega} + \frac{\tau}{2} + \frac{2}{\omega} \right) \right]^{1/2} \tag{6.18b}$$

$$\phi_{q,\text{rms}} = \frac{\phi_q}{2} \sqrt{\frac{3}{2}} \tag{6.18c}$$

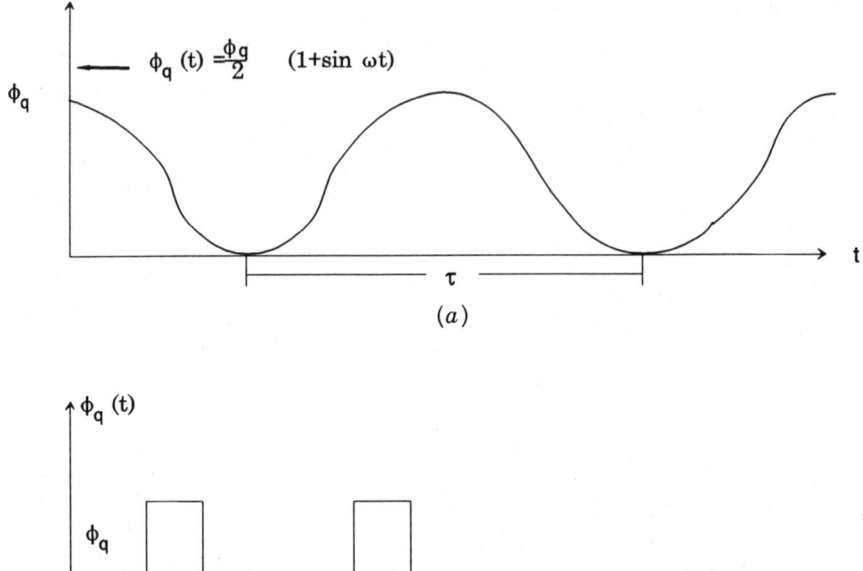

Figure 6.5. (a) Signal-flux sinusoidal variation in time; (b) signal-flux pulses in time.

The rms shot-noise current from an ideal photodiode is

$$i_s = \sqrt{2_q \bar{i} \, \Delta f} \tag{6.19}$$

where the average current is

$$\bar{i} = \eta_q \frac{\phi_q}{2} \tag{6.20}$$

which is the signal current produced by the detector for an average flux of $\phi_q/2$. Substituting Eq. (6.20) in Eq. (6.19)

$$i_s = \sqrt{q^2 \eta \phi_q \, \Delta f} \tag{6.21}$$

The photogenerated signal is $i_g = \eta q \phi_{q,\mathrm{rms}}$. The SNR is

$$\frac{i_g}{i_s} = \frac{q \eta \phi_{q,\mathrm{rms}}}{\sqrt{q^2 \eta \phi_q \, \Delta f}} \tag{6.22a}$$

Substituting for $\phi_{q,\text{rms}}$ from Eq. (6.18) and Eq. (6.22) for a SNR of 1:

$$1 = \frac{\eta \phi_{q,\text{rms}}}{\left[\eta \, \Delta f 2 \phi_{q,\text{rms}} \sqrt{\frac{2}{3}} \right]^{1/2}} \qquad (6.22b)$$

Squaring and simplifying,

$$\eta^2 \phi_{q,\text{rms}}^2 = \eta \, \Delta f 2 \phi_{q,\text{rms}} \sqrt{\frac{2}{3}} \qquad (6.23a)$$

$$\phi_{q,\text{rms}} = \frac{2 \Delta f}{\eta} \sqrt{\frac{2}{3}} \qquad (6.23b)$$

This noise-equivalent photon flux is required for SNR = 1. To find NEP from the noise-equivalent photon flux, multiply each photon by the energy of a photon hc/λ:

$$\phi_{e,\text{rms}} = \phi_{q,\text{rms}} \frac{hc}{\lambda} \qquad (6.24a)$$

$$\text{NEP} = \frac{hc \, \Delta f}{\eta \lambda} \sqrt{\frac{8}{3}} \qquad (6.24b)$$

which can be interpreted as the NEP ultimate limit for the case where the photon flux is the dominant noise. The NEP is on the order of magnitude of a single photon energy per measurement interval.

The corresponding $D*$ using Eq. (6.7) is

$$D* = \frac{\eta \lambda}{hc} \sqrt{\frac{3 A_d}{8 \, \Delta f}} \qquad (6.25)$$

Note that the $D*$ is proportional to quantum efficiency.

If we assume the photons are discrete quanta, as shown in Fig. 6.5b, the ultimate NEP is slightly different from Eq. (6.24)

$$\text{NEP} = 2\sqrt{2} \, \frac{\Delta f hc}{\eta \lambda} \qquad (6.26)$$

6.6.2. BLIP Performance

When the background photon flux is much larger than the signal flux and is the dominant noise source, we have the case called *BLIP: background-limited* *infrared photodetector.* We will derive an expression for the spectral D^* under BLIP operating conditions for a photovoltaic detector. Background-limited performance is commonly the case in infrared scanning systems and sometimes in staring cameras. We consider the photon irradiance from the signal source and background as $E_{q,\,\text{signal}}$ and $E_{q,\,\text{background}}$, respectively.

The corresponding signal-current and noise-current expressions can be developed. The rms noise current is developed assuming that shot noise in the detector is being caused by dc-photogenerated current $\bar{i}$ flowing across a potential barrier,

$$i_n = [2q\bar{i}\,\Delta f]^{1/2} \tag{6.27}$$

where the photogenerated current is caused by both background and signal photons

$$\bar{i} = \eta\,(E_{q,\,\text{background}} + E_{q,\,\text{signal}})A_d q \tag{6.28}$$

However, typically the background irradiance is much larger than the signal irradiance. Substituting Eq. (6.28) into Eq. (6.27), and assuming $E_{q,\,\text{background}} \gg E_{q,\,\text{signal}}$, we get for noise current, i_n,

$$i_n = [2q\,(\eta E_{q,\,\text{background}}A_d q)\,\Delta f]^{1/2} \tag{6.29}$$

An SNR can be written

$$\frac{S}{N} = \frac{\eta E_{q,\,\text{signal}}A_d q}{[2q^2\eta E_{q,\,\text{background}}A_d\,\Delta f]^{1/2}} \tag{6.30}$$

Setting the SNR equal to 1 determines the minimum-detectable signal irradiance $E_{q,\,\text{signal}}$:

$$E_{q,\,\text{signal}} = \left[\frac{2E_{q,\,\text{background}}\,\Delta f}{\eta A_d}\right]^{1/2} \tag{6.31}$$

Equation (6.31) can be interpreted as the noise-equivalent irradiance on the detector surface. From this expression we can obtain the spectral NEP, which is in units of watts. The signal power corresponding to this noise-equivalent irradiance is

$$\phi_{e,\,\text{signal}} = E_{q,\,\text{signal}}\frac{hc}{\lambda}A_d \tag{6.32}$$

If $E_{q,\,\text{signal}}$ is from Eq. (6.31), then $\phi_{e,\,\text{signal}}$ in Eq. (6.32) is just the spectral noise-equivalent power:

$$\text{NEP}(\lambda, f) = \left[\frac{2E_{q,\text{background}} A_d \, \Delta f}{\eta} \right]^{1/2} \frac{hc}{\lambda} \tag{6.33}$$

The spectral D^* can now be obtained from the definition of D^* in terms of NEP

$$D^*(\lambda, f) = \frac{\sqrt{A_d \, \Delta f}}{\text{NEP}(\lambda, f)} \tag{6.34}$$

Solving

$$D^*(\lambda, f) = \frac{\lambda}{hc} \sqrt{\frac{\eta}{2E_{q,\text{background}}}} \tag{6.35}$$

Figure 6.6 is a plot of $D^*(\lambda, f)$ versus wavelength. For the ideal detector shown, the wavelength of peak response (λ_p) and cutoff wavelength (λ_c) are the same wavelength. In real detectors, λ_p and λ_c are not the same. In general, the peak-response wavelength is shorter than the cutoff wavelength.

If we had analyzed a photoconductive detector instead of Eq. (6.27) for the shot noise of a photovoltaic detector, we would have used the expression for the rms current fluctuation caused by generation–recombination noise Eq. (5.83):

$$i_{\text{gr}} = 2qG[E_q A_d \eta \, \Delta f]^{1/2} \tag{6.36}$$

By substituting Eq. (6.36) into Eq. (6.4) for the NEP and Eq. (6.9) for the current responsivity, we obtain for $G = 1$ the spectral NEP for a photoconductor

$$\text{NEP}(\lambda, f) = 2 \left[\frac{E_{q,\text{background}} A_d \, \Delta f}{\eta} \right]^{1/2} \frac{hc}{\lambda} \tag{6.37}$$

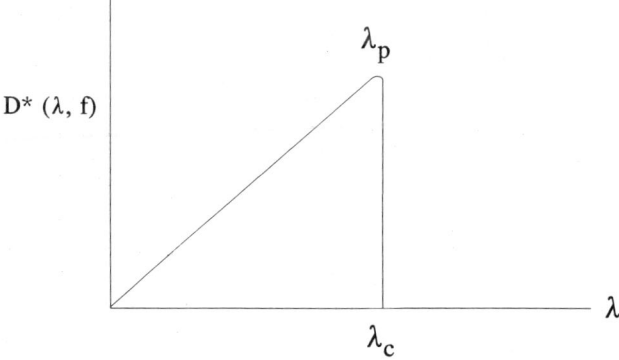

Figure 6.6. D^* vs. wavelength.

with a resulting spectral D^* of

$$D^*(\lambda, f) = \frac{\lambda}{2hc}\sqrt{\frac{\eta}{E_{q,\,\text{background}}}} \qquad (6.38)$$

The difference between the D^* expressions for a photovoltaic, Eq. (6.35), and a cooled photoconductor, Eq. (6.38), is caused by the additional recombination noise in a photoconductor. The variances of generation noise and recombination noise add, leading to an increase in the rms noise voltage by a factor of the square root of 2. For photoconductive detectors, the temperature of operation affects the D^* and NEP achieved under BLIP operation. For a cooled photoconductive detector, D^* is given by Eq. (6.38). However, if a photoconductor is uncooled (not the typical case, because cryogenic cooling is usually required), then the detector temperature $T_d = T_{\text{background}}$. In this case, photons are coming from 4π steradians, so an additional noise degradation occurs. For room-temperature photoconductors, the expression for D^*_{BLIP} for cooled photovoltaics is divided by 2. For an uncooled photovoltaic, the expression for D^*_{BLIP} for a cooled photovoltaic is divided by the square root of 2. To summarize:

$$D^*_{\text{BLIP}}(\lambda, f) = \frac{\lambda}{2hc}\sqrt{\frac{\eta}{E_{q,\,\text{background}}}} \qquad \text{(Cooled PC)} \qquad (6.39a)$$

$$D^*_{\text{BLIP}}(\lambda, f) = \frac{\lambda}{hc}\sqrt{\frac{\eta}{2E_{q,\,\text{background}}}} \qquad \text{(Cooled PV)} \qquad (6.39b)$$

$$D^*_{\text{BLIP}}(\lambda, f) = \frac{\lambda}{2hc}\sqrt{\frac{\eta}{2E_{q,\,\text{background}}}} \qquad \text{(Ambient Temperature PC)} \qquad (6.40a)$$

$$D^*_{\text{BLIP}}(\lambda, f) = \frac{\lambda}{2hc}\sqrt{\frac{\eta}{E_{q,\,\text{background}}}} \qquad \text{(Ambient Temperature PV)} \qquad (6.40b)$$

In each case, for BLIP operation, the D^* is inversely proportional to the square root of the background irradiance ($E_{q,\,\text{background}}$). By controlling background irradiance, we can improve the SNR achieved by a detector. The noise is reduced by decreasing the number of background photons incident on the detector, each of which contributes to either shot noise or generation–recombination noise. To calculate the background irradiance, we integrate over the entire spectral response of the detector, including all background-photon wavelengths to which the detector can respond. The background irradiance on the detector is controlled by the geometry of the detector and cold shield, as shown in Fig. 6.7. The detector views the outside scene (sources plus background)

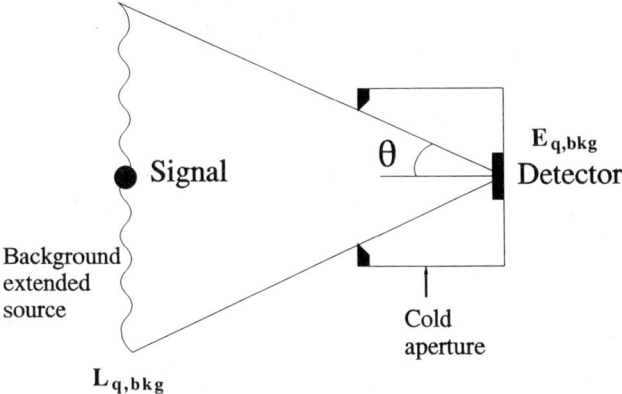

Figure 6.7. Detector enclosure limiting photon cone onto detector.

through a hole in the cold-shield structure that encloses the detector. Radiation reaches the detector within a cone that subtends a planar angle θ, shown in Fig. 6.7.

From the geometry of Fig. 6.7, the background irradiance, E_q, on the detector can be expressed as

$$E_{q,\text{background}} = \pi L_{q,\text{background}} (\text{NA})^2 \qquad (6.41a)$$

where NA is numerical aperture, $\text{NA} = \sin^2 \theta$, assuming that the surrounding medium is air:

$$E_{q,\text{background}} = \pi L_{q,\text{background}} \sin^2 \theta \qquad (6.41b)$$

The peak spectral $D^*(\lambda_p, f)$, Eq. (6.35) for photovoltaic detectors or Eq. (6.38) for photoconductive detectors, can now be evaluated for various cutoff wavelengths (assume $\lambda_c = \lambda_p$). The background irradiance on the detector, $E_{q,\text{background}}$, can also be expressed in terms of a background exitance [$M_q(\lambda, T)$] from the source as

$$E_{q,\text{background}} = \left\{ \int_0^{\lambda_p} M_q(\lambda, T_{\text{background}}) \, d\lambda \right\} \sin^2 \theta \qquad (6.42)$$

In the D^* equations, Eqs. (6.35), (6.38), (6.39), and (6.40), the background photon irradiance has been integrated over the spectral response of the detector as in Eq. (6.42).

Figure 6.8 shows a peak-spectral-D^*-versus-peak-cutoff-wavelength plot for ideal photovoltaic and photoconductor detectors, assuming that a 300-K background fills the field of view (assumed to be hemispherical) of the detectors. The dashed line (higher D^*) is for a photovoltaic detector that is evidence of

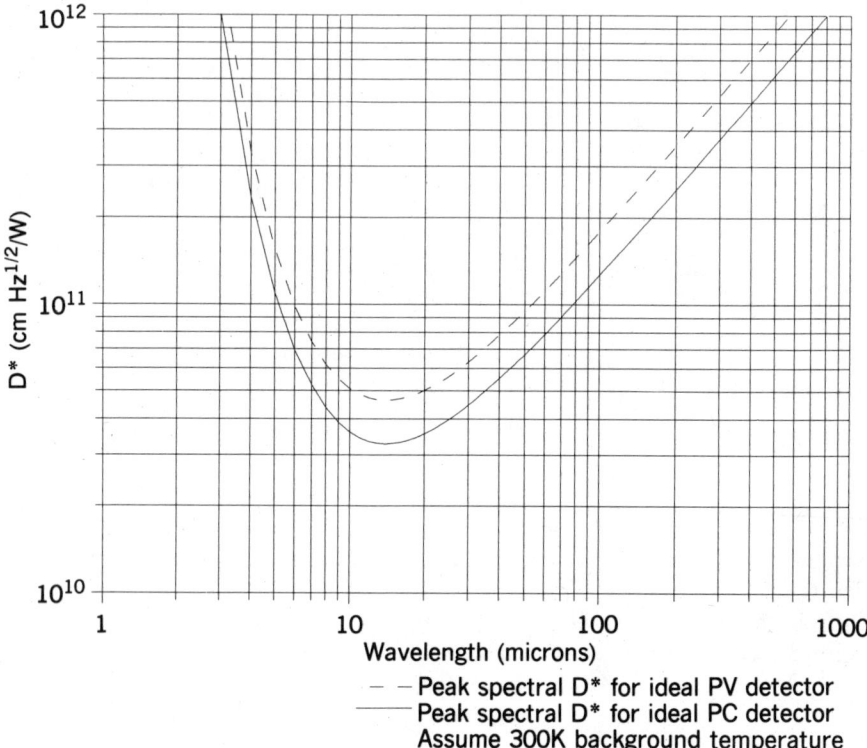

Figure 6.8.   Locus of peak spectral $D^*(\lambda_p, f)$ vs. peak wavelength (λ_p).

the smaller rms fluctuation associated with shot noise, as compared to the generation–recombination noise that occurs in photoconductors. Figure 6.9 plots the $D^*(\lambda, f)$ expression of Eq. (6.38), the photoconductive-detector case, with E_q modified to include the Bose factor. This shows the effect of Bose–Einstein statistics on D^*, compared to the Poisson-statistics approximation. Bose–Einstein statistics result in a clumping of photons at longer wavelengths (beyond 25 μm), which causes a reduction in D^* values at these long wavelengths (see Chap. 5, Sec. 5.2.2).

If the background temperature drops but still fills the field of view (assumed hemispherical) of 2π steradians, the curves for peak spectral-D^*-versus-cutoff-wavelength plots shift toward higher values, as shown in Fig. 6.10 for photovoltaic detectors (similar plots can be made for photoconductors). Figure 6.10 plots only the peak D^* value using Eq. (6.35) for background temperatures of 150, 200, 250, and 300 K. We assume here that the detector temperature is sufficiently low that self-emission processes do not radiate an appreciable amount of photons. However, at any nonzero temperature, the self-emission of the detector produces photon flux and hence photon noise. The highest temperature at which a detector can operate and remain within BLIP levels was discussed by Kinch (1989) and Jensen (1990). The results of their

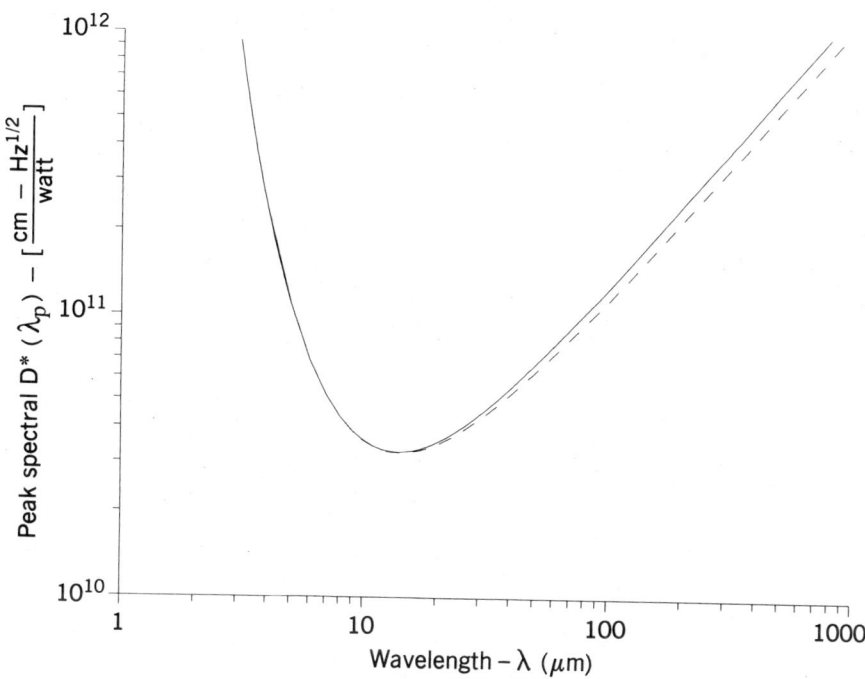

Figure 6.9. Peak spectral D^* vs. cutoff wavelength for a photoconductor without (solid line) and with (dashed line) the Bose factor.

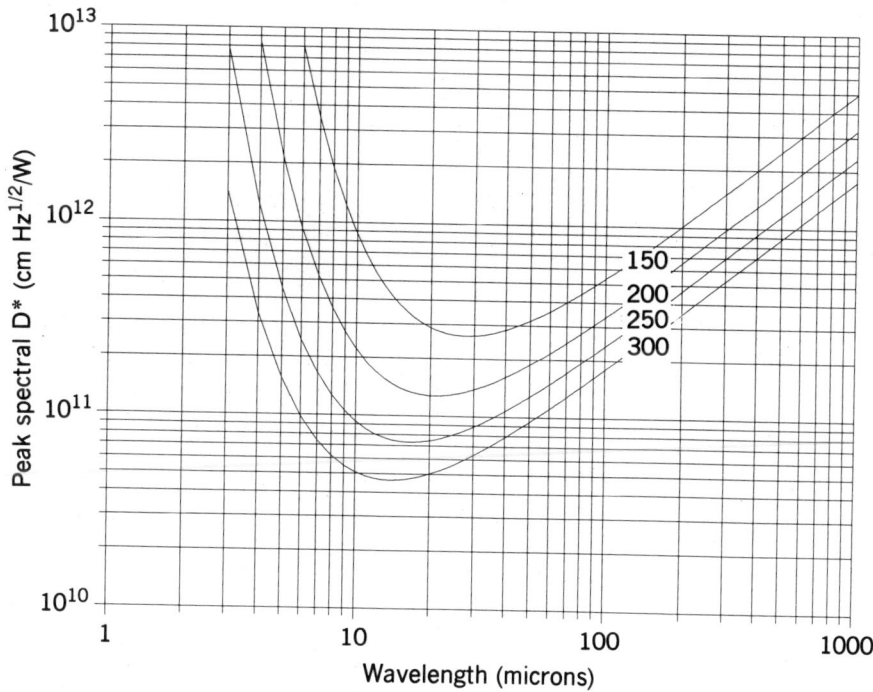

Figure 6.10. Peak spectral D^* vs. cutoff wavelength for various background temperatures with 2π steradians for a photovoltaic detector.

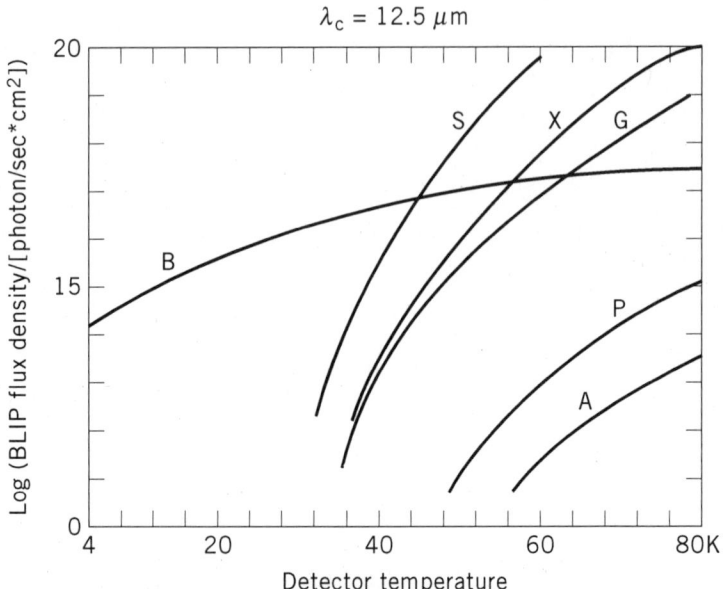

Figure 6.11. Log (BLIP flux density) vs. detector temperature, 12.5-μm cutoff wavelength. *Key:* A = absolute minimum; P = HgCdTe photovoltaic detector; C = GaAs/AlGaAs super-lattice; X = extrinsic silicon detector; S = superconducting detector; B = bolometric detector. (Originally appeared in A. S. Jensen, "Temperature Limitations of Infrared Detectors," *SPIE Proc.* **1308**, 1990. Reprinted by permission of SPIE.)

analysis of the self-radiative generation of carriers is shown in Fig. 6.11 for various detector materials, assuming a quantum efficiency of 1 and a wavelength-independent index of refraction of the detector material. These plots have great practical significance because they illustrate the performance limits for various detector technologies. If a particular D^* is desired for a given temperature of operation, it must be consistent with these limits if BLIP operation is to be achieved.

The spectral detectivity for any wavelength can be expressed, given the peak spectral D^*, assuming that the detector's responsivity (volt/watt) is linear with wavelength:

$$D^*(\lambda, f) = \left(\frac{\lambda}{\lambda_p}\right) D^*(\lambda_p, f) \tag{6.43}$$

The peak spectral $D^*(\lambda_p, f)$ is the maximum point on the D^*-versus-λ curve shown in Fig. 6.8. Equation (6.43) is often useful when finding the integrated response of a blackbody and the D^* expression, because it facilitates the multiplication of the spectral D^* and the Planck equation so that the product can be integrated.

6.6.3. D^{**}: D Double Star

As was shown by Eq. (6.41b) and Fig. 6.7, the background irradiance depends on the planar half-angle θ over which the detector views the background through the cold shield. Hence, the D^* expression for the BLIP case is also a function θ. To eliminate this dependence under BLIP operation, the D^{**} parameter was introduced. The D^{**} parameter only applies to the background-limited case where the background subtends 2θ linear angle as seen from the detector.

Substituting Eq. (6.43) in Eq. (6.35), the BLIP D^* is

$$D^*_{\text{BLIP}}(\lambda, f) = \left(\frac{\lambda}{hc}\right) \sqrt{\frac{\eta}{\left\{2\int_0^{\lambda_p} M_q(\lambda, T)\, d\lambda\right\} \sin^2 \theta}} \qquad (6.44)$$

Rearranging

$$D^*_{\text{BLIP}}(\lambda, f, \theta) = \left(\frac{1}{\sin \theta}\right) D^*_{\text{BLIP}}(\lambda, f, \pi) \qquad (6.45)$$

The definition of D^{**} is the D^*_{BLIP} corresponding to a hemispherical background $D^*_{\text{BLIP}}(\lambda, f, \pi)$ is

$$D^{**}(\lambda, f) = [\sin \theta] D^*_{\text{BLIP}}(\lambda, f) \qquad (6.46)$$

D^{**} has normalized the dependence on $\sin \theta$ ($D^{**} < D^*$). The figure of merit D^{**} allows a comparison of detectors normalized to hemispherical background, removing the θ dependence. This relation can also be expressed in terms of the solid angle Ω_{bkg} subtended by the background

$$D^{**} = D^* \sqrt{\frac{\Omega_{\text{bkg}}}{\pi}} \qquad (6.47)$$

Because Fig. 6.10 is a plot of peak spectral D^* for a hemispherical field of view, the D^*s plotted are also the peak spectral D^{**}s for the background temperatures 300 K, 250 K, 200 K, and 150 K.

6.6.4. Actual D^* Performance

In the previous section, we developed theoretical plots of spectral D^* versus wavelength for background-limited operation. In fact, real detectors only approach this result, because their quantum efficiency is not 100%. Figure 6.12 shows some D^* plots for different photoconductive detector materials (at given operation temperatures). The ratio of the peak spectral D^* of a real detector to

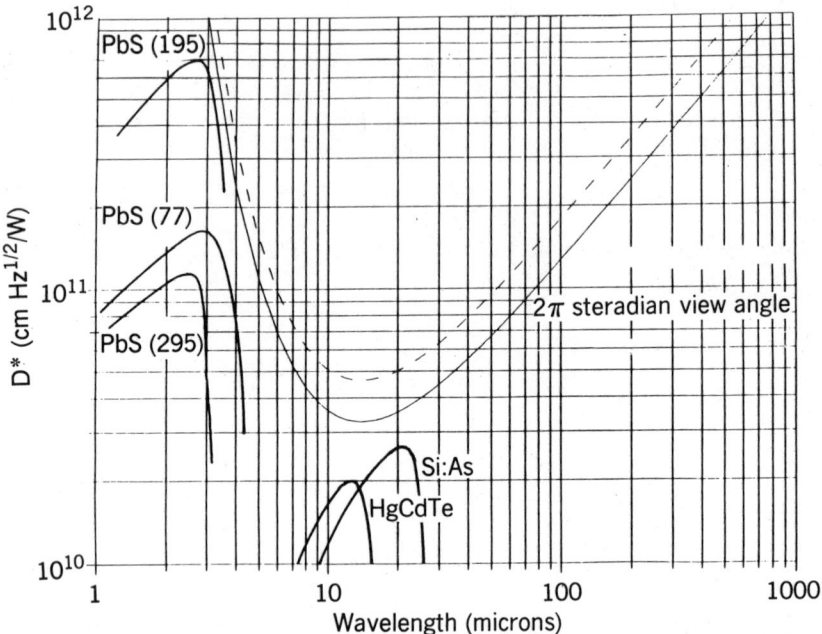

Figure 6.12. Spectral $D^*(\lambda, f)$ vs. wavelength for various photoconductors, compared to ideal peak spectral D^* for photoconductors (solid curve) and for photovoltaics (dashed curve).

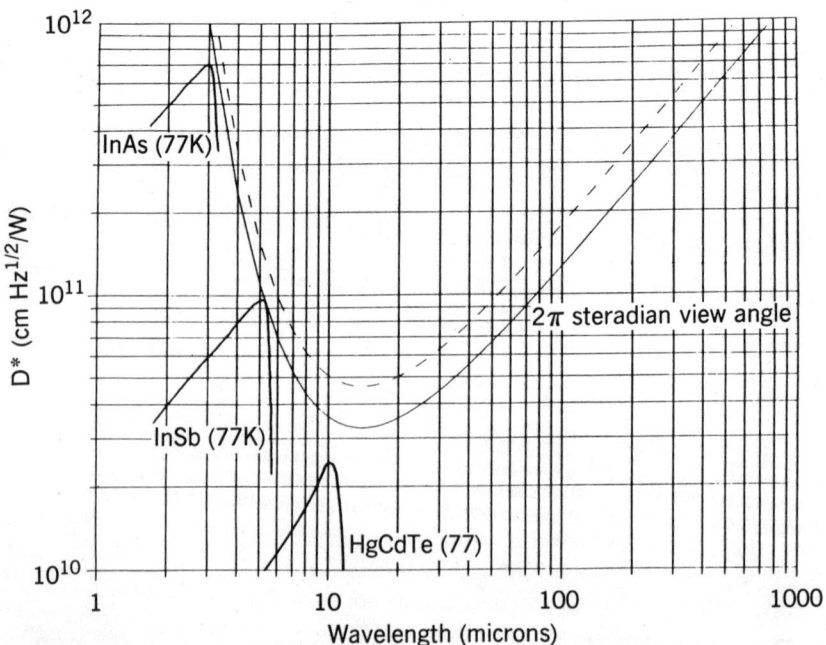

Figure 6.13. Spectral $D^*(\lambda, f)$ vs. wavelength for various photovoltaic detectors, compared to ideal peak spectral D^* for photoconductors (solid curve) and for photovoltaics (dashed curve).

the theoretical peak spectral $D*$ (at the same wavelength) is the square root of the quantum efficiency. For example, consider the HgCdTe detector with a peak spectral $D*$ of 2×10^{10} at 12 μm. As shown in Fig. 6.12, the theoretical limit is 3.35×10^{10}, which indicates a quantum efficiency of 36%. Figure 6.13 shows some real detector $D*$s for photovoltaic-detector materials.

6.7. JOHNSON-NOISE-LIMITED PERFORMANCE

For any system configuration, the best situation would be photon-noise limited, with signal-shot-noise limited being preferred to background-shot-noise limited. However, in situations where the photon flux reaching the detector is small, the photon noise of the signal or background is not the limiting detector noise. This occurs especially for deep space applications, where the system designer chooses cryogenically cooled filters and optics. Deep space has a temperature of about 2.7 K. In cases such as these, the photon-flux levels on the detector are very low and the best SNR is achieved when the Johnson noise of the detector is the largest noise contribution. Johnson-limited (JOLI) performance occurs when the photon (shot) noise has been reduced and is considerably less than the Johnson noise, and the Johnson noise dominates:

$$i_J^2 \gg i_s^2 \tag{6.48}$$

Recalling the equation for Johnson noise,

$$i_J = \left[\frac{4kT_d \, \Delta f}{R_d} \right]^{1/2} \tag{6.49}$$

where

T_d = detector temperature
R_d = detector resistance

The SNR of the photovoltaic detector for this condition is

$$\frac{S}{N} = \frac{q\eta\phi_{q,\text{signal}}}{\sqrt{\left[\dfrac{4kT_d \, \Delta f}{R_d} \right]}} \tag{6.50}$$

Solving for the noise-equivalent photon flux for an SNR = 1 is

$$\phi_{q,\text{signal}} = \frac{\sqrt{4kT_d \, \Delta f}}{q\eta \sqrt{R_d}} \tag{6.51}$$

Converting to NEP in watts,

$$\phi_{e,\,signal} = \frac{hc}{\lambda}\, \phi_{q,\,signal} = \frac{hc}{\lambda q \eta} \sqrt{\frac{4kT_d\,\Delta f}{R_d}} \qquad (6.52)$$

Substituting numerical values for the constants in Eq. (6.52) yields

$$NEP \cong 9 \times 10^{-12}\, \frac{1}{\lambda\,[\mu m]\eta} \sqrt{\frac{T_d[K]\,\Delta f\,[Hz]}{R_d[Ohm]}} \qquad (6.53)$$

Recalling that smaller is beter for NEP, as quantum efficiency, detector resistance, or wavelength increase, NEP gets better. By cooling the detector (T_d), the NEP also improves.

Substituting the NEP expression into Eq. (6.7), the spectral $D*$ for a photovoltaic diode that is Johnson-noise limited can be expressed as

$$D* = \frac{\lambda \eta q}{2hc} \sqrt{\frac{R_d A_d}{kT_d}} \qquad (6.54a)$$

Similarly, for a photoconductor,

$$D* = \frac{\lambda \eta q}{2hc}\, G \sqrt{\frac{R_d A_d}{kT_d}} \qquad (6.54b)$$

The two most important results that this analysis shows are

1. $D*$ is directly proportional to quantum efficiency (not square root as in BLIP)
2. $D*$ is proportional to the detector resistance-area product (RA product)

The second conclusion just cited is very important in photovoltaic detectors, because the RA product is an inherent characteristic of the detector material and fabrication process. More desirable materials have higher RA products, yielding a higher $D*$ and thus a higher output SNR for a given application. It is not possible to increase the detector area without reducing its resistance, leading to the observation that the $R_d A_d$ product is a constant for a given detector material. The optimum situation is a material with the largest RA value for a given cutoff wavelength (i.e., HgCdTe $\approx 10^7$ ohm-cm^2; InSb $\approx 10^{10}$ ohm-cm^2). Often, manufacturers will quote an RA product instead of $D*$ or NEP values.

Knowledge of the RA product provides the user with the range of back-

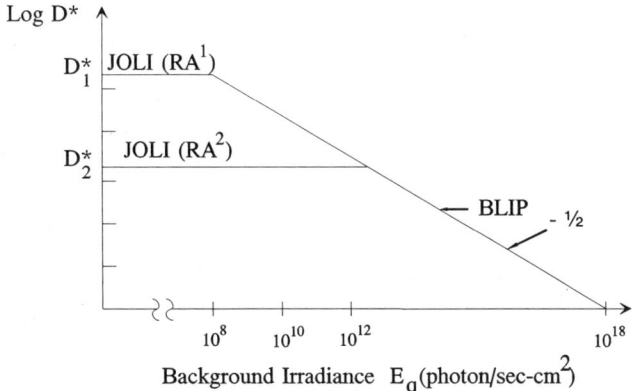

Figure 6.14. D^* vs. background irradiance on the detector.

ground photon flux levels over which the detector is either BLIP or JOLI. A plot of D^* versus background irradiance is shown in Fig. 6.14. If the detector is photon-noise limited (BLIP), the plot of D^* versus background irradiance follows Eq. (6.35) with a slope of negative one half ($-1/2$) on the log–log plot (Fig. 6.14). Consider two detectors, with RA products RA^1 and RA^2. Because the detector background irradiance is lowered (at approximately 10^{12} on the curve), the detector that has an RA product of RA^2 does not improve D^* with decreasing background irradiance (no improvement in D^* is possible above D_2^*). However, for the detector with RA product $RA^1 > RA^2$, the ultimate D^* obtainable is higher by a factor of 100 (square root of 10^4). The higher RA product allows D^* to continue to improve with decreasing background irradiance, down to the 10^8 point on the curve. The higher RA product puts the transition between BLIP and JOLI D^* performance at a lower background-irradiance level, increasing the ultimate D^* eventually obtained, according to Eq. (6.54).

6.8. TESTING DETECTORS AND DESCRIBING THEIR PERFORMANCE

We have reviewed and developed figures of merit common to all detectors. How do we measure these parameters? A schematic of a standard laboratory test set to measure detector performance is shown in Fig. 6.15. Shown in the figure are various components that require further discussion or description. The radiation source is a blackbody simulator. Assume $\varepsilon = 1$ and the temperature T to be one of the standards, either 500 K or 2800 K. The chopper blade is at room temperature. Assume the modulation is a 50% duty cycle, at f_0. The detector bias depends on the type of detector being tested. This is usually the variable the researcher uses to optimize performance. The calibration circuit

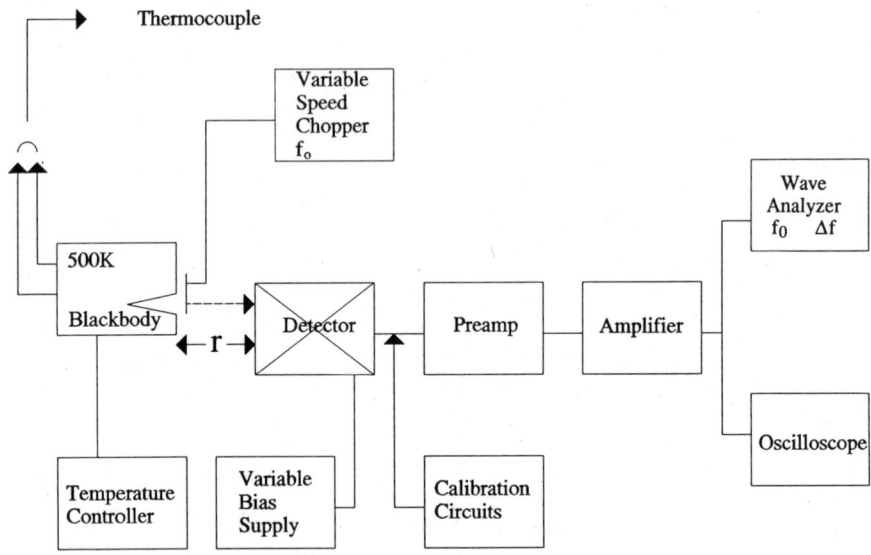

Figure 6.15.   Detector measurement setup.

is used to test the preamplifier, to verify noise level and frequency response. The preamplifier is typically the biggest problem, because it must provide signal amplification while introducing no appreciable noise above the detector noise level. The wave analyzer is an rms voltmeter over an electrical bandwidth Δf. Ideally you want to make true rms measurements. However, most meters are average reading with a scale factor adjustment, rather than true rms. The wave analyzer could also be replaced by a lock-in voltmeter that uses synchronous detection at the chopper frequency f_0.

Table 6.2 lists typical parameters that must be measured and recorded to evaluate the performance of a detector. The bias voltage is varied to obtain maximum SNR (maximum sensitivity). After this procedure has been completed, parameters relating to this best-performance condition are recorded. Now, we can calculate the rms power on the detector from Fig. 6.15 and Table 6.2:

$$\phi_{e,\,\mathrm{rms}} = (L_e A_s A_d F_F t)/r^2 \tag{6.55}$$

Table 6.2   Required Data for Detector Performance Evaluation

Bias Volt	Signal (rms)	Noise $\Delta f = 10$	Range	Blackbody Temperature	F/#	Frequency f_0	Room Temperature	Detector Area
90	3 mV	10 μV	500 cm	500 K	1	1000 Hz	300 K	1 mm^2

where

$$L_e = \frac{1}{\pi} [\sigma_e T^4_{source} - \sigma_e T^4_{background}]$$

T_{source} = blackbody temperature
$T_{background}$ = background temperature
A_s = blackbody source area
A_d = detector area
t = optical transmission between blackbody and detector
r = distance between blackbody and detector
F_F = form factor for conversion of peak-to-peak to rms values

The form factor F_F allows conversion of the peak-to-peak values of radiant power on the detector that arise from the mechanical chopper to the rms value of radiant power. The standard equipment seen in Fig. 6.15 only measures the rms voltage of the fundamental frequency (f_0), so we must find the rms of the fundamental frequency of the radiant signal. Recall the square waveform of 50% duty cycle. The square wave can be decomposed into its Fourier components (fundamental, third, fifth, etc.). Solving for the rms of the fundamental of the Fourier series yields 0.45 of the peak-to-peak value or

$$F_F = \frac{\sqrt{2}}{\pi} = 0.45 \qquad (6.56)$$

Equation (6.56) assumes that the blackbody-aperture dimension is much smaller than the aperture of the chopper, so that the radiant power signal produced is a good approximation to a square-wave signal (assuming a 50% duty-cycle chopper). If blackbody aperture just equals the chopper aperture, then a tri-angle-wave radiant signal is produced, and the rms of the fundamental is 0.286 ($F_F = 0.286$). The form factor results from the use of wave analyzer in the setup of Fig. 6.15, of bandwidth Δf around the center frequency f_0. If a wide bandwidth measurement were employed instead, the rms would be 0.707 of the peak-to-peak value of a square wave. Assuming an optical transmission of 90%, and substituting the other parameters from Table 6.2 into Eq. (6.55) yields

$$\phi_{e,rms} = \frac{9.8 \times 10^{-2}}{500^2} \frac{\pi (1)^2}{4} 10^{-2}(0.45)(0.9) \qquad (6.57a)$$

$$\phi_{e,rms} = 1.25 \times 10^{-9} \text{ watts} \qquad (6.57b)$$

This is flux from a blackbody, not spectral or monochromatic radiation. Therefore, blackbody-voltage responsivity is

$$\Re_v(500 \text{ K, } 1000 \text{ Hz}) = \frac{3.0 \times 10^{-3} \text{ volt}}{1.25 \times 10^{-9} \text{ watts}} = 2.4 \times 10^6 \text{ V/W} \qquad (6.58)$$

Using Eqs. (6.5) or (6.6), the blackbody NEP can be obtained:

$$\text{NEP} = \frac{v_n}{\Re_v} = \frac{\phi_{e,\text{rms}}}{S/N} \tag{6.59}$$

Substituting for noise and responsivity yields

$$\text{NEP}(500 \text{ K}, 1000 \text{ Hz}) = \frac{10.0 \times 10^{-6} \text{ V}}{2.4 \times 10^6 \text{ V/W}} = 4.17 \times 10^{-12} \text{ W} \tag{6.60}$$

Using Eq. (6.7), the corresponding expression for blackbody D^* is

$$D^*(T, f) = \frac{\sqrt{A_d \, \Delta f}}{\text{NEP}} = \frac{\sqrt{A_d \, \Delta f}}{\phi_{e,\text{rms}}} \frac{S}{N} \tag{6.61}$$

$$D^*(500 \text{ K}, 1000 \text{ Hz}) = \frac{\sqrt{10^{-2} \, 10}}{1.25 \times 10^{-9}} 300 = 7.59 \times 10^{10} \frac{\text{cm} \sqrt{\text{Hz}}}{\text{watt}} \tag{6.62}$$

The peak spectral $D^*(\lambda_c, f)$ can now be found from blackbody $D^*(T, f)$ using Eq. (6.16) for the conversion ratio, u:

$$D^*(\lambda_c, f) = \frac{\sigma T^4_{\text{source}}}{\dfrac{hc}{\lambda_c} \displaystyle\int_0^{\lambda_c} M_q(T_{\text{source}}, \lambda) \, d\lambda} D^*(T_{\text{source}}, f) \tag{6.63}$$

For this example, we now need to know additional information about the detector measurement conditions, such as

$$\lambda_c = 23 \text{ } \mu\text{m}; \text{ (Si:As photoconductor; see Fig. 6.16)}$$
$$u = 2.7 \text{ (conversion ratio, Figure 6.3)}$$
$$\text{Cold shield } \theta = 10°$$
$$\text{Background} = 300 \text{ K}$$
$$E_{q,\text{bkg}} = 6 \times 10^{16} \text{ photons/(s-cm}^2)$$

Therefore,

$$D^*(23 \text{ } \mu\text{m}, 1000) = 2.7(7.59 \times 10^{10})$$
$$= 2.05 \times 10^{11} (\text{cm Hz}^{1/2})/\text{W} \tag{6.64}$$

Is the detector operating under BLIP conditions? A convenient test is to put a spherical mirror in front of the detector, at a distance equal to the radius of curvature. This will image the detector onto itself. If the noise decreases when the detector views a cold background (the image of itself), then the detector

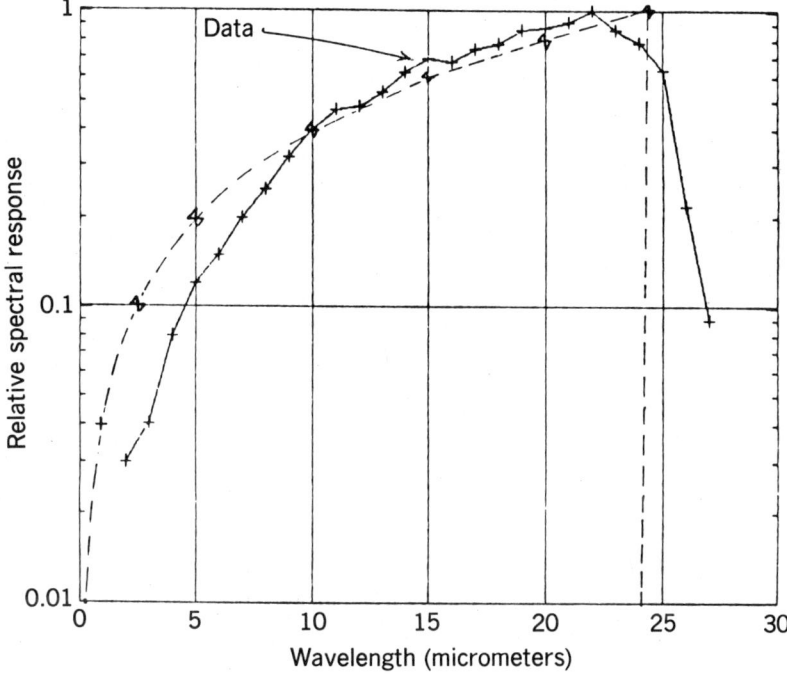

Figure 6.16. Relative spectral response of detector.

was originally BLIP. Assuming BLIP operation, the quantum efficiency can be calculated using Eq. (6.38) for a photoconductor [for a photovoltaic, we would use Eq. (6.35)]:

$$\eta = 75\% \qquad (6.65)$$

6.8.1. Relative Spectral-Response Tests

The setup needed to measure the spectral response of a detector is shown in Fig. 6.17. With a monochromator producing monochromatic radiation, the detector responsivity to spectral radiation can be compared to a broad-spectral-band thermal detector. The monochromatic radiation is simultaneously detected by both detectors using a beam splitter. The ratio of the two detector signals yields the relative spectral response of the detector under test. This result is a plot, similar to the one shown in Fig. 6.16. The ideal spectral responsivity for a photodetector (a triangle on a linear–linear plot) is shown as the dashed line. Figure 6.16 also illustrates the distinction between the peak-response wavelength (λ_p) and cutoff wavelength (λ_c). The cutoff wavelength is defined to be where the spectral responsivity of the detector falls below some arbitrary criterion. In general, the wavelength of peak response is lower than the cutoff wavelength.

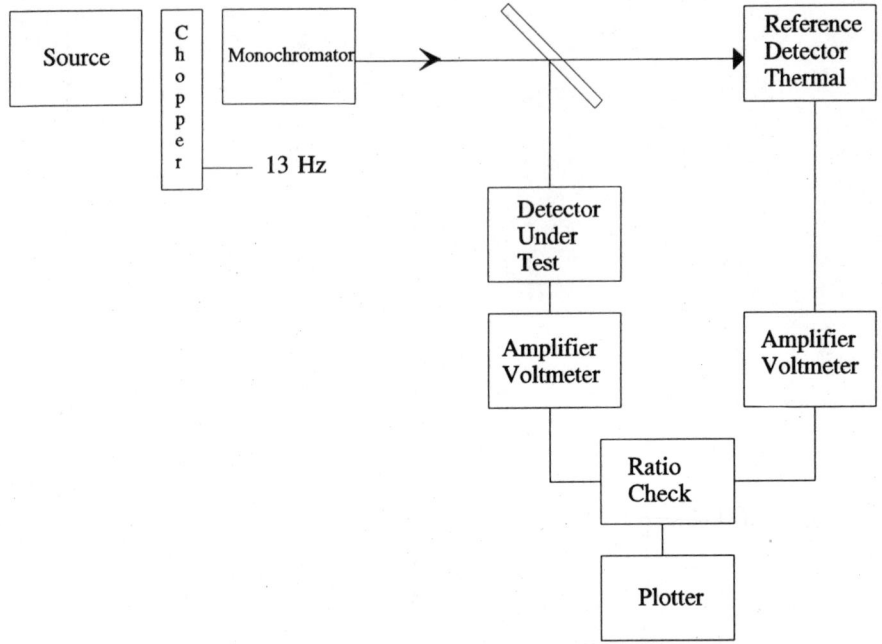

Figure 6.17. Measurement of spectral response.

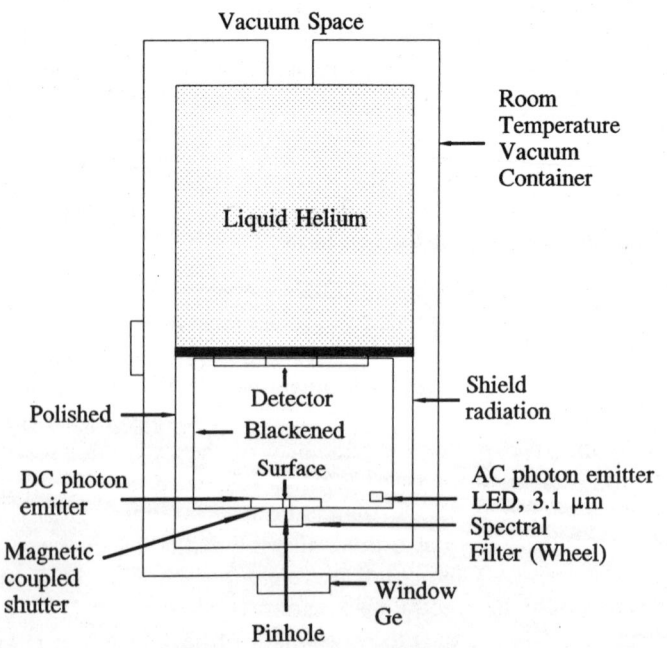

Figure 6.18. Cryogenic pinhole camera.

The other important parameter, discussed in Chapter 5, is the temporal response. Conceptually, to find the frequency response, the chopper frequency is increased to the frequency where the signal drops in amplitude versus frequency at 3 dB/octave. For photon detectors, this is not practical because the chopper blade cannot be rotated fast enough with conventional equipment. Therefore, a pulsed LED or laser is used to measure the time response. The pulse rise time is measured between the 10% and 90% points to calculate the frequency response (see Appendix D).

6.8.2. Low-Background-Flux Texts

Testing the performance of infrared detectors for their ultimate performance limits at very low background-irradiance levels requires cryogenically cooled pinhole-camera enclosures (Sayanagi, 1967). This is especially true for very-long-wavelength detectors because they respond to photons from very-low-temperature blackbodies. A typical cryogenically cooled pinhole-camera chamber is shown in Fig. 6.18.

REFERENCES

Kinch, M. A., "Katloy," *Appl. Phys. Lett.* **55** (20) (1989).

Jensen, A. S., "Temperature Limitations of Infrared Detectors," *SPIE Proc.* **1308**, 26 (1990).

Sayanagi, K., "Pinhole Imagery," *J. Opt. Soc. Am.* **57**, 1091 (1967).

Seib, D. H., and L. W. Aukerman, in *Advances in Electronic and Electron Physics* (L. Marton, ed.), Academic Press, New York, 1973, p. 212.

BIBLIOGRAPHY

Kruse, P. W., L. D. McGlauchlin, and R. B. McQuistan, *Elements of Infrared Technology: Generation, Transmission, and Detection*, Wiley, New York, 1962.

Petroff, M. D., M. G. Stapelbroek, and W. A. Kleinhans, "Detection of Individual 0.4–28 μm Wavelength Photons via Impurity-Impact Ionization in a Solid-State Photomultiplier," *App. Phys. Lett.* **51**, 406 (1987).

Wolfe, W. L., and G. J. Zissis, *The Infrared Handbook*, Optical Engineering Press, Bellingham, WA, 1990.

PROBLEMS

6.1. Compare and contrast each pair of quantities below, and write the defining equation for each. Define your symbols.

(a) responsivity and sensitivity

(b) responsive quantum efficiency and detective quantum efficiency

(c) transmission versus transmissivity

(d) absorptance versus absorptivity

6.2. Plot the current responsivity versus electrical frequency for a photo-conductor with the following characteristics:

$$\eta = 30\%$$

$$\lambda = 10 \ \mu m$$

$$G = 0.5$$

$$\tau = 10^{-7} \ s$$

What can you say about the noise spectrum ignoring $1/f$ noise?

6.3. Relate the figures of merit of responsivity, noise equivalent power (NEP), and detectivity (D^*). Equations are necessary.

6.4. The specific detectivity of an infrared detector is

$$D^* = \frac{\sqrt{A_d \ \Delta f}}{\phi_e} \frac{v_s}{v_n}$$

where A_d is the detector area, Δf is the effective noise bandwidth, v_s is the rms signal voltage, v_n is the rms noise voltage, and ϕ_e is the power on the detector.

(a) Compare the use of D^* as a figure of merit with NEP, quantum efficiency, and responsivity.

(b) Explain why and when D^* is proportional to $\sqrt{A_d}$ and $\sqrt{\Delta f}$.

(c) Discuss the use of D^* for photon and thermal detectors.

(d) Derive an expression for D^* for a photovoltaic detector when the limiting noise is shot noise.

6.5. You are testing a square detector, 1 mm on a side, with a resistance of 1 megohm. Your source is a HeNe laser (0.6328 μm) with a power of 0.5 μW. Your preamp readout is in terms of input current.

Step A: With the laser off and the detector in place, you measure 5 picoamps rms with a 100-Hz effective noise bandwidth filter.

Step B: You place a 1-megohm resistor in place of the detector and measure 2-picoamps rms.

Step C: You turn the laser on and use a 1000-Hz square-wave chopper to measure 0.23-$\mu amps$ rms from the detector.

Find the following detector parameters:

(a) responsivity $\Re$ (0.6328, 1000)

(b) noise-equivalent power NEP (0.6328, 1000)

(c) detectivity D^* (0.6328, 1000)

6.6. A photon detector that has a quantum efficiency of 0.5 from $\lambda = 0$ to 6 μm, has dimensions of 100×100 μm. It is placed without any optics 1 m from a 500-K blackbody source that has a 1-cm circular aperture. The light from the blackbody is 100% modulated sinusoidally at 50 Hz. Assume the chopper temperature to be 300 K. The detector output is 10-μV rms. When the signal (blackbody) is turned off, the detector rms output voltage is 1-nV rms. Both signal and noise are measured with a 9-Hz bandpass filter (equivalent noise bandwidth). What is the blackbody D^*?

6.7. What is the value of peak spectral D^* and D^{**} for a photoconductor and photovoltaic detector having a 40% quantum efficiency, cutoff wavelength of 25 μm, and photon-noise limited by a 196-K background (assume background subtends 2π steradian at the detector).

6.8. A 1-μm beam of radiation of wavelength 1.5 μm falls on an ideal germanium photodiode detector 1 mm $\times$ 1 mm $\times$ 1 mm operated at zero bias. What is the approximate quantum efficiency? What photocurrent is generated? If the noise-equivalent bandwidth is 10 kHz, estimate the noise current (consider only photon noise). What current would be generated under the same conditions in a silicon detector?

	BANDGAP	REFRACTIVE INDEX
Si	1.04 eV	3.4
Ge	0.63 eV	4.0

6.9. Consider the following 500-K blackbody (1-cm^2) signal source (see figure) as detected by a 1-mm^2 photovoltaic detector with a blackbody

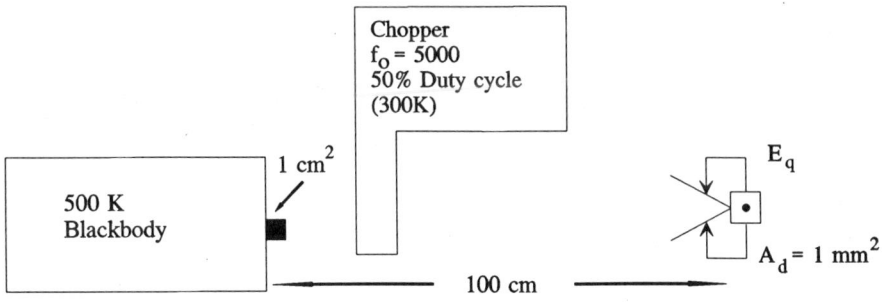

Figure P6.9

responsivity, $\mathcal{R}_i$ (500, 100), of 3.3 amp/W, a quantum efficiency (η) of 50%, and a spectral response from 0 to 20 μm. The detector temperature is 77 K, resistance is 1000 Ω, and the background photon irradiance in the spectral band on the detector is 10^{15} photon/s-cm^2 (cold shielded to view a 2° full field of view centered on the blackbody). The noise-equivalent bandwidth (Δf) for this setup is 100 Hz.

(a) What is the rms signal radiant power incident on the detector? (Assume 100% transmission.)

(b) What is the rms signal current out of the detector?

(c) What is the Johnson-noise current?

(d) What is the shot-noise current (with and without signal)?

(e) What is the SNR?

6.10. What is the analytical expression for the best NEP (λ, f) and $D^*(\lambda_p, f)$ for an ideal photovoltaic detector when the dominant noise is caused by the fluctuations in the radiant signal flux (ϕ_p), that is, signal photon-noise limited. Assume electrical bandwidth and quantum efficiency equal to 1.

6.11. A photovoltaic HgCdTe detector ($\lambda_c = 20$ μm) was measured in a standard blackbody laboratory setup with a full view angle of 10°. The following data were obtained for a *triangle* chopped waveform (refer to Table 6.3). What is the best SNR for 1 nanowatt of 3.39-μm HeNe-laser radiation on this detector in the configuration just cited, if the laser replaces the blackbody source?

6.12. What is the NEP of a bolometer (detects all λs) detector that measures an SNR of 10, $\Delta f = 1$, looking at a 500-K blackbody, and chopped with a 300-K, 50% duty cycle as shown in the figure?

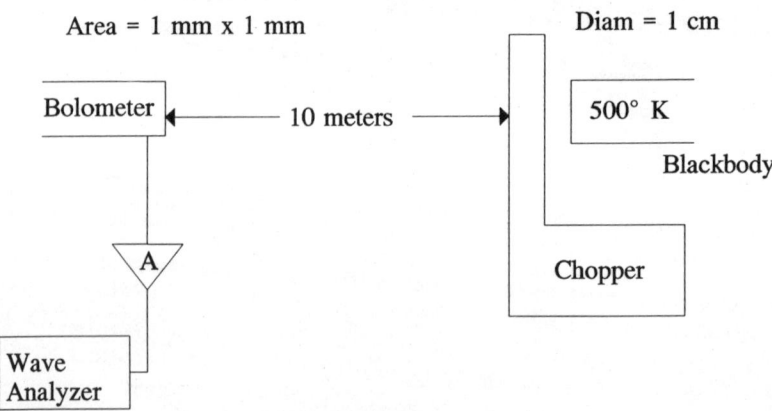

Figure P6.12

Table 6.3 Data Table for Problem 6.11

V_b Bias	Signal (mV)	Noise (μV)	R (cm)	Blackbody A_b (cm^2)	Temperature (K)	f_0 (Hz)	Δf (Hz)	Room Temperature (K)	Detector Area (mm^2)	Total Trans
10	1	1.5	150	1	500	1000	10	300	1	50%
20	2	1.5	150	1	500	1000	10	300	1	50%
30	3	1.5	150	1	500	1000	10	300	1	50%
40	4	1.9	150	1	500	1000	10	300	1	50%
50	5	2.5	150	1	500	1000	10	300	1	50%
60	6	3.0	150	1	500	1000	10	300	1	50%

6.13. Derive the expression for the background-limited NEP for a photovoltaic detector.

6.14. Derive the $D^*(\lambda, f)$ expression when BLIP (background-limited infrared photodetector) condition exists for a photovoltaic detector.

6.15. Typically, we assume that photons follow Poisson statistics; however, for some cases, they follow Bose–Einstein statistics. Plot a background-limited D^* (λ, f) versus wavelength for 300-K background temperature, but show approximately where in wavelength and how (+ or −) this effect causes peak spectral D^* to change for a photovoltaic.

6.16. Derive the expression for the background-limited NEP for a thermal detector that is radiation-noise limited. The noise power is given by

$$\overline{\Delta\phi_e^2} = 4kT_d^2 K\Delta f$$

where

k = Boltzmann's constant
K = thermal conductance
T_d = detector temperature

Assume the lead conductance is zero and the detector is in a vacuum (no conduction or convection); only radiative exchange takes place with the environment. Assume the detector temperature T_d is equal to the background temperature T_d and ϵ_b is the background emissivity.

6.17. Calculate the conversion factor to convert blackbody D^*, $D^*(T, f)$, to peak spectral $D^*(D^*(\lambda, f))$ for a detector with a cutoff wavelength of 25 μm, and measured using a 500-K blackbody.

6.18. What is the peak spectral D^* for a photovoltaic detector that has a spectral response from 0 to 10 μm with the following conditions?

Detector resistance: $R_d = 10^6\ \Omega$
Detector area: $A_d = 1\ mm^2$
Background irradiance: $E_q = 10^{10}$ photons/s-cm^2
Quantum efficiency: $\eta = 50\%$
Noise-equivalent bandwidth: $\Delta f = 10$ Hz
Detector temperature: $T = 77$ K
Blackbody responsivity: $\mathcal{R}_i (500, 500) = 0.5$ amp/watt

What could you do to improve upon this D^* value if you were only trying to detect 8–10-μm radiation?

6.19. Plot peak spectral D^*, that is, $D^*(\lambda_p, f)$ versus wavelength (1–100 μm) for BLIP performance for a photovoltaic detector for various background temperatures (T_B = 25 to 300 K in 25-K increments). What can be said about the minima of these curves?

6.20. The noise-equivalent power (NEP) of a detector whose noise is determined by the background flux can be derived for a photodetector (see Problem 6.13). A professor at the Optical Sciences Center, University of Arizona, has concluded that a better figure of merit for a photodetector would be noise-equivalent photon flux, NEQ, (photon/s). Derive an expression for a background-limited photodetector, NEQ. Discuss the significance of integration time for a charge-integrating photodetector. Can you come up with a more appropriate figure of merit for such a photodetector?

6.21. Discuss photon-noise-limited detection versus Johnson-noise-limited detection. When would each apply for a typical long-wavelength infrared photodetector (LWIR) detector?

6.22. A cooled detector with a spectral responsivity independent of wavelength and a diameter of 1 mm is located at the focal point of a 1-m-diameter $F/1$ paraboloidal telescope mirror (see figure). The mirror has a reflectance of 0.95. The detector is surrounded by a cold shield whose radiance is negligible and just allows the detector to see the full aperture of the mirror. The target is a star with a spectral distribution identical with that of a blackbody at a temperature of 10^4 K, a distance of 5×10^{13} km, and a radiant intensity of 2.5×10^{26} W/sr. Ignoring the intervening atmosphere, determine the temperature of the mirror if its

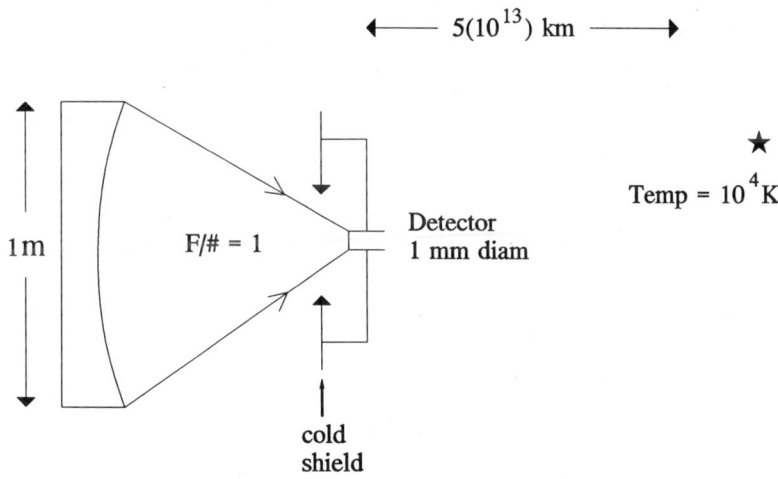

Figure P6.22

contribution to the detector output is 10% of the contribution from the target. State how you might further decrease the contribution of the mirror to the signal.

6.23. What is the effect of placing a hemisphere of transparent optical material of index n directly on a detector as shown in the figure? How much? What happens and how much change is there to the detectivity, D^*, if the detector is background limited? If it is limited by Johnson noise?

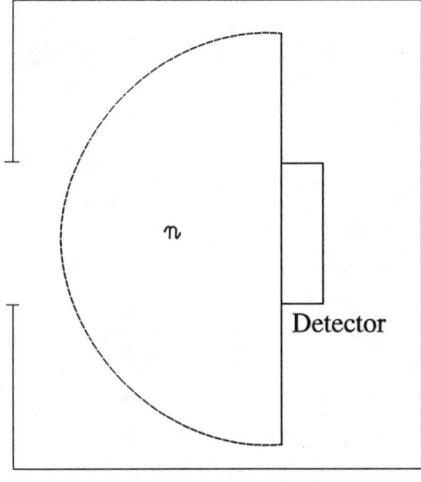

Figure P6.23

6.24. Derive the expression for D^*_{BLIP} for a background-limited photoconductive detector. When a detector that is background-limited for a 300-K background is operated under reduced background conditions (low E_q), the D^* changes as shown in the figure.

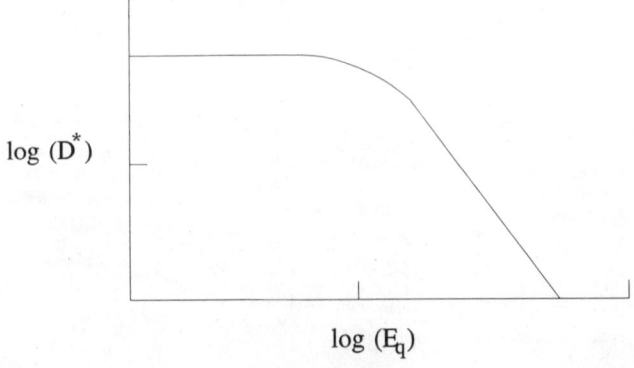

Figure P6.24

(a) Explain and label this curve.

(b) If a superconductor (resistance $= 0$) was used as an IR detector, assuming η is the same, how would this curve change and why?

6.25. Derive the D_λ^* expression in terms of RA product (resistance $\times$ area) for a photodetector. (Disregard $1/f$ and preamplifier noise.) State all assumptions. Discuss the importance of the RA product under conditions of fixed background noise.

6.26. The resistance-area product is often used as a measure of quality of a photovoltaic detector. ($RA = 10^7$ ohm-cm^2 for this detector; detector temperature, $T_d = 100$; quantum efficiency, $\eta = 1.0$; noise-equivalent bandwidth, $\Delta f = 1$.)

(a) What is the ultimate D^* for a 12-μm cutoff detector with the characteristics just given?

(b) How can the D^* in part (a) be achieved?

(c) About what value of photon irradiance (E_q) is this detector BLIP limited?

(d) If a photoconductor was used instead of a photovoltaic, how would parts (a), (b), (c) change? Assume a photoconductive gain of 0.5.

6.27. A 1-mm^2 ideal photovoltaic detector (quantum efficiency $= 100\%$) views a hemisphere 2π-steradian 300-K background. The detector material is a variable bandgap such that its cutoff wavelength response varies with detector temperature over a limited spectral range (0.1 to 10 μm).

$$\epsilon_g(\text{eV}) = 1.376 \times 10^{-3} \times T(\text{K})$$

The detector resistance is 350 Ω and the noise-equivalent bandwidth is 1000 Hz. What is the optimum temperature for maximum peak spectral D^*, $D^*(\lambda_c, f)$? What is the value for this peak spectral $D^*(\lambda_c, f)$ at this temperature?

6.28. Explain the relationship between blackbody responsivity, blackbody noise-equivalent power, and blackbody D^*. Use of equations in the explanation is appropriate. Blackbody $D^*(T, f)$ relates to peak spectral D-star [$D^*(\lambda_p, T, f)$] as derived; however, the equation below is somewhat different. Please relate the two expressions, if possible:

$$D^*(T, f) = \frac{\int_0^\infty D^*(\lambda, f)\phi_e(T, \lambda)\, d\lambda}{\int_0^\infty \phi_e(T, \lambda)\, d\lambda}$$

6.29. The BLIP $D*$ for a 15° view angle was 2×10^{11}. What is $D**$ for this detector?

6.30. A cooled photodetector has a $D**$ (10, 500) given to be 10^{10} at 10 μm. However, it was used with a cold-stop full view angle of 20° for a HeNe laser at 3.39 μm, chopped at 500 Hz. What is the expected $D*$ (3.39, 500) for this configuration?

6.31. Plot the peak spectral $D*$ versus wavelength peak (λ_p) for 3-K space background over the 1- to 1000-μm region. Why can't these values be realized from a practical detector?

6.32. Calculate the (spectral $D*$) (Planck equation) product for a detector with a cutoff wavelength (λ_c) of 14 μm, over the following spectral regions of a 500-K blackbody.
   (a) 0–14
   (b) 8–14
   (c) 3–5
   Assume linear responsivity and $D*$ (14, f) = 1.

6.33. Considering self-emission of the detector material from a photon-flux point of view, what is the absolute minimum temperature versus BLIP flux a detector will have if the cutoff wavelength is 14 μm?

6.34. A thermal detector has a signal incident on it of 5 μW chopped with a 50% duty cycle at 50 Hz. The detector responsivity is

$$\mathcal{R}_v(\lambda, f) = \frac{3 \times 10^6}{\sqrt{1 + [(1.333 \times 10^{-2})f]^2}}$$

What is the rms of the output signal? Explain why it is not 0.707 times peak-to-peak voltage.

6.35. Consider an optical system that forms an image of a point source. The system is diffraction limited and thus forms an irradiance profile of a first-order Bessel function squared, for example.
   (a) Plot a curve that shows that the signal varies as a function of detector radius. Assume a round detector for convenience.
   (b) Plot on the same graph a curve that shows how rms noise varies with detector radius (assume noise is caused by background photons).
   (c) What radius of the detector would provide an optimum SNR?

7

Photovoltaic Detectors

7.1. INTRODUCTION

Photovoltaic (PV) detectors, or photodiodes, are both the most common and the best detectors available. These detectors are made by building a *p-n* junction in a semiconductor. The photodiode is a special photon-sensitive diode, producing current or voltage output in response to optical-radiation input. In a photodiode, the optical radiation is absorbed at the junction of the two materials. The absorbed photons thus produce an electron–hole pair separated by the internal electric field producing an external current or voltage from the diode. Table 7.1 lists some examples of photodiodes and their expected energy gap, $\mathcal{E}_g$, and cutoff wavelength (λ_c).

We discuss general characteristics of photodiodes in the first part of this chapter, and then discuss the more important examples of photodiodes and their characteristics in engineering terms.

Before we can consider the physics of photodiodes, we must review some basic semiconductor theories and concepts to understand the required nomenclature for photodiodes. Figure 7.1 shows the thermal-equilibrium concentration in an intrinsic semiconductor.

The energy gap ($\mathcal{E}_g$) is only a slight function of temperature; however, in thermal equilibrium, the hole concentration (p) equals the number of electrons (n) and their product is the intrinsic carrier concentration squared:

$$n_i^2 = np = 4\left(\frac{2\pi kT}{h^2}\right)^3 \exp\left\{-\mathcal{E}_g/kT\right\}(m_h^* m_e^*)^{3/2} \qquad (7.1)$$

where m_h^* and m_e^* are the effective mass of holes and electrons, respectively.

Because the conductivity (σ_c) is proportional to n or p, σ_c increases as $\mathcal{E}_g$ decreases. Thus, the resistivity (ρ) is lower with a smaller energy gap, because of $\exp\left\{-\mathcal{E}_g/kT\right\}$ in Eq. (7.1). Low-leakage devices have a wide energy gap.

Recall that the Fermi level ($\mathcal{E}_f$) is an energy level where the probability of finding a filled energy state is 50%. Thus, an equal number of filled and vacant levels are both above and below the Fermi level. For an intrinsic semiconductor, the Fermi level is in the center of the forbidden energy gap, as shown in Fig. 7.1.

For an *n*-type semiconductor (e.g., silicon), a donor level lies in the forbid-

Table 7.1   Examples of Photovoltaic Detectors

Material	$\mathcal{E}_g$ (eV)	λ_c (μm)
GaAs	1.4	0.89
Si	1.12	1.1
Ge	0.68	1.82
InAs	0.29	4.2
InSb	0.21	5.9
$Hg_{1-x}Cd_xTe$	0.1	12.3
GaP	2.4	0.51
InAsSb	0.13	0.91

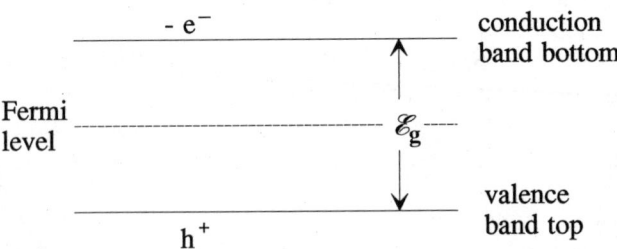

Figure 7.1.   Intrinsic semiconductor-band diagram.

den gap as shown in Fig. 7.2. The silicon material (valence 4) is doped with As, P, or Sb (valence 5). The Fermi level has now risen between the donor level and the conduction band, as shown in Fig. 7.2. The majority carriers are electrons and are mobile, whereas the holes are trapped. The concentration of donors in the silicon is denoted n_d (number/cm^3).

The p-type semiconductor (e.g., silicon) is made by doping the semiconductor with a material of valence 3, thus making an acceptor in the forbidden gap (Fig. 7.3). The acceptor ions lack an electron (valence $+3$, such as B, Al,

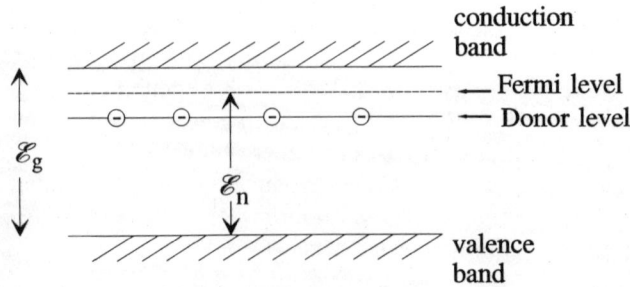

Figure 7.2. n-Doped silicon.

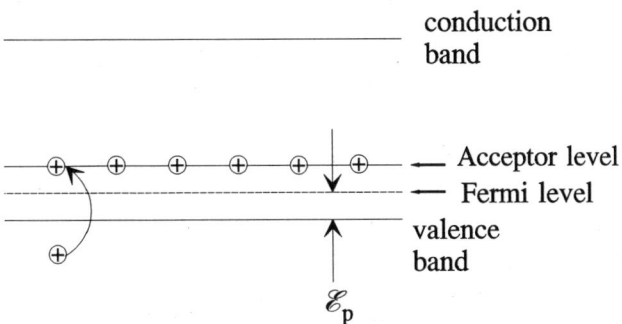

Figure 7.3. *p*-Doped silicon.

In, or Ga) in a silicon lattice. The acceptor doping concentration is represented in n_a (number/cm^3). Transition of an electron from the valence band to the acceptor level leaves ionized holes in the valence band that are mobile. Thus, the acceptor levels trap electrons.

7.2. *P-N* PHOTODIODE

A *p-n* photodiode is formed by joining a *p*-type and an *n*-type material. The junction between the two materials forms a diode, as shown in Fig. 7.4. Conceptually, when we join the two materials, a diffusion current flows across the junction immediately upon contact (Fig. 7.5). Diffusion currents ($J_{d,e}$ and $J_{d,h}$ as current densities) cause an electrical imbalance in the charge distribution because holes annihilate electrons, leaving a net charge. When a *p-n* junction is formed, some free electrons in the *n* region are attracted to the *p* region (drift current flow), and holes drift to the *n* region. The *n* region is left with a net positive charge, and the *p* region is left with a net negative charge, because both were initially (before contact) neutral (Fig. 7.5b). Therefore, a potential barrier is developed. The Fermi levels in each material align to the same energy, and, in Fig. 7.6, are represented with a dahsed line after steady state has been established.

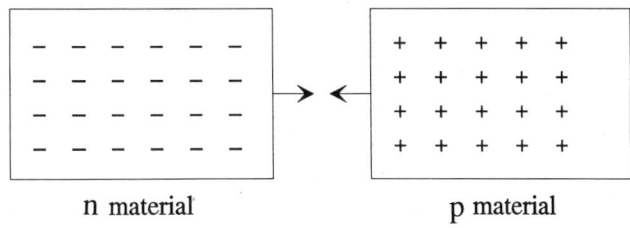

Figure 7.4. Two oppositely doped semiconductor substrates prior to contact.

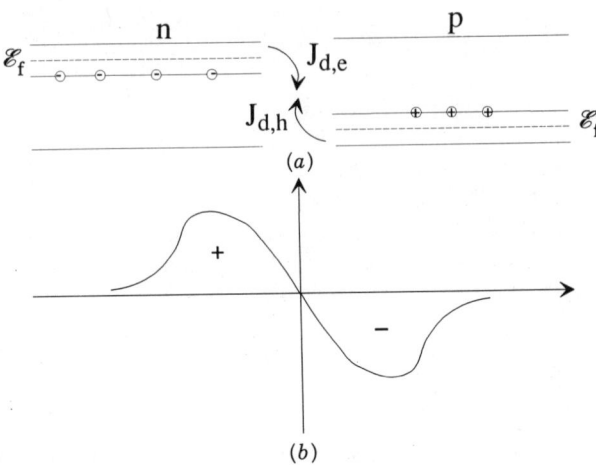

Figure 7.5. Carrier flow when two oppositely doped semiconductors are joined.

Figure 7.6 shows that the sum of diffusion (drift) and field currents must be zero for the holes:

$$J_{f,h} + J_{d,h} = 0 \qquad (7.2)$$

Substituting for drift and field currents:

$$q\mu_h \mathsf{p}(x)\mathsf{E} + qD_h \frac{d\mathsf{p}(x)}{dx} = 0 \qquad (7.3)$$

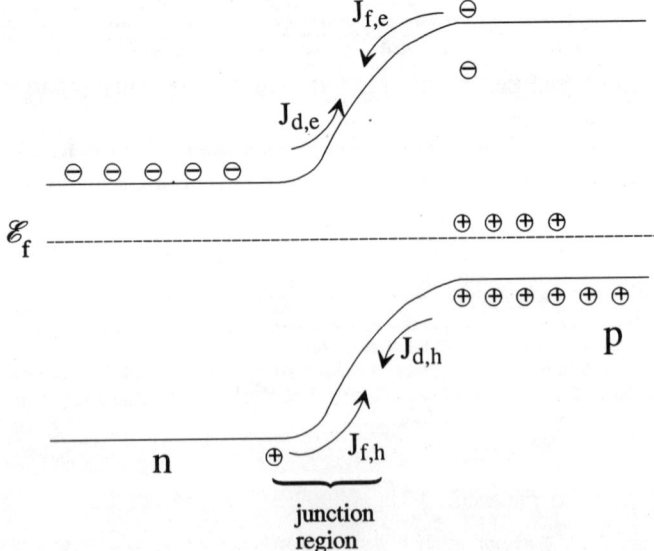

Figure 7.6. Energy-band diagram of a *p-n* junction.

where

$$E = \text{electric field } (d\theta_0(x)/dx, \text{ in transition region})$$
$$\theta_0 = \text{contact potential (volts)}$$
$$D_h = \text{hole diffusion constant } (-\mu_h kT/q)$$
$$p(x) = \text{hole concentration across transition region}$$
$$\mu_h = \text{hole mobility}$$

The electron drift and field currents can be used to develop an analog to Eq. (7.3) for electrons. Rearranging the differential equation and substituting for the electric field:

$$\frac{\mu_h}{D_h} \frac{d\theta_0(x)}{dx} = -\frac{1}{p(x)} \frac{dp(x)}{dx} \tag{7.4}$$

Using the Einstein relation,

$$\frac{\mu_h}{D_h} = \frac{q}{kT} \tag{7.5a}$$

and then substituting into Eq. (7.4), dropping the x dependence, $p = p(x)$, yields

$$\int_0^d d\theta_0(x) = \frac{kT}{q} \int_{p_n}^{p_p} \frac{dp}{p} \tag{7.5b}$$

where p_n is the hole concentration in the n-type side, p_p is the hole concentration in the p-type side, and d is the width of the depletion region. By integrating across the depletion region, W, the potential across it, θ_0 is found:

$$\theta_0 = \frac{kT}{q} \ln \left(\frac{p_p}{p_n} \right) \tag{7.6}$$

using the intrinsic carrier concentration, (Eq. 7.1.):

$$n_i^2 = np = n_n p_n \tag{7.7}$$

but n_n is the doping concentration of the n region. Recall

$$p_p = n_a \qquad \text{(which is the doping concentration in the } p \text{ region)}$$

and

$$n_n = n_d \qquad \text{(which is the doping concentration in the } n \text{ region)}$$

Thus,

$$n_i^2 = n_d p_n \qquad (7.8)$$

We substitute into Eq. (7.6) to obtain an expression for contact potential:

$$\theta_0 = \frac{kT}{q} \ln \left(\frac{n_a n_d}{n_i^2} \right) \qquad (7.9)$$

The contact potential is $\mathcal{E}_n - \mathcal{E}_p$, as defined in Figs. 7.2 and 7.3. Note that a diode can be used as a cryogenic-temperature sensor by measuring contact potential, because it is a linear function of temperature.

The transition region across the abrupt junction, depleted of holes and electrons (shown in Fig. 7.7a), is a region of interest in an optical detector. The width of the depletion region (depletion meaning the dopant levels are not occupied) is denoted by d. We can analyze and derive the depletion region, d, by the one-dimensional model for this junction and using Poisson's equation:

$$\nabla^2 \theta = -\rho_c / \epsilon \qquad (7.10)$$

where θ is the potential, ρ_c is the charge distribution, and ϵ is the dielectric constant.

Integrating the charge distribution from $-x_n$ to 0 or 0 to x_p shown in Fig. 7.7b yields the effective electric field $\mathbf{E}$:

$$\nabla \theta = -\mathbf{E} = \int -\frac{\rho_c}{\epsilon} dx \qquad (7.11a)$$

$$\mathbf{E} = \int_{x_n}^{0} \frac{q n_d}{\epsilon} dx = \frac{q n_d}{\epsilon} x_n \qquad (7.11b)$$

Because an equal charge exists on both sides of $x = 0$ in Fig. 7.7b,

$$\mathbf{E} = \frac{q n_d}{\epsilon} x_n = \frac{q n_a}{\epsilon} x_p$$

The contact potential can be found by integrating the electric field:

$$\nabla \theta = -\mathbf{E} \qquad (7.12a)$$

$$\theta = \int_{x_n}^{x_p} \mathbf{E} \, dx \qquad (7.12b)$$

Recalling that the integration can be interpreted as an area under the curve (in

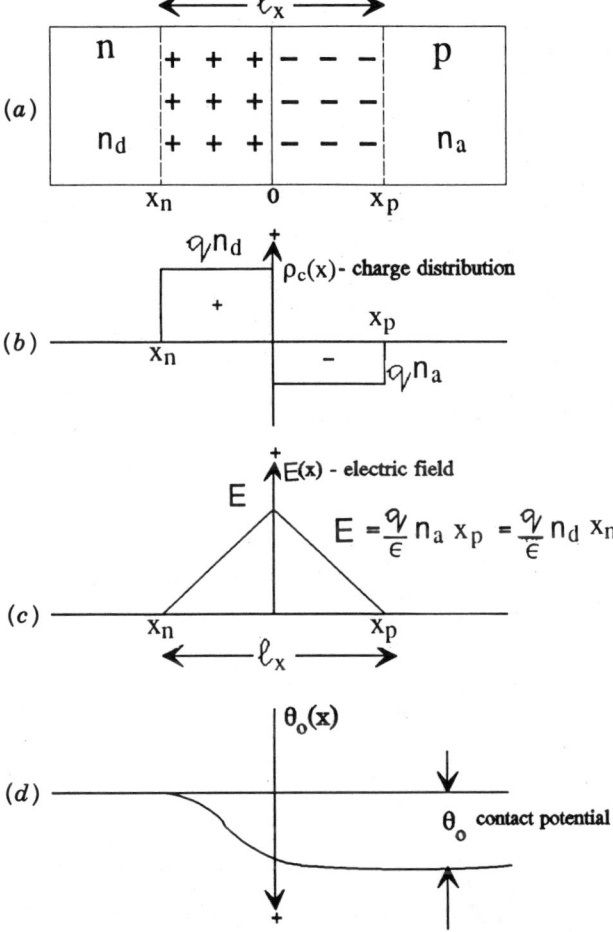

Figure 7.7. *p-n* Diode analysis.

this case a triangle), we can simplify the integration by using the area of a triangle as one-half the base times the height:

$$\theta_0 = \frac{1}{2} d \frac{q}{\epsilon} n_a x_p \qquad (7.13)$$

where d is the width of the depletion region, or

$$d = x_p + x_n \qquad (7.14)$$

and x_p and x_n are the distances from the center of the *p-n* junction to the edge of the depletion region on the p and n side, respectively. Recall that

$$n_a x_p = n_d x_n \qquad (7.15a)$$

$$d = x_p + \left(\frac{n_a}{n_d}\right) x_p \qquad (7.15b)$$

$$d = \left(\frac{n_d + n_a}{n_d}\right) x_p \qquad (7.15c)$$

Substituting for x_p into Eq. (7.13) for potential:

$$\theta_0 = \frac{1}{2} d \frac{q}{\epsilon} n_a \left(\frac{n_d}{n_d + n_a}\right) d = \frac{1}{2} \frac{q}{\epsilon} \left(\frac{n_a n_d}{n_d + n_a}\right) d^2 \qquad (7.16)$$

where the contact voltage θ_0 is a function of doping concentration and depletion width. Solving for depletion width:

$$d = \left[\frac{2\epsilon\theta}{q} \left(\frac{1}{n_a} + \frac{1}{n_d}\right)\right]^{1/2} \qquad (7.17)$$

where θ is the total bias, the sum of the self-induced contact potential θ_0 (Eq. (7.9)), and the applied bias voltage external to the diode (v_B):

$$\theta = \frac{kT}{q} \ln \left(\frac{n_a n_d}{n_i^2}\right) - v_B \qquad (7.18)$$

We can model the diode as a plane-parallel-plate capacitor where d is a function of voltage, in other words, a voltage-variable capacitor. Two concepts from this analysis are important with respect to photodiodes. First, the frequency response caused by the RC time constant is a function of voltage. Second, the speed of response depends on the velocity of carriers across the width of the junction.

7.2.1. P-N Junction under Bias Conditions

If the p-n junction is in equilibrium, the diffusion current equals drift-field current for the holes and electrons (in steady state):

$$J_{d,e} + J_{d,h} = J_{f,e} + J_{f,h} \qquad (7.19)$$

This equilibrium becomes unbalanced with the application of voltage (forward or reverse) or exposure to optical radiation. Figure 7.8a shows the applied voltage that causes the diode to be forward biased. The potential barrier

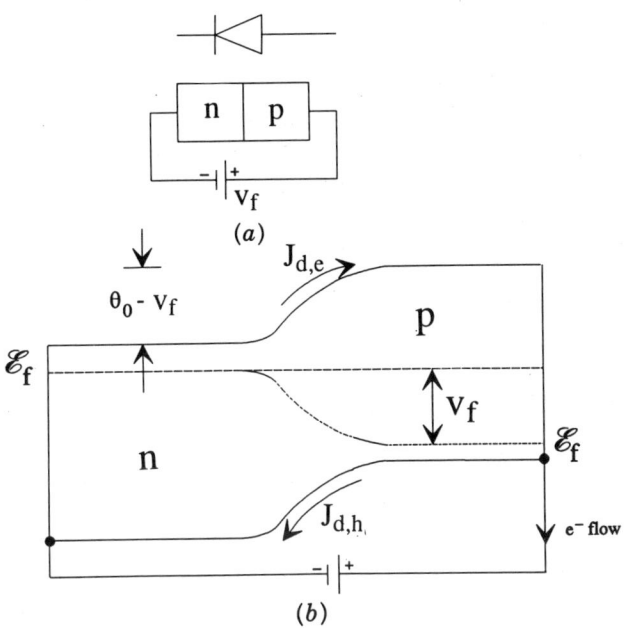

Figure 7.8.   Forward-biased diode.

shown in Fig. 7.8*b* is reduced by the applied voltage (v_f) from equilibrium so electrons and holes see a lower potential barrier, resulting in a higher current flow.

If a reverse-bias voltage is applied to the photodiode as shown in Fig. 7.9*a*, a very small current flows. Figure 7.9*b* shows that the potential barrier has increased by v_r. The reverse current is low because the minority carriers are controlled by a larger potential barrier. Tunneling can also occur between holes in the valence band and electrons in the conduction band laterally across the depletion region. The temperature dependence on reverse saturation current flow through the depletion region is very strong. For silicon, the current approximately doubles for every 10°C in temperature, as seen in Eq. (7.1).

7.2.2.   Optical Detection

Optical radiation incident on a *p-n* junction causes an external current to flow or a voltage to develop, depending on whether the diode is in a short-circuit or an open-circuit configuration, respectively. The photon must have sufficient energy to cause an electron to jump the entire energy gap of the substrate materials, because no dopant levels exist within the depletion region. The *p-n* junction must remain at equilibrium during exposure, but because no external current flows (assuming an open circuit), the barrier potential decreases, as shown in Fig. 7.10. An open-circuit voltage (v_{oc}) is generated across the diode.

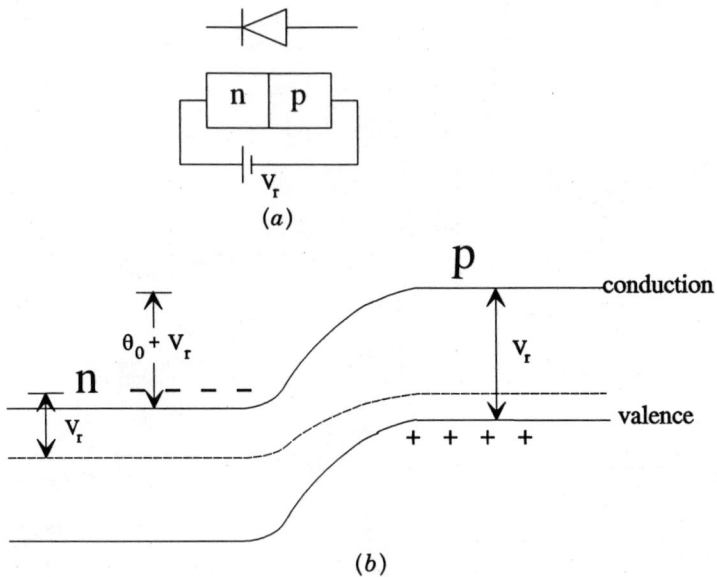

Figure 7.9.   Reverse-biased diode.

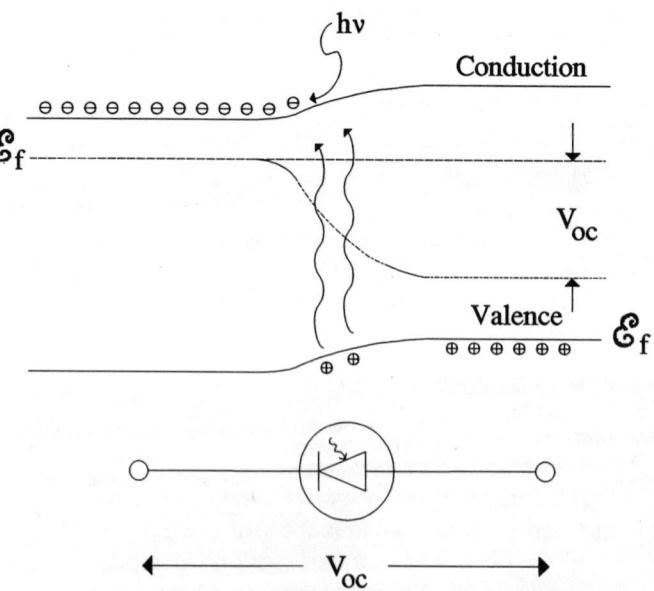

Figure 7.10.   Optical-radiation input on photodiode junction.

From an optical-detection point of view, the radiation must be absorbed at the junction. If the electron–hole pair is generated too far from junction, it recombines before being separated by the depletion field. Detection occurs at a region near the junction. Outside this region, recombination centers cause loss of the photo-generated carriers. Only photons detected within one diffusion length of the depletion region are separated, and therefore produce some detectable parameter change.

The photoactive region in a *p-n* diode acts as a detector, as shown in Figure 7.11*a*. The shaded region is where the photons are detected.

The depletion region appears as a narrow slit in cross section but, from before, is the optically active detector area (A_d). The width of the *p-n* junction (d), given in Eq. (7.17), is approximately 10^{-4} cm for silicon and germanium. The sensitivity to the light region extends beyond d by the diffusion length of the electrons (L_e) and holes (L_h) (see Fig. 7.11*b*). The diffusion lengths L_e and L_h have different values because of their respective mobility differences, so the profiles shown in Fig. 7.11*b* are asymmetric or skewed. The diffusion lengths for electrons and holes are

$$L_h = \sqrt{\frac{kT\mu_h\tau_h}{q}} \tag{7.19a}$$

$$L_e = \sqrt{\frac{kT\mu_e\tau_e}{q}} \tag{7.19b}$$

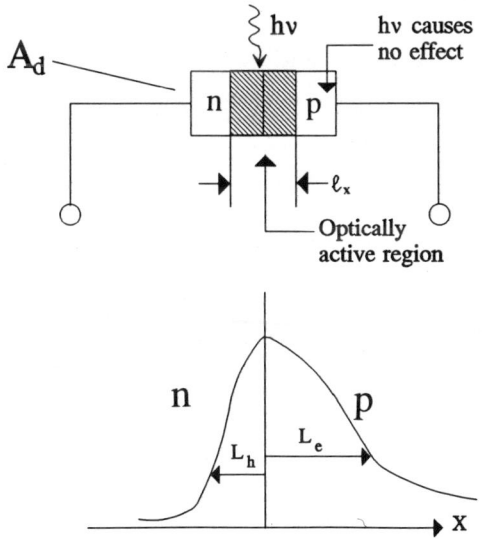

Figure 7.11. Photodiode optically active area and effects of diffusion lengths in the detection process.

where

μ_e = mobility of electrons
μ_h = mobility of holes
τ_e = carrier lifetime of electrons
τ_h = carrier lifetime of holes
k = Boltzmann's constant
T = temperature
q = charge of the electron

7.2.3. Photodiode *i-v* Curve

The continuity equation can be written for an abrupt *p-n* junction with concentration gradients and using Fick's law as

$$\frac{d\Delta p}{dt} = g + \frac{D_e}{q} + \frac{\partial^2 \Delta p}{\partial x^2} - \frac{\Delta p}{\tau_e} \qquad (7.20)$$

where

Δp = number/cm^3—change in holes above thermal equilibrium
D_e = diffusion constant for electrons $(-(kT\mu_e/q))$
L_e = diffusion length for electrons $\sqrt{D_e \tau_e}$
τ_e = carrier lifetime for electrons
g = rate of generation (number/cm^3-s)

Solving the continuity equation for steady-state conditions and no photo-generated carriers (i.e., $d\Delta p/dt = 0$ and $g = 0$), the diode equation for holes is derived as

$$i_h = \frac{q n_p D_e}{L_e} (e^{qv/kT} - 1) A_d \qquad (7.21)$$

We can add the electron contribution to the current after solving the continuity equation for electrons. This yields the diode equation

$$i = i_0 (e^{qv/kT} - 1) \qquad (7.22a)$$

where i_0 is the reverse saturation current:

$$i_0 = q \left(\frac{n_p D_e}{L_e} + \frac{p_n D_h}{L_h} \right) A_d \qquad (7.22b)$$

For silicon with a 10-mm^2 cross-sectional area, the reverse saturation current $i_0 \approx 10^{-8}$ amperes for a temperature of 300 K. This is the diode equation or

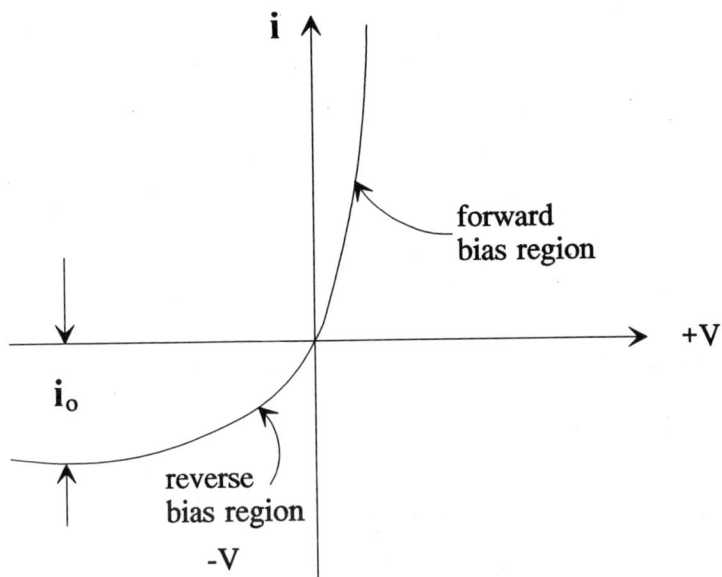

Figure 7.12.   Diode curve illustrating current flow vs. voltage bias for a *p-n* diode.

Shockley equation for a diode. An *i-v* plot of the current (i) versus voltage (V) is shown in Fig. 7.12.

Examining the diode curve near zero bias, we see that the curve does not follow the ideal exponential curve as derived. In fact, an additional term to correct for this empirical observation is the diode nonideality factor, β_d, which is put in the exponent of Eq. (7.22a):

$$i = i_0(e^{qv/\beta_d kT} - 1) \qquad (7.23)$$

The nonideality factor (β_d) lies between 1 and 2. When the diffusion current is the dominant current through the junction, β_d is 1, and when recombination current dominates, β_d is 2.

An electron–hole pair is produced whenthe incident photons have sufficient energy (i.e., $h\upsilon > \mathcal{E}_g$). If the pair are within a diffusion length of the *p-n*-junction depletion region, they are separated; the electron goes into the *p*-type region and the hole goes into the *n*-type region, causing current to flow. Recall that the photo-generated current (i_g) is

$$i_g = \eta q E_q A_d = \eta q E_e \frac{\lambda}{hc} A_d \qquad (7.24)$$

Adding this photo-generated current (i_g) to the diode equation, we get

$$i = i_0(e^{qv/\beta_d kT} - 1) - i_g \qquad (7.25)$$

Photodiodes are typically used in various electrical configurations that effectively make them operate with different desirable characteristics. The operating regions are depicted on the i-v curve as the various quadrants shown in Fig. 7.13. The photodiode operating in each quadrant has quite different optical and electrical characteristics, depending on its operating point. Four operating regions have special significance: (1) reverse bias in quadrant 3, shown in Fig. 7.13; (2) solar cell; (3) short-circuit operation; and (4) open-circuit operation.

7.2.3.1. Reverse Bias. Reverse-bias operation (negative voltage, see Fig. 7.12) widens the junction region that causes a decrease in capacitance ($C = \epsilon A/d$) so that the resistance–capacitance time constant decreases, or the detector response to a radiation pulse is faster. In addition, the minority carrier lifetime is shorter than that of the majority carriers, which enables quick response. Also, the increased electric field sweeps out carriers faster for high-frequency response.

7.2.3.2. Solar Cell. Operation in the first quadrant of the i-v curve corresponds to the forward-biased condition. This situation is not optimum for detector use; however, it will work. Because R_d (inverse slope of i-v curve) is small, the RA product is small. Also, the current is large, so the shot noise is high.

Figure 7.13 shows the effect of optical radiation on a photodiode. Fourth-quadrant operation is the solar-cell region, resulting in power generation. Both

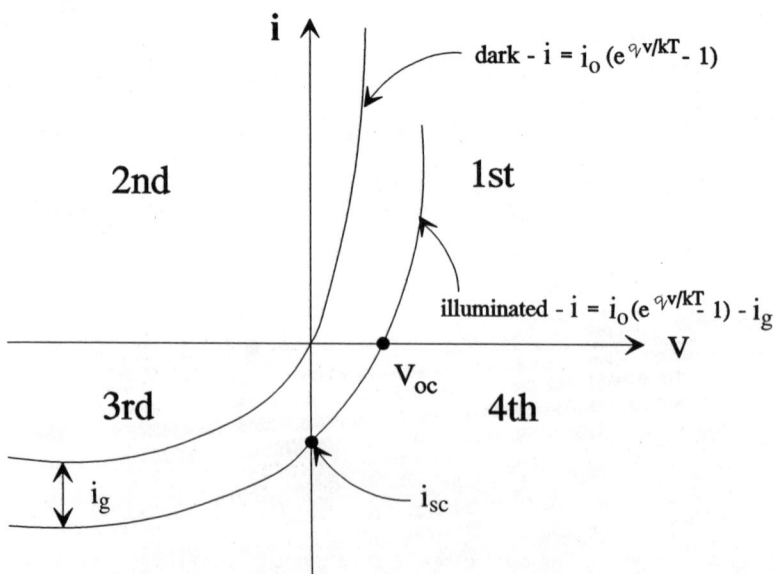

Figure 7.13. Photodiode curve with and without radiation.

a current and voltage are produced by the diode, generating power (the product of voltage and current). Two interesting regions of operation for an optical photodiode are the short-circuit current (i_{sc}) and the open-circuit voltage (v_{oc}), as delineated in Fig. 7.13.

7.2.3.3. Short-Circuit Current Operation. Under this operation, no voltage exists ($v = 0$) across the detector and all the photo-generated current is forced to flow into an electrical short circuit:

$$i = i_g = \eta q A_d E_q \tag{7.26}$$

The photo-generated current is linearly related to photon irradiance (E_q).

7.2.3.4. Open-Circuit Voltage Operation. This voltage is generated by optical radiation when no current is allowed to flow through the diode ($i = 0$):

$$v_{oc} = \frac{kT}{q} \ln \left\{ \frac{i_g}{i_0} + 1 \right\} \tag{7.27}$$

The open-circuit voltage varies with the logarithm of E_q as Eq. (7.27) indicates. Dynamic range compression of the signal exists when $i_g/i_0 \gg 1$, so a direct logarithmic response occurs.

7.2.4. Photodiode Resistance

The slope on the i-v curve determines the ac resistance, $\Delta v/\Delta i$. Mathematically, taking the partial derivative of the diode equation of Eq. (7.23), we obtain

$$R_d = \frac{1}{(\partial i/\partial v)} = \frac{\beta_d kT}{q i_0} e^{qv/\beta_d kT} \tag{7.28}$$

As the temperature of the photodiode is decreased, the reverse saturation current i_0 decreases. Thus the resistance is seen to increase with decreasing temperature. Note that the reverse saturation current is a function of the intrinsic carrier concentration.

Figure 7.14a shows the effect of the nonideality factor β_d. If all other parameters were equal, the diode curve shown with $\beta_d = 2$ would give a larger *RA* product than the diode with $\beta_d = 1$. Therefore the diode with the larger β_d should be the detector of choice. To better understand the influences that affect detector resistance, recall Eq. (7.21), the expression for reverse saturation current:

$$i_0 = q \left(\frac{n_p D_e}{L_e} + \frac{p_n D_h}{L_h} \right) A_d \tag{7.29}$$

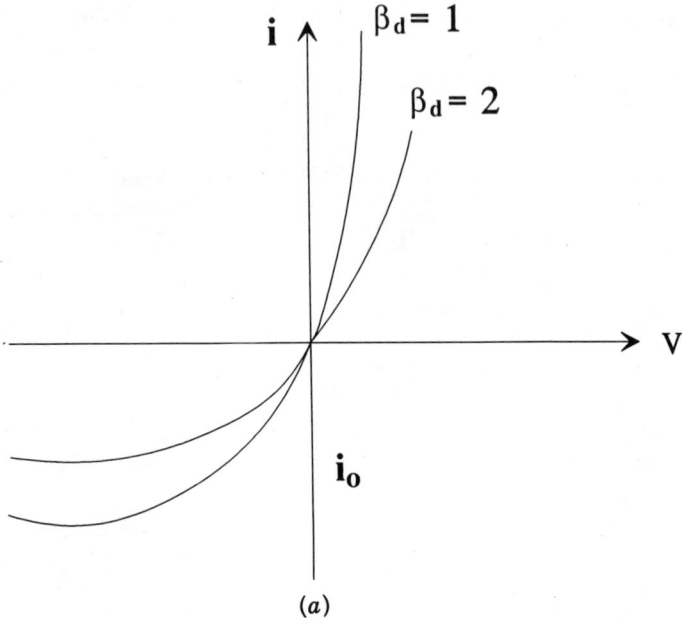

Figure 7.14. (a) Diodes with different nonideality factors.

Assuming that the second term in the parenthesis (corresponding to the hole influence) is small, the reverse saturation current, i_0, is

$$i_0 = q\left(\frac{n_p D_e}{L_e}\right) A_d \tag{7.30}$$

Recall the expressions for diffusion constant (D_e) and diffusion length (L_e) relationships:

$$L_e = \sqrt{D_e \tau_e} \tag{7.31a}$$

$$D_e = \frac{kT}{q}\mu_e = \frac{L_e^2}{\tau_e} \tag{7.31b}$$

The intrinsic carrier concentration can be related to doping concentration and minority density similar to Eq. (7.8) by

$$n_i^2 = n_p p_p \tag{7.32a}$$

$$n_p = \frac{n_i^2}{p_p} \tag{7.32b}$$

$$p_p = n_a \tag{7.32c}$$

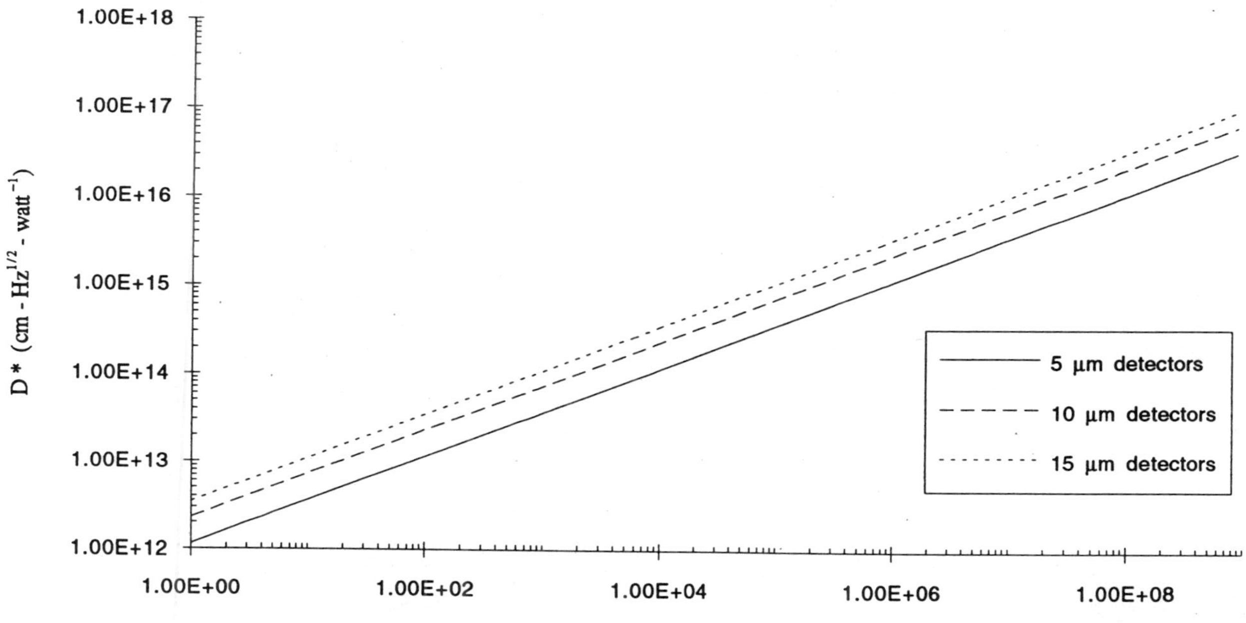
D * (cm - Hz^1/2 - watt^-1)
1.00E+18
1.00E+17
1.00E+16
1.00E+15
1.00E+14
1.00E+13
1.00E+12
1.00E+00
1.00E+02
1.00E+04
1.00E+06
1.00E+08
Resistance Area Product / Temperature (ohm - cm^2 / K)
5 μm detectors
10 μm detectors
15 μm detectors

(b)

Figure 7.14. (*b*) D^* vs. resistance area/temperature.

where n_p is the concentration of electrons in the p-type material, which can be written as

$$n_p = \frac{n_i^2}{n_a} \qquad (7.33)$$

Substituting into the reverse saturation Eq. (7.30) along with Eq. (7.31b) we obtain

$$i_0 = A_d q \frac{L_e^2}{\tau_e} \frac{n_i^2}{n_a} \left(\frac{1}{L_e}\right) \qquad (7.34)$$

which can be substituted into Eq. (7.28) for detector resistance. Now the resistance-area (RA_d) product can be found for the case of zero voltage (otherwise the exponential would necessarily be in the expression). The subscript indicates the voltage is zero ($v = 0$), $R_{d,0}$, written as R_0:

$$R_0 A_d = \frac{\beta_d k T n_a \tau_e}{q^2 L_e n_i^2} \qquad (7.35)$$

If $\beta_d = 1$ (ideal),

$$R_0 A_d = \frac{k T n_a \tau_e}{q^2 L_e n_i^2} \qquad (7.36)$$

where

τ_e = carrier lifetime
n_a = doping $\sim 4\,(10^{16})$
L_e = diffusion length (L_e is $\approx$ the thickness of the p-material)
kT/q^2 = a constant
n_i = intrinsic carrier concentration

We find that the $R_0 A_d$ product is proportional to the carrier lifetime and inversely proportional to the square of the intrinsic carrier concentration.

The resistance-area product is often used as a figure of merit of the detector material: the larger the RA is, the better. In Fig. 7.14b, we plotted D^* versus RA product divided by the detector temperature (RA/T) because the RA product is a function of detector temperature and larger RA products might be achieved by cooling to a lower temperature, causing a misleading conclusion. This presentation eliminates that confusion, and predicts the D^* for a given operating temperature. Figure 7.14b is for three of the detector cutoff wavelengths most commonly used. This prediction is theoretical, and should be taken as such, because D^*s of greater than 10^{14} are seldom realized.

7.2.5. Electrical Interface to Photodiodes

Photovoltaic detectors are interfaced to electronic circuits in several ways. Each configuration is application dependent. The circuit creates a load resistance (or impedance) with which the detector must work in concert. Maximum power will be transferred when the detector resistance equals the load resistance. Even for situations where an exact match is not required, it is generally advantageous to make the impedance of the detector and the impedance of the load approximately matched.

If the detector is used as a solar cell to produce power, the circuit shown in Fig. 7.15 is used. The diode is connected to a load resistance (representing the electrical system or component that requires the power for its operation) that dissipates the power generated by the detector. The voltage generated as a function of the optical flux is a nonlinear process. Using Kirchhoff's law for Fig. 7.15, the voltages around the loop must equal zero:

$$v_d - iR_L = 0 \qquad (7.37)$$

If we plot Eq. (7.37) on a typical *i-v* diode curve as shown in Fig. 7.16, we have a load line that goes through the origin, with a slope of $1/R_L$. From Eq. (7.37):

$$i/v_d = 1/R_L \qquad (7.38)$$

The power generated is determined by the area ($i \times v$). The larger the area, the greater the power generated. Note that the power generated is a function of the load resistor, and should be chosen to match the diode resistance at the operating point to maximize power.

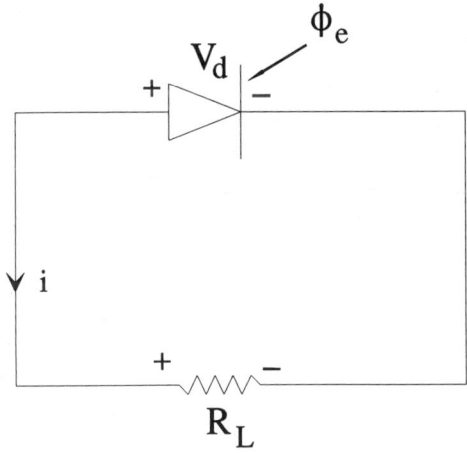

Figure 7.15. Power generator (V_d) and current (i).

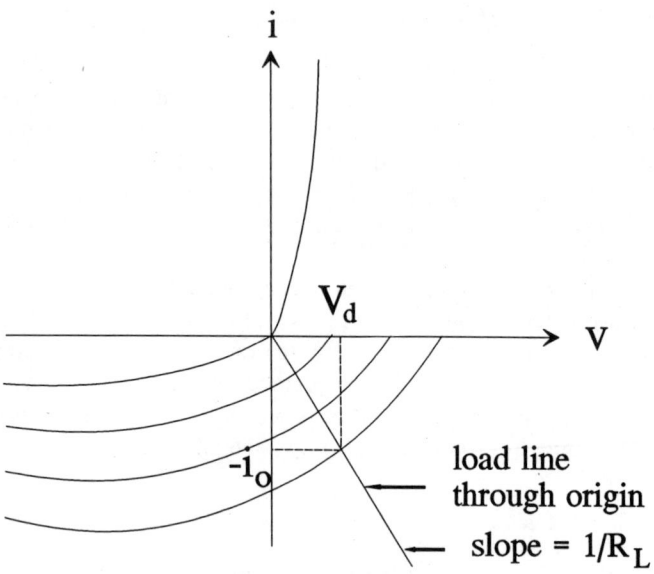

Figure 7.16.   Photodiode (solar cell) i-v curve for power generation.

Next, let us consider the reverse-biased PV illustrated in Fig. 7.17a. Again, using Kirchhoff's law for the loop voltage,

$$-v_B + v_d + iR_L = 0 \tag{7.39}$$

If this equation is plotted on the i-v diode curve, we have a load line as shown in Fig. 7.17b. The line has a slope of $1/R_L$, but the intercepts are at $-v_B$ and i. This is the same as the power-generation case, except the load line has been shifted to the left by the applied bias $(-v_B)$. If we rewrite Eq. (7.39) in a standard slope-intercept form,

$$i = \frac{v_d}{R_L} - \frac{v_B}{R_L} \tag{7.40}$$

where the slope is $1/R_L$ and its current-axis intercept is $i = -V_b/R_L$.

Consider a reverse-biased photodiode ($v_b = 20$ V) with a load resistor ($R_L = 100$ kΩ) as shown in Fig. 7.18a. To obtain the responsivity from the curves shown in Fig. 7.18b we must (1) draw the load line; (2) the mark the intercepts of the diode curves corresponding to optical fluxes, ϕ_1, ϕ_2, and ϕ_3; and (3) read the voltage output for a given optical-flux input. Voltage responsivity, $\Re_v$, from Fig. 7.18b is

$$\Re_v = \frac{\Delta v}{\Delta \phi} = \frac{(16 - 6)\ \text{V}}{(300 - 100)\ \mu\text{W}} = 10^5\ \text{V/W} \tag{7.41}$$

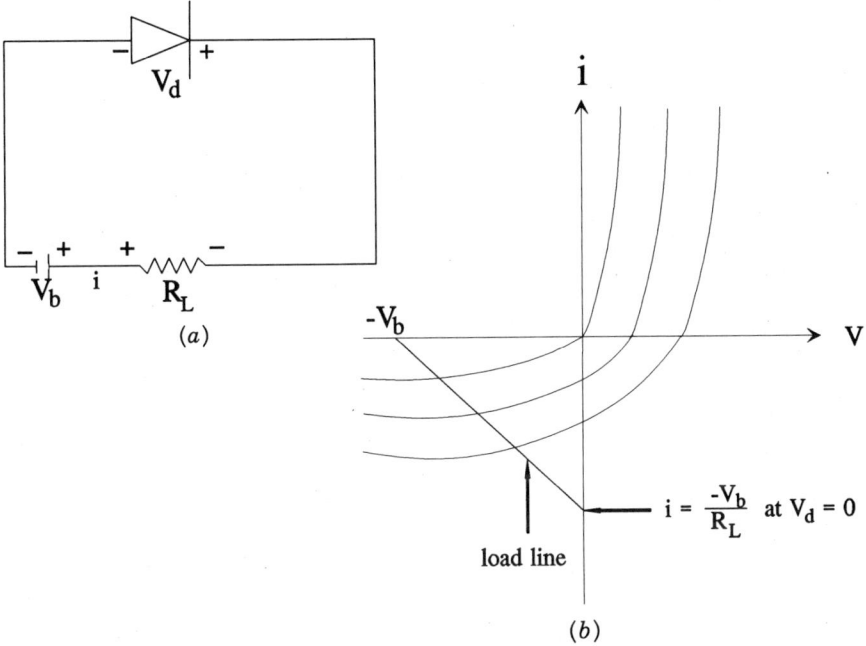

Figure 7.17.   Reverse-biased diode: (*a*) circuit configuration; (*b*) diode curves with load line.

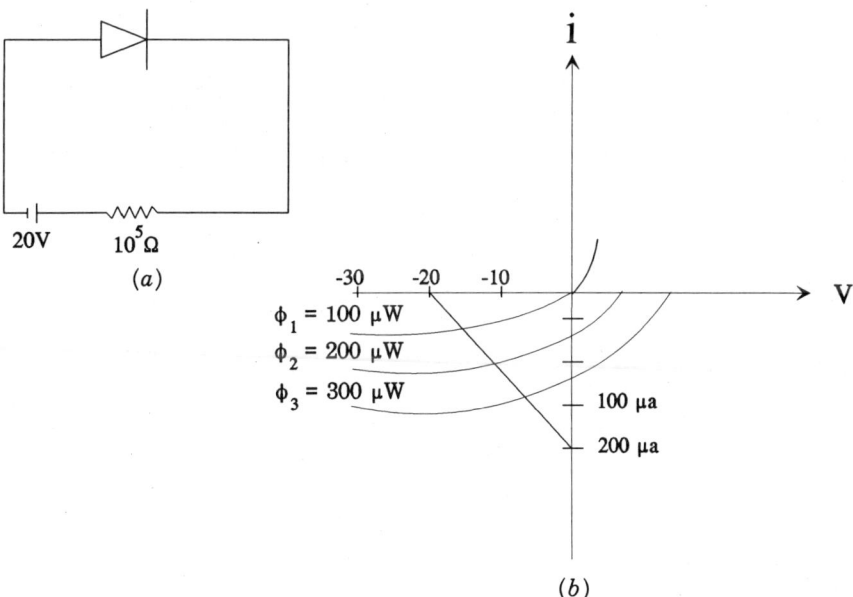

Figure 7.18.   Reverse-biased photodiode: (*a*) circuit; (*b*) diode curves for various amounts of incident radiation.

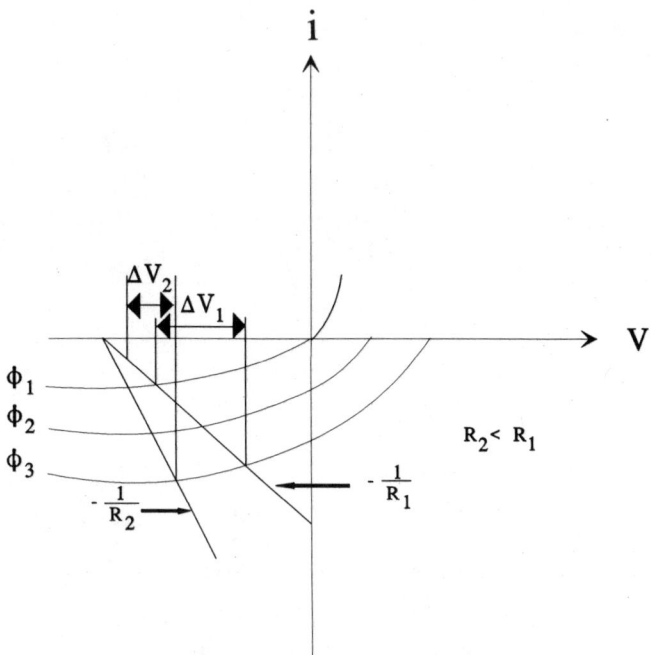

Figure 7.19.   Effect of load resistance on voltage responsivity.

Note that the voltage responsivity is a function of the load-line slope and the load resistance, R_L.

Figure 7.19 illustrates the effect of the load resistor for exactly the same detector, thus giving two different values of voltage responsivity for different load resistors:

$$\mathcal{R}_{v,1} = \frac{\Delta v_1}{\phi_3 - \phi_1} \tag{7.42a}$$

and

$$\mathcal{R}_{v,2} = \frac{\Delta v_2}{\phi_3 - \phi_1} \tag{7.42b}$$

where the value of load resistor R_1 is greater than R_2 ($R_2 < R_1$).

The two limiting cases for the load-resistance value are zero and infinity ($R \to 0$, $R \to \infty$). The load line in these two limiting cases would be the abscissa and ordinate, respectively, for the standard diode i-v axes for short-circuit or open-circuit operation. These limiting cases can be approximately created by the use of an operational amplifier (opamp). A typical electronic preamplifier using an operational amplifier to interface a PV detector as shown in Fig. 7.20,

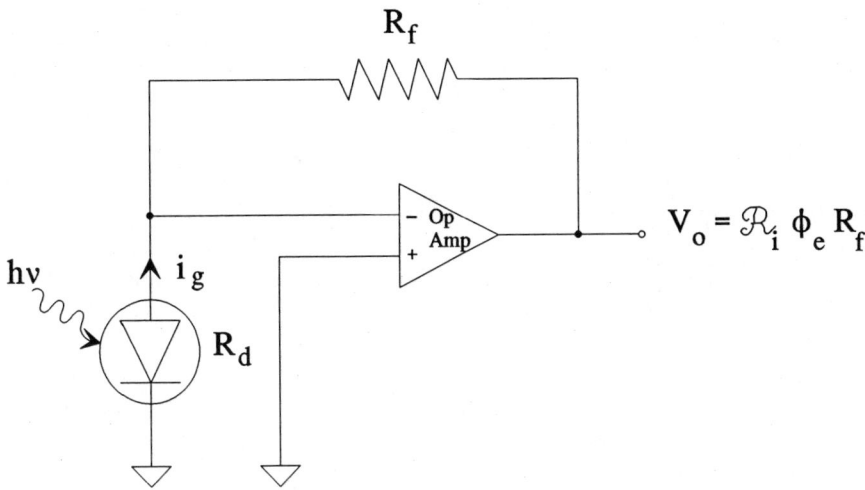

Figure 7.20.   Transimpedance amplifier.

where R_f is the value of the feedback resistor. An ideal operational-amplifier symbol and its circuit equivalent are shown in Fig. 7.21. An opamp in Fig. 7.21 is a differential amplifier with gain, and its input–output characteristic is written as

$$v_o = -G_{OL} v_i \qquad (7.43)$$

where

   v_i = input voltage
   v_o = output voltage
   G_{OL} = open-loop gain

Ideally, the operational amplifier exhibits no temperature effects, and adds no noise of its own to the amplified signal. Other ideal properties and character-

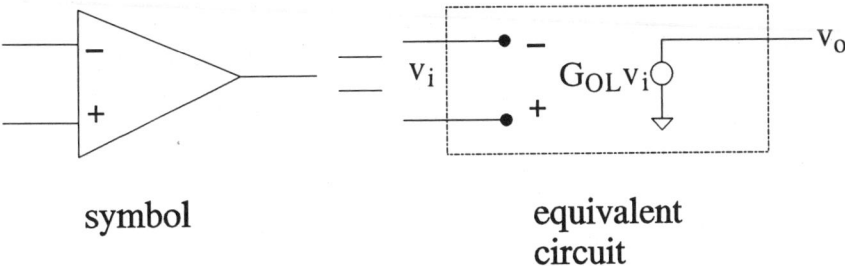

symbol   equivalent
                                          circuit

Figure 7.21.   Operational-amplifier equivalent circuit.

istics are:

$$\text{input impedance} = \text{open circuit } (\infty)$$
$$\text{output impedance} = 0$$
$$\text{open-loop gain } (G_{\text{OL}}) = \infty$$
$$\text{zero output for zero input} = (v_o = 0 \text{ for } v_i = 0)$$
$$\text{infinite frequency response} = (f_{3\text{dB}} = \infty)$$

The positive and negative terminals in Fig. 7.22 indicate inputs, noninverting (+) and inverting (−). Certain input constraints result because of the ideal conditions assumed previously:

1. No current flows into or out of either input terminal (opened circuit)
2. When negative feedback is applied, no voltage difference exists between input terminals (positive and negative terminals)

Figure 7.22 shows the effects of feedback, β_f, on the operational amplifier. The output voltage is

$$v_o = G_{\text{OL}}(v_i - \beta_f v_o) \tag{7.44a}$$

$$v_o + G_{\text{OL}}\beta_f v_o = G_{\text{OL}} v_i \tag{7.44b}$$

Rearranging to obtain the ratio of output-to-input voltage:

$$\frac{v_o}{v_i} = \frac{G_{\text{OL}}}{1 + G_{\text{OL}}\beta_f} \tag{7.45}$$

If $(G_{\text{OL}}\beta_f) \gg 1$, then

$$\frac{v_o}{v_i} \approx \frac{1}{\beta_f} \tag{7.46}$$

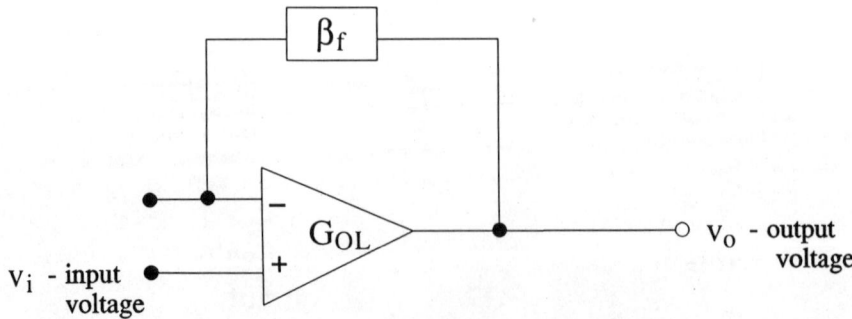

Figure 7.22. Feedback with operational amplifier.

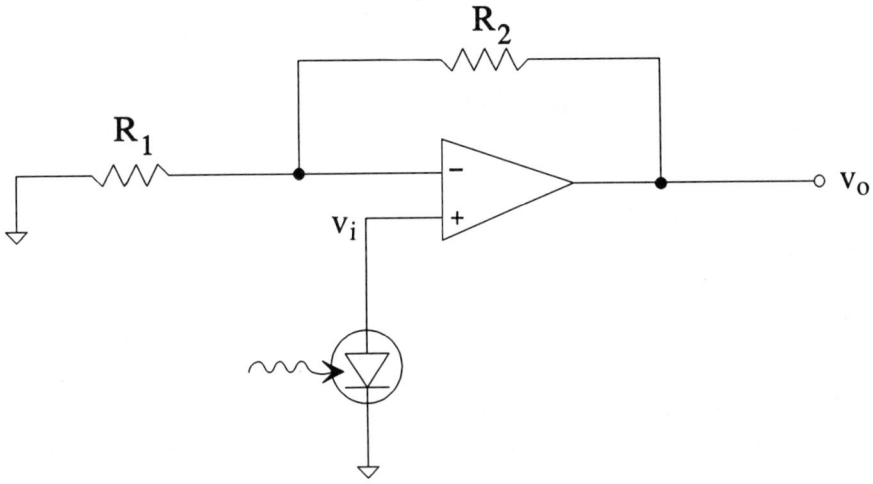

Figure 7.23. Photodiode connected to noninverting operational amplifier with gain.

Figure 7.23 shows a circuit where the detector is connected to the noninverting side (an open-circuit case). What is the output voltage of the preamplifier? From the voltage divider network in Fig. 7.23, the feedback factor is

$$\beta_f = \frac{R_1}{R_1 + R_2} \tag{7.47}$$

so that

$$\frac{v_o}{v_i} = 1 + \frac{R_1}{R_2} \tag{7.48}$$

and the output voltage is expressed as

$$v_o = v_i \left(1 + \frac{R_1}{R_2} \right) \tag{7.49}$$

A PV detector can be modeled as a current source (Fig. 7.24) with photo-generated current i_g and shunt resistance R_s.

Ideally, one would want all of the photo-generated current driven through the load resistor. This condition is $i_g \approx i_L$ (Fig. 7.24). Therefore, one should choose a small R_L compared to the shunt resistance ($R_L \ll R_s$) to get most of the photo-generated current to the load. The load resistor represents the input impedance of the preamplifier. What is the input resistance of a transimpedance amplifier shown in Fig. 7.20 with a photodiode? The equivalent input

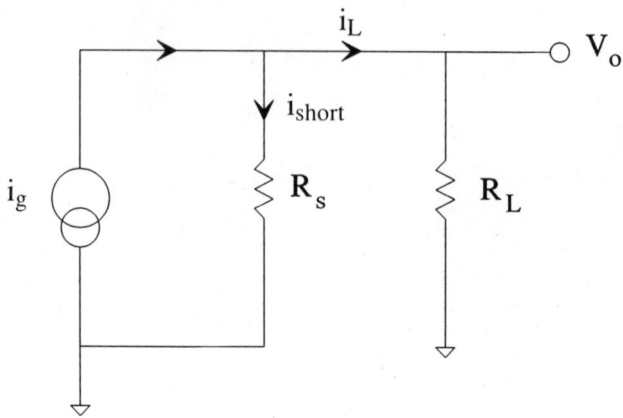

Figure 7.24. Electrical equivalent circuit for photodiode connected to operational amplifier.

impedance of the amplifier (R_L) is

$$R_L = \frac{R_f}{(1 + G_{OL})} \tag{7.50}$$

Therefore, because G_{OL} is so large, R_L in Fig. 7.24 is small. In fact, the detector operates into a virtual short circuit (virtual ground), and photo-generated current flows through R_f. So the output voltage v_o is $i_g R_f$. This forces the detector into short-circuit operation (as discussed in Section 7.2.3).

7.2.6. Sources of Noise

The primary sources of noise in the combination of a PV detector and preamplifier are: shot noise of the reverse saturation current; photon noise; Johnson noise of the detector resistance; Johnson noise of the load resistor; $1/f$ noise; and preamplifier noise. Because these noise sources are independent, the resulting noises can be summed in quadrature by adding the noise powers (noise voltages squared). The total noise (squared) expression for any PV detector is

$$i_n^2 = 4q^2 i_0 \,\Delta f + 2q^2 \eta (\phi_{q,\text{sig}} + \phi_{q,\text{bkg}}) \,\Delta f + \frac{4kT_d \,\Delta f}{R_d} + \frac{4kT_f \,\Delta f}{R_f}$$

$$+ \frac{\beta_0 \bar{i} \Delta f}{f} + \overline{i_{\text{pa}}^2} + \overline{v_{\text{pa}}^2} \left(\frac{R_f + R_d}{R_f R_d} \right)^2 \tag{7.51}$$

where

i_0 = reverse saturation current
q = electron charge

$\phi_{q,\text{sig}}$ = signal photon flux
η = quantum efficiency
$\phi_{q,\text{bkg}}$ = background photon flux
R_d = detector resistance
R_f = feedback resistance
Δf = electrical bandwidth
k = Boltzmann's constant
T_f = feedback-resistor temperature
β_0 = $1/f$ noise proportionality factor
i = dc current through detector
f = electrical chopping frequency
i_{pa} = preamplifier current noise
v_{pa} = preamplifier voltage noise

The first term in Eq. (7.51) has a factor of 4 instead of 2 because the reverse saturation current, i_0, consists of two parts at low injection levels. Assuming a p-type diffused layer on an n-type bulk substrate, a current (i_0) flows because of holes injected into the n region from the p region. In addition, a current ($-i_0$) flows because of holes injected into the p region from the n region. Both currents have shot noise associated with them, giving rise to the commonly seen form of the shot-noise equation having $2 i_0$ as the saturation current through the diode.

The second term is the contribution of the photon noise of the background and the signal. The third and fourth terms are the Johnson noise of the detector and feedback resistor, respectively. The fifth term is the $1/f$ noise associated with the dc current flow, and the last two terms are the squared-current contributions of the preamplifier current and voltage noises.

Some noises can be eliminated; we have some control over others as detector users. The preamplifier noise can be controlled by a judicious choice of components and operation. The selection of field-effect transistors (FETs) or bipolar transistors with low voltage noise can make the preamplifier noise negligible; examples of low noise devices are 2N6484 and 2N4403. If necessary, low values of current preamplifier noise can also be obtained by cooling the preamplifier to cryogenic temperature. Recall the dependence of $1/f$ noise on the total current flowing in the sensor. If open-circuit conditions apply, the detector current is nearly zero, and thus the $1/f$ noise can be approximately eliminated, for the open-circuit case at least.

The remaining noises are photon and Johnson noises in Eq. (7.51). If we are photon-noise limited, usually background-limited infrared photodetector (BLIP), from Eq. (7.51):

$$2q^2\eta(\phi_{q,\text{sig}} + \phi_{q,\text{bkg}})\,\Delta f \gg 4k\Delta f\left(\frac{T_d}{R_d} + \frac{T_f}{R_f}\right) \tag{7.52}$$

Recall that the signal current (i_{sig}) is

$$i_{\text{sig}} = \eta \phi_{q,\text{sig}} q \tag{7.53}$$

so the signal-to-noise ratio (SNR) is

$$\frac{S}{N} = \frac{q\eta(\phi_{q,\text{sig}})}{[2q^2\eta(\phi_{q,\text{sig}} + \phi_{q,\text{bkg}})\Delta f]^{1/2}} \tag{7.54}$$

The BLIP spectral D^* can be derived directly from

$$D^*(\lambda, f) = \frac{\sqrt{A_d \Delta f}}{\phi_e} \frac{S}{N} \tag{7.55}$$

Substituting Eq. (7.54) into (7.55), with the added conversion of photon signal (ϕ_q) to signal radiant power (ϕ_e) and the assumption that the background photon flux was much greater than signal flux, we get

$$D^*(\lambda, f) = \frac{\sqrt{A_d \Delta f}}{\left(\dfrac{hc}{\lambda}\right)\phi_{q,\text{sig}}} \frac{q\eta(\phi_{q,\text{sig}})}{[2q^2\eta(\phi_{q,\text{bkg}})\Delta f]^{1/2}} = \frac{\lambda}{hc}\sqrt{\frac{\eta}{2\phi_{q,\text{bkg}}/A_d}} \tag{7.56}$$

Further, assuming uniform background across the detector so that the denominator of the radical can be approximated as background irradiance ($E_{q,\text{bkg}}$):

$$D^*(\lambda, f) = \frac{\lambda}{hc}\sqrt{\frac{\eta}{2E_{q,\text{bkg}}}} \tag{7.57}$$

the same result as shown previously (in Chap. 6) for a BLIP condition of a photovoltaic detector.

If Johnson noise is limiting the performance of a PV detector, the "much-greater-than" symbol is reversed in Eq. (7.52). Under conditions of reduced levels of photon flux from background and signal sources, the photon noise is small compared to Johnson noise, so for a PV detector the total noise is

$$i_n = \left[4k\Delta f\left(\frac{T_d}{R_d} + \frac{T_f}{R_f}\right)\right]^{1/2} \tag{7.58}$$

The load resistor is assumed to be at detector temperature, so $T_d = T_L$, because reduced microphonics-induced noise will result from small distances between the detector and load resistor. The load resistor is usually chosen to be much larger than the detector resistance ($R_d \ll R_L$) so

$$i_n = \left[\frac{4kT_d\Delta f}{R_d} \right]^{1/2} \tag{7.59}$$

The SNR for Johnson-noise-limited operation (JOLI) is

$$\frac{S}{N} = \frac{q\eta\phi_{q,\text{sig}}}{\sqrt{4k(T_d/R_d)\Delta f}} \tag{7.60}$$

The spectral D^* for JOLI performance, substituting Eq. (7.60) into Eq. (7.55) is

$$D^*(\lambda, f) = \frac{\sqrt{A_d\Delta f}}{(hc/\lambda)\phi_{q,\text{sig}}} \frac{q\eta\phi_{q,\text{sig}}}{\sqrt{4k(T_d/R_d)\Delta f}} = \frac{\eta\lambda q}{2hc} \sqrt{\frac{R_d A_d}{kT}} \tag{7.61}$$

This is the exact expression obtained in Chapter 6. Desirable characteristics for a good high-performance photodiode are high quantum efficiency, large RA product, and low operating temperature. Recall the temperature dependence of the RA product in Eq. (7.33). The RA product determines the range of background-irradiance levels over which the detector is background limited.

The photovoltaic detector noise typically has a minimum versus bias voltage as shown in Fig. 7.25a. The minimum in noise occurs at a slightly negative voltage because the i-v slope is negative (as seen in the i-v characteristic of the photodiode shown in Fig. 7.25b), so the resistance increases with decreasing negative voltage. Thus the rms Johnson-noise current decreases, according to Eq. (7.59). A counteracting trend is that as the voltage is decreased from zero, the current through the detector becomes nonzero, so shot noise increases. The overall noise thus has a minimum at a slightly negative voltage.

7.3. SILICON PHOTOVOLTAIC DETECTOR

The silicon photodiode is the most common PV detector in use. Its spectral response covers approximately from 0.2 to 1.15 μm, with a peak response occurring at approximately 0.9 μm. They are constructed as either planar-diffused or Schottky-barrier photodiodes (Figs. 7.26 and 7.27, respectively).

The planar-diffused construction is an n-type diffusion on a p-type substrate, thus producing a p-n junction. The time and temperature of the diffusion process are controlled to obtain the proper thickness of the n-type diffusion. The proper thickness maximizes absorption in the junction region. The Schottky diode is a junction formed by deposition of metal on a semiconductor. We can consider the metal as a very highly doped n-type material, producing the junction shown in Fig. 7.27.

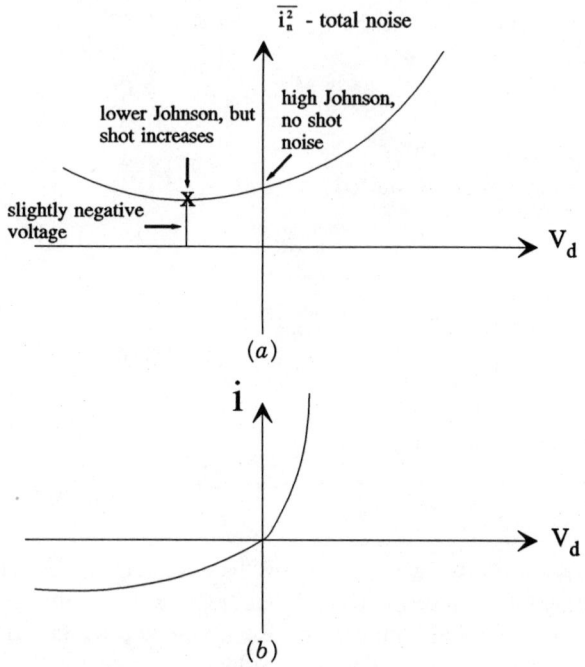

Figure 7.25. (a) Total noise vs. bias voltage for a photodiode. (b) i-v Characteristic for a photodiode.

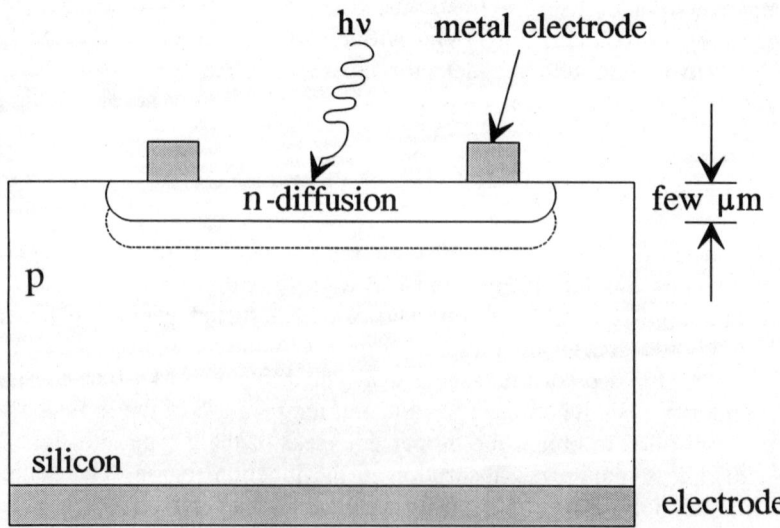

Figure 7.26. Planar photodiode construction.

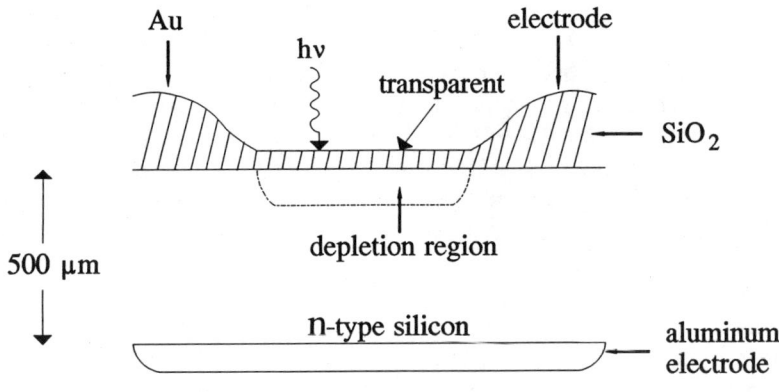

Figure 7.27. Schottky-barrier photodiode.

The thickness of gold (Au) is critical in the Schottky-barrier device, because the maximum transmission of light is necessary along with maximum electrical conductivity. The thickness is typically on the order of 150 Å. The energy-band diagram for the Schottky diode is shown in Fig. 7.28. The work function for metal ($\psi_{wf,m}$) is larger than the work function for the n-type semiconductor. Therefore, electrons moving from left to right in Fig. 7.28 must overcome a larger barrier than electrons moving from right to left. Therefore, we have a one-way preferred direction for electron flow in the Schottky-barrier diode.

A transparent electrode can be produced using a very heavy doping region (n^+) over an n region. This configuration is similar to a Schottky diode, but the work function requirement is reversed compared to Fig. 7.28 ($\psi_{wf,m} < \psi_{wf,s}$). The very high doping produces a material with high electrical conductivity, but which is optically transparent. An aluminum contact must be deposited on this n^+ region to provide bonding and ohmic contact. The spectral

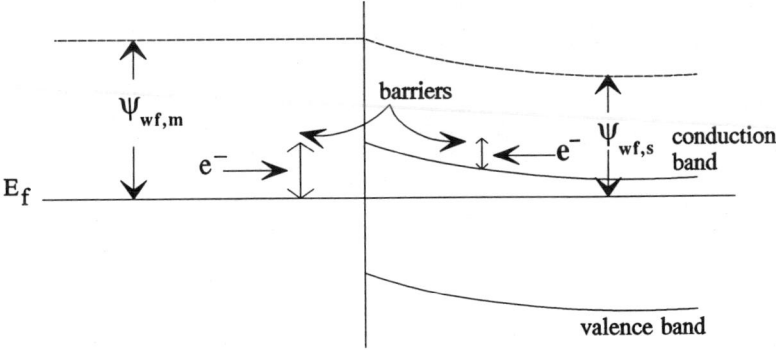

Figure 7.28. Schottky-diode energy-band diagram. Work functions shown are $\psi_{wf,m}$ for the metal and $\psi_{wf,s}$ for the semiconductor.

response (~ 0.2 μm to ~ 1.15 μm) is not constant through the spectral region, as shown in Fig. 7.29. The ultraviolet (UV) response is determined by the relatively large absorption coefficient for the short-wavelength photons ($> 10^5$ cm^{-1}). These photons get absorbed near the surface; and surface traps, which cause a short recombination lifetime, lead to an increased probability that the photogenerated charge will be lost before reaching the junction. Therefore, the short-wavelength quantum efficiency is reduced (Fig. 7.29). Another factor contributing to reduced quantum efficiency at short wavelengths is the increase in refractive index, leading to a higher Fresnel-reflection coefficient at short wavelengths. As wavelength increases, the absorption coefficient decreases (~ 100 cm^{-1}), so photons can penetrate closer to the junction-depletion region, and they travel a shorter distance to be detected. Hence the mid-wave quantum efficiency is higher than at short wavelengths. The quantum efficiency at the long-wavelength end falls off compared to the mid-wave region because there is a smaller absorption coefficient as the band edge is approached. The energy gap from the conduction to the valence band is not sharp, because of thermal agitation in the lattice at room temperature. Thus, some photons at long wavelengths will see a slightly larger energy gap that allows them to pass through the detector material without absorption, even though their wavelength is below the cutoff wavelength. The quantum efficiency at the long-wave end can be increased by increasing the reverse bias. By increasing the reverse bias, one increases the depletion width (d) as was shown by Eq. (7.17). This provides a larger volume for the photons to be captured and increasing quantum efficiency. Also, carrier sweep-out increases with increasing reverse bias, decreasing the recombination rate, and again leading to a higher quantum efficiency.

In summary, the quantum-efficiency losses from the ideal case are caused by Fresnel-reflection losses, trapping by surface recombination centers, diffu-

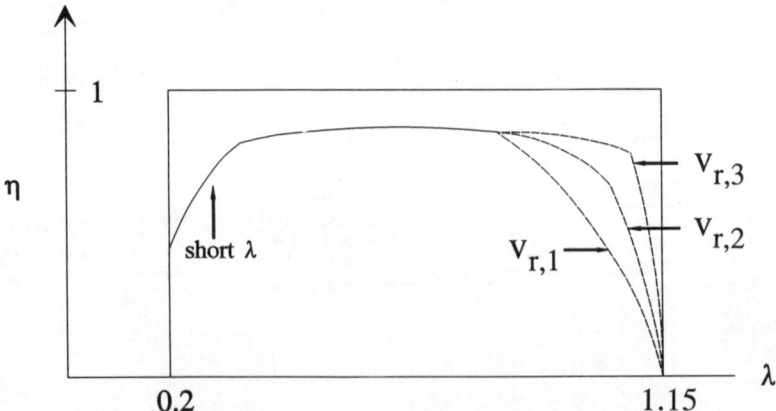

Figure 7.29. Silicon-photodiode spectral quantum efficiency where reverse bias is increased ($v_{r,3} > v_{r,2} > v_{r,1}$).

sion-length effects (L_e), and the wavelength dependence of the absorption coefficient.

The typical spectral responsivities for planar-diffused silicon and Schottky photodiodes are shown in Fig. 7.30. The size and geometry of the detector do not affect spectral response. Significant differences exist between the Schottky-barrier and planar-diffused type. The planar-diffused type has an improved response for wavelengths longer than 8000 Å. In this spectral region, the thin gold film of the Schottky barrier exhibits a reflection coefficient of greater than 30%, which reduces the number of photons entering the Si, thus reducing the responsivity. Antireflection coatings can be used to improve Schottky barriers in this region. The Schottky barrier has a considerably improved response for wavelengths shorter than 8000 Å.

An equivalent circuit for the silicon photodiode is shown in Fig. 7.31. A good rule of thumb for junction capacitance per unit area is about 0.01 μ farad/cm^2 for silicon at an applied voltage of zero. This value varies with applied reverse-bias voltage (v_r). As shown in Eq. (7.62), as the reverse bias goes up, the capacitance goes down:

$$(v_r \, C_d)^{1/2.5} = \text{constant} \tag{7.62}$$

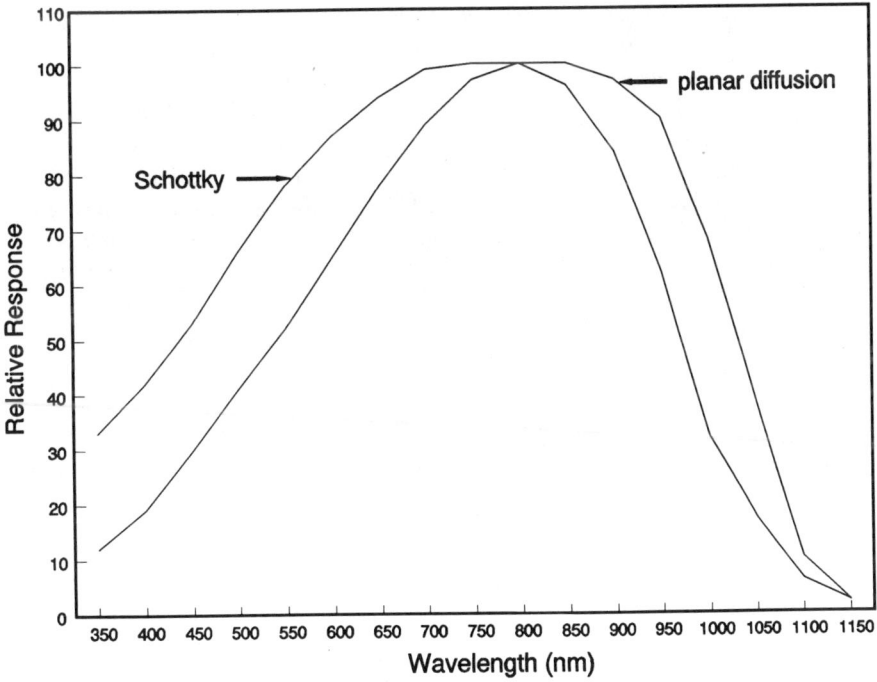

Figure 7.30. Relative spectral responses of planar-diffusion and Schottky photodiodes.

Consider an example of a 1-mm²-active-area silicon photodiode in parallel with a load resistor of 50 Ω. To find the time constant,

$$C_d = (0.01 \ \mu \ \text{farad/cm}^2) \times 1 \ \text{mm}^2 = 10^{-10} \ \text{farad} \qquad (7.63a)$$

$$R = R_d \| 50 \ \Omega = 50 \ \Omega \qquad (7.63b)$$

$$\tau = RC_d = 50 \times 10^{-10} \ \text{s} \qquad (7.63c)$$

Then, we apply 100-V reverse bias to the diode, using Eq. (7.62a)

$$C_d = 1.6 \times 10^{-11} \ \text{F} \qquad (7.64)$$

$$RC = 9 \times 10^{-10} \ \text{s} \qquad (7.65)$$

By increasing the depletion width, the frequency response has increased.

7.3.1. PIN Photodiodes

A photodiode that has a built-in depletion region, with an intrinsic layer of silicon between the p and n materials, is called a PIN photodiode. This intrinsic layer is tailored in thickness to optimize quantum efficiency, frequency response, and spectral response. The depletion region width has been substantially increased over the p-n junction depletion width. Figure 7.32 illustrates the increased depletion width in the PIN diode. The addition of this intrinsic layer lowers the capacitance and increases the volume of the absorption (depletion) region. Three forms of construction for PIN photodiodes are shown in Fig. 7.33: the Schottky-barrier, planar-diffused, and side-illuminated PIN.

Because the p layer is very thin, negligible absorption occurs in this region; however, the p layer is also fully depleted. The entire depletion region (mainly intrinsic material) is thick enough to capture most of the radiation, but not so thick that transit time for the carrier to cross the region is too long compared to the carrier lifetime. In each case, an n^+ region is used to make ohmic contact to the aluminum electrode. In the case of the Schottky construction, the gold

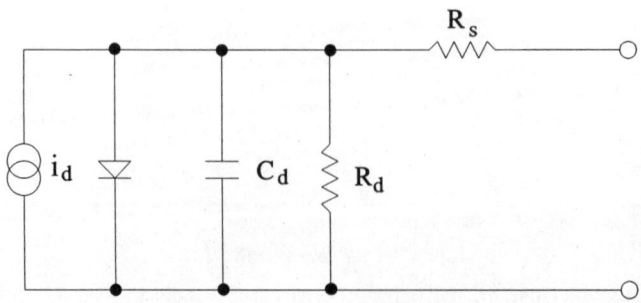

Figure 7.31. Silicon-photodiode equivalent circuit.

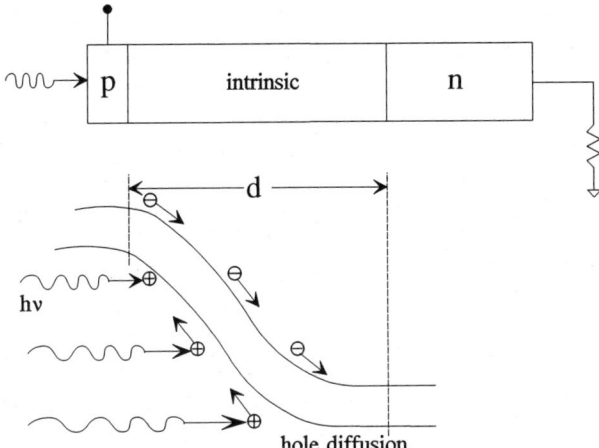

Figure 7.32. PIN construction and corresponding energy-level diagram showing depletion width.

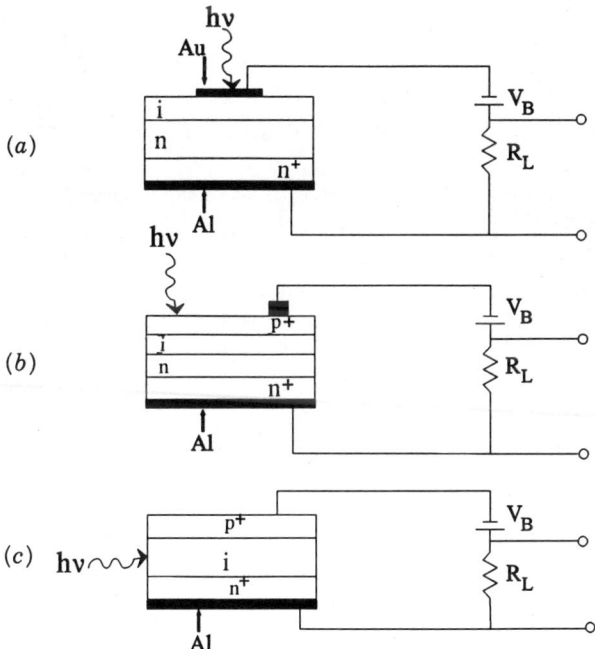

Figure 7.33. Construction of PIN photodiodes: (a) Schottky-barrier PIN; (b) planar-diffused PIN; (c) side-illuminated PIN.

is transparent to light because it is so thin. The incident radiation is absorbed in the depletion region, where it produces electron–hole pairs. This depletion region exists because of the built-in field produced by the diffusion of minority carriers across the junction between the p-type and intrinsic region.

By applying an external voltage V_B as seen in Fig. 7.33, the depletion region will extend throughout the intrinsic layer. The intrinsic layer has no free charge, so its resistance is high. This causes most of the voltage drop to appear across the intrinsic region and creates a strong electric field. This high electric field will enhance the separation of photo-generated electron–hole pairs and then cause these carriers to have a high drift velocity through the region.

The reverse current of the device in the dark (no photons incident) arises from two sources: (1) generation–recombination dark currents in the depletion region, and (2) surface-leakage currents. Currents generated in bulk silicon can be kept low by using high-purity silicon and using care in processing to avoid introducing crystalline defects. Bulk current densities as low as 10^{-11} amp/mm^2 can be achieved. Surface leakage on silicon detectors can be reduced by passivation techniques. For this discussion, we assume that good passivation techniques will eliminate this source of current.

Dark current depends on temperature. The exact variation of dark current with temperature is determined by the device-structure fabrication process and follows Eq. (7.1). The approximate equation is

$$i_0 \propto \exp\left\{ -\frac{0.7 \text{ eV}}{kT} \right\} \tag{7.66}$$

In the range of room temperature to 70°C, this corresponds to about an order of magnitude increase in dark current for each 30°C rise in temperature.

7.3.1.1. Quantum Efficiency of PIN. The intrinsic layer has a much higher resistivity than the doped layer. Therefore, the reverse-biasing voltage will mainly fall across the intrinsic layer. Because the electric E varies as dv/dx, E is contained within the intrinsic region. The natural existence of the unbiased, steady-state electric field (attributable to the coulomb-induced diffusion across the junction) creates a sufficiently high electric field to guarantee that any appreciable reverse-bias voltage will cause drift current to dominate in the intrinsic region. Conversely, because there is very little voltage across the p- and n-type regions, the lack of an electric field causes diffusion current to dominate in those regions. Because absorption can occur in the p region, the p region is always made very thin so that photo-generated carriers will not have to rely on the (slower) diffusion currents to carry them to the intrinsic region. The total current is the sum of the drift and diffusion currents, given by

$$i = (1 - \imath)\phi_q q \left(1 - \frac{e^{-ad}}{1 + aL_h} \right) \tag{7.67}$$

where

ι = Fresnel reflectance
a = absorption coefficient
L_h = diffusion length of holes
d = total depletion width

The quantum efficiency can be written as

$$\eta = \frac{i}{q\phi_q} = (1 - \iota)\left(1 - \frac{e^{-ad}}{1 + aL_h}\right) \qquad (7.68)$$

For Si, with $\lambda \leq 1.1\ \mu m$ and $aL_h \ll 1$, the denominator can be ignored. If antireflection coatings are applied to the surface of the detector (effectively making ι negligible), the depletion width would be the most important factor in the quantum efficiency: $\eta \approx 1 - e^{-ad}$. If the depletion width is much greater than the absorption depth ($1/a$), very high quantum efficiencies can be attained (around 85%).

7.3.1.2. Time Response. Three basic phenomena control detector-response time: (1) transit time of photogenerated carriers; (2) trapping of photogenerated carriers; and (3) equivalent-circuit RC time constant.

The transit time is the sum of the time taken for the photo-generated minority carrier to move out of the junction and the time taken for the photo-generated majority carrier to move to the ohmic contact. Because the PIN-type photodiodes collect photo-generated carriers in a strong electric in PIN photodiodes. Thus a 5-mm-thick PIN photodiode could have a transit time of 5 ns if the applied external voltage were sufficient to fully deplete the detector. Comparing this to the transit time of a PN photodiode, which is in the microsecond region (photoconductors in the range of milliseconds to microseconds and thermodetectors in the millisecond range), the PIN photodiode has an extremely short transit time.

The trapping effect of photo-generated carriers by deep-lying energy states in Si can cause long rise times and decay times. This problem can be alleviated by proper selection of an extremely pure starting material, and a final selection in the choice of device used. These procedures produce time constants of less than a microsecond.

The equivalent-circuit-response time constant will be $\tau = R_L C_d$, which is the time it takes for the current to reach 67% of a steady-state value after the light is turned on or off. If the diode capacitance was 100 pF and the load resistance has 1 kΩ, the RC time would be 10^{-7} s. Although the transit time within the cell would be much faster, this RC response time would dominate, and thus would be the observable time constant of the device.

There is a tradeoff to be made in the selection of load resistance for both fast response time and high sensitivity. A fast response requires that the load resistance be small, while a high sensitivity requires a high load resistance. D^* is a function of $(A_d R_L)^{1/2}$, where R_L dominates the Johnson noise instead of the detector Johnson noise. A compromise must be made for each application.

The intrinsic region width affects the transit time, which affects the electrical bandwidth of the PIN diode. As usual, a sinusoidally varying input photon irradiance $E_q e^{j\omega t}$ is assumed where $\omega = 2\pi f$ is the electrical radian frequency. According to Sze (1981), the photons are to be absorbed at the edge of the intrinsic layer ($d \gg 1/a$). It is also assumed that the bias voltage within the intrinsic region is sufficient to make the average drift velocity reach its saturation value (v_s). If the incident irradiance is uniform, then the current density created at the surface becomes a plane wave:

$$J(x) = q(1 - r)E_q e^{j(\omega t - kx)} = q(1 - r)E_q e^{j\omega(t - x/v_s)} \qquad (7.69)$$

and the total current is the average current density throughout the intrinsic region:

$$J = \frac{1}{d} \int_0^d J(x)\, dx$$

$$= q(1 - r)E_q \frac{1 - e^{j\omega d/v_s}}{j\omega d/v_s} e^{j\omega t} \qquad (7.70a)$$

$$J = q(1 - r)E_q e^{j\omega t} \operatorname{sinc}\left(\frac{fd}{v_s}\right) \qquad (7.70b)$$

The sinc function determines the temporal response of the system. The point at which the squared current (J^2) response has decreased to half of its maximum ($\operatorname{sinc}^2(fd/v_s) = 1/2$), its 3-dB point, is $fd/v_s = 0.44$, so the 3-dB frequency caused by drift is

$$f_{\text{drift}} = 0.44\, v_s/d \qquad (7.71)$$

For Si, the saturation drift velocity $v_s = 10^7$ cm/s, and assuming d in cm,

$$f_{\text{drift}} = 0.44\, v_s/d = (4.4 \times 10^6)/d \qquad (7.72)$$

For a 10-μm depletion width in a silicon PIN, the 3-dB frequency is 4.4 GHz.

The width of the intrinsic region has a very simple effect on the RC time constant. The width affects the capacitance by $C = \epsilon A/d$. For Si, $\epsilon = 11.8 \times \epsilon_0$, where ϵ_0 is the dielectric constant of free space. Assuming the lengths and areas are in cm units, the RC-limited bandwidth is

$$f_{RC} = \frac{1}{2\pi RC} = \frac{1}{2\pi \epsilon R} \frac{d}{RA} = 1.5 \times 10^{11} \frac{d}{RA} \qquad (7.73)$$

We can see that, in addition to its effect on noise properties already investi-gated, the RA product is an important factor in the response speed as well.

7.3.1.3. Performance Expected. The shot noise of the PIN photodiode de-pends on the dc current flowing through the PIN photodiode and the bandwidth of the device. The noise effects of the PIN diode are only indirectly affected by the intrinsic region width (through quantum efficiency). The primary noise sources are shot noise and Johnson noise. When considering the dc current caused by signal flux (i_{sig}), current resulting from background flux (i_{bkg}), and dark current (i_0) (for a sufficiently large reverse bias), the net noise current is

$$i_n^2 = \sqrt{2q(i_{sig} + i_{bkg} + i_0)\Delta f + \frac{4\,kT\Delta f}{R_L} + \frac{4\,kT\Delta f}{R_d}} \qquad (7.74a)$$

$$i_n^2 = \sqrt{2q(i_{sig} + i_{eq})\Delta f} \qquad (7.74b)$$

where $i_{eq} = i_{bkg} + i_0 + 2kT/(qR_{eq})$ and $R_{eq} = R_L \| R_d$.

At low frequencies ($f < 1$ kHz), $1/f$ noise appears. This is a surface-effect-related noise that varies from detector to detector. In general, Schottky pho-todiodes have lower $1/f$ noise than the planar-diffused type.

The SNR is then

$$\frac{S}{N} = \frac{i_{sig}}{\sqrt{2q(i_{sig} + i_{eq})\Delta f}} \qquad (7.75)$$

Then, $i_{sig, NEP}$, the signal current flowing in response to an incident flux equal to the noise-equivalent power (NEP), is found by setting the SNR equal to one:

$$i_{sig, NEP}^2 = (2q\Delta f)i_{sig, NEP} + (2q\Delta f)i_{eq} \qquad (7.76a)$$

$$\cdot\ i_{sig, NEP}^2 - (2q\Delta f)i_{sig, NEP} - (2q\Delta f)i_{eq} = 0 \qquad (7.76b)$$

We solve the preceding for $i_{sig, NEP}$ using the quadratic equation

$$i_{sig, NEP} = (q\Delta f)\left(1 + \sqrt{1 + \frac{2i_{eq}}{q\Delta f}}\right) \qquad (7.76c)$$

The radiant flux necessary to produce the signal current $i_{sig, NEP}$ is the NEP, which is found from

$$i_{sig, NEP} = q\eta\,\frac{\lambda}{hc}\,\text{NEP} = (q\Delta f)\left(1 + \sqrt{1 + \frac{2i_{eq}}{q\Delta f}}\right) \qquad (7.77)$$

Therefore

$$\text{NEP} = \left(\frac{hc\Delta f}{\eta\lambda}\right)\left(1 + \sqrt{1 + \frac{2i_{eq}}{q\Delta f}}\right) \tag{7.78}$$

For the frequent case where noise processes other than the signal (e.g., background, thermal, or dark current) dominate, the noise $i_{eq} \gg (q\Delta f)/2$ and

$$\text{NEP} = \frac{hc}{\eta\lambda}\sqrt{\frac{2i_{eq}\Delta f}{q}} \tag{7.79a}$$

The specific detectivity is

$$D^* = \frac{\eta\lambda}{hc}\sqrt{\frac{qA_d}{2i_{eq}}} \tag{7.79b}$$

For the limiting cases of background noise (for the BLIP case, $i_{eq} = q\eta E_q A_d$) and Johnson noise (for the JOLI case, $i_{eq} = 2kT/qR_d$), the same equations as those obtained in Chapter 6 for general PVs are found. Substituting for i_{eq}, we obtain

$$D^* = \frac{\eta\lambda}{hc}\sqrt{\frac{qA_d}{2\,(i_0 + i_{bkg} + 2\,kT/(qR_{eq}))}} \tag{7.80}$$

Figure 7.34 shows the cases for BLIP performance when the background current (i_{bkg}) approaches the dark current (i_0), for a typical Si PIN ($\eta = 75$, $\lambda = 0.77\ \mu\text{m}$, $i_0 = 150$ pa, $T = 300$ K, and $\Delta f = 1$).

Here the reverse saturation currents will limit the detector D^* for lower background-irradiance values. For instance, if the dark current i_0 of a PIN photodiode is 10^{-7} amp, the responsivity is 0.3 amp/W, and the bandwidth of the measurement system is 1 Hz, then the critical resistance (resistance beyond which we are Johnson-noise limited) is 50 kΩ. The NEP for all values of load resistance larger than this would be 6×10^{-13}. For example, UDT-PIN-8LC has a detectivity: $D^*(\lambda,\ f) = D^*\ (0.850\ \mu\text{m},\ 900\ \text{Hz}) = 5.6 \times 10^{12}$ cm Hz$^{1/2}$/W. Although he PIN photodiode can be operated at temperatures greater than 50°C (323 K), cooling as shown in Table 7.2 is required (~ 77 K, liquid nitrogen temperature) to make NEP as low as possible ($\sim 5 \times 10^{-16}$ W).

The responsivity will vary with changes in wavelength of the incident light and will also vary slightly with changes in applied voltage and changes in temperature. The responsivity will change with wavelength because the reflection and absorption coefficients of Si change with wavelength. The applied

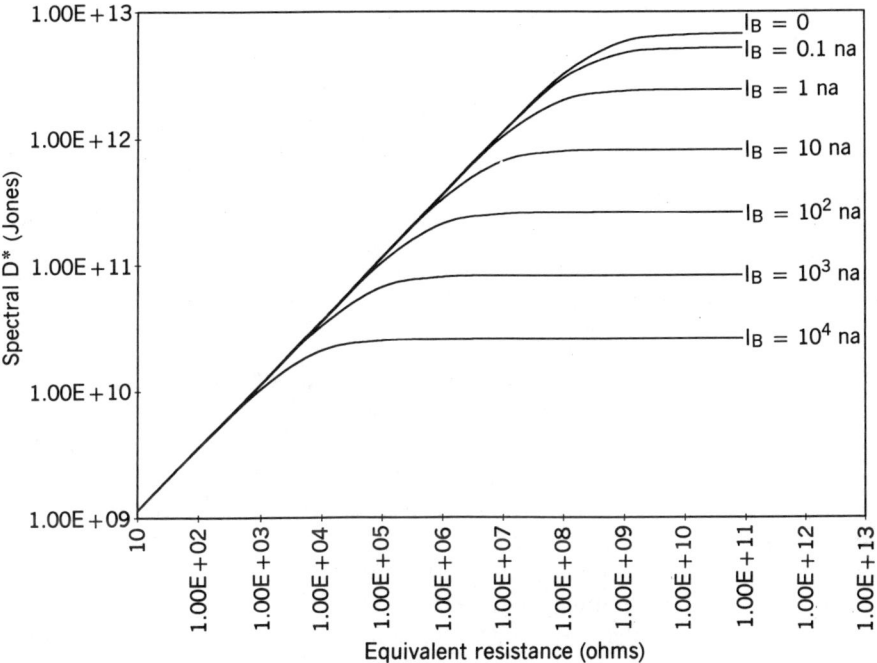

7.34. D^* vs. equivalent resistance for various values of background current.

voltage affects the collection process of the photo-generated electron–hole pairs within the Si, and thus causes the responsivity to vary. Temperature changes affect both the optical constants of Si and the collection process, thus affecting the responsivity.

A design example will show how the intrinsic region width affects some parameters of the detector, along with the RA product. Suppose we want to operate the detector at $T = 300$ K and $\lambda = 0.5$ μm. Also, assume that the detector is Johnson-noise limited, and that the reverse bias is sufficiently high to cause the electron velocity to saturate. Using a refractive index of 3.44 and

Table 7.2 Dark Currents of 1-mm-Diameter Pin Diodes

Technology	i_0 at 25°C	i_0 (doubles for) (°C)
Silicon	5 nA	7
Germanium	2 μA	8
InGaAs	12 μA	10
InGaAsP	6 μA	10

$a(\lambda, T) = 10^4 \text{ cm}^{-1}$ (Eq. 7.68),

$$D^* = \frac{\eta q \lambda}{2\,hc} \sqrt{\frac{R_d A_d}{kT}} = 2.2 \times 10^9 \, (1 - e^{-10^4 d}) \sqrt{R_d A_d} \qquad (7.81)$$

Recall Eqs. (7.71) and (7.73):

$$f_{\text{drift}} = (6 \times 10^6)/d \qquad (7.82a)$$

$$f_{RC} = (1.5 \times 10^{11})d/(R_d A_d) \qquad (7.82b)$$

In Eqs. (7.82a) and (7.82b), d is in units of cm, A_d is in units of cm^2, and R_d is in units of ohms. From this set of equations, it is easy to see how d is important to the operation of the PIN detector. The two parameters, d and the RA product, trade off to determine the performance of the detector. If we were to specify a particular bandwidth target, the depletion width d could be found from the f_{drift} equation, and the RA could be found from the f_{RC} equation (which could be as large as possible, enabling the D^* to be as large as possible).

If we wanted to optimize the width of the depletion region for the best frequency response, the two bandwidths should be set equal to avoid having one mechanism dominate ($f_{\text{drift}} = f_{RC}$). This will allow for as large a D^* as possible. In this case,

$$(6 \times 10^6)/d = (1.5 \times 10^{11})d/(R_d A_d) \qquad (7.83a)$$

$$d = 6.32 \times 10^{-3} \sqrt{R_d A_d} \qquad (7.83b)$$

For widths greater than this, transit-time effects will dominate, and for narrower widths, the RC time constant will dominate.

7.3.1.4. PIN Interface.

The PIN photodiode functions in a manner similar to a constant current generator. The amount of current is proportional to the incident light intensity. Any variation in responsivity with light intensity represents a variation in linearity. Maximum deviation from linearity over the range of irradiances from the limit of detectivity, 10^{-13} W/m^2 to 10^{-3} W/cm^2, is about 5%. At input light intensity greater than 10 mW/cm^2, major deviations in linearity will begin to occur in both the Schottky-barrier and planar-diffused devices. At higher light levels, the photocell must be back biased for best linearity. Bias voltages greater than 50 V should be used with light intensities greater than 10 mW/cm^2.

Figure 7.31 shows the equivalent circuit for both the Schottky-barrier and planar-diffused devices. There are a junction depletion layer capacitance C_d, a junction parallel resistance R_d (which is the effective resistance of the depletion region itself), and a series resistance (which is the resistance of the undepleted silicon and any lead resistance). The light flux incident on the cell generates a constant current, i_g.

If the device is working into a short circuit (corresponding to a vertical load line, $R_L = 0$), the current is linear with incident illumination. As the load resistance increases, operation becomes nonlinear until the open-circuit condition is obtained ($R_L = \infty$). At this point, the open-circuit voltage output is proportional to the log of the incident radiation. Minimal dark current exists in this photovoltaic mode of operation, making this mode ideal for low-level measurements where changing dark currents are a significant portion of the total current output.

In the reverse-bias mode, linear operation is maintained if the saturation power is not exceeded (about 10 mW/cm^2), and the bias level remains higher than the product of the maximum signal current and the load resistance ($i_g R_L < V_B$).

To obtain a large signal power from a photodiode, the photodiode must be connected to a very large load resistance, at the expense of frequency response (signal bandwidth). This is an example that the gain–bandwidth product of a device is constant, independent of any external factors. The gain–bandwidth product of the photodiode is inversely proportional to its junction capacitance. A typical value of gain–bandwidth product is 10^{19}.

The reverse-bias PIN photodiode behaves as a current source with very large shunt impedance, small junction capacitance, and small series resistance. For signal amplification, the photocurrent should be transformed into a voltage with a moderate output impedance. If the rather large load resistance used with most photodiodes were connected directly to the amplifier input, the impedance mismatch would greatly reduce the signal. The solution to this problem is to use a transimpedance amplifier (TIA), as in Fig. 7.35a, which converts input current into output voltage. Because of the high junction resistance of the reverse-bias photodiode, the operational amplifier should be a FET type with very high input impedance. Consequently, the photocurrent flows through the feedback resistance, R_L, which acts as the load resistance for the photodiode. Because the negative input of the operational amplifier acts as a virtual ground, the output voltage of the circuit is

$$v_o = -i_g R_L \qquad (7.84)$$

The transimpedance amplifier in Fig. 7.35a also increases the power available to auxiliary electronics. The maximum power available from the photodiode preamplifier is obtained when the impedance of the additional electronics matches the output impedance of the operational amplifier (R_{out}):

$$P_{av} = \frac{v_o^2}{R_{out}} = \frac{i_g^2 R_L^2}{R_{out}} \qquad (7.85a)$$

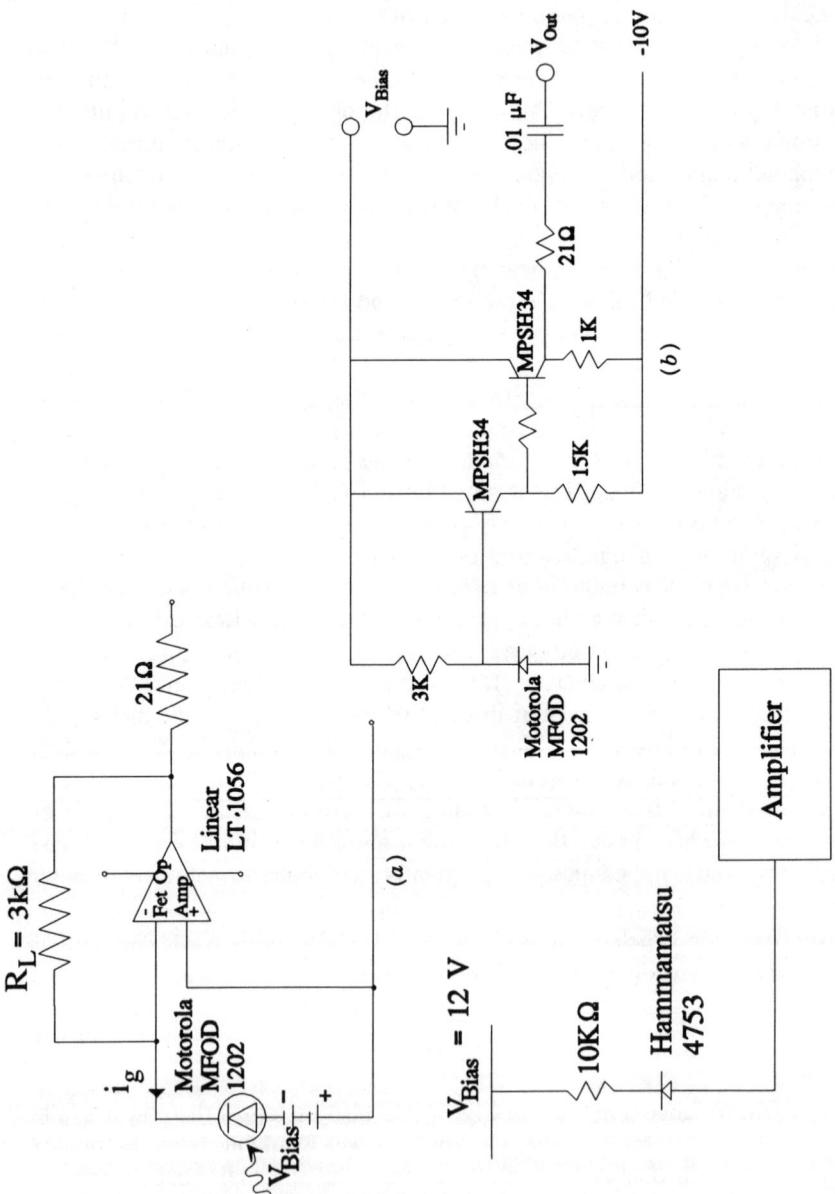

Figure 7.35. PIN electrical interfaces: (*a*) transimpedance circuit for a linear response; (*b*) discrete high-speed circuit; (*c*) bias circuit using high-speed modular amplifier (0.1–1.5 GHz).

282

compared with the power delivered to the load resistor (R_L):

$$P_L = R_L|i_s|^2 = \frac{i_g^2 R_L}{1 + \omega^2(R_s + R_L)^2 C_d^2} \qquad (7.85b)$$

The output power has been increased by the factor of $R_L/4R_{out}$. Typically R_L = 1 MΩ and R_{out} = 100 Ω, giving an increase in output power of 2500 times.

When the photodiode is used to detect light intensity as, for example, in a laser power meter or photometer, the dc output voltage ideally should be proportional to the photocurrent. Because in this case signal bandwidth is not needed, a very large feedback resistance may be used; values as high as 100 MΩ are typical in practice.

When the photodiode is used as a detector in an optical-communication system or digital-data link, ideally the bandwidth should be large and the signal gain high. In this case, the size of the feedback resistance is restricted by the requirement of signal bandwidth. Further signal amplification is provided by a second capacitor-coupled stage, which acts as an ac amplifier.

Figure 7.35b shows a discrete amplifier constructed with low-input-capacitance bipolar transistors. The decreased input capacitance allows the detection system to be limited in frequency response by the junction capacitance of the detector. Because the junction capacitance is inversely proportional to bias voltage, the frequency response of the system will increase only if the detector bias voltage is increased. This result is readily observed up to frequencies of 350 MHz, at which point the frequency response of the amplifier begins to roll off. The PIN diode used in this amplifier scheme demonstrated an upper 3-dB point of 80 MHz at a bias of 3.8 V, and increased to 100 MHz at a bias of 6.0 V. At frequencies above 100 MHz, the response of the light-emitting-diode (LED) source used in the evaluation circuit rolls off. If a laser diode were used, it would modulate at frequencies in the gigahertz range.

An example of a truly "high-speed" detector circuit is shown in Fig. 7.35c. The PIN detector has a terminal capacitance of only 1.5 pF at a bias of 12 V. The amplifier used in this circuit is a multistage discrete amplifier with a gain of 100 and a frequency response that approaches 1.5 GHz. In this detector system, all component leads must be kept as short as possible to minimize stray capacitance. The disadvantage of this detector is its low-end frequency response. The detector system will not respond to radiation below 100 kHz.

Table 7.3 compares PIN and avalanche photodiodes (APD) at 25°C. From these data, we can see that PIN devices are used for speed and APD diodes are used to improve the signal-to-noise ratio.

7.3.2. Avalanche Photodiodes

Avalanche photodiodes are photon detectors that operate under a large reverse bias and produce large currents attributable to impact ionization resulting in carrier multiplication. This phenomenon leds to avalanche breakdown for or-

Table 7.3 Photodiode Performance Comparison

	λ_p (μm)	$\Re_i$ (A/W)	D^* (λ_p)	Rise Time	i_0 (A)
Silicon	0.85	0.45	5 (10^{13})	400 ns	5 (10^{-12})
PIN (Si)	0.9	0.5	2 (10^{12})	50 ps	150 (10^{-12})
APD ($G = 75$)	0.9	45	4 (10^{13})	8 ns	100 (10^{-9})

dinary reverse-biased *PN* junction diodes. This section presents the physics of conventional avalanche photodiodes, as well as gain, noise, frequency response, and temperature characteristics.

Avalanche multiplication was first discovered in the early 1950s with silicon and germanium by McKay and McAfee (1953), and developed by McIntyre (1966) and MacGregor (1991). This mechanism is also an important consideration in bipolar transistors, MOSFETs, and IMPATT diodes (Sze, 1985). Yariv (1985) provided the simplest conceptual explanation of the avalanche multiplication process. The construction of an APD is shown in Fig. 7.36*a*, with the corresponding energy-band diagram depicting the avalanche process shown in Fig. 7.36*b*.

Be reversing a *PN* junction, the electric field in the depletion layer can increase to a point at which carriers accelerated across the depletion layer can gain enough kinetic energy to kick new electrons from the valence band to the conduction band by impact ionization (avalanche multiplication), often visualized as a reverse breakdown of a classic diode. If the limiting noise source is the photogenerated current, a PIN photodiode will be adequate. The PIN can exceed the APD in speed, but if the preamplifier is the major source of noise, a device with internal gain is necessary. In general, avalanche photodiodes are more economical and easier to use than photomultiplier tubes, and thus are suitable for most of these applications.

Detection systems are usually limited by either signal photon noise, Johnson noise, or preamplifier noise. If preamplifier noise is the limiting noise source, it is desirable to have an internal gain mechanism in the detector that multiplies, as noiselessly as possible, both the detector signal and the detector noise until the latter is greater than the preamplifier noise. The photomultiplier tube satisfies these criteria uniquely with low-noise amplification. The APD, although not having noiseless gain, may be better suited for certain applications by virtue of its ruggedness, size, speed, cost, reduced operating voltage, and greater quantum efficiency in the near infrared.

The electrical signal current is multiplied by an avalanche multiplication factor (gain) *G*, which increases the signal:

$$i_{sig}^2 = (\eta q \phi_q)^2 G^2 \tag{7.86}$$

However, the shot noise arising from both photon-induced current and dark current are also multiplied by the gain plus an effective excess noise factor

$\mathfrak{f}(G)$, which is a function of gain. The noise currents present are given by

$$i^2_{s,\text{photon}} = 2q^2\Delta f\eta\phi_q G^2\mathfrak{f}(G) \tag{7.87}$$

$$i^2_{s,\text{dark-current}} = 2q\Delta f\frac{kT}{qR_d}[\exp(qv/kT) - 1]\,G^2\mathfrak{f}(G) \tag{7.88}$$

$$i^2_J = 4kT\Delta f\left(\frac{1}{R_L} + \frac{1}{R_d}\right) \tag{7.89}$$

We can obtain the SNR by adding the noises in quadrature:

$$\frac{S}{N} = \frac{\eta q\phi_q G}{\sqrt{2q\Delta f\left\{q\eta\phi_q + \dfrac{kT}{qR_d}[\exp(qv/kT) - 1]\right\}G^2\mathfrak{f}(G) + 4kT\Delta f\left(\dfrac{1}{R_L} + \dfrac{1}{R_d}\right)}} \tag{7.90}$$

With ideal electron-current multiplication, $\mathfrak{f}(G)$ would be exactly 1.0 and the signal-to-noise ratio could be characterized by a Johnson-noise-limited region for low values of gain and a shot-noise-limited region for high values of gain. However, because of the statistical nature of the gain mechanism (a Markovian process), the multiplication is not ideal and must be characterized statistically. There is an average gain and corresponding fluctuation about that average, thus the multiplication process increases overall noise. This increased noise is accounted for by the excess-noise factor $\mathfrak{f}(G)$. The origin of the non-ideal behavior stems from the statistical behavior of the ionization rates or coefficients of the electrons and holes. As described in Fig. 7.36b, the electrons and holes travel an average distance before impact ionization occurs. The probability of impact ionization is a strong function of electric field, which in turn is a function of bias voltage and position in the depletion layer. Therefore, the gain is a function of position and bias voltage also. The probability that an impact ionization will occur in a distance dx is $\alpha_i\,dx$ for electrons and $\beta_i\,dx$ for holes, where α_i and β_i are carrier-impact-ionization coefficients. The mean gain (G) is a function of the carrier-ionization coefficients. Different materials have particular carrier-ionization coefficients. Hence, the value of $\mathfrak{f}(G)$ is different for each material and is a function of the mean gain (G), and the ratio of the ionization coefficients (β_i/α_i).

In avalanche photodiodes, current gain or multiplication is obtained from the electric field, which supplies the photo-generated carriers with sufficient energy, typically 1.5 to 1.8 times $\mathcal{E}_g$ to create secondary electron–hole pairs. These secondary electron–hole pairs can be accelerated until they also ionize pairs of free carriers, which continues the process and thus provides multiplication. This multiple electron–hole-pair creation occurs for electric-field magnitudes of approximately 10^5 V/cm.

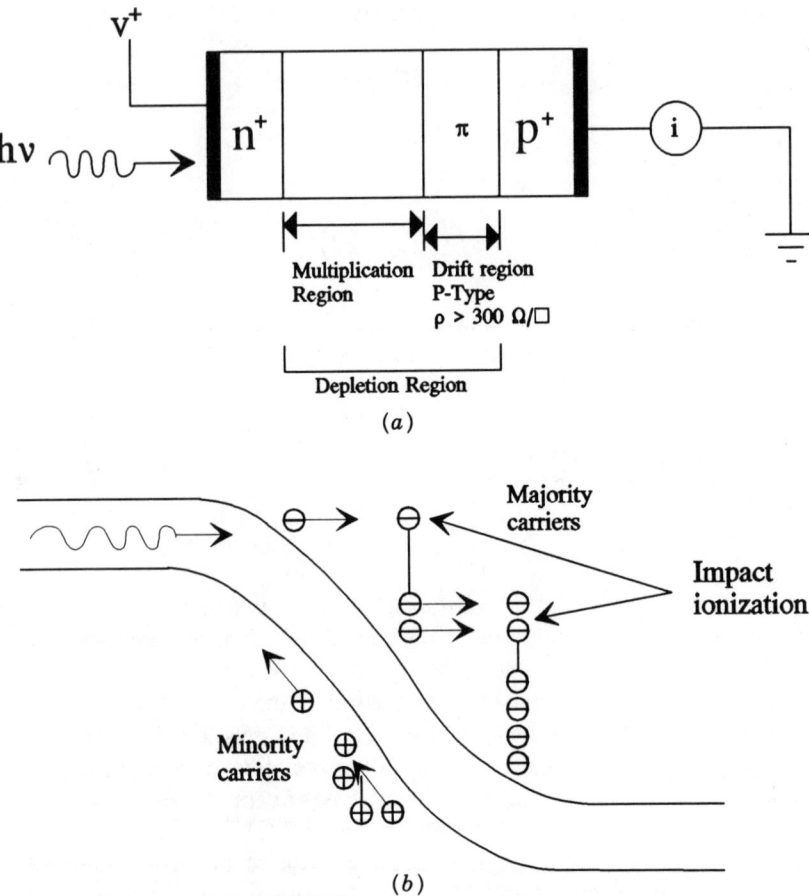

Figure 7.36. Avalanche photodiode (*a*) structure and (*b*) energy-band diagram.

The generation of the electron–hole pairs can be described in terms of the impact-ionization coefficients, α_i and β_i, for electrons and holes, respectively. The coefficients have units of cm^{-1} and are exponential functions of electric-field strength. Two important cases for α_i and β_i illustrate the best- and worst-possible behavior of the APD: the best situation is when one of the ionization coefficients is zero, and the worst case is when the ionization coefficients are equal.

Figure 7.37 shows the electron and hole currents and the space-charge (depletion) region in the photodiode. The electrons move in a direction opposite the direction of current flow. Let ℓ_x be the distance across the space-charge region measured from the p to the n side, where avalanching of the (n_e) initial electrons occurs. Consider the case where n_e electrons enter the space-charge region at $x = 0$ (assuming no holes enter the space-charge region at $x = \ell_x$).

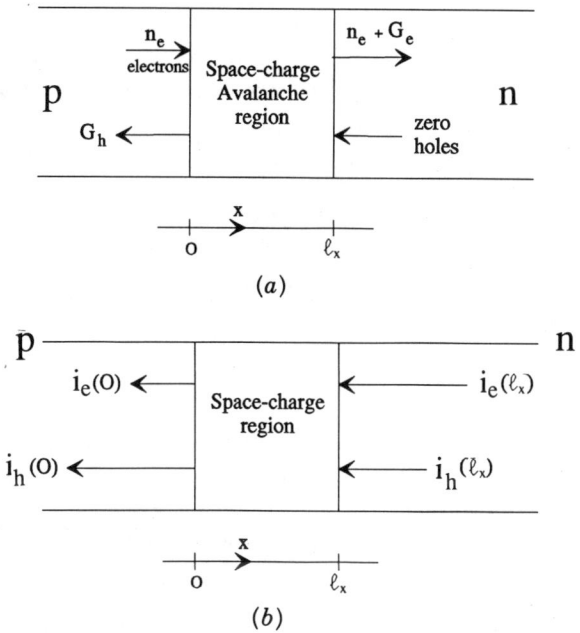

Figure 7.37. (a) Avalanching of electrons; (b) space-charge region in the photodiode.

As the electrons move a distance dx into the space-charge region, they produce additional electrons; in fact, n_e is continuously increasing in value to $n_e + G_e$, where G_e represents the number of electrons produced by the avalanching process, so the gain is $(n_e + G_e)/n_e$ as shown in Fig. 7.37a. If ℓ_x is the distance across the space-charge region measured from the n-type material to the p-type material, then we can write expressions for the hole and electron currents in differential volumes in the gain region as a function of distance in the x direction:

$$\frac{di_e}{dx} = \alpha_i i_e - \beta_i i_e \tag{7.91a}$$

$$\frac{di_h}{dx} = -\alpha_i i_h + \beta_i i_h \tag{7.91b}$$

Equation (7.91a) assumes that no holes are entering the depletion region from the n-type region and the current is being produced by the avalanching process of electrons and holes. Similarly, Eq. (7.91b) assumes no electrons are entering the depletion region from the p-type region.

The subscript e refers to electrons and the subscript h pertains to holes. We see in Eq. (7.91a) that the ionized carriers add a positive amount to the electron-current magnitude and add a negative amount to the hole current.

The first case we consider is the one for an impact ionization coefficient β_i = 0, that is, no ionization of holes. For simplicity we can assume that the hole current entering the right side of the space-charge region is zero: $i_h(\ell_x) = 0$. Solving Eq. (7.91a) for i_e when $\beta_i = 0$ gives us

$$i_e(x) = i_e(0) \exp \left(\int_0^x \alpha_i \, dx \right) \tag{7.92}$$

If we call the electron current leaving the left side $i_e(0)$, then, by Kirchhoff's law, we can write

$$i \equiv i_e(\ell_x) = i_e(x) + i_h(x) \tag{7.93}$$

For $\beta_i = 0$, it is important to note that the current in the device is always finite, and it is a strong function of electric field (therefore applied voltage), noting the hole current is negligible. When $\alpha_i = \beta_i$, the avalanching process results in the worst-case performance of an avalanche photodiode. For this case, solving Eq. (7.91a) for i_e leads to

$$i \equiv i_e(\ell_x) = \frac{i_e(0)}{1 - \exp \left(\int_0^{\ell_x} \alpha_i \, dx \right)} \tag{7.94}$$

which becomes infinite when the integral is equal to one. This occurs when each injected carrier (on the average) generates only one electron–hole pair during the time it takes to travel through the space-charge region. With few carriers in the space-charge region, statistical variations in the impact-ionization process can cause large fluctuations in the gain, resulting in considerable excess noise for this case. In the case of unequal (but both nonzero) ionization coefficients, the process is more difficult to treat mathematically, but a breakdown condition still arises.

The statistical fluctuation in the number of carriers in the space-charge region causes an uncertainty in the magnitude of the gain. Also, background-generated current and dark current both experience gain. The net result is that the gain mechanism generates excess noise that degrades the SNR by the multiplication process. However, under conditions where the overall SNR is limited by thermal noise or other electronics noise, the internal-gain device can produce a significant improvement in overall SNR.

The current gain for the charge transversing the entire depletion region is G, which is just $i/i_e(0)$. $G(x)$ is the current gain, or change in output current caused by a change in the electron current initiating at x, $i_e(x)$. Thus the mean-square shot-noise current is

$$i_n^2 = 2q \left(G^2 i_e(0) + \int_{i_e(0)}^{i} G^2(x) \, di_e(x) \right) \Delta f \tag{7.95}$$

This infers that a given increase in current (di) at x produces a shot-noise component that is amplified by the square of the remaining gain as the carriers travel from x (absorption location) to ℓ_x (the edge of the gain region).

Consider again the case where $\beta_i = 0$. $G(x)$ is equal to $i/i_e(x)$, because we can write from Eq. (7.91a) integrated from x to ℓ_x where the amplified current exits:

$$i \equiv i_e(\ell_x) = i_e(x) \exp \left\{ \int_x^{\ell_x} \alpha_i \, dx \right\} \tag{7.96}$$

Substituting for $G(x) = i/i_e(x)$ and solving Eq. (7.95) yields

$$i_n^2 = 2qi_e(0)G^2 \left(2 - \frac{1}{G} \right) \Delta f \tag{7.97}$$

An ideal multiplier would exhibit a noise current multiplied by a factor of G and therefore a mean-square noise multiplied by a factor of G^2. We have shown here an extra noise factor of 2 for high gain when $\beta_i = 0$.

Equation (7.97) gives the best possible noise performance for an avalanche photodiode. The situation is much worse for the case of equal impact-ionization coefficients. This is because for nonzero β_i, a small increase in current now produces both extra electrons that travel to the right and extra holes that travel to the left through the space-charge region. The sum of the contributions gives a constant net increase in current independent of the position of the initial fluctuation in current. So now $G(x) = G$ and Eq. (7.95) can be rewritten as

$$i_n^2 = 2q \left(G^2 i_e(0) + \int_{i_e(0)}^{i_e(\ell_x)} G^2 di_e(x) \right) \Delta f \tag{7.98a}$$

Solving the integral yields

$$i_n^2 = 2q[G^2 i_e(0) + G^2(i_e(\ell_x) - i_e(0))] \Delta f \tag{7.98b}$$

and

$$i_n^2 = 2qG^2 i_e(\ell_x) \Delta f \tag{7.98c}$$

Substituting $i_e(\ell_x) = Gi_e(0)$ gives us

$$i_n^2 = 2qG^3 i_e(0) \Delta f \tag{7.98d}$$

Now we have an extra factor of G multiplying the noise, causing the signal-to-noise ratio to be much worse than it was in the $\beta_i = 0$ case. Of course, this is using an idealized model for the gain mechanism in the device. It has been shown that the excess noise factor $\mathfrak{f}(G)$ is bounded (McIntyre, 1966) by

$$\left(2 - \frac{1}{G}\right) < \mathfrak{f}(G) \le G \tag{7.99}$$

We have shown that the two extremes for the performance of an avalanche photodiode occur for $\beta_i = 0$ and $\alpha_i = \beta_i$. Generally α_i and β_i are not equal, so the fabrication of the device should be done in such a way as to take advantage of this difference. This is because the avalanche gain will depend on where the carriers are generated in the multiplication region. For higher gain the device structure should be designed so that multiplication is initiated by the carrier with the higher ionization coefficient, typically electrons. The maximum multiplication obtainable is limited by the gain–bandwidth product limitations (breakdown) and gain–saturation effects. Gain–saturation can be caused by the voltage drop across resistive elements in the circuit, resistive heating (which can reduce α_i and β_i), and reduction of the electric-field strength as the carriers travel through the depletion region. When the photocurrent is smaller than the dark current, maximum multiplication is limited by the dark current. Thus the dark current must be as low as possible to attain maximum multiplication from gain–saturation (GaAs has a problem here). High dark current should also be avoided because of the associated shot noise.

The gain is a function of applied voltage and temperature; therefore both must be held very stable. If the excess-noise factor $\mathfrak{f}(G)$ is lumped with the exponent of the gain (G^γ), the limits on this exponent would be between 2 and 3 ($2 < \gamma \le 3$). Note that it cannot ever equal two ($\gamma \ne 2$). This analysis is useful for instructional purposes (MacGregor 1991). The equation for total mean-square noise would then become

$$i_n^2 = 2q^2\Delta f(i_{\text{sig}} + i_{\text{bkg}})G^\gamma$$
$$+ 2q\Delta f \frac{kT}{qR_d}\left[\exp(qv/kT) - 1\right]G^\gamma + 4\,kT\,\Delta f\left(\frac{1}{R_L} + \frac{1}{R_d}\right) \tag{7.100}$$

(*Note:* When $G = 1$, we have a classic *p-n* junction photodiode.) The $G = 2$ case is the ideal case (where either electrons or holes are causing the avalanching); and in the $G = 3$ case, both electron and holes are equally effective in the multiplication process. For the two limited cases, these approximations are correct.

For instructional purposes, let's consider the case of $G = 2$ for a small background. The SNR would be

$$\frac{S}{N} = \frac{\eta q \phi_q G}{\sqrt{2q\Delta f\left\{\eta\phi_q + \dfrac{kT}{qR_d}\left[\exp(qv/kT) - 1\right]\right\}G^\gamma + 4kT\Delta f\left(\dfrac{1}{R_L} + \dfrac{1}{R_d}\right)}}$$

$$\tag{7.101}$$

Consider the SNR as a function of gain (G). When Johnson noise dominates, the gain increases the SNR until shot noise dominates, as shown in Fig. 7.38. Observe the effects of $\gamma > 2$ in Fig. 7.38, for example, $\gamma = 2.5$. The SNR-versus-gain plot passes through a maxima, so further increasing of G does not enhance SNR, but actually lowers it. As G increases, the excess noise expressed in the exponent dominates the total noise.

Rewriting the SNR with the excess-noise factor again:

$$\frac{S}{N} = \frac{\eta q \phi_{q,\text{sig}} G}{\sqrt{2q(i_{\text{sig}} + i_{\text{bkg}} + i_d)\Delta f G^2 \mathcal{f}(G) + \left(\frac{4\,kT\,\Delta f}{R_{\text{eq}}}\right)}} \tag{7.102}$$

where

i_d = dark current
R_{eq} = parallel equivalent resistance $(R_d \| R_L)$

The PIN-diode signal-to-noise equivalent is found by setting $G = 1$ and $\mathcal{f}(G) = 1$; therefore the SNR for the two cases is

$$\frac{(S/N)_{\text{PIN}}}{(S/N)_{\text{APD}}} = \sqrt{\mathcal{f}(G)} \tag{7.103}$$

as seen from Eq. (7.102). A general expression for the excess-noise factor as a function of the ionization coefficient lies between the extremes of one-carrier amplification and the case where both holes and electrons have equal ionization coefficients. Noting that (β_i/α_i) is the ratio of ionization coefficients, we then rewrite the expression for shot noise for the case of either holes or electrons

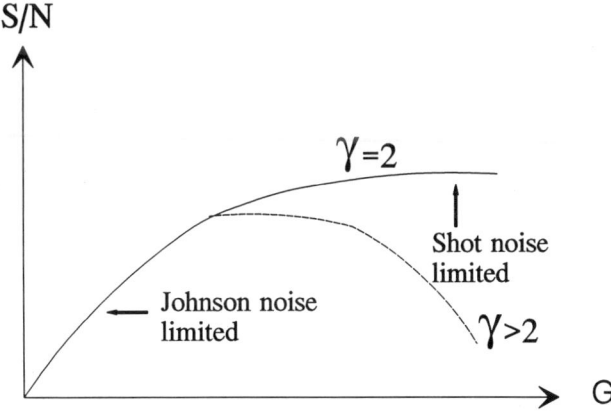

Figure 7.38. SNR vs. gain.

initiating the avalanching (McIntyre, 1966):

$$i_s^2 = 2q\bar{i}G^3 \left[1 + \frac{1 - (\beta_i/\alpha_i)}{(\beta_i/\alpha_i)} \left(\frac{G - 1}{G} \right)^2 \right] \qquad (7.104a)$$

$$i_s^2 = 2q\bar{i}G^3 \left[1 - (1 - (\beta_i/\alpha_i)) \left(\frac{G - 1}{G} \right)^2 \right] \qquad (7.104b)$$

When electrons are injected or initiate avalanche, these shot-noise expressions are more complex when both electrons and holes are injected (McIntyre, 1966):

$$i_s^2 = 2q\bar{i}G^2 \left[(\beta_i/\alpha_i)G + (1 - (\beta_i/\alpha_i)) \left(2 - \frac{1}{G} \right) \right] \qquad (7.105)$$

So the excess noise factor for electron avalanching is

$$\beta(G) = (\beta_i/\alpha_i)G + (1 - (\beta_i/\alpha_i)) \left(2 - \frac{1}{G} \right) \qquad (7.106)$$

Figure 7.39 is a plot of excess-noise factor versus gain for various values of the ionization coefficient ratio (β_i/α_i). For $\beta_i = 0$, $(\beta_i/\alpha_i) = 0$, so $\beta(G) = 2 -$

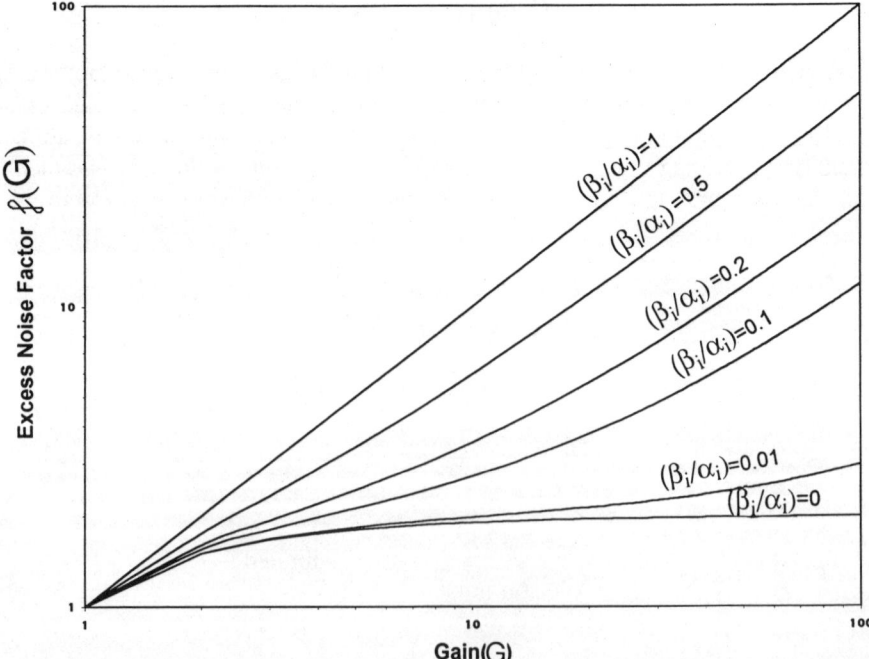

Figure 7.39. Plot of excess-noise factor vs. gain for different values of the ionization–coefficient ratio (β_i/α_i).

$1/G$. As stated earlier, the excess-noise factor lies between $(2 - 1/G)$ and G. The key elements to obtain low-noise signal-carrier amplification are:

- A large ratio of ionization coefficients.
- The carrier with the largest ionization coefficient starts the multiplication process.
- There exists a high probability of ionization at the localized point.
- The carrier energy is greater than the bandgap.

As the temperature of the device is decreased, the ionization ratio, (β_i/α_i), decreases and the allowable gain, G, increases as shown in Fig. 7.40. The quantum efficiency, η, is also a function of device temperature. Webb et al. (1974) calculated that the quantum efficiency increases with increasing temperature. This quantum efficiency is a function of the absorption coefficient that is dependent on wavelength and temperature. Because η is an important parameter in the SNR (Eq. 7.90), an optimum operation wavelength exists that yields maximum quantum efficiency, and both η and the operation wavelength are commonly specified in commercial data sheets. A useful figure of merit to compare various ADPs is $\eta/(\beta_i/\alpha_i)^{1/2}$, because η includes structural qualities and (β_i/α_i) includes material qualities.

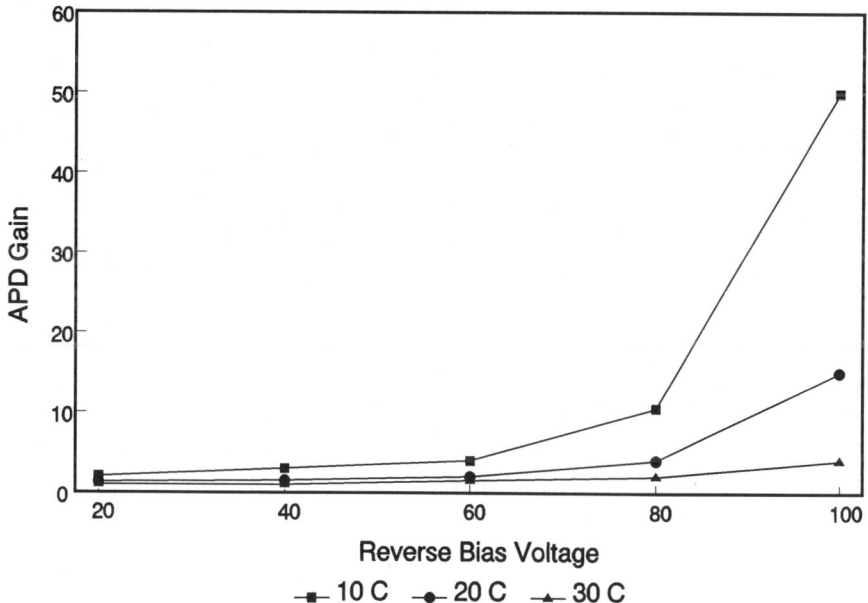

Figure 7.40.   Temperature dependence of gain.

7.4. GERMANIUM PHOTODIODES

Silicon is the most-used and best-understood detector material. The second most well-known semiconductor material is germanium, which was researched heavily during the 1950s and 1960s. Germanium has the highest refractive index of any elemental semiconductor ($n \sim 4$). Pure germanium is considered an intrinsic semiconductor, but doping it creates a $p\text{-}n$ junction photodiode.

Germanium photovoltaic detectors are used in the long-wavelength visible and the near infrared, approximately from 0.6 μm to 1.8 μm. Few other detectors operate in this range, and germanium is the least expensive of these. Other advantageous properties of germanium are:

- Relatively high sensitivity (NEP $\sim 10^{-13}$ watt).
- High speed (65 MHz for a 0.8-mm^2 area).
- Ruggedness; good overload recovery (except that native oxide is water soluble).
- Wide dynamic range with linear response (10^7).
- High refractive index makes antireflection coating easier to design for selective wavelengths.

7.4.1. Construction of Ge Photodiode

A simplified version of the construction of a "passivated mesa germanium photodiode" is shown in Fig. 7.41. A piece of p-type germanium (gallium doped to a 10^{15}-cm^{-3} concentration or 0.8-Ω-cm) is placed in an environment of an element (such are arsenide) from the fifth (V) column of the periodic table of the elements. The arsenide diffuses a small distance (~ 1 μm) into the p-type germanium, filling in the acceptor levels and creating donor levels until there is effectively an n-type region, and therefore a graded (as opposed to abrupt) $p\text{-}n$ junction is created. Note that the doping is not important in the detection process because, as pointed out earlier, the intrinsic energy gap determines the spectral response. The center of the newly created n-type is masked off and a portion of the surrounding semiconductor material is etched away, leaving a raised flat spot (hence the term *mesa*) that contains a $p\text{-}n$ junction as shown in Fig. 7.41b. Subsequently, an oxide film is deposited on the mesa structure. The function of this "passivation," or oxide film, is to reduce surface conductivity in the vicinity of the $p\text{-}n$ junction, which could ultimately destroy the detector efficiency by introducing a shunt resistance that is lower than the detector resistance (lowering the RA product). Finally, an antireflection coating can be deposited on the surface upon which the radiation will be incident to preserve the signal.

Having fabricated the $p\text{-}n$ junction, electrodes are connected to a chip carrier and the package is mounted onto a thermoelectrically temperature-controlled surface. The diode is then hermetically sealed in a canlike package with a glass window. The ramifications of using a thermoelectric cooler will become apparent.

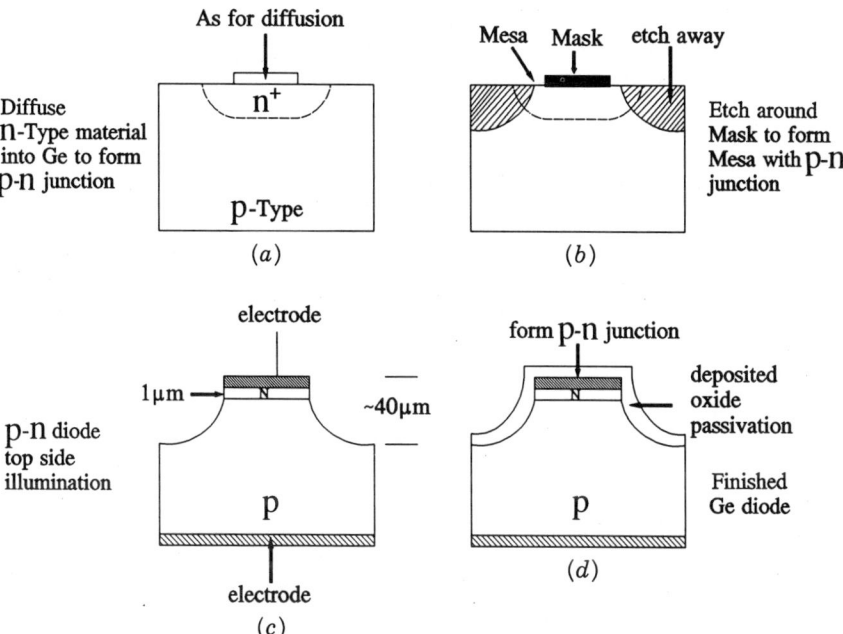

Figure 7.41. Mesa construction of a Ge photodiode.

Germanium photodiodes are readily made in areas ranging from 0.05 to 3 mm^2, and the capability exists to make the area as small as 5×10^{-4} mm^2 or as large as 500 mm^2. The lower limitation results from the mesa structure (the diffusion, masking, etching process), and the upper limit is imposed by raw-material uniformity. If other techniques are employed, different geometries can be attained.

7.4.2. Operating Characteristics of Ge Photodiodes

The basic operating features of any photovoltaic detector can be determined from its v-i curve (see Fig. 7.3). When the irradiance on the detector is zero, the diode curve is the standard form:

$$i = i_0 \left(\exp \left\{ \frac{qv}{kT} \right\} - 1 \right) \tag{7.106}$$

where i_0, the reverse saturation current, is specific to a given detector material. When the detector is exposed to radiation, the characteristic curve is shifted down an amount i_g (Chap. 3), which is given by

$$i_g = \eta q \phi_q \tag{7.107}$$

The lower limit of spectral response for germanium can be explained by understanding how the quantum efficiency is affected by wavelength. Quantum efficiency is related to the absorption coefficient by

$$\eta = (1 - e^{-ax}) \frac{4n}{(n + 1)^2} \qquad (7.108)$$

The absorption coefficient becomes large for shorter wavelengths (see Fig. 3.1), making the quantum efficiency very nearly equal to one minus the Fresnel reflection. Thus, photons are easily absorbed, freeing electrons very near the surface of the detector. Defects and impurities in the surface serve to "trap" these electrons, slowing their movement, and causing them to recombine with their respective holes before becoming effective current. Thus, the dynamic resistance at zero volts (from Eq. 7.28) is

$$R_0 = \left. \frac{\partial v}{\partial i} \right|_{v=0} = \frac{kT}{qi_0} \qquad (7.109)$$

Figure 7.42 shows the dynamic resistance at zero volts plotted as a function of temperature for several detector diameters.

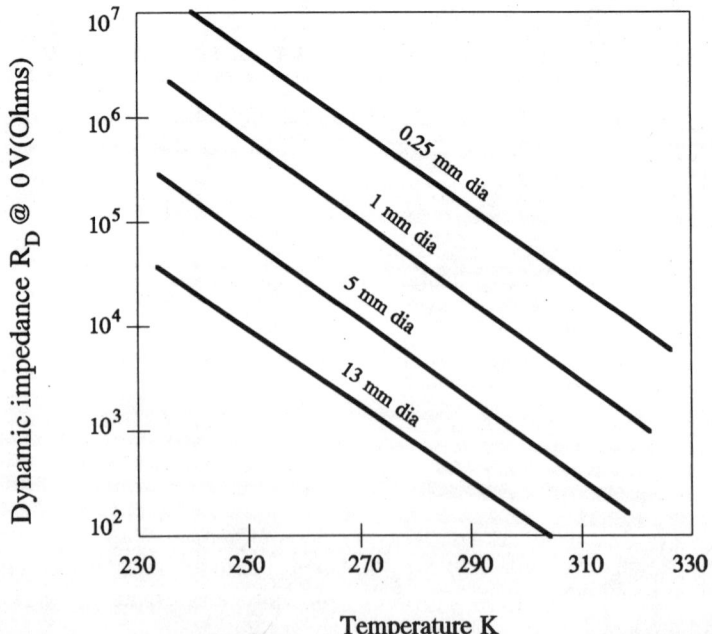

Figure 7.42. Resistance of Ge photodiode vs. temperature for various detector diameters.

The reverse saturation current, i_0, is the maximum current present in a PV detector when a negative voltage is applied to the diode. This saturation effect is observed (rather than ohmic behavior) because few mechanisms enable the charge carriers to go "backwards" in a photodiode, and these mechanisms do not necessarily depend on the applied voltage. The reverse saturation current is often written as

$$i_0 = J_d A_d + J_{gr} A_d + J'S \tag{7.110}$$

where

J_d = diffusion-current density in amps/area
J_{gr} = generation–recombination current density in amps/area
J' = surface leakage current density
A_d = area of the p-n junction
S = surface area of the perimeter of p-n junction

The "passivation" or oxide film mentioned earlier, plus good engineering practice, high-quality starting material, uncontaminated surfaces, and hermetic packaging, make J' small enough to neglect for purposes of our discussion. Now i_0 is written

$$i_0 = (J_d + J_{gr})A_d \tag{7.111a}$$

In germanium, the diffusion current overwhelms the generation–recombination current because

$$J_d \propto n_i^2 \quad \text{whereas} \quad J_{gr} \propto n_i \tag{7.111b}$$

and germanium has a relatively high intrinsic-carrier concentration (see Table 7.4; Streetman, 1980). We can now write

$$i_0 = J_d A_d = \left(\frac{n_p D_e}{L_e}\right) A_d q = A_d q n_p \sqrt{\frac{D_e}{\tau_e}} \tag{7.112}$$

where we recall that n_p is the electron concentration in the p-type material. It is observed that the p-type extrinsic makes the most significant contribution to the diffusion current, J_d, so we have ignored that portion attributable to the n-type.

From Table 7.4 we find that, for room-temperature germanium, $D_e = 101$ cm^2/s. The carrier lifetime, τ_e, is given by τ_e (μs) $= 50\ \rho$, where ρ is the resistivity in ohm-cm. The resistivity is a function of the substrate doping levels, typically, $\tau_e = 2 \times 10^{-4}$ s. With this information, we refer to Fig. 7.43, which shows the reverse saturation current plotted as a function of temperature

Table 7.4 Properties of Semiconductors at Room Temperature

Property	Si	Ge	InSb	HgCdTe	GaAs	SiO$_2$
Energy gap, $\mathcal{E}_g$ @ 300 K [eV]	1.107	0.67	0.163	0.102	1.35	~8
Temperature coefficient of $\mathcal{E}_g$, $d\mathcal{E}_g/dT$ [eV/K]	-2.3×10^{-4}	-3.7×10^{-4}	-2.8×10^{-4}	3.0×10^{-4}	-5×10^{-4}	
Electron mobility μ_e [cm^2/V s]	1900	3800	78,000	10^5 3×10^5 @ 77 K	8800	
Hole mobility μ_h [cm^2/V s]	500	1820	750	150 1000 @77 K	400	
Temperature coefficient of μ_e, $d\mu_e/dT$ [cm^2/V sk]	-2.6	-1.66	-1.6		-1.0	
Temperature coefficient of μ_h, $d\mu_h/dT$ [cm^2/V $\cdot$ s $\cdot$ k]	-2.3	-2.33	-2.1		-2.1	
Electron diffusion coefficient [cm^2/s]	35	100			220	
Hole diffusion coefficient [cm^2/s]	12.5	50			10	
Intrinsic carrier concentration [/cm^3]	1.38×10^{10}	2.5×10^{13}		2×10^{13} ~10^{15}	2×10^6	
Intrinsic resistivity [Ω-cm]	2.3×10^5	43	0.06	0.03	4×10^8	10^{14}–10^{16}
Breakdown field [V/cm]	3×10^5	8×10^4				5–10×10^6
Density of states (cb) [/cm^3]	3.22×10^{19}	1.0×10^{18}		1.91×10^{15}		

Density of states (vb) [/cm³]	1.83×10^{19}	6.0×10^{18}		1.15×10^{18}		
Effective mass—e^-	$1.1\ m_0$	$0.55\ m_0$				
Effective mass—h^+	$0.56\ m_0$	$0.37\ m_0$				
Dielectric constants	11.8	16	17.7		13.2	3.8–3.9
Refractive index (near bandgap)	3.42	4.00	4.00		3.65	1.46
Molecular weight	28.09	72.59	236.57	310.11	144.64	60.08
Lattice constant [Å]	5.43072	5.65754	6.47877	6.47	5.65315	~3.5163
Density [g/cm³]	2.3283	5.3234	5.775	7.6195	5.316	2.27
Melting point [K]	1685 ± 2	1231	798	700	1510	~1700
Specific heat [J/kgK]	702	321.9	144	60.74		1000
Thermal conductivity [W/cm K]	1.240	0.640	0.160	0.201	0.560	0.014
Thermal diffusivity [cm²/sec]	0.9	0.36				0.01
Linear thermal expansion coefficient [/K]	4.68×10^{-6}	6.1×10^{-6}	4.7×10^{-6}	2.0×10^{-6}	5.4×10^{-6}	5×10^{-7}
Elastic constants [dyne/cm²] C_{11}	1.66×10^{12}	1.30×10^{12}	1.15×10^{12}	5.46×10^{11}	1.18×10^{13}	
C_{12}	0.64×10^{12}	0.49×10^{12}	6.26×10^{11}	3.77×10^{11}	5.31×10^{12}	
C_{44}	0.80×10^{12}	0.67×10^{12}	3.02×10^{12}	2.03×10^{11}	5.94×10^{12}	
Volume compressibility [cm²/dyne]	3.06×10^{-12}	7.68×10^{-12}	4.42×10^{-12}		7.71×10^{-12}	
Young's modulus ⟨111⟩ [dyne/cm²]	1.69×10^{12}		-1.34×10^{12}		-3.31×10^{12}	
Bulk modulus [dyne/cm²]	7.7×10^{11}		4.65×10^{12}		7.47×10^{12}	$2–4 \times 10^{9}$
Rupture modulus (bending) [g/cm²]	$0.7\text{–}3.5 \times 10^{6}$					
Break strength (compression) [g/cm²]	$4.9\text{–}5.6 \times 10^{6}$					

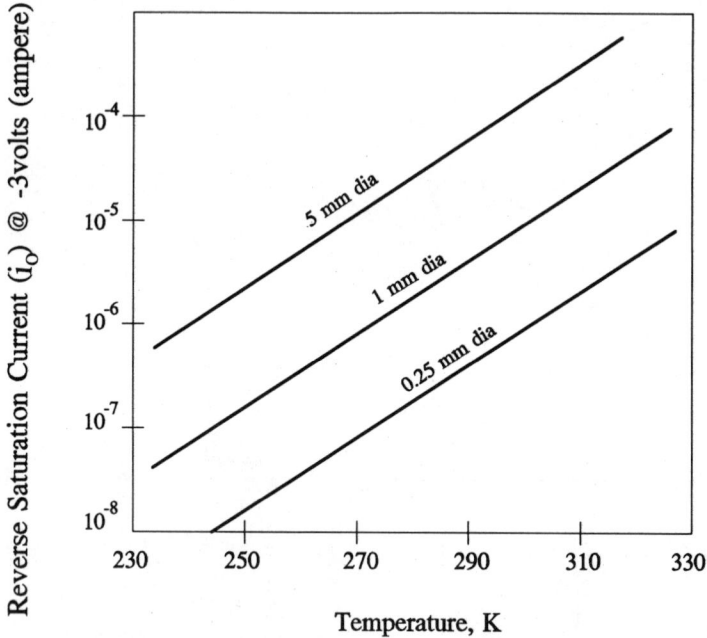

Figure 7.43. Reverse saturation current, i_0, as a function of temperature.

for various detector areas. Note that for a 1-mm-diameter photodiode at 250 K, $i_0 \approx 1 \times 10^{-6}$ amps. Thus, we can deduce the impurity concentration or the number of acceptors in the p region, n_a

$$n_a \approx n_p = \frac{i_0}{A_d q} \sqrt{\frac{\tau_e}{D_e}} = 1.12 \times 10^{12} \text{ cm}^{-3} \qquad (7.113)$$

7.4.3. Figures of Merit

First, let us consider the responsivity. Assuming that we operate into a virtual short circuit (such as a transimpedance amplifier), the current responsivity of a photovoltaic detector is given by

$$\mathscr{R}_i = \frac{\eta q \lambda}{hc} \qquad (7.114)$$

for the theoretical photodetector. The cutoff wavelength (λ_p) from Chapter 3, using the energy gap of 0.67 eV, infers a theoretical value of 1.85-μm cutoff wavelength. In practice, thermal (kT) effects cause some rounding in the neighborhood of the cutoff wavelength. The actual responsivity of a germa-

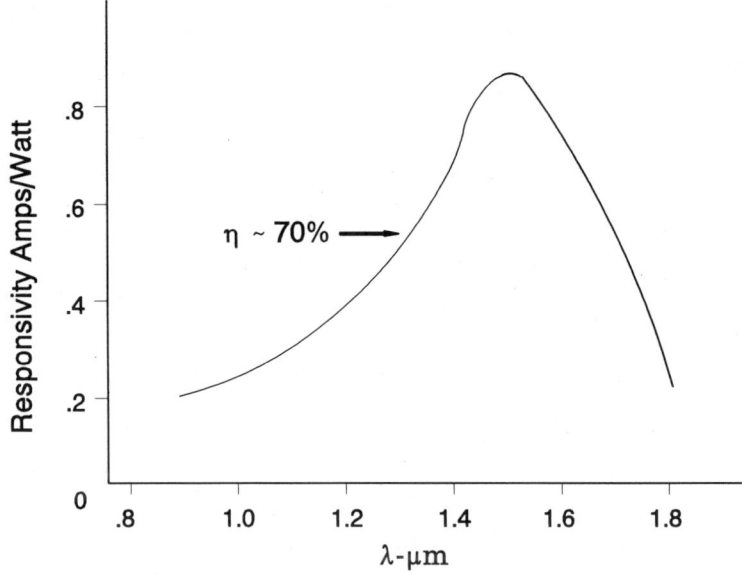

Figure 7.44. Representative current responsivity for a Ge photodiode.

nium photodiode is shown in Fig. 7.44. This is only a representative response curve, as fabrication parameters can affect the specific shape of the curve.

The energy gap in germanium is a function of temperature, which means that λ_p is also a function of temperature. The energy gap increases slightly as germanium is cooled. This effect is usually ignored, but is represented by

$$\mathcal{E}_g(T) = \mathcal{E}_g(T = 0) - \frac{\alpha_i T^2}{T + \beta_i} \tag{7.115}$$

where $\mathcal{E}_g(T = 0) = 0.741$, $\alpha_i = 4.56E{-}4$, and $\beta_i = 210$.

Figure 7.45 illustrates an interesting effect. A well-meaning experimenter might naively cool his detector to reduce the noise, but if the wavelength he is trying to detect is near the cutoff wavelength (λ_p) for room-temperature operation, he might have no response at all to that wavelength after cooling. For example, it appears that for a 1.5-μm signal, using a thermoelectric cooler to lower the temperature to 250 K would have a desirable effect, whereas at 1.8 μm, it would be a disaster. This effect, as well as the smoothing (kT) effect, can readily be seen in Fig. 7.45. It even seems that in some cases where noise is not a major concern, it might be advantageous to heat the detector slightly.

We are concerned fundamentally with detector performance and sources of detector noise. Therefore preamplifier noise is not discussed here, but its presence must be accounted for prior to any attempt at system design. Fortunately, 1/f noise does not seem to trouble germanium PVs because of a low electron-

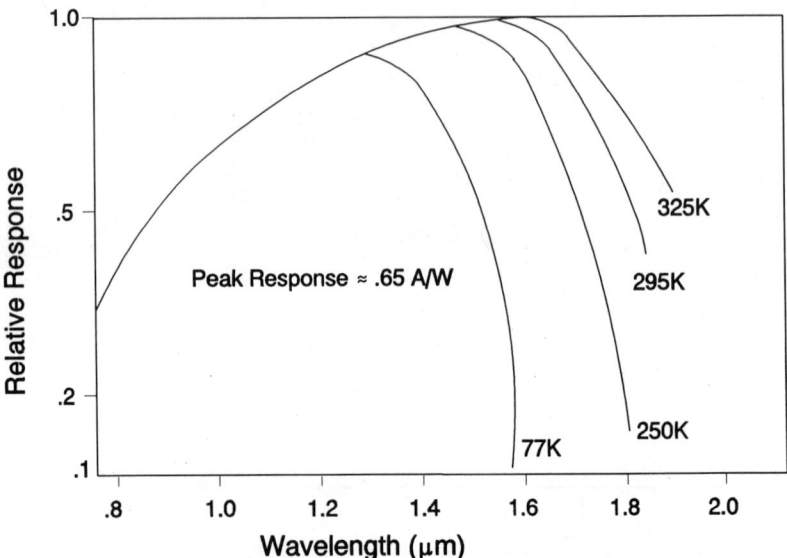

Figure 7.45.   Response vs. wavelength for a Ge photodiode for various temperatures.

to-hole ionization ratio. The $1/f$ noise is effectively proportional to the direct current in the diode, and this is or can be made small for PV detectors. This fact coupled with the fact that techniques exist to reduce $1/f$ noise sufficiently that it can be neglected in most cases.

Johnson and shot noise are the remaining noise sources we will investigate. The Johnson-noise current is written as

$$i_J = \sqrt{4\ kT\Delta f \left(\frac{1}{R_L} + \frac{2}{R_d} \right)} \tag{7.116}$$

As usual, it was assumed that the feedback/load resistor is at the same temperature as the detector.

Shot noise or photon noise is given by

$$i_s = \sqrt{2q(\bar{i} + i_g)\ \Delta f} = \sqrt{2q(\bar{i} + \eta q \phi_q)\ \Delta f} \tag{7.117}$$

If we are using an opamp, then we operate into a virtual short circuit, and $V = 0$, so

$$\bar{i} = i_0[\exp(qv/kT) - 1] = 0 \tag{7.118a}$$

and

$$i_s = \sqrt{2q^2 \eta \phi_q \Delta f} \tag{7.118b}$$

It is important to remember that if the load on the detector does have any appreciable resistance, then one must include the resultant dc current in the shot-noise calculation. For BLIP operation, the photon noise must overwhelm the Johnson noise, so to be photon-noise limited, photon noise >> Johnson noise:

$$q^2 \eta \phi_q \gg 2\,kT \left(\frac{1}{R_L} + \frac{1}{R_d} \right) \tag{7.119}$$

or

$$\phi_{e,\text{min}} = E_{e,\text{min}} A_d \phi_q \gg \frac{2\,kT}{\eta q^2} \left(\frac{hc}{\lambda} \right) \left(\frac{1}{R_L} + \frac{1}{R_d} \right) \tag{7.120}$$

This irradiance, $E_{e,\text{min}}$, is the minimum that will result in BLIP operation. The corresponding flux in watts is $\phi_{e,\text{min}}$, and the area of the detector is A_d. If the detector resistance dominates the Johnson noise, and we agree that a factor of 4 will satisfy the "much greater than" in the BLIP condition given earlier, then we find that the minimum irradiance for BLIP operation for 1.5 μm depends on parameters that we already know:

$$E_{e,\text{min}} = \frac{8\,hc}{\eta \lambda_p} n_p \sqrt{\frac{D_e}{\tau_e}} = 1.2 \times 10^{-7}\ \text{W/cm}^2 \text{ at room temperature} \tag{7.121}$$

This result is sufficiently high that we, unfortunately, will almost always be Johnson-noise limited.

Because we have the Johnson noise, and we can get the responsivity from either the equation or the actual data (Fig. 7.44), we can calculate the spectral NEP and normalized detectivity for a Johnson-noise-limited system. Generally, the spectral NEP normalized detectivity is

$$\text{NEP}(\lambda) = \frac{i_n}{\mathcal{R}_i} \quad \text{and} \quad D^*(\lambda) = \frac{\sqrt{A_d \Delta f}}{\text{NEP}(\lambda)} \tag{7.122}$$

where i_n is the dominant root-mean-square-noise (rms) current. Some interesting points may be brought out by combining equations already presented to calculate the Johnson-noise-limited D^*:

$$D^*(\lambda) = \frac{\mathcal{R}_i}{2} \sqrt{\frac{A_d R_d}{kT}} = \frac{\mathcal{R}_i}{2} \sqrt{\frac{A_d}{q i_0}} = \frac{\mathcal{R}_i}{2q \sqrt{n_p}} \left(\frac{\tau_e}{D_e} \right)^{1/4} \tag{7.123}$$

We see that D^* depends only on responsivity, carrier lifetime, doping levels, and the diffusion coefficient. Because we know these quantities (Table

7.4), we can calculate peak spectral $D^* = 9.5 \times 10^{10}$ cm Hz$^{1/2}$ W^{-1} for room temperature, assuming an electrical bandwidth of 1 Hz. Another observation is that for Johnson-noise-limited systems that have the detector resistance as the major noise source, the peak spectral NEP seems to decrease as the temperature is raised. This surprising result occurs because the temperature dependence of the detector resistance cancels that of the Johnson noise, and we are left only with the temperature dependance of the cutoff wavelength through the changing energy gap. Thus, to first order, as temperature increases, so does the cutoff wavelength, the current responsivity, and peak spectral D^*; hence the NEP is reduced.

Another important point is that, because we are Johnson-noise limited, we can improve detector performance by immersion in a hemispherical lens. The effective area of the detector increases by n^2, where n is the refractive index of the medium. Germanium, with $n = 4$, is an excellent candidate material for the immersion lens because of its high refractive index.

One more operating characteristic must be addressed. Because the p-n junction acts like a capacitor—there is some load resistance (regardless of system configuration)—we have an RC time constant that results in an electrical bandwidth of the system. The resistance used in this calculation must be the parallel combination of the detector resistance and the load resistance. Whether using an opamp or not, the minimum realistic value is usually around 50 Ω. There is also a cutoff frequency for the photodiode all by itself, but in most systems this cutoff frequency is much higher than that imposed by the interface, and so the resistance of the interface is included in the RC time constant, as earlier.

The capacitance per unit area is

$$\frac{C}{A} \text{ [pF/cm}^2\text{]} = \sqrt{\frac{q\epsilon n_b}{2(v_{bi} - v_r)}} = 0.011 \sqrt{\frac{n_b}{(v_{bi} - v_r)}} \qquad (7.124)$$

where ϵ is the dielectric constant for germanium, and $n_b \approx n_a$ is the bulk impurity level in cm^{-3}. Note that we can lower the capacitance by applying a reverse bias (v_r). This practice often occurs in high-speed applications. The built-in voltage v_{bi} common to all PVs is approximately

$$v_{bi} = \frac{kT}{q} L_e \left(\frac{n_d n_a}{n_i^2}\right) \approx \frac{kT}{q} L_e \left(\frac{n_a^2}{n_i^2}\right) = 0.16 \text{ V} \qquad (7.125)$$

for room-temperature germanium. The number of donors in the n region is n_b $\approx n_a$. The built-in voltage, or work function, is particularly important to know when operating into a virtual short circuit so that $v = 0$. Having a means to calculate the capacitance, and presuming we know our load resistance, we can proceed to calculate the cutoff frequency for a given situation:

$$f_{RC} = \frac{1}{2\pi C} \left(\frac{1}{R_d} + \frac{1}{R_L}\right) \qquad (7.126)$$

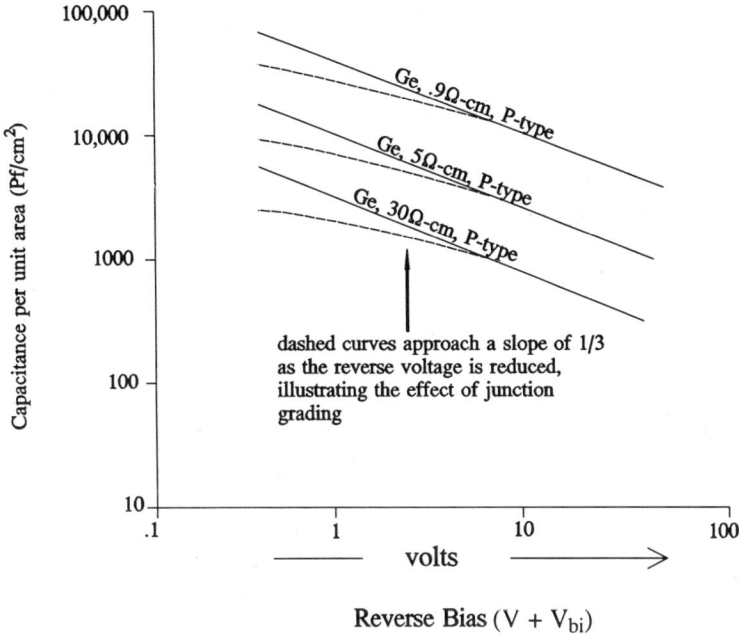

Figure 7.46.   Germanium-photodiode capacitance as a function of reverse voltage.

If we are operating without a bias voltage, then the cutoff frequency depends only on the area of the detector:

$$f_{RC}[\text{Hz}] \doteq \frac{2.2 \times 10^5}{A_d[\text{cm}^2]} \tag{7.127}$$

assuming a resistance of 50 Ω. A bias voltage can greatly extend the frequency response; however, Fig. 7.46 shows capacitance for germanium photodiodes of different resistivities plotted as a function of voltage. The solid curve is predicted by the preceding equation, and the dashed curve represents actual data. The departure of the curves occurs because the *p-n* junction is gradual rather than abrupt. Note that the minimum voltage is determined by the built-in voltage of 0.16 V.

7.5. InSb PHOTODIODE

Indium antimonide (InSb) is one of the most highly developed and widely used detector materials for the near-infrared portion of the spectrum (1 to 5 μm). With recent advances in material preparation and preamp design, the InSb photovoltaic detector is one of the most sensitive detectors available for use in the 3- to 5-μm atmospheric transmission "window."

InSb was first considered for infared-detector applications because it could be prepared in single-crystal form using conventional techniques. The ability to analytically determine is bulk crystalline solid-state properties was a great design advantage compared to competing detector materials such as PbSe (Kruse, 1970). By the late 1950s high-performance diffused-junction InSb detectors were available (Lesser et al., 1958). The InSb photodiode has been a very reliable and useful detector because it closely follows a simple theoretical model.

7.5.1. Physical Properties

Indium antimonide (InSb) is a member of the class of photodetective materials known as III-V semiconductors. The crystal structure of pure InSb is zincblende cubic, similar to the compound InAs and the alloy system $Hg_{1-x}Cd_xTe$. Physically, this structure is composed of two interpenetrating face-centered cubic sublattices displaced by (a/4, a/4, a/4) along the cubic body diagonal, where a is the lattice constant. The p-type material is formed if more indium is used in the compound and an n-type material is formed if more antimony is used. Thus, a p-n junction can be formed by the control of the InSb ratio.

Indium antimonide is an intrinsic semiconductor with a direct energy bandgap, because both the conduction-band minimum and valence-band maximum are located at the center of the Brillouin zone. At room temperature (300 K) the energy gap of 0.17 eV corresponds to a long-wavelength response cutoff of around 7 μm. If cooled to a liquid-nitrogen temperature of 77 K, the energy gap increases to 0.23 eV, shifting the cutoff wavelength to 5.5 μm. The relationship between energy gap and operating temperature from 4 K to 300 K is

$$\mathcal{E}_g[eV] = 0.233 - (6.498 \times 10^{-5})T - (5.0826 \times 10^{-7})T^2 \quad (7.128)$$

The overall spectral response of InSb at 77 K is typically 1 to 5.5 μm, shown by the relative spectral response curve in Fig. 7.47. The optical-absorption coefficient is very sharply turned at approximately 5.2 μm, as can be inferred from Fig. 7.47.

A plot of the energy bandgap $\mathcal{E}_g$ versus temperature (Fig. 7.48) shows the conventional negative temperature coefficient. As was shown, it is desirable to achieve the highest RA product possible to maximize the D^* of the detector. The RA product was derived by Long (1980) using Melngailis and Harman (1970) for a geometric model of a p-n junction shown in Fig. 7.49. The RA product may be written as

$$R_dA_d = \frac{kT}{q}\left(\frac{qD_e p_p}{L_e}\coth\frac{\ell_{z,n}}{L_e}\right)^{-1} \quad (7.129)$$

This equation assumes that the back contact of the detector is a high-recombination-rate interface, that is, no space-charge layer is formed at the contact.

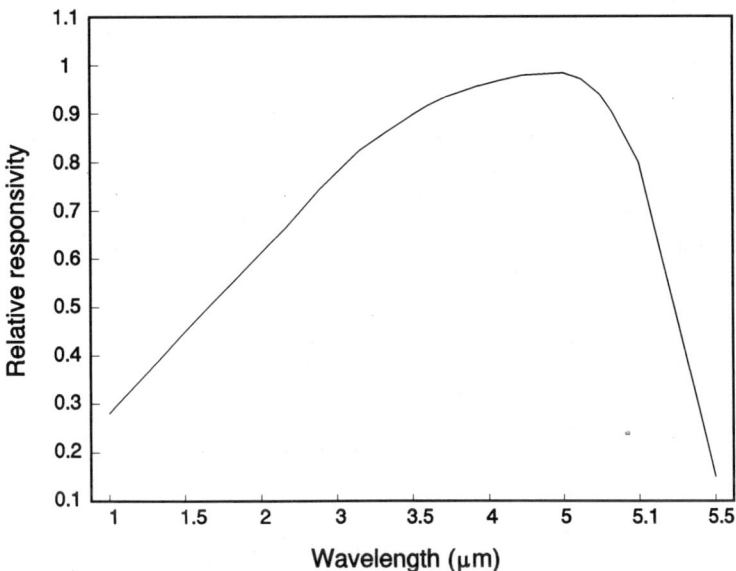

Figure 7.47.   Relative spectral response of InSb PV detector.

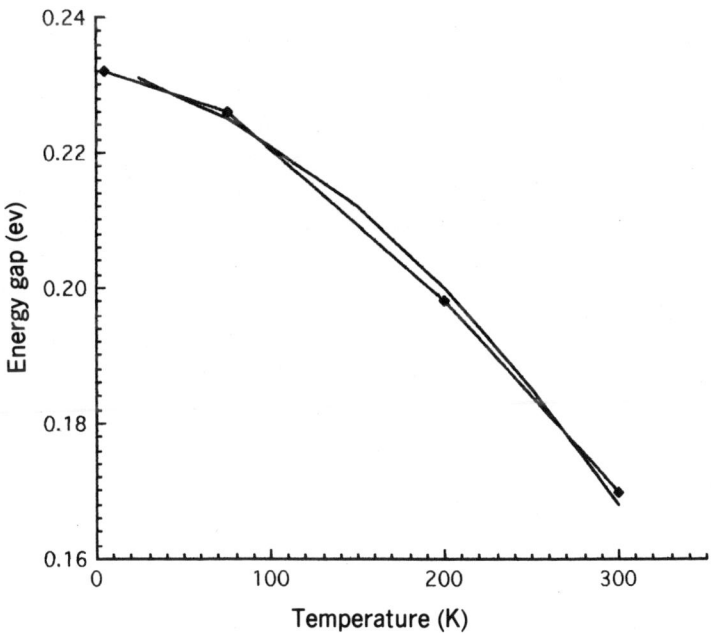

Figure 7.48.   Dependence of the energy gap of InSb upon temperature.

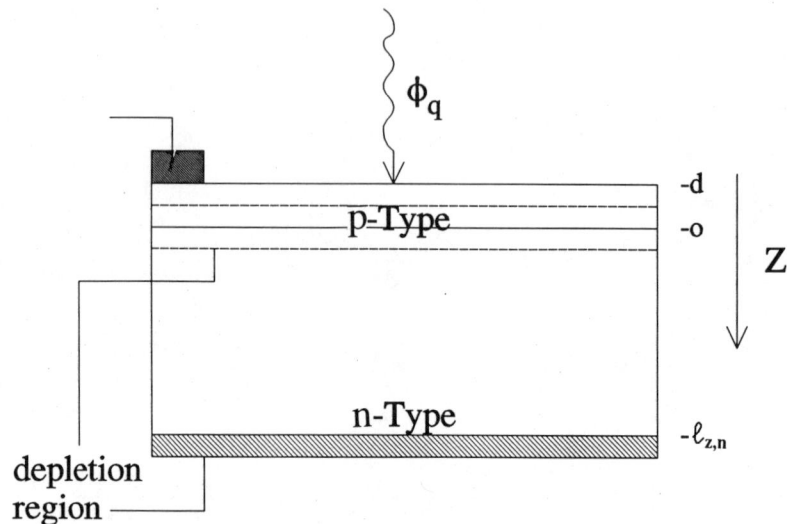

Figure 7.49. InSb photodiodes with n-region thickness, $\ell_{z,n}$.

If the thickness of the n-type substrate is much larger than the diffusion length ($\ell_{z,n} \gg L_e$) then the RA product is

$$R_d A_d = \frac{kT}{q^2} \frac{\tau_h n_p}{L_e n_i^2} \tag{7.130}$$

where $n_i^2 = p_p n_p$ and n_p is the doping of the p-type material. Depending on the particular detector, it is often experimentally observed that a decrease in detector temperature causes a corresponding increase in detector resistance, that is, the RA product. Recall that the n_i^2 varies at least as fast as T^3. An often-achieved value of 2×10^6 Ω-cm^2 at 77 K can be improved by decreasing the detector temperature to 60 K and obtaining values of 2×10^7 Ω-cm^2. For infrared detectors for low-background astronomy, Richards and Greenberg (1982) report an RA product greater than 10^7 Ω-cm^2 for InSb at 77 K. After cooling the detector to a liquid-helium temperature of 4.2 K, the RA product increased to greater than 10^{10} Ω-cm^2. The quantum efficiency is typically about 65%.

One property of InSb PV detectors that is not yet fully understood is the change in detector quantum efficiency with changes in detector temperature. The refractive index is about 3, so $\eta = 75(1 - e^{-az})$. Hall et al. (1975) measured the quantum-efficiency temperature dependence for a number of InSb detectors and have found that the quantum efficiency is independent of temperature to 60 K. However, at lower temperatures, the quantum efficiency exhibits a significant decrease, especially at longer wavelengths.

Figure 7.50 shows this phenomenon by plotting quantum efficiency versus

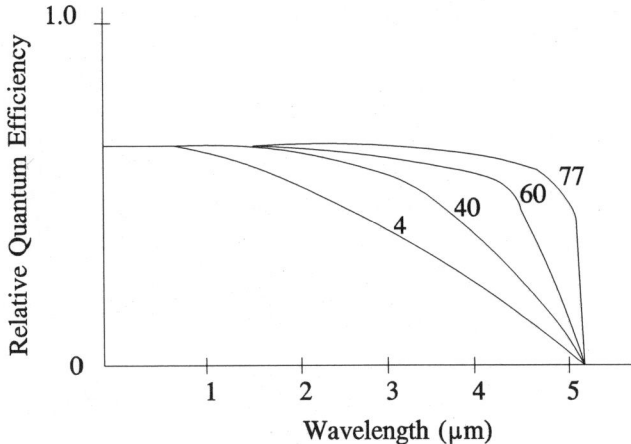

Figure 7.50. Quantum efficiency vs. wavelength for various operating temperatures.

wavelength for various detector temperatures. The reduction in quantum efficiency for device operation at 4 K (where the RA product is highest) causes a decrease in detector D^*. Future improvements in InSb material technology and/or fabrication techniques may help eliminate this problem. This loss at longer wavelengths may result because the carrier diffusion length is a function of temperature. Therefore, long-wavelength photons may not be absorbed sufficiently near the depletion region where they can be detected.

7.5.2. InSb Detector Fabrication

The processing of high-purity single crystals of InSb has been discussed by Liang (1962), Hulme and Mullin (1962), and Hulme (1965). Indium antimonide is formed by reacting commercially available, high-purity (0.999999) indium and antimony above the compound melting point of 523°C. The reaction can take place at atmospheric pressure because the partial pressures of indium and antimony are low. However, a reducing atmosphere is used to prevent the formation of indium and antimony oxides, which would react with the quartz vessel.

Harman (1956) showed that the dominant foreign-atom impurities are those of zinc (an acceptor) and tellurium (a donor). Horizontal zone refining in a reducing atmosphere is used to remove these impurities from the InSb compound (tellurium being the more difficult to remove). Once these zone-refined polycrystalline ingots are formed, single crystals of InSb are produced using either the Czochralski method or by zone leveling. As before, crystal growth is done in a reducing atmosphere in a quartz vessel. Initial testing of the bulk crystal is done using a variety of techniques. Hall coefficient and resistivity measurements are used to determine the majority carrier concentration and mobility, while thermoprobing reveals the n- and p-type regions in the bulk crystal.

Most extensive zone refining has been done on n-type InSb, for which the stoichiometric content of antimony is greater than indium. A consequence of this more improved technology for n-type InSb is a lower intrinsic carrier concentration than in p-type material. As a result, most PV InSb detectors are made by diffusing p-type InSb directly into a mesa of n-type InSb, forming a top p-type mesa of typically 5- to 10-μm thickness. (Joyce, 1983).

7.5.3. Operational Characteristics of InSb

Indium-antimonide photovoltaic detectors exhibit current-versus-voltage characteristics typical of any diode junction. This behavior is described by

$$i = i_0(e^{qv/kT} - 1) - i_g + G_s v \qquad (7.131)$$

Pruett and Petritz (1959) point out that the conductance term G_s represents an additional current path in diffused, alloyed, or grown junctions. One such path may be surface leakage current that causes a loss in RA product. Recall that the photo-generated current i_g is

$$i_g = \eta q A_d E_q \qquad (7.132)$$

For a detector with $\eta = 1$ viewing a 180° background at $T = 300$ K, the photo-generated current density is 3200 μA/cm^2. Hall et al. (1975) indicated that if the detector is operated by holding the junction voltage constant and measuring changes in current, the detector responsivity will be linear over a wide range of flux and will be independent of the operating point (the particular value of voltage). Also, if the current is held constant and changes in voltage are measured, this will not be true.

The electrical properties of an InSb PV detector may be modeled by the equivalent circuit shown in Fig. 7.51, where the shunt resistance $R_s = 1/G_s$

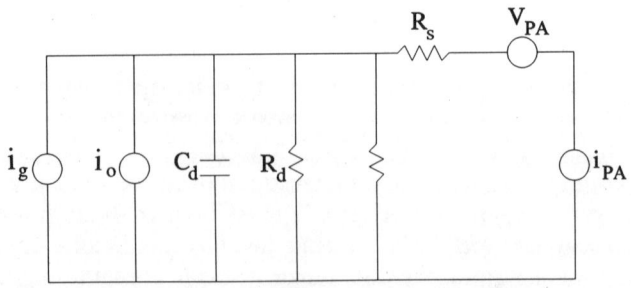

Figure 7.51.   Diode equivalent circuit with simplified preamplifier and noise sources.

(leakage-current term). InSb detectors have typical capacitances of 50 nF/cm^2. The various noises affecting these detectors are the same as those for PV detectors in general: $1/f$, Johnson, photon, and preamp noises. By operating the detector in a zero-bias mode ($v = 0$), $1/f$ noise may be eliminated, and by using cooled, low-noise FETs as the preamp-input stage, preamp noise can be reduced below the Johnson- and photon-noise limits. Under these conditions, the detector $D*$ is

$$D*(\lambda) = \frac{\eta\lambda}{hc\sqrt{2\eta E_{q,\text{bkg}} + \dfrac{2kT}{q^2 R_d A_d}}} \tag{7.133}$$

In the case of BLIP, this expression reduces to

$$D*(\lambda) = \frac{\lambda}{hc}\sqrt{\frac{\eta}{2E_{q,\text{bkg}}}} \tag{7.134}$$

where a typical value is D^*_{BLIP} ($\lambda = 5.3$ μm, 500 Hz) $= 1.2 \times 10^{11}$ cm Hz$^{1/2}$W^{-1} for a background temperature of 300 K.

For operation at very low background levels, as in the case of infrared astronomy of faint objects, the detector may become Johnson-noise limited. In this limit, the expression for $D*(\lambda)$ becomes

$$D*(\lambda) = \frac{\lambda\eta q}{2hc}\sqrt{\frac{R_d A_d}{kT}} \tag{7.135}$$

For a given wavelength, the sensitivity of the detector thus depends on the detector's quantum efficiency, operating temperature, and resistance. Techniques for increasing the detector's resistance are typically related to cooling to lower temperatures. A typical RA product at 60 K is 10^{10} Ω-cm^2 and as high as 10^{12} at 4 K.

To operate these detectors under Johnson-noise-limited conditions, preamp noise must be kept as low as possible. Several designs of low-noise preamps specifically tailored to the operating characteristics of InSb have appeared in the literature (Levan, 1980; Zhang and Williamson, 1982; Rieke et al., 1981). Perhaps the most noticeable of these is that of Hall et al. (1975) shown in Fig. 7.52, which uses a transimpedance amplifier compensated to prevent dc offset and drift. Using a pair of matched, cooled JFETs as the input to an opamp, the circuit allows for very large values of feedback resistance (10^{12} Ω) to be used to match or equal the high values of detector resistance commonly encountered.

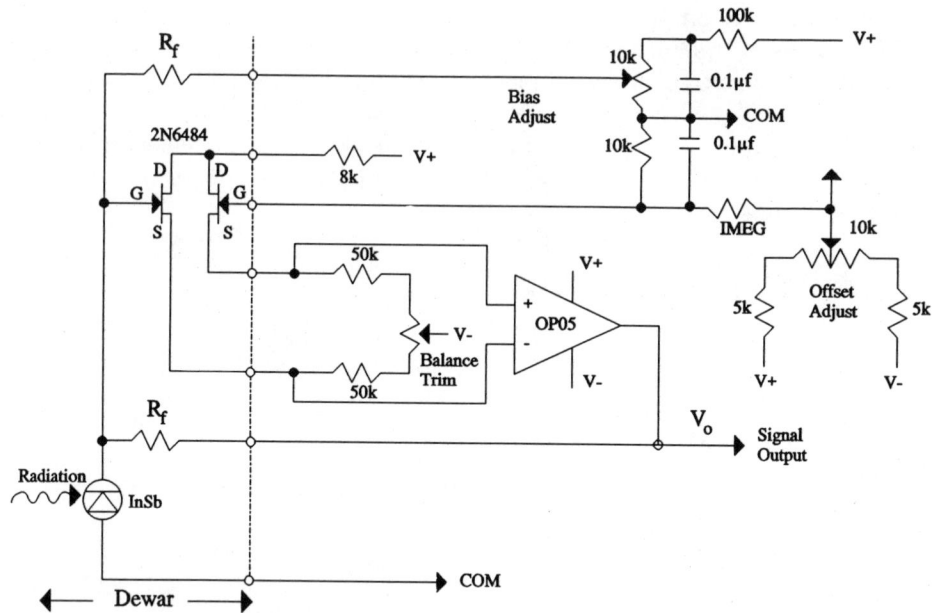

Figure 7.52. InSb preamp with cooled dual JFETs.

7.6. GaAs PHOTODIODES

Gallium arsenide (GaAs) is the genesis of many different optical detectors that can be developed by proper alloying and fabrication. GaAs is the base material that can be used to sense the same spectral region covered by the combination of silicon and germanium photodiodes. GaAs has many advantages over silicon, including speed of response, lower power dissipation, controlled variable-cutoff wavelength, low junction capacitance, and large shunt resistance. In performance comparison, the three measures are quantum efficiency, response time, and detectivity.

The speed of response is higher because a weaker electric field can set electrons in motion through a GaAs lattice more readily than is the case with silicon. The velocity of electrons is about five times faster in GaAs than in silicon for electric fields less than 1000 V/cm. The electrons in GaAs have less effective mass, so they can attain terminal velocity much more quickly than in silicon. This leads to circuits that are subject to much less joule heating and therefore less heat loss. There are fewer ricochets because fewer residual impurities scatter them and cause loss of energy. The less inhibited motion of an electron through a GaAs lattice because of the uniform crystalline structure also lends itself to less noise. The velocity distribution of electrons is more compact than in silicon or germanium, and therefore the variance on the signal electrons is less. Thus, a lower noise level is seen within the detector.

Another inherent advantage of GaAs is the possibility of controlling the energy gap by alloying the material. The energy gap for GaAs (1.43 eV), which is larger than that of silicon, can be narrowed by judicious alloying (substitution of gallium atoms with other elements). The bandgap is directly proportional to the fraction of material in the alloy. Typical alloying materials are indium, phosphor, and aluminum. By stacking such material in layers (a few atoms thick) a superlattice can be formed. This is discussed further in Chapter 11.

Unlike silicon, GaAs can emit photons as well as detect them. The transition from the conduction to the valence band can produce electromagnetic radiation. The same process in silicon only produces heat by means of phonons. Recall that a GaAs photodiode is reverse biased, an emitter is forward biased.

Unfortunately, several problems and material-handling considerations are associated with the manufacturing and processing of gallium-arsenide materials. The first is the lack of a good insulator. The native oxide of silicon is an insulator; for gallium arsenide it is not. Silicon oxide is electronically, mechanically, and chemically suited for detectors. Large areas of GaAs are not readily available with typical values less than 5 mm^2. The defects in ion-implanted GaAs cannot be annealed out like they can be in silicon. Vaporization of arsenic is seen with elevated temperature.

GaAs forms the basis of all materials used in the III-V photodiode; however, it is typically alloyed with a third element. It is the most prominent of the nine III-V compounds listed in Table 7.5. We see that InSb is also in this list, which was discussed in Section 7.3.

The various III-V compounds and their corresponding alloys can be quantified as to their intersolubility by an energy-gap-versus-lattice-constant plot and connected lines. Figure 7.53 shows the nine III-V compounds plotted in this manner (Ferry, 1985). The lines connecting two materials indicate that a ternary compound can be formed (i.e., GaAs–AlAs can be made by varying the ratio of the number of atoms of gallium to aluminum, and thus the energy gap can be controlled from 1.42 to 2.16 eV). Therefore, we can tailor the

Table 7.5   III-V Detector Materials

Material	ε_g (eV)
GaAs	0.42
GaSb	0.73
GaP	2.26
AlAs	2.16
InAs	0.36
AlP	2.45
AlSb	1.58
InP	1.35
InSb	0.17

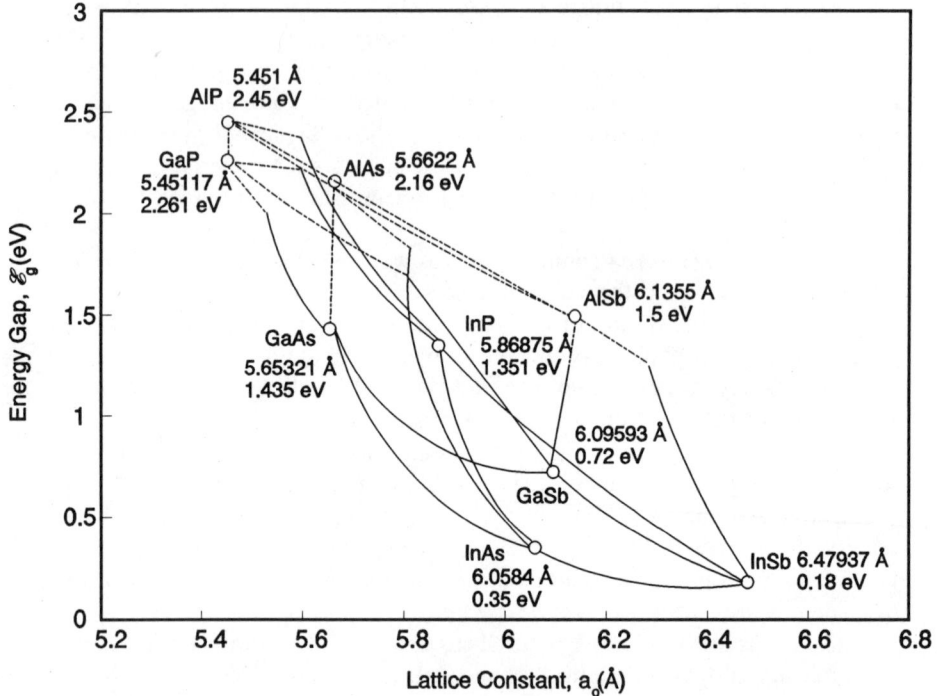

Figure 7.53. A plot of energy bandgap vs. lattice constant for major III-V compounds. Ternary alloys are denoted by lines between the binaries. The solid lines and the dashed lines represent the direct-bandgap and indirect-bandgap regions, respectively.

energy gap of the alloy to the wavelength of interest by controlling the mixing. The corresponding energy gap and lattice constant is 1.11 eV and 5.43 Å for silicon and 0.6 eV and 5.69 Å for germanium, respectively.

$Ga_{1-x}Al_xAs$ is an alloy, represented in Fig. 7.53 as a line joining the two compounds with x varying from 0 to 1 (1 meaning AlAs, 0 meaning GaAs). The lattice constants are almost identical, introducing no strain in the structure. A quaternary compound can also be formed by combining four elements (two III and two V elements). The material is represented as an area in Fig. 7.53, considered InGaAsP, or $(In_{1-x}Ga_x)(As_{1-y}P_y)$. The most common photodiode materials in use today are compounds of InGaAs, GaAsP, and InGaAsP.

7.6.1. GaAsP

This material has a spectral response from about 300 nm to 800 nm, depending on whether it is of Schottky construction or a diffusion type (see Fig. 7.54). For operation at room temperature D^* varies from 3×10^{13} to 10^{14}. A room-temperature-operated GaAsP detector has the following typical (approximate)

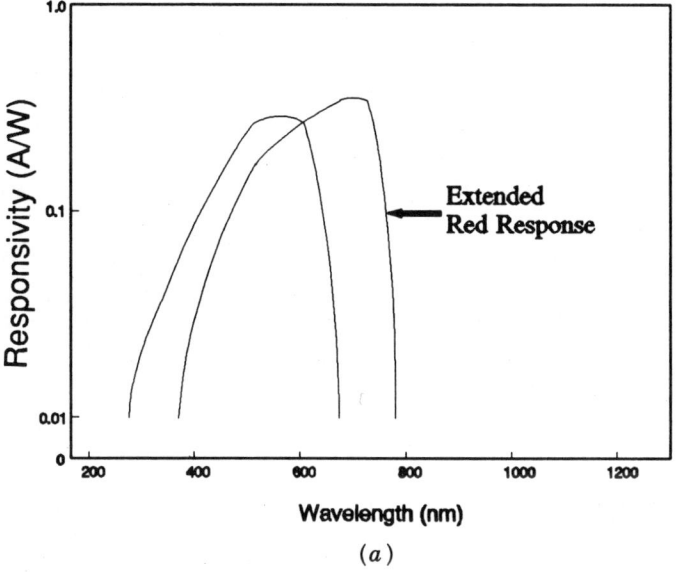

(a)

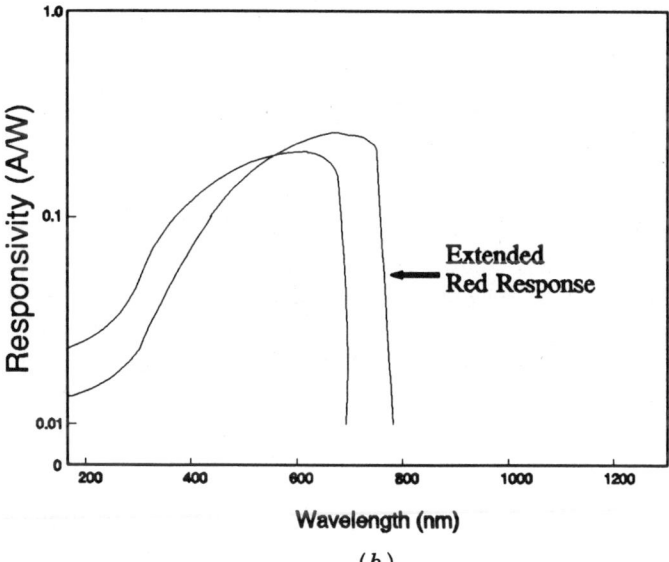

(b)

Figure 7.54. GaAsP relative responsivity vs. wavelength: (a) diffusion junction, (b) Schottky junction.

parameters:

$$R_d A_d = 2 \times 10^{19} \ \Omega \ cm^2$$
$$C_d = 250 \ pF/mm^2$$
$$i_0 = 3 \ pA/cm^2$$
$$\tau = 5 \ \mu s$$

As indicated earlier, the n- or p-type doping concentration does not determine the wavelength response. However, the doping concentration does effect the dark current. An essential feature of a GaAsP photodiode is its extremely dark current, which is 10 to 100 times less than in silicon. Dark current is also a much weaker function of temperature in GaAsP than it is in silicon. In addition, its large energy gap produces a larger open-circuit voltage than does silicon.

The linearity of the device extends over 10 orders of magnitude; however, detector areas are always less than 5 mm². Care must be given to handling these devices to prevent fingerprints on the window (UV loss) or too much bias on the detector. These detectors can be damaged by measuring their resistance with a digital voltmeter (DVM) across the terminals.

7.6.2. InGaAs

The indium-gallium-arsenide (InGaAs) photodiode is a high-speed broad-wavelength detector. It responds in the range from 0.5 to 1.7 μm, as shown in Fig. 7.55. Because optical fibers transmit best (lowest insertion loss around 1.3 μm), this detector is an ideal choice for fiber-optic communication. Manufacturers often eliminate the low-end wavelength response ($\lambda < 0.9 \ \mu$m) by using an opaque substrate to obtain a response between 0.9 μm and 1.7 μm.

The quantum efficiency at 1.5 μm is 55% without an antireflection coating (refractive index $n \approx 3.65$). Different forms of the InGaAs photodetector provide various quantum efficiencies. Efficiencies of $\eta = 38\%$ at 1.3 μm for high-speed tailored photodetectors have been measured. Quantum efficiency decreases with decreasing thickness of the undiffused ternary layer.

The response bandwidth is a major benefit of InGaAs. In theory, bandwidths of 75 GHz or higher are possible, provided that thin absorbing layers and small detector areas are used. To obtain maximum bandwidth of a PIN detector, the transit-time and parasitic-capacitance limitations must also be considered. Minimizing transit time requires thinner intrinsic layers. However, minimizing the inherent parasitic capacitance of the detector junction requires thicker intrinsic layers. Also, with thin-layer detectors, quantum efficiency decreases. Consequently, for any given detector-junction area, an optimum intrinsic-layer thickness exists that yields the maximum bandwidth.

Good sensitivity is possible with low dark current and high quantum efficiency. Leakage currents are low, on the order of 0.2 to 10 nA, even near the

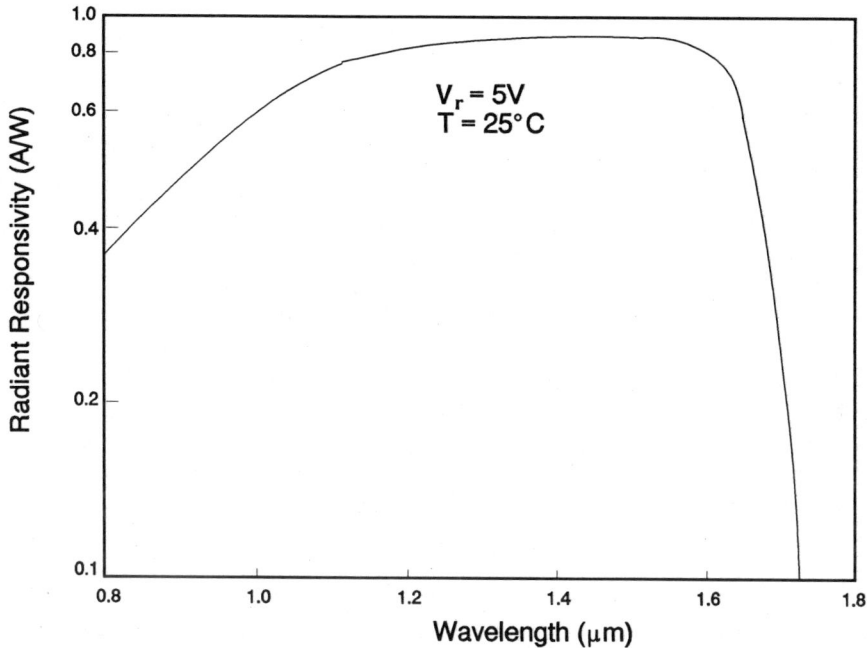

Figure 7.55. Spectral response of InGaAs photodiode.

onset of the avalanche multiplication. For a 25-μm device, a fourfold reduction in the dark current occurs, when compared with a 75-μm device. This is consistent with the decrease in device perimeter. For the same device, the ninefold decrease in device capacitance is consistent with the decrease in device area. Chip capacitances are about 0.03 pF (at -10 V). These photodetectors are fully depleted at a bias level of only 4 to 6 V.

7.7. HgCdTe

Mercury-cadmium-telluride PV detectors, which operate to 13 μm, have the longest wavelength response. This material is a II-VI compound that is an alloy composition of a wide energy-gap semiconductor CdTe, with a semimetallic compound, HgTe, that is considered a semiconductor with a ''negative energy gap.'' By controlling the relative proportions of HgTe and CdTe, the energy gap could be controlled for the alloy.

In 1959, Lawson et al. reported that $Hg_{1-x}Cd_xTe$ was a semiconductor whose bandgap energy varied from -0.3 to 1.605 eV, depending on the x value, which is the mole fraction ratio of Cd to Hg. This energy-gap control gives HgCdTe considerable versatility. By controlling the proportion of CdTe mixed with HgTe, a compound of the form $Hg_{1-x}Cd_xTe$ results. Pure CdTe ($x = 1$) will detect radiation at wavelengths out to about 0.83 μm and will not respond to longer wavelengths. The greater the percentage of Hg in the compound, however, the longer the wavelength response. Hence, detectors can be tailor-made to respond to the desired wavelength in the infrared.

The HgCdTe material is a zinc-blende crystal structure composed of two face-centered cubics displaced by one-half-lattice spacing along the diagonal (Fig. 7.56). Covalent bonds exist between II and VI ions. The cations (Hg and Cd) are mixed in the mole ratio of x, but the zinc-blende crystal structure is maintained.

In $Hg_{1-x}Cd_xTe$ compounds, the bandgap can be controlled by adjusting the proportion of HgTe and CdTe. As is typical of most semiconductors, CdTe has a bandgap that decreases with temperature. HgTe, however, is a semiconductor and its bandgap energy increases with temperature (Schmit and Stelzer, 1969). Near absolute zero, the bandgap energy of CdTe is about 1.6 eV and for HgTe it is -0.3 eV. The negative sign for HgTe shows that the bottom of the conduction band is lower than the top of the valence band; however, this bandgap is indirect.

Various empirical expressions have been developed relating the bandgap energy to the temperature T and the fraction x of cadmium. Hansen et al. (1982) give perhaps the most accurate expression, relying on the data of 22 different studies. They note that the relationship between x and $\mathcal{E}_g$ is nearly linear for values of x between 0.2 and 0.6. Almost all HgCdTe detectors fall in this range. However, for the best fit to the data, particularly for $x = 1$, an equation cubic in x and linear with temperature was necessary:

$$\mathcal{E}_g = -0.302 + 1.93\ x + (5.35 \times 10^{-4})T\ (1 - 2x)$$
$$- 0.310x^2 + 0.832x^3 \tag{7.136}$$

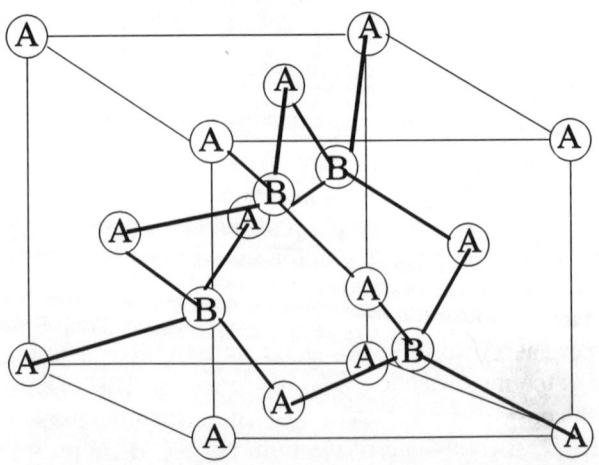

Figure 7.56. Zinc-blende crystal structure.

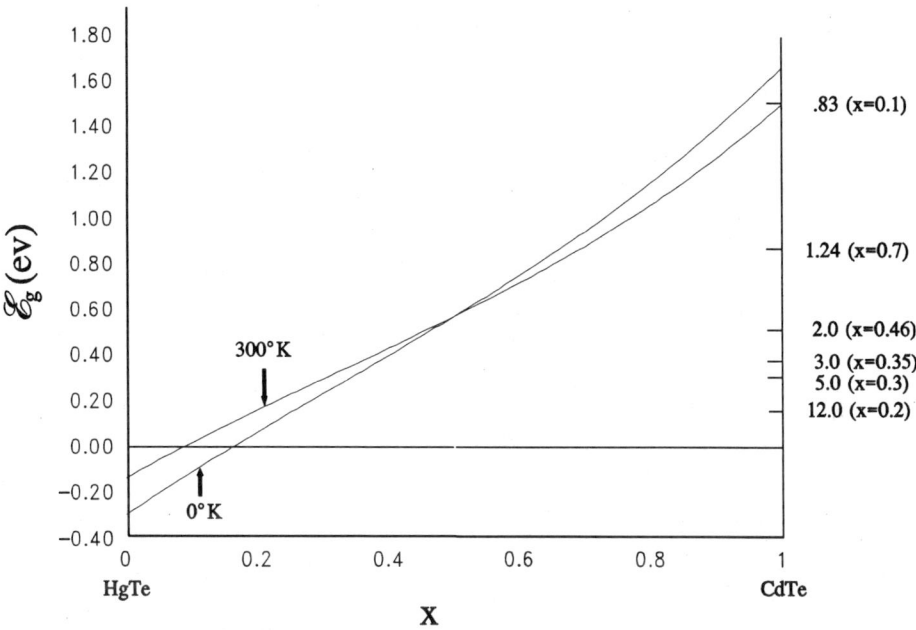

Figure 7.57. $Hg_{1-x}Cd_xTe$: x vs. $\mathscr{E}_g$ (corresponding cutoff wavelength shown on right).

Equation (7.136) is plotted as a function of x in Fig. 7.57 for temperatures of $T = 4.2$ K and 300 K. The right-hand side of the figure shows corresponding values of cutoff wavelength.

The usual method of creating the *p-n* regions of the material is by deliberately introducing impurities. Where easy-to-excite electrons are desired, we add a "donor" type of impurity (such as boron). This region of the material is called "*n*-type." For the other region we add an "acceptor" impurity (such as copper) and call that region "*p*-type." The boundary between the two regions is called the *depletion region*. The *n*-type side of the depletion region will have a net positive charge because some of its electrons get trapped or captured by impurity acceptor atoms on the *p*-type side of the depletion region.

If a photon with sufficient energy is absorbed in or near the depletion region, an electron will gain sufficient energy to move into the conduction band. The electron will then be attracted to the *n*-type side. The electron leaves behind a vacancy or "hole" that moves toward the *p*-type side. Electrodes attached to the detector will see a current flow or difference in voltage as a result of the separation of the electron–hole pair. A PV detector is still "intrinsic," because the absorption of the radiation is due to the bulk material, HgCdTe, not the impurities.

This material is unique as an infrared detector for two major reasons: (1) the wavelength response can be tailored by means of the mixing ratio x; and (2) because the intrinsic energy gap determines the wavelength response, the cooling requirements (how low a temperature is necessary to eliminate dark current) are not as severe as for extrinsic-semiconductor-based detectors.

7.7.1. Quantum Efficiency for HgCdTe

The refractive index of HgCdTe is approximately 3.7 (see Table 7.1), giving a Fresnel reflectivity of about 30%. The optical absorption coefficient ranges from 10^3 to 10^4 cm^{-1}. This high absorption causes all photons to be absorbed within 10 μm of the incident surface, which means that the length of the junction collection region is considerably less than the minority diffusion length, leading to effects such as minority-carrier sweepout. This high absorptivity also means that the detector thickness can be much less than the thickness required of other detector materials for similar wavelengths. The quantum efficiency seen for these detectors can be as high as 65%.

7.7.2 Control of the Mixing Ratio in Hg$_{1-x}$Cd$_x$Te

An early problem with Hg$_{1-x}$Cd$_x$Te detectors was producing a homogeneous material. Because of the high vapor pressure of free Hg, early fabrication techniques sealed the elements in a quartz ampoule and heated them to a certain temperature above the melting point. The alloy was then cooled through a temperature gradient. This is known as the Bridgman technique. Early ingots grown in this way exhibited variations in the x fabrication parameter of ±0.03 over about 20 cm (Schmit and Stelzer, 1969). The value of x could be found using an electron microprobe, which measured the electron diffraction through the material. The ingot would then be sliced and the resulting isocomposition contours used to obtain a uniform x material. In practice, the interface form is roughly conical, with the point of the cone pointing toward the end of the ingot that solidified first. By controlling radial and axial solidification rates independently, Nelson et al. (1980) reported obtaining ±0.002 compositional homogeneity over 95% of wafers up to 25 mm in diameter.

Because of the difficulties in obtaining highly homogeneous Hg$_{1-x}$Cd$_x$Te using crystal-growth techniques, epitaxial-growth techniques were investigated (Micklethwaite, 1981). Epitaxial techniques are favorable because near-ideal substrates are readily available. Both cadmium telluride and mercury telluride have only a 0.3% variation in the lattice parameter. Cadmium telluride is a semi-insulator and transparent to IR radiation. However, problems still exist with epitaxal growth of Hg$_{1-x}$Cd$_x$Te. The most important of these is the variation of x with depth in the growth layer. Recall that the variation of x can be related to the cutoff wavelength by

$$\lambda(x) \ [\mu m] = 1.24/\mathcal{E}_g(x) \ [eV] \tag{7.137}$$

where $\mathcal{E}_g$ is given by Eq. (7.136). Substituting and rearranging terms yields

λ_p [μm]

$$= \frac{1}{-0.244 + 1.556x + (4.31 \times 10^{-4})\, T\,(1 - 2x) - 0.65x^2 + 0.671x^3}$$

$$(7.138)$$

where x is the molar ratio and T is the temperature. To determine the relationship between cutoff wavelength and x uncertainty, we will take the differential of Eq. (7.138). This will relate x variations in manufacturing to variation in the cutoff wavelength:

$$d\lambda_p = \lambda_p^2[1.556 - 8.62 \times 10^{-4}T - 1.3x + 2.013x^2]\, dx \quad (7.139)$$

Table 7.6 shows some uncertainty in cutoff wavelength for x variations of 0.2%. This variation in x value is typical of a good material. For short-wavelength (~ 3 μm) and mid-wavelength (~ 5 μm) materials, the variation in cutoff wavelength is not large. However, for the long-wavelength materials (~ 14 μm) the uncertainty in the cutoff wavelength is large and cannot be neglected. For $x = 0.196$, the uncertainty as to the cutoff wavelength is 0.51 μm. That is, for a 14-μm wavelength cutoff, the value may be at 13.5 or 14.5 μm. This response variation causes radiometric-calibration problems in that the radiation is detected over a different spectral region than expected. For example, integrating the exitance of a blackbody at temperature T from 0 to 13.5 μm will yield the same output response as integration from 0 to 14 μm of a source at the same $T - \Delta T$. Therefore, the response variation inherent to HgCdTe makes it impossible to uniquely determine a source's temperature by this approach.

If a linear array was built of contiguous pixels, a different output voltage for a uniform 300-K background would appear just because of variations in x. The fabrication of highly pure HgCdTe is both an art and a science. In every case, the starting material must be very pure, better than 1 ppm.

7.7.2.1. Hg (Atomic Weight = 200.6). Mercury occurs in many minerals, but is invariably purified by roasting the red sulfide, "cinnabar." Mn, Zn, Pb,

Table 7.6 Cutoff Wavelength for x Variations of 2%, and the Corresponding Cutoff Wavelength Shift for $Hg_{1-x}Cd_xTe$

x Value	Cutoff Wavelength [μm]	Temperature	Uncertainty
$x = 0.395$	$\lambda_p = 3$	$T = 77$ K	$\Delta\lambda_p = 0.023$
$x = 0.295$	$\lambda_p = 5$	$T = 77$ K	$\Delta\lambda_p = 0.064$
$x = 0.21$	$\lambda_p = 10$	$T = 77$ K	$\Delta\lambda_p = 0.26$
$x = 0.196$	$\lambda_p = 14$	$T = 77$ K	$\Delta\lambda_p = 0.51$

and Cu are removed by oxidation by adding acids. The metallic oxides are insoluble and nonwetable by Hg. They rise to the surface and are skimmed. Multiple distillation achieves parts per billion (ppb) level of purity.

7.7.2.2. Cd (Atomic weight = 112.4). Cadmium occurs naturally in zinc materials and is relatively easy to segregate. Cd boils at 767°C at 1 atm pressure, which is 141°C below the boiling point of zinc. Cd is therefore purified by distillation in an inert (oxygen-free) atmosphere. Multiple distillations at 10^{-5} torr at 450°C commonly result in six-nines-purity (0.999999) material. Zone refining in dry high-purity hydrogen excludes oxygen and aids the "volatilization" of remaining impurities. Up to 50 zone passes of a distillation-processed bar of Cd at a zone speed of 1 mm/min yields the highest purity material. Of this, the center of the zone-refined bar is the purest and is used.

7.7.2.3. Te (Atomic weight = 127). Te occurs in copper sulfide, nickel sulphide, and lead sulphide areas. Te is most often recovered from copper refinery slims. The chemical purification of Te is complicated. It is followed by low-pressure distillation at roughly 500°C. Se, As, Na and other metals are best removed by successive zone refining in a vacuum.

7.7.3. RA Product

As discussed in Chapter 6, the RA product for photodiodes is an important figure of merit under Johnson-noise-limited conditions. The RA product is usually low for $Hg_{1-x}Cd_xTe$ photodiodes because the RA product is inversely proportional to the square of the intrinsic-carrier concentration (see Eq. 7.35). For a $Hg_{0.8}Cd_{0.2}Te$ photodiode operating at 77 K, the intrinsic-carrier concentration is $9 \times 10^{13}/cm^3$ (Schmit, 1970). This can be compared to the value for silicon of $1.5 \times 10^{13}/cm^3$ at 300 K. The theoretical upper limit of RA products can be calculated from the diffusion current (Long, 1980). Typical maximum RA products are about 10^5 Ω-cm^2. The intrinsic-carrier concentration is

$$n_i(x, T, \mathcal{E}_g) = [9.908 - 5.21x + 3.07 \times 10^{-4} T + 5.94 \times 10^{-3} Tx]$$

$$\cdot [10^{14} \mathcal{E}_g^{0.75} T^{1.5} e^{-\mathcal{E}_g/2kT}] \tag{7.140}$$

where $\mathcal{E}_g$ is also a function of x and T.

RA-product-versus-inverse-temperature plots are shown in Fig. 7.58. The RA-product values for cooled detectors are quite good for short-wavelength and mid-wavelength detectors. However, for long wavelength, that is, in the 8- to 12-μm region, the RA products are relatively low. Recall that this means that the BLIP-operation range is limited.

7.7.4. 1/f Noise

Tobin et al. (1980) has studied 1/f noise in $Hg_{1-x}Cd_xTe$ n^+-on-p photodiodes. The authors varied background photon flux, reverse-bias voltage, and potential

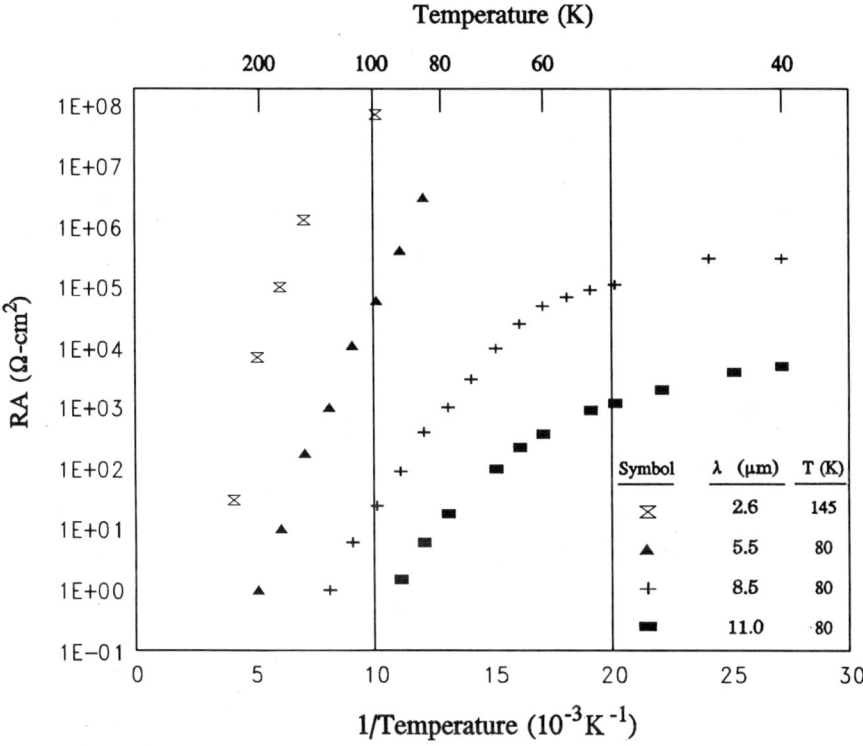

Figure 7.58. Plots of *RA* products for various α-valued Hg$_{1-x}$Cd$_x$Te detectors.

applied to an insulated-gate electrode while they measured 1/f noise current. They concluded that 1/f noise was independent of photocurrent and diffusion current, and linearly dependent on surface-leakage current. The result is similar to that found for gate-controlled silicon junctions observed by Hsu et al. (1968). However, the mechanism of surface-leakage current in Hg$_{1-x}$Cd$_x$Te and its relationship to 1/f noise is still poorly understood.

The effects on 1/f noise plots by bias voltage are shown in Fig. 7.59. We can see that slight bias causes a large change in 1/f noise. As stated earlier, the 1/f noise is a function of current, but what current is the dominant contributor to 1/f noise? Figure 7.60 shows the equivalent circuit of a HgCdTe PV detector. The equivalent circuit for an Hg$_{1-x}$Cd$_x$Te photodiode is an ideal diode in parallel with a capacitance, a diffusion-current resistance, a generation–recombination current resistance, a tunneling-current resistance, and a surface-leakage-current resistance. The resistances all add in parallel, so the lowest resistance will dominate the current flow. The surface-leakage current can be controlled by better passivation techniques, which will increase the resistance. The test samples used to evaluate surface leakage should be equal-area detectors with different perimeter values. By testing several perimeter values we can determine if the current is proportional to the area or the perimeter.

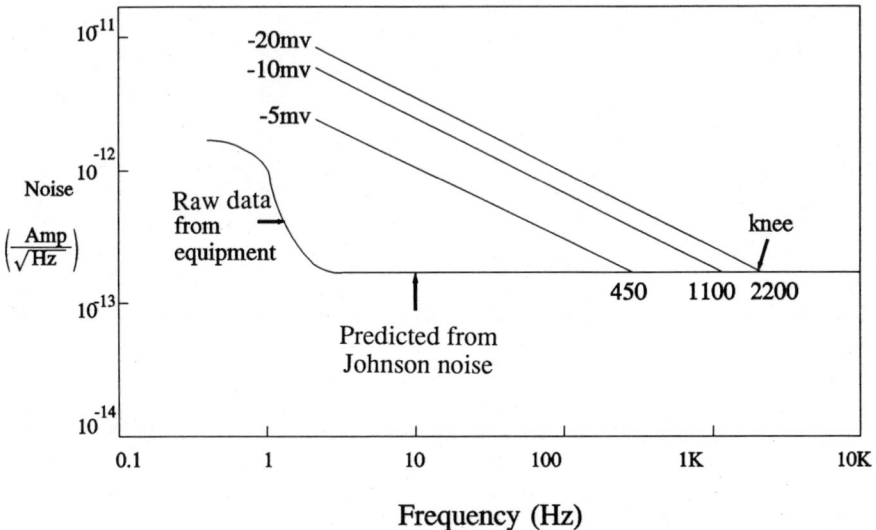

Figure 7.59.   Frequency vs. $1/f$ noise for various bias levels.

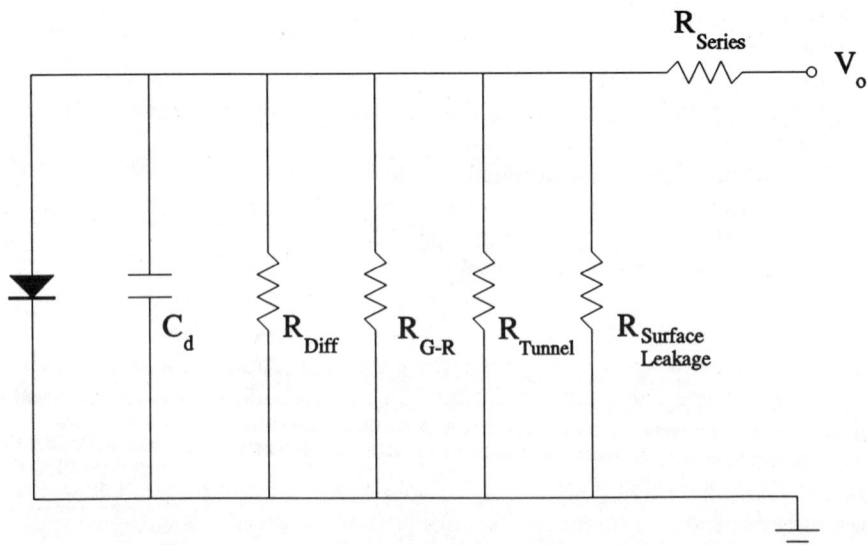

Figure 7.60.   Equivalent circuit for various currents in a HgCdTe diode.

A curious effect occurs in $1/f$ noise at about 73 K. The noise dips to a minimum value. This minimum in $1/f$ occurs approximately at the temperature where the material linear coefficient of expansion goes from positive to negative (Dornhaus and Nimitz, 1976). In addition, the electron mobility reaches a maximum at this temperature (Xu et al., 1989). From our discussion, we can surmise that $1/f$ noise may be orders of magnitude above the BLIP limit at low frequencies.

7.7.5 Manufacturability

Modern fabrication methods that use molecular-beam epitaxy (MBE) or metal-organic chemical-vapor deposition (MOCVD) are applied to the building of superlattice structures, which consist of 200 to 500 thin CdTe/HgTe layers stacked. This approach controls x values better and reduces $1/f$ noise contributions.

7.7.6. Summary of $Hg_{1-x}Cd_xTe$

$Hg_{1-x}Cd_xTe$ photodiodes have found many uses in infrared-detector applications. By tailoring the composition ratio of the material, the intrinsic energy gap can be varied to provide cutoff wavelengths specific to a certain application. Because $Hg_{1-x}Cd_xTe$ is an intrinsic material, higher operating temperatures are possible. One use for $Hg_{1-x}Cd_xTe$ photodiodes has been in high-frequency detection of 10.6-μm CO_2 laser radiation. The high-frequency response of $Hg_{1-x}Cd_xTe$ photodiodes results from the low dielectric constant of the material, which provides a low junction capacitance. Another more recent use has been to join $Hg_{1-x}Cd_xTe$ junctions to silicon charge-coupled device (CCD) arrays. These hybrid mosaic focal-plane arrays are generally used in the 3-to-5-μm and 8-to-12-μm wavelength regions for direct IR imaging.

However, problems with $Hg_{1-x}Cd_xTe$ detectors still exist. Because of the difficulties associated with producing homogeneous material, detectors arrays are small and different detectors have slightly different cutoff wavelengths, resulting in nonuniform detector response between individual detector elements in two-dimensional detectors. Also, the RA product of $Hg_{1-x}Cd_xTe$ is low in the long-wavelength IR, which lowers the D^* for low background irradiances. $Hg_{1-x}Cd_xTe$ photodiodes are also sensitive to $1/f$ noise, which is linearly dependent on surface-leakage current. To control $1/f$ noise in $Hg_{1-x}Cd_xTe$, the mechanism that produces surface-leakage current and its relationship to $1/f$ noise requires more investigation.

In practice, the D^* of HgCdTe photodiodes approaches the ideal BLIP values, but at the present time D^* is still limited by a low RA product. A typical spectral detectivity is about 2.5×10^{10} cmHz$^{-1/2}$/W for a 77-K detector, 295-K background, 60° field of view, at a frequency of 1 kHz and at a maximum wavelength of 10 μm. For comparison, the ideal photovoltaic detectivity under these conditions would be 5×10^{10} cmHz$^{1/2}$/W.

REFERENCES

Dornhaus, R., and G. Nimitz, *Springer Tracts in Modern Physics*, Springer-Verlag, Berlin/New York, 1976, pp. 1–119.

Dornhaus, R., and G. Nimitz, *Narrow Gap Semiconductors*, Springer-Verlag Tracts in Modern Physics, Vol. 98, Springer-Verlag, Berlin/New York, 1983.

Ferry, Dave K., *Gallium Arsenide Technology*, Sams of MacMillan, Howard W. Sams & Co., Indianapolis, 1985.

Hall, D. B., R. S. Aikens, R. Joyce, and T. W. McCurnin, "Johnson Noise Limited Operation of Photovoltaic InSb Detectors," *Appl. Opt.* **14** (2), 450 (1975).

Hansen, G. L., J. L. Schmit, and T. N. Casselman, "Energy Gap Versus Alloy Composition and Temperature in Hg$_{1-x}$Cd$_x$Te," *J. Appl. Phys.* **53** (10), 7099 (1982).

Harman, T. C., "Effect of Zone-Refining Variables on the Segregation of Impurities in Indium-Antimonide," *J. Electrochem. Soc.* **103**, 128 (1956).

Hsu, S. T., D. J. Fitzgerald, and A. S. Grove, "Surface State 1/f Noise in p-n Junctions and MOS Transistors," *Appl. Phys. Lett.* **12**, 287 (1968).

Hulme, K. F., in *Materials Used in Semiconductor Devices* (C. A. Hogarth, ed.), Wiley, New York, 1965, p. 115.

Hulme, K. F., and J. B. Mullin, "Indium Antimonide—A Review of Its Preparation, Properties, and Device Applications," *Solid-State Electron* **5**, 211 (1962).

Joyce, R. R., "Indium Antimonide Detectors for Ground-Based Astronomy," *SPIE Proc.* **443**, 50 (1983).

Lawson, W. D., S. Nielsen, E. H. Putley, and A. S. Young, "Preparation and Properties of HgTe and Mixed Crystals of HgTe-CdTe," *J. Phys. Chem. Solids* **9**, 325 (1959).

Lesser, M. E., P. Cholet, and E. C. Wurst, Jr., "High-Sensitivity Crystal Infrared Detectors," *J. Opt. Soc. Am.* **48**, 468 (1958).

Levan, P. D., "Contemporary infrared sensors and instruments," *SPIE Proc.* **246**, 34 (1980).

Liang, S. C., in *Compound Semiconductors*, vol. 1 (R. K. Willardson and H. L. Goering, eds.), Reinhold, New York, 1962, p. 227.

Long, D., "Photovoltaic and Photoconductive Infrared Detectors," in *Optical and Infrared Detectors*, 2d ed. (R. J. Keyes, ed.), Springer-Verlag, New York, 1980, pp. 111–112.

MacGregor, A. D., "39 Photon/Bit Direct Detection Receiver," *SPIE Proc.* **1511**, 349 (1991).

McIntyre, R. J., "Multiplication Noise in Uniform Avalanche Diodes," *IEEE Trans. Electron. Devices* **ED-13**, 164 (1966).

McKay, K. G., and K. B. McAfee, "Electron Multiplication in Silicon and Germanium," *Phys. Rev.* **91**, 1079 (1953).

Melngailis, I., and T. C. Harman, "Single-Crystal Lead-in Chalcogenides," in *Semi-

conductors and Semimetals (R. K. Willardson and A. C. Beer, eds.), Academic Press, New York, 1970, pp. 111–174.

Micklethwaite, W. F. H., "The Crystal-Growth of Cadmium Mercury Telluride," in *Semiconductors and Semimetals*, (R. K. Willardson and Albert C. Beer, eds.), Academic Press, New York, 1981, p. 85.

Nelson, D. A., W. M. Higgins, and R. A. Lancaster, "Advances in (Hg, Cd)Te Materials Technology," *SPIE Proc.* **225,** 48 (1980).

Pruett, G. R., and R. L. Petritz, "Detectivity and Preamplifier Considerations for Indium Antimonide Photovoltaic Detectors," *Proc. I.R.E.*, paper 4.1.9, p. 1524, Sept. (1959).

Richards, P. L., and L. T. Greenberg, "Infrared Detectors for Low-Background Astronomy: Incoherent and Coherence Devices from One Micrometer to One Millimeter," in *Infrared and Millimeter Waves: Systems and Components* (K. J. Button, ed.), Academic Press, New York, 1982, p. 165.

Rieke, G. H., E. F. Montgomery, M. J. Lebofsky, and P. R. Eisenhardt, "High Sensitivity Operation of Discrete Solid State Detectors at 4 K," *Appl. Opt.* **20,** 814 (1981).

Schmit, J. L., "Intrinsic Carrier Concentration of Hg1-CH1-CDCHI-TE as a Function of CHI and T Using K-P Calculations," *J. Appl. Phys.* **41,** 2876 (1970).

Schmit, J. L., and E. L. Stelzer, "Temperature and Alloy Compositional Dependences of Energy Gap of HG1-XCDXTE," *J. Appl. Phys.* **40,** 4865 (1969).

Streetman, B. G., *Solid State Electronic Devices*, 2d ed., Prentice Hall, Englewood Cliffs, NJ, 1980.

Sze, S. M., *Physics of Semiconductor Devices*, Wiley, New York, 1981.

Sze, S. M., *Semiconductor Devices: Physics and Technology*, Wiley, New York, 1985, p. 285.

Tobin, S. P., S. Iwasa, and T. J. Tredwell, "1-F Noise in (Hg, Cd)Te Photodiodes," *IEEE Trans. Electron. Devices* **Ed-27,** 43 (1980).

Webb, P. P., R. J. McIntyre, and J. Conradi, "Properties of Avalanche Photodiodes," *RCA Review* **36,** 234 (1974).

Xu, X., X. Hu, J. Shen, and J. Fang, "The Influence of Dislocation Scattering on N-type HgCdTe Mobility," *SPIE Proc.* **1107,** 236 (1989).

Yariv, A., *Optical Electronics*, 3rd ed., Holt, Rinehart, & Winston, New York, 1985, p. 376.

Zhang, Y. X., and F. O. Williamson, "Evaluation of an InSb Infrared Detector at Liquid N_2 and Liquid He Temperatures," *Appl. Opt.* **21,** 2036 (1982).

BIBLIOGRAPHY

Ando, H., H. Kanbe, T., Kimwa, T. Yamaska, and T. Kaneda, "Characteristic of Germanium, Avalanche Photodiode in the Wavelength Region 1-.6 μm," *IEEE J. Quant. Electron.* **QE-14,** 804 (1978).

Conradi, J., "The Distribution of Gains in Uniformity Multiplying Avalanche Photodiodes: Experimental," *IEEE J. Electron. Devices* **ED-19,** 6 (1972).

Conradi, J., "Temperature Effects in Silicon Avalanche Diodes," *Solid State Electron.* **17**, 99–106 (1974).

Crowell, C. R., and S. M. Sze, "Temperature Dependence of Avalanche Multiplication in Semiconductors," *Appl. Phys. Lett.* **9**, 242 (1966).

Forrest, S. R., "Gain-Bandwidth Limited Response in Long-Wavelength Avalanche Photodiodes," *J. Lightwave Tech.* **LT-2**, 34 (1984).

Forrest, S. R., R. G. Smith, and O. K. Kim, "Performance of $In_{0.53}Ga_{0.47}As/InP$ Avalanche Photodiodes," *IEEE J. Quant. Electron.* **QE-18**, 2040 (1982).

Hughes, R. C., "III-V Compound Semiconductor Superlattices for Infrared Photodetector Applications," *Opt. Eng.* **26** (3), 249 (1987).

Hurwitz, C. E., and J. J. Hsich, "GaInAsP/InP Avalanche Photodiodes," *Appl. Phys. Lett.* **22**, 487 (1978).

Kanbe, H., N. Susa, H. Nakagome, and H. Ando, "InGaAs Avalanche Photodiode with InP p-n Junction," *Electron. Lett.* **16**, 163 (1980).

Keyes, R. J., *Optical and Infrared Detectors II*, Springer-Verlag, New York, 1980.

Kingston, R. H., *Detection of Optical and Infrared Radiation*, Springer-Verlag, New York, 1978.

Kruse, P. W., "Indium Antimonide Photoconductive and Photoelectro Magnetic Detectors," in *Semiconductors and Semimetals*, (R. K. Willardson and A. C. Beer, eds.), Academic Press, New York, 1970, pp. 15–83.

Kruse, P. W., L. D. McGlauchlin, and R. B. Quistan, *Elements of Infrared Technology*, Wiley, New York, 1962.

Law, H. D., K. Nakano, and L. R. Tomasetta, "III-V Alloy Heterostructure High Speed Avalanche Photodiodes," *IEEE J. Quant. Electron.* **EQ-15**, 549 (1979).

Lee, C. A., R. A. Logan, R. L. Batdorf, J. J. Kleimack, and W. Weigmann, "Ionization Rates of Holes and Electrons in Silicon," *Phys. Rev. A* **134**(3A), A761, (1964).

Li, Tingye, "Lightwave Telecommunication," *Physics Today* **38**(5) (1985).

Lindley, W. T., R. J. Phelan, C. M. Wolfe, and A. J. Foyt, "GaAs Schottky Barrier Avalanche Photodiodes," *Appl. Phy. Lett.* **14**, 197 (1969).

Lucovsky, G., and R. B. Emmons, "High Frequency Photodiodes," *Appl. Opt.* **4**, 697 (1965).

Major, L. D., Jr., R. W. Olmsted, and R. R. Kusner, "Performance Characteristics of Diffused Junction Silicon Photodiodes," *SPIE Proc.* **190**, 355 (1979).

Mathur, D. P., R. J. McIntyre, and P. P. Webb, "A New Germanium Photodiode with Extended Long-Wavelength Response," *Appl. Opt.* **9**, 1842 (1970).

McKay, K. G., "Avalanche Breakdown in Silicon," *Phys. Rev.* **94**, 877 (1954).

Melchior, H., and J. Lumin, "Sensitive High Speed Photodetectors for the Demodulation of Visible and Near Infrared Light," *Bell Syst. Tech. J.* **52**, 390 (1973).

Melchior, H., A. R. Hartman, D. P. Schinke, and T. E. Seidel, "Planar Epitaxial Silicon Avalanche Photodetector," *Bell Syst. Tech. J.* **57**, 1791 (1978).

Mikawa, T., S. Kagawa, T. Kaneda, T. Sakwai, H. Ando, and O. Mikami, "A Low Noise n^+np Germanium Avalanche Photodiode," *IEEE J. Quant. Electron.* **QE-17**(2), 210 (1981).

Murray, L. A., K. Wang, and K. Hesse, "A Review of Avalanche Photodiodes, Trends, and Markets," *Optical Spectra* **14**(4), 54 (1980).

Nishida, K., K. Taguchi, and Y. Matsumoto, "InGaAsP Heterostructure Avalanche Photodiodes with High Avalanche Gain," *Appl. Phys. Lett.* **35,** 251 (1979).

Personick, S. D., "Statistics of a General Class of Avalanche Detectors with Applications to Optical Communications," *Bell Syst. Tech. J.* **50,** 167 (1971).

Ross, Douglas A., "Solid State Photodetectors—The Photodiode and Phototransistor," in *Optoelectronic Devices and Optical Imaging Techniques*, Macmillian, Great Neck, NY, 1979.

Schinke, D. P., R. G. Smith, and A. R. Hartman, "Photodetectors," in *Semiconductor Devices for Optical Communication* (H. Kressel, ed.), Springer-Verlag, Berlin/New York, 1980, pp. 63–87.

Senior, J. M., *Optical Fiber Communications*, Prentice Hall, Englewood Cliffs, NJ, 1985, p. 340.

Shockley, W., *Electrons and Holes in Semiconductors*, D. Van Nostrand, Princeton, NJ, 1950.

Stillman, G. E., and C. M. Wolfe, "Avalanche Photodiodes," in *Semiconductors and Semimetals*, (R. K. Willardson and A. C. Beer, eds.), Academic Press, New York, 1977, p. 291–391.

Teich, M. C., K. Matsuo, and B. E. A. Saleh, "Excess Noise Factors for Conventional and Superlattice Avalanche Photodiodes and Photomultiplier Tubes," *IEEE J. Quant. Electron.* **QE-22,** 1184 (1986).

Temkin, H., R. E. Frahm, N. A. Olsson, C. A. Burrus, and R. J. McCoy, "Very High-Speed Operation of Planar INGAAS/INP Photodiode Detectors," *Electron. Lett.* **22,** 1267 (1986).

Temkin, H., M. B. Panish, and S. N. G. Chu, "InGaAs/InP Superlattice Avalanche Photodetectors Grown by Gas Source Molecular Beam Epitaxy," *Appl. Phys. Lett.* **49,** 859 (1986).

Willardson, R. K., and A. C. Beer (eds.), *Infrared Detectors II, Semiconductors and Semimetals*, Academic Press, New York, 1977.

Willardson, R. K., and A. C. Beer (eds.), *Semiconductors and Semimetals*, Academic Press, New York, 1980.

Wolf, H. F., ed., *Handbook of Fiber Optics*, Garland STPM Press, New York, 1979.

PROBLEMS

7.1. Discuss the various modes of operation for a photovoltaic (PV) detector. List advantages and disadvantages for each mode.

7.2. Why is the cutoff wavelength of a PV detector independent of the energy gap of the impurity levels used in its fabrication (unlike a photoconductor)? What effect, if any, does the doping concentration have on the detector?

7.3. A PV detector ($A_d = 1$ mm^2, $R_d = 10^{10}$ Ω, $\lambda_c = 10$ μm, $\eta = 1.0$) is used in a cryogenic optical system, with $f/1$ optics ($t = 100\%$) (see figure). The system is in a rotating satellite; the detector views the earth (300 K) for one-half revolution and deep space (~ 5 K) for the remain-

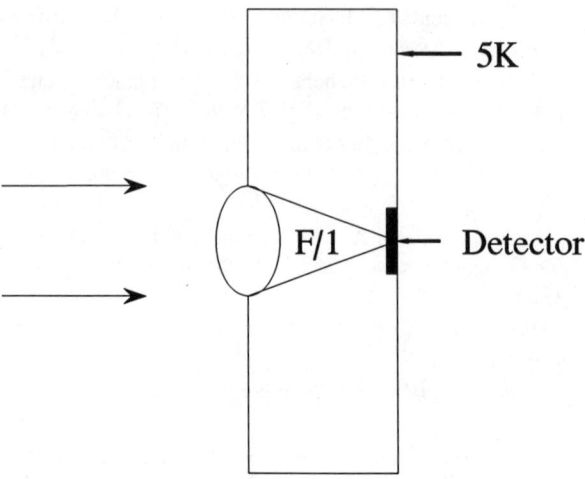

Figure P7.3

der of the revolution. How does the D^* vary qualitatively during one revolution (plot one period as you expect it)? Calculate its maximum and minimum values of peak spectral D^*.

7.4. Explain the silicon PIN photodiode mechanism of operation. Discuss the spectral sensitivity characteristic one might expect versus wavelength, which might be different from a simple *p-n* detector. What is the smallest pulse width one could detect with 99% accuracy in radiometry?

7.5. For a Markovian process, prove that the excess noise factor is

$$\mathcal{E}(G) = 2 - 1/G$$

where G is the average gain across the gain region.

7.6. The product of gain squared and the excess noise factor for an avalanche photodiode is approximated by gain raised to a power between 2 and 3:

$$G^2\mathcal{E}(G) \approx G^\gamma$$

If $\gamma = 2.5$, what is the best guess at the ionization coefficient ratio α_i/β_i for $G = 20$? What is the worst-case error for any G between 10 and 100 using this approximation?

7.7. Compare a PIN and an avalanche photodiode (APD). Why doesn't the PIN avalanche?

7.8. Derive the temperature dependences on the reverse saturation current (i_0) for a PV detector using the equation given for reverse saturation current (i_0) and the intrinsic carrier concentration ($n_i^2 = np$). What temperature change is required for a factor of 2 change in reverse saturation current ($2x$) at 300 K for silicon?

7.9. For a silicon 1-mm^2 PV detector, with equal doping concentration, $n_a = n_d = 10^{15}$/cm^3, calculate the detector capacitance with 10-V reverse bias and with zero bias.

7.10. Relate the SNR from a PV detector coupled to a voltage-mode preamplifier ($i = 0$) to a current mode preamplifier ($v = 0$) given the same signal level.

7.11. Sketch the BLIP peak spectral D^* (λ_c, f) versus wavelength (1–100 μm, every micrometer) of an ideal PV detector for a 2π-sr background of temperature 250 K (assume $\eta = 100\%$).

7.12. Plot the BLIP peak spectral detectivity ($D^*(\lambda_c, f)$) versus cutoff wavelength from 1 to 100 μm for ideal PV operation with and without the Bose factor effect for a 300-K, 2π-sr background.

7.13. The expression for the current from a PV detector is where i_g is the photo-generated current, and the nonideality factor (β) is found empirically. If two photodetectors were exactly the same, except $\beta = 1$ for one and $\beta = 2$ for the other, which detector would you choose to use? Why? What operating condition could be varied so that a single photodiode would yield these two situations?

7.14. Derive the NEP expression for a PV detector in which Johnson noise is the dominant noise.

7.15. Discuss the sources of noise that may be present in a PV detector. Describe in detail an experiment to determine both the existence of and magnitude of each type of noise you have identified.

7.16. A photovoltaic detector has a characteristic i-v curve given by the equation:

$$i = i_0 \left(\exp\left(qv/bkT\right) - 1\right) - i_g$$

for the following values,

$i_0 = 1 \times 10^{-9}$ amps

$T = 200$ K

$b = 2$

$A_d =$ detector area, 1×10^{-5} sq cm

$\Delta f =$ noise electrical bandwidth, 10 Hz

(a) Calculate the RA product at zero bias.

(b) What is the Johnson-noise current at zero bias?

(c) What is the "dark" shot noise at "large" reverse bias (i.e., where $i = i_0$)?

(d) Discuss the relationship between shot noise and Johnson noise at zero bias.

(e) Given the choice of b as 1 or 2, which would you prefer, and why?

(f) Calculate the RA-product-limited D^*.

7.17. What is the peak spectral D^* for a PV detector that has a spectral response from 0 to 10 μm width the following conditions?

$$\text{detector resistance}-R_d = 10^6 \ \Omega$$
$$\text{detector area}-A_d = 1 \ \text{mm}^2$$
$$\text{background irradiance}-E_p = 10^{10} \ \text{photon/s-cm}^2$$
$$\text{quantum efficiency}-\eta = 50\%$$
$$\text{noise equivalent bandwidth}-\Delta f = 10 \ \text{Hz}$$
$$\text{detector temperature}-T = 77 \ \text{K}$$
$$\text{blackbody responsivity}-\mathcal{R}_i \ (500, 500) = 0.5 \ \text{amp/watt}$$

What could you do to improve upon this value?

7.18. Explain the energy band and cutoff wavelength dependence in $Hg_{1-x}Cd_xTe$ using equations and physical description of the material.

7.19. Plot the spectral cutoff wavelength, λ_c, for $Hg_{1-x}Cd_xTe$ detectors versus the molar ratio of Hg and Cd (x).

7.20. For a $Hg_{1-x}Cd_xTe$ photovoltaic detector ($T = 77$ K), plot the uncertainty in cutoff wavelength (λ_c) versus cutoff wavelength (5, 10, 15, and 20 μm) for an x uncertainty of 0.1%, 0.2%, and 0.5%.

7.21. Explain the energy-band and cutoff-wavelength dependence in $Hg_{1-x}Cd_xTe$ using equations and some physical description of the material.

7.22. Discuss the various problems with $Hg_{1-x}Cd_xTe$ (PV) detectors in the LWIR region. List them relating to some equations, if possible, and give possible solutions to them.

7.23. HgCdTe has a linear coefficient of expansion ($\Delta l/l$) that changes sign at 72 K (Dornhaus and Nimitz, 1983) (see figure). Also, the $1/f$ noise has a minimum and the mobility of a maximum at this temperature. From our understanding of $1/f$ noise, what conclusion can you draw from these data?

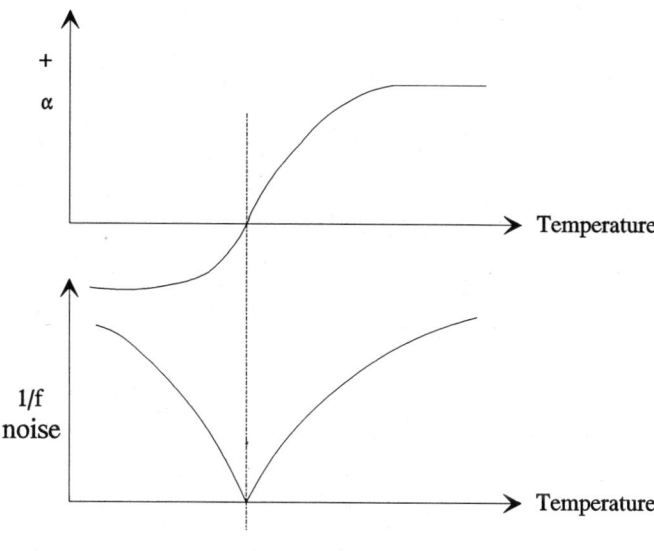

Figure P7.23

7.24. Plot $\varepsilon_g(x, T)$ for a $Hg_{1-x}Cd_xTe$ (PV) detector at a temperature of 77 K. How can this energy gap be negative yet be called a semiconductor? (Hansen et al., 1982.)

7.25. For a HgCdTe PV detector, show that the total noise (Johnson, $1/f$, and shot) has a minimum for some value of detector bias voltage. Find an analytical expression for it. What is optimum bias if $i_0 = 10^{-10}$ A, $B_{1/f} = 2(10^{-5})$, $f = 100$ Hz, $\beta_{1/f} = 2$ for $1/f$ noise, and the nonideality factor is 1.

8

Photoconductive Detectors

8.1. INTRODUCTION

The most readily understood photodetector is the photoconductor, in which absorbed optical radiation causes a corresponding conductance or resistance change. The photoconductor is a piece of semiconductor with ohmic contacts to form a two-port electrical device. When optical radiation falls on the device, free-electron carriers are generated by band-to-band transitions of electrons (intrinsic photoconductors) or by transitions involving energy levels deliberately formed in the forbidden energy gap by appropriate doping materials (extrinsic). The wavelengths of radiation that produced this response were limited to photons with energies greater than the effective energy gap ($\mathcal{E}_g$).

Photoconductors became prominent during World War II with the advent of lead salts (PbS, in particular). The great discovery was made that the resolution of a surveillance system scaled with wavelength. Microwave antennas were huge, bulky, and slow because of mass constraints, while infrared system at shorter wavelengths scaled system mass proportionally and improved the resolution proportionally to wavelength. In addition, photoconductors had a much quicker response than thermal detectors to optical radiation; therefore, they were the detector of choice.

The difference between intrinsic and extrinsic photoconductors is the long-wavelength response. The extrinsic photoconductor increases the spectral response far into the infrared. Table 8.1 lists some common intrinsic-photoconductor detector materials. Extrinsic photoconductors require much colder operation to avoid thermal carrier generation, because the effective energy gap is reduced. Recall that to detect photo-generated carriers, we must freeze most of the thermally generated carriers. In addition, we can expect the quantum efficiency of an extrinsic to be approximately half the intrinsic-detector value. In either type, the conductance or resistance is changed by the presence of incident radiation. This change in resistance is sensed by a preamplifier circuit.

8.2. GENERAL ANALYSIS OF PHOTOCONDUCTORS

For the case of an intrinsic photoconductor, the conductivity is given by (Kittel, 1976)

$$\sigma_c = q\mu_e n + q\mu_h p \qquad (8.1)$$

where

$$\mu_e, \mu_h = \text{mobility of electron, hole}$$
$$n = \text{concentration of electrons}$$
$$p = \text{concentration of holes}$$

For illumination, the conductivity increases with the increase in the number of carriers.

$$\Delta\sigma_c = q\mu_e \Delta n + q\mu_h \Delta p \qquad (8.2)$$

For an extrinsic photoconductor, only the majority carrier (holes for p-type, electrons for n-type) causes the conductance changes. The photoexcitation is produced by photons causing a carrier move between a band edge and the dopant level within the energy gap. Because this energy gap is smaller than the valence-to-conduction bandgap, the spectral response has been increased. However, because the energy gap is smaller, thermally generated carriers become a larger problem. The location of the impurity energy levels within the forbidden bandgap determine the wavelength sensitivity.

After each photon–electron interaction has occurred, the electron recombines with a hole to return to equilibrium or initial conditions. The recombination time or carrier lifetime contributes significantly to the response time of the detector. Cryogenic cooling is often required to keep thermally generated carriers from masking the photoconductive changes. Obviously, the smaller the effective energy gap, the longer the wavelength response and the colder the detector temperature needed to prevent thermal electrons from being generated. For a photoconductor to operate at wavelengths longer than 4 μm, cooling is required. For an extrinsic photoconductor, the conductivity can be expressed as

$$\sigma_c = \mu q n \qquad (8.3)$$

Table 8.1   Intrinsic Photoconductors

Material	$\mathcal{E}_g$	λ_c
PbS	0.42	2.9
PbSe	0.23	5.3
CdS	2.4	0.52
CdSe	1.8	0.69
Ge	0.67	1.85
Si	1.12	1.1
$Hg_{1-x}Cd_x Te$	function of x	25 μm (max)
InSb	0.23	5.3

where

μ = majority carrier mobility
n = concentration of carriers

and the change in conductivity results from the change in the number of majority carriers. Recall that electrons are typically more mobile than holes.

To simplify the analysis, only extrinsic photoconductors are discussed. (A parallel discussion applies to the intrinsic-photoconductor case when substituting $\mu_e + \mu_h$ for μ in Eq. 8.3.) At steady state, the photo-generation rate must equal the recombination rate. Figure 8.1 shows the physical layout of an illuminated photoconductor chip. The conductance is

$$\sigma_{c,\text{illuminated}} = \sigma_c + \Delta\sigma_c \tag{8.4}$$

The photocurrent flowing between the electrodes can be expressed as

$$i = \Delta\sigma_c E \ell_x \ell_y \tag{8.5}$$

where

E = electric field parallel to current flow
$\ell_x \ell_y$ = cross-sectional area to current flow.

$$\Delta\sigma_c = q\mu\Delta n \tag{8.6}$$

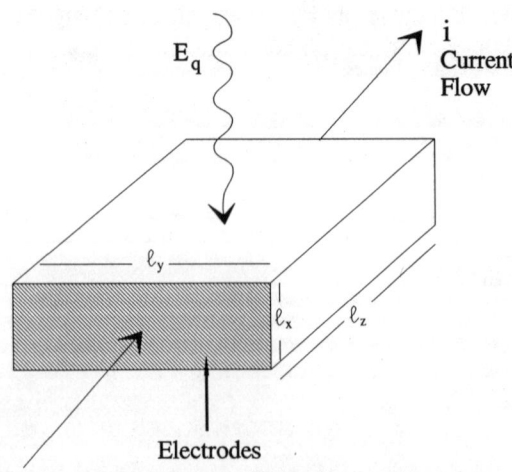

Figure 8.1. Photoconductive chip.

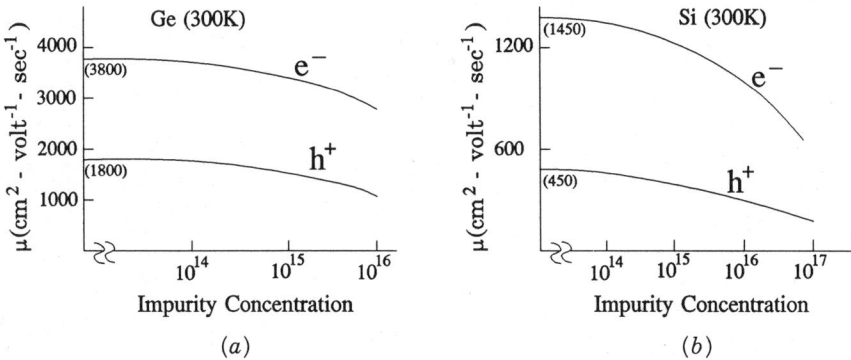

Figure 8.2. (a) Carrier mobility in germanium for various doping concentrations. (b) Carrier mobility for various doping concentrations in silicon.

Substituting into Eq. (8.5) and recalling that the drift velocity of a charge particle is related to the mobility and electric field:

$$v_d = \mu E \tag{8.7}$$

$$i = q\ell_x\ell_y\,\Delta n\mu E = q\ell_x\ell_y\,\Delta n v_d \tag{8.8}$$

The mobility for silicon and germanium is shown in Fig. 8.2a and b. The effect of doping these semiconductors is to decrease the mobility. The value for the average number of carriers generated by the incident irradiance must be evaluated. To obtain the photo-generation rate of carriers per unit volume (g), we must evaluate the absorption process.

Typically, the photoconductor absorption path or detector thickness in the direction of optical propagation is much larger than the reciprocal of the absorption coefficient ($\ell_z \gg 1/a$). The rate of carrier generation varies exponentially with distance from the irradiated surface. The excess carrier concentration therefore must be evaluated by integration of the charge versus distance as shown in Fig. 8.3. Most carriers are generated within a thickness of the value of $1/a$, which is much smaller than the thickness (ℓ_z). Therefore, a carrier-concentration increase near the surface causes a diffusion of carriers (like charges repel) from this layer into the bulk. The concentration falls off exponentially as shown in Fig. 8.3 and approaches zero at the back surface. Remember that the current flow is in the orthogonal direction.

The continuity equation for carrier flow is

$$\frac{\partial \Delta n}{\partial t} = g(z) - \frac{\Delta n}{\tau} + \frac{1}{q}\,(\nabla \cdot \mathbf{J}_e) \tag{8.9}$$

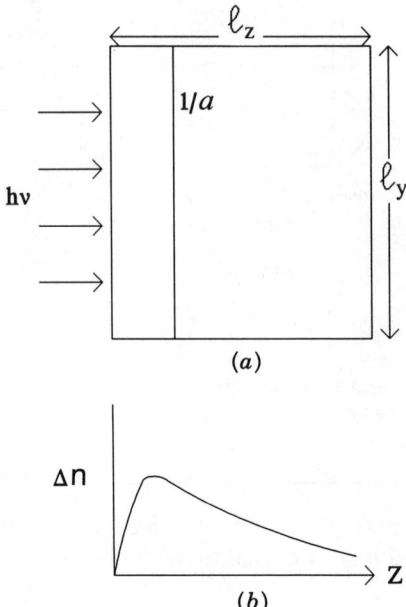

Figure 8.3. Carrier distribution in the direction of optical propagation.

where

Δn = change in carrier concentration caused by illumination
$g(z)$ = generation rate as a function of z
$\Delta n / \tau$ = recombination rate

$$J_e = q\mu_e n\mathbf{E} + qD_e \frac{\partial n}{\partial x} \qquad (8.10)$$

Substituting into the continuity Eq. (8.10), rearranging, and assuming steady state and a constant electric field ($\mathbf{E}$):

$$g(z) = D_e \frac{\partial^2 n}{\partial z^2} - \frac{\Delta n}{\tau} \qquad (8.11)$$

where τ = carrier lifetime. Solving this differential equation,

$$\Delta n = \mathfrak{A}e^{-z/D_e \tau} - \frac{\tau g(z)}{D_e \tau - 1} \qquad (8.12)$$

Recall from Chapter 7 that $L_e = \sqrt{D_e \tau}$. Using the boundary condition that no carriers are present at $z = 0$, ($\Delta n = 0$):

$$\mathfrak{A} = \frac{\tau g(z = 0)}{D_e \tau - 1} \tag{8.13}$$

To find the mean value of the excess carrier concentration in the detector, Eq. (8.10) must be integrated over the thickness:

$$\overline{\Delta n} = \frac{\displaystyle\int_0^{\ell_z} \Delta n(z) \, dz}{\ell_z} \tag{8.14}$$

$$\overline{\Delta n} = \frac{\dfrac{\tau g(z = 0)}{D_e \tau - 1} \displaystyle\int_0^{\ell_z} (e^{-z/L_e} - e^{z/a}) \, dz}{\ell_z} \tag{8.15}$$

Integrating out,

$$\overline{\Delta n} = \frac{\tau g(z = 0) \, L_e}{(L_e^2 - 1)\ell_z} \cong \frac{\tau g(z = 0)}{\ell_z} \tag{8.16}$$

Also taking into account the other two dimensions to obtain volume density, the carrier concentration per unit volume is

$$\overline{\Delta n} = \frac{\tau g(z = 0) \, L_e}{\ell_x \ell_y \ell_z} \tag{8.17}$$

Realizing that $g(0)$ is just the incident photon flux:

$$g(0) = \eta E_q A_d \tag{8.18}$$

The photo-generated carrier density that is generated in the steady state is given by

$$\overline{\Delta n} = \frac{\tau \eta E_q A_d}{\ell_x \ell_y \ell_z} \tag{8.19}$$

The density of photo-generated carriers is related to carrier lifetime and incident photon irradiance. When we substitute Eq. (8.19) into the current Eq. (8.8), the photo-generated current is

$$i = q\mu E \ell_x \ell_y \times \frac{\tau \eta E_q A_d}{\ell_x \ell_y \ell_z} \tag{8.20}$$

Recall some radiometric relations between the radiant power on the detector and the photon irradiance:

$$\phi_e = \phi_q \frac{hc}{\lambda} = E_q A_d \frac{hc}{\lambda} \tag{8.21}$$

Therefore, we can rewrite the current in Eq. (8.20) in terms of incident radiant power:

$$i = \frac{\eta \lambda q}{hc} \phi_e \frac{\mu E \tau}{\ell_z} \tag{8.22}$$

Equation (8.22) has the form of the responsivity figure of merit, the ratio of current output to radiant power input. Terms rearranged yield

$$\Re_i = \frac{i}{\phi_e} = \frac{\eta \lambda q}{hc} \frac{\mu E \tau}{\ell_z} \tag{8.23}$$

The responsivity is exactly the form for the photodiode except for the second term in Eq. (8.23). It is proportional to the electric field, which is the voltage across the detector (V_d) divided by its length:

$$E = \frac{v_d}{\ell_z} \tag{8.24}$$

The responsivity increases linearly with the applied voltage because the carriers are moving with a larger velocity, so the current goes up. Therefore, we must optimize the bias of the photoconductor. The electric field cannot be increased without limit. The typical biasing circuit for a photoconductor is shown in Fig. 8.4.

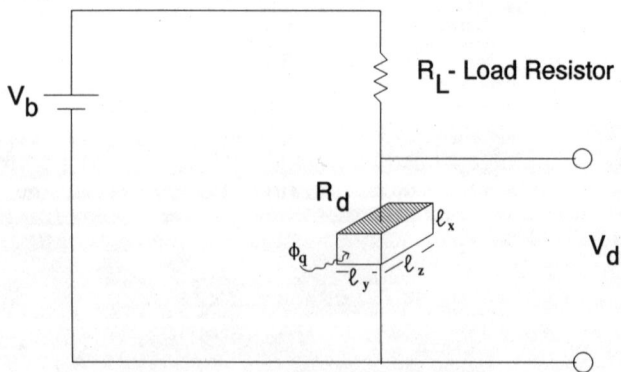

Figure 8.4. Photoconductor biasing circuit.

The second term in Eq. (8.23) is called the *photoconductive gain* (*G*). Note the term is unitless. This gain is a function of the applied voltage across the photoconductor.

$$\mathcal{R}_i = \frac{\eta \lambda q}{hc} G \tag{8.25}$$

8.2.1. Photoconductive Gain

We can rewrite Eq. (8.23) using the carrier-velocity relationship with mobility and electric field:

$$\mathcal{R}_i = \frac{\eta \lambda q}{hc} \frac{v_d \tau}{\ell_z} \tag{8.26}$$

where v_d = drift velocity of carrier. The second term is the photoconductive gain written in terms of carrier velocity. The carrier transit time between electrodes can now be determined. It is the interelectrode spacing (ℓ_z) divided by the velocity (v_d):

$$\tau_t = \frac{\ell_z}{v_d} \tag{8.27}$$

The second term in Eq. (8.26) can be expressed as the ratio of the carrier lifetime to the transit time. This ratio is called the photoconductive gain:

$$G = \frac{\tau}{\tau_t} \tag{8.28}$$

An internal photoconductive gain results because the recombination lifetime and transit time have different values. Because an electron travels faster than a hole, the electron passes through a semiconductor faster. Therefore, for current continuity, an external electrode must supply another electron from the opposite side's electrode to cause charge neutrality. This new electron may move more quickly across the detector than the old hole can recombine with it. A gain is produced that is proportional to the number of times an electron can transit the detector electrodes in its lifetime. This internal photoconductive gain is less than one if the recombination lifetime is short, such that the carrier does not reach the electrode before recombining. The transit time for a typical 1-mm detector is 10^{-8} s for a velocity of 10^7 cm/s. Depending on the detector size, material, and doping, the photoconductive gain can vary from less than one to 10^5. The upper limit is restricted by space-charge effects, dielectric breakdown, and ionization effects.

Rewriting the responsivity expression using Eqs. (8.26), (8.27), and (8.28), which is just Eq. (8.25) with an alternative definition of photoconductive gain, we obtain

$$\mathcal{R}_i = G\frac{\eta\lambda q}{hc} \tag{8.29}$$

To find quantum efficiency, we must know the photoconductive gain. This makes the measurements of quantum efficiency more difficult; what is often quoted is the gain-quantum-efficiency product.

8.2.2.   Quantum Efficiency

The interpretation of quantum efficiency for a photoconductor is somewhat different than in the photovoltaic. The photon can be absorbed anywhere in the semiconductor and be detected because a field exists across the entire chip. The same problems that cause a loss of quantum efficiency apply to a photo-conductor as were discussed with a photovoltaic, however, some improvements in absorption can be forced. The reflections from the second surface (back surface) are important. The expression for quantum efficiency takes the form of

$$\eta = (1 - \imath_1 - \imath_2)(1 - e^{-a\ell_z}) \tag{8.30}$$

where

$\imath_1$ = Fresnel reflection on incident surface
$\imath_2$ = back-surface reflection
ℓ_z = detector thickness in direction of propagation

The back-surface can be made (and often is formed) to cause total internal reflection to increase the optical path length.

One method used to trap the light in the backside is to taper the detector as shown in Fig. 8.5a. Recall from geometrical optics, a tapered light pipe can cause total internal reflection. Figure 8.5b illustrates how to determine which rays pass through the light pipe (Wolfe and Zissis, 1990). The ray shown will be returned through the detector; by putting the taper on the detector, we increase the absorption length by greater than two times the length ($>2\,\ell_z$). The quantum efficiency is also more uniform across the wavelength response because the absorption process will always be in a fully active material region.

8.2.3.   Temporal Response

The response of a photoconductor is determined by the carrier lifetimes, RC time constants, and transit times. The rise-and-fall time of a detector in response to a pulse of radiation was demonstrated in Chapter 5. The responsivity

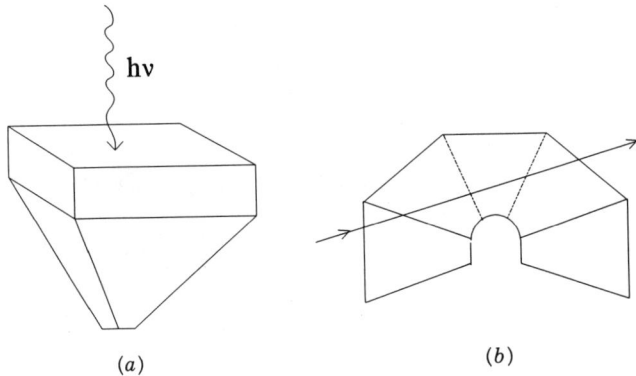

Figure 8.5. (a) Photoconductor geometry to obtain total internal reflection. (b) Light-pipe design to get acceptance angle for total internal reflection.

expression, Eq. (8.25), can be rewritten to include the temporal response:

$$\mathcal{R}_i = \frac{\eta \lambda q}{hc} G \frac{1}{(1 + \omega^2 \tau^2)^{1/2}} \tag{8.31}$$

where τ is an aggregate time constant.

8.2.4. Noise Processes in Photoconductors

The four major noise sources in photoconductors are Johnson noise, $1/f$ noise, generation–recombination (G–R) noise, and preamplifier noise. The typical noise-power spectral density is plotted in Fig. 8.6. The dominant noise at low frequencies is $1/f$ noise, and at mid frequencies is G–R noise. The G–R noise

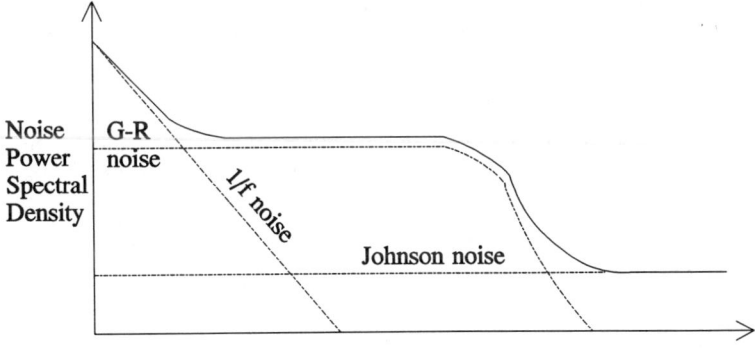

Figure 8.6. Photoconductor typical noise spectrum.

rolloff results from the carrier lifetime and has the form

$$i_{gr}(f) = \frac{i_{gr}}{\sqrt{1 + (2\pi f \tau)^2}} \tag{8.32}$$

At high frequencies, the G–R noise has rolled off, as shown in Fig. 8.6, and the Johnson noise is dominant.

The Johnson noise is generated in both the photoconductor and load resistor used to bias the detector:

$$i_J = \sqrt{\frac{4kT_d\Delta f}{R_d} + \frac{4kT_L\Delta f}{R_L}} \tag{8.33}$$

Let us further assume that the detector temperature and the load-resistor temperature are equal (i.e., $T_d = T_L = T$). A biasing circuit containing a load resistor is required (Fig. 8.4). The equivalent resistance of the load resistor and detector resistor in parallel can be expressed as

$$R_{eq} = \left(\frac{1}{R_d} + \frac{1}{R_L}\right)^{-1} = \frac{R_d R_L}{R_d + R_L} \tag{8.34}$$

The Johnson-noise expression now becomes

$$i_J = \sqrt{\frac{4kT\Delta f}{R_{eq}}} \tag{8.35}$$

The exact cause of $1/f$ noise in photoconductors is not known. Generally, it is attributed to contacts or electrodes. Control of fabrication processes ensures a smooth transition between the semiconductor and metal. Construction of transparent electrodes by means of degenerate doping at the contact has proved successful. The conductivity modulation that occurs at the metal-contact-to-semiconductor transition has been thought to influence $1/f$ noise. Here, the mobility of the carrier undergoes a velocity transition, thus giving off bremsstrahlung radiation. Another effect known to influence the $1/f$ noise is the existence of surface-state traps at the electrodes. The surface-state traps have varying lifetimes that give a nonwhite noise spectrum. To avoid $1/f$ noise, photoconductors are usually operated at frequencies above the $1/f$-noise knee. The $1/f$-noise expression is given by (Van der Ziel, 1976)

$$i_{1/f} = (B_{1/f} i^{\alpha_{1/f}} f^{-\beta_{1/f}} \Delta f)^{1/2} \tag{8.36}$$

where $i_{1/f}$ is the root-mean-square (rms) $1/f$ noise current, $\alpha_{1/f} \cong 2$, $\beta_{1/f} \cong 1$,

$B_{1/f}$ is a proportionality constant, f is the electrical frequency, Δf is the electrical bandwidth, and i is the dc current. Because a bias current is always present in a photoconductor, the $1/f$ noise is always present. To minimize this noise, the detector must have good contacts and must not be operated at low frequencies. Under these conditions, $1/f$ noise can become negligible, compared to the other noises present.

The remaining noise is G–R and/or photon noise that results from the random number of free-charge carriers that exist at succeeding instances of time. This leads to random changes in conductivity of the crystal, causing fluctuations in current flow. The distinction between the G–R and photon noise is that in G–R noise the free carriers are generated and recombined as a result of crystal-lattice vibrations, while in photon noise they are generated by external photons and recombined at minority impurity states.

The G–R noise can be expressed as (Long, 1980)

$$i_{gr} = \sqrt{4q^2\eta E_q A_d G^2 \Delta f + 4q^2 g_{th} G^2 \Delta f} \qquad (8.37)$$

where E_q is background irradiance on the detector and g_{th} is the thermal-generation rate of carriers. The photoconductive gain, G, is the number of excess signal electrons through the photoconductor per absorbed signal photon (Reine and Broudy, 1977). The first term in the preceding expression is the noise produced by background photons and the second term is the noise that results from thermal excitations in the detector. If the detector is cryogenically cooled to a sufficiently low temperature, the second term becomes negligible and the G–R noise current becomes

$$i_{gr} = 2qG\sqrt{\eta E_q A_d \Delta f} \qquad (8.38)$$

For a photon noise such that $hc/\lambda kT \leq 1$, the boson factor must be retained because the photon statistics are more accurately described by Bose–Einstein statistics rather than Poisson statistics. The equation for photon noise then becomes

$$i_{gr} = 2qG(\eta A_d \Delta f E_q(1 + b_{Bose}))^{1/2} \qquad (8.39)$$

where b_{Bose} = Bose factor. For background temperatures of 300 K, b_{Bose} is negligible; however, at 50 μm, $b_{Bose} = 0.60$, and thus must be taken into account. Thus, the bose factor is important to consider at longer wavelengths. The total noise current is

$$i_n = (i_{gr}^2 + i_j^2 + i_{1/f}^2 + i_{pa}^2)^{1/2} \qquad (8.40)$$

Normally, the preamplifier is carefully chosen such that its noise is negligible compared to the other sources of noise.

Now, let us assume that the detector operates in the G–R-noise BLIP-limited mode:

$$i_{gr}^2 \gg i_j^2 \qquad (8.41)$$

Substituting Eqs. (8.38) and (8.35), the constraints and impacts are

$$4q^2G^2\eta E_{q,\text{bkg}}A_d\Delta f \gg 4k\Delta f\left(\frac{T}{R_{\text{eq}}}\right) \qquad (8.42)$$

where the first term represents noise from the background photons and the second term represents the Johnson noise of the combination of the load resistor and the photoconductive detector. Rearranging the preceding inequality,

$$\frac{T}{R_{\text{eq}}} \ll \frac{q^2G^2\eta E_{q,\text{bkg}}A_d}{k} \qquad (8.43)$$

This relationship sets a requirement on the choice of operating temperature and load resistor that should be used in a particular system design to assure BLIP operation. The right side of Eq. (8.43) is determined by radiometry (E_q), detector area, and quantum efficiency. To maintain BLIP operation, the user must choose the load resistor and temperature required to assure Eq. (8.43) inequality is met.

8.2.5. D^* and NEP Relations

The figures of merit for a photoconductor can now be derived for both the (BLIP) background-limited infrared photodetector and Johnson-limited (JOLI) cases. The fundamental limit to sensitivity of a photoconductor depends on the operation temperature, background photon irradiance, and load resistor. BLIP performance is achieved when the limiting noise is the G–R type.

For BLIP operation, the noise-equivalent power (NEP) is

$$\text{NEP}(\lambda, f) \equiv \frac{i_{gr}}{\mathcal{R}_i} = \frac{2qG(\eta E_{q,\text{bkg}}A_d\Delta f)^{1/2}}{(\eta\lambda q/hc)G} \qquad (8.44)$$

Canceling terms, and noting the gain (G) should not effect the sensitivity (it increases signal and noise equally):

$$\text{NEP}(\lambda, f) = \frac{2hc}{\lambda}\left(\frac{E_{q,\text{bkg}}A_d\Delta f}{\eta}\right)^{1/2} \qquad (8.45)$$

Using the definition of D^* relating to NEP for the corresponding $D^*(\lambda, f)$ for BLIP operation of a photoconductor:

$$D^*(\lambda, f) = \frac{\lambda}{2hc} \sqrt{\frac{\eta}{E_{q,\text{bkg}}}} \tag{8.46}$$

As the temperature of the detector is increased, thermal generation becomes important, and $D^*(\lambda, f)$ decreases rapidly. No attempt is generally made to operate a photoconductor at elevated temperature. Ideal BLIP detectivity curves are usually calculated and plotted assuming $\eta(\lambda) = 1$ to the cutoff wavelength and that the Bose factor is negligible. This is seldom absolutely true in practice and is one reason why theoretical D^*s are never achieved.

The background-limited D^* was derived, but as the background is reduced to zero ($E_{q,\text{bkg}} \approx 0$), the Johnson noise becomes dominant. The expression for Johnson noise is

$$i_J^2 = \frac{4kT\Delta f}{R_{\text{eq}}} \tag{8.47}$$

assuming the load resistor and detector are mounted on the same heat sink as before. Again, the equivalent effective resistance is

$$R_{\text{eq}} = \frac{R_L R_d}{R_L + R_d} = \left(\frac{1}{R_L} + \frac{1}{R_d}\right)^{-1} \tag{8.48}$$

For the NEP expression relating responsivity and noise:

$$\text{NEP} = \frac{\sqrt{i_J^2}}{\mathfrak{R}_i} \tag{8.49}$$

Substituting

$$\text{NEP} = \frac{\sqrt{\dfrac{4kT\Delta f}{R_{\text{eq}}}}}{\dfrac{\eta\lambda q}{hc} G} \tag{8.50}$$

This is the JOLI limit (Johnson-noise limit). Again the parallel combination of the load resistor and detector resistor should be as large as possible. Substituting the NEP into the D^* relationship yields

$$D^*(\lambda, f) = \frac{\eta\lambda q}{2hc} G \sqrt{\frac{R_{\text{eq}} A_d}{kT}} \tag{8.51}$$

For the photoconductor, it is important for both the $R_{eq}A_d$ product and the photoconductive gain (G) to be high.

For JOLI operation to be high, the quantum efficiency and photoconductive gain affect D^* directly. The resistance-area product has two special cases that can be used because the load resistor can be selected

1. if $R_L \gg R_d$; $R_{eq} \approx R_d$
2. if $R_L = R_d$ (maximizes $\Re_i$); $R_{eq} = R_d/2$

In either case, $R_d A_d$ needs to be as large as possible. The $R_d A_d$ product is not an inherent material property, as it was in the case of the photodiode. (For the photodiode, if the area was increased the resistance decreased proportionally so that the product was a constant; in other words, a figure of merit of the detector material used for the photodiode.)

For the photoconductor, increasing the area by increasing ℓ_z (see Fig. 8.3) causes both resistance and area increase, so the RA product *is not* an inherent material figure of merit. However, resistivity of the material, ρ, is a material parameter for a photoconductor. If we consider the detector resistance

$$R_d = \frac{\rho \ell_z}{\ell_x \ell_y} \tag{8.52}$$

If $\ell_z = \ell_y$ (a square detector), the resistance is

$$R_d = \frac{\rho}{\ell_x} \tag{8.53}$$

This value is often referred to as ohms/square. Therefore the aspect ratio of the optically active area determines the detector resistance. For example, a detector geometry of $3:1$ has a resistance of 3 times the ohms/square.

8.3. MERCURY CADMIUM TELLURIDE PHOTOCONDUCTORS

The photoconductive mercury cadmium telluride (HgCdTe) detector has the same genesis as the photovoltaic, in that it can be made to operate over different spectral regions simply by controlling the Cd/Hg ratio in the material, which in turn controls the wavelength. However, the photoconductor bandgap can be varied to have a spectral range that is further in the infrared, to approximately 30 μm. The photoconductive $Hg_{1-x}Cd_xTe$ detector is important because detectors in this range are scarce, it possesses a higher operational temperature, and its performance is excellent (Broudy and Mazurczyk, 1981).

The photoconductive $Hg_{1-x}Cd_xTe$ detector is a good intrinsic infrared detector and can achieve background-limited sensitivities at operating tempera-

Table 8.2 Extrinsic Photoconductors

Material	Impurity Concentration	$\mathcal{E}_g$	λ_c	Operating Temperature (K)
Si : As	3×10^{16}	0.052	24	10
Si : P	3×10^{16}	0.045	28	4
Si : Ga	3×10^{16}	0.073	19	18
Si : In		0.155	8	45
Ge : Cu	3×10^{16}	0.041	29	4
Ge : Hg	8×10^{15}	0.088	14	30
Ge : In		0.033	38	4
Ge : Zn		0.0104	118	2

ture substantially higher than the extrinsic detectors. Table 8.2 listed typical material parameters (Long and Schmit, 1970) for extrinsic. The zinc-blende lattice structure (two faced-centered cubic (FCC) interpenetrating) was shown in Chapter 7 (Fig. 7.56). The Hg (cations) forms a weak bond in this structure, so the Hg atom is often dislocated.

The $Hg_{1-x}Cd_x Te$ energy gap is a direct bandgap and is a function of the mole fraction of cadmium to mercury, x, and of the temperature, T, given by (Hansen et al., 1982)

$$\mathcal{E}_g(x, T) = -0.302 + 1.93x + (5.35 \times 10^{-4})(1 - 2x)T$$
$$- 0.810x^2 + 0.832x^3 \tag{8.54}$$

This equation is fit for the energy-gap data for $0 \le x \le 0.6$ (plus $x = 1$) and for $4.2\ K < T \le 300\ K$. Figure 8.7 shows the plot of energy gap and mole fraction. The most important region of interest is the 8- to 14-μm region, which corresponds to x values of 0.203 to 0.245.

A unique characteristic of $Hg_{1-x}Cd_x Te$ is that it is an infrared intrinsic semiconductor, which leads to two very important features. First, no impurity levels are present in its band structure, so its temperature need only be such that it cannot cause electron transitions from valence to conduction band. This means that the $Hg_{1-x}Cd_x Te$ detectors can operate at higher temperatures than the equal cutoff-wavelength extrinsic semiconductors (Broudy and Mazurczyk, 1981) because extrinsic semiconductors have impurity levels in their forbidden region, and they require much lower operational temperatures to reduce the thermal (kT) excitation to or from these levels. The second important and advantageous feature is that the absorptivity of $Hg_{1-x}Cd_x Te$ is better than that of an extrinsic semiconductor. It has more atoms capable of capturing a photon, and therefore, it has a larger capture cross section; hence, these semiconductors can be made thinner. Typical operating values of the optical absorption coefficient are (Reine and Broudy, 1977) $a = 10^5\ cm^{-1}$, so virtually all of the

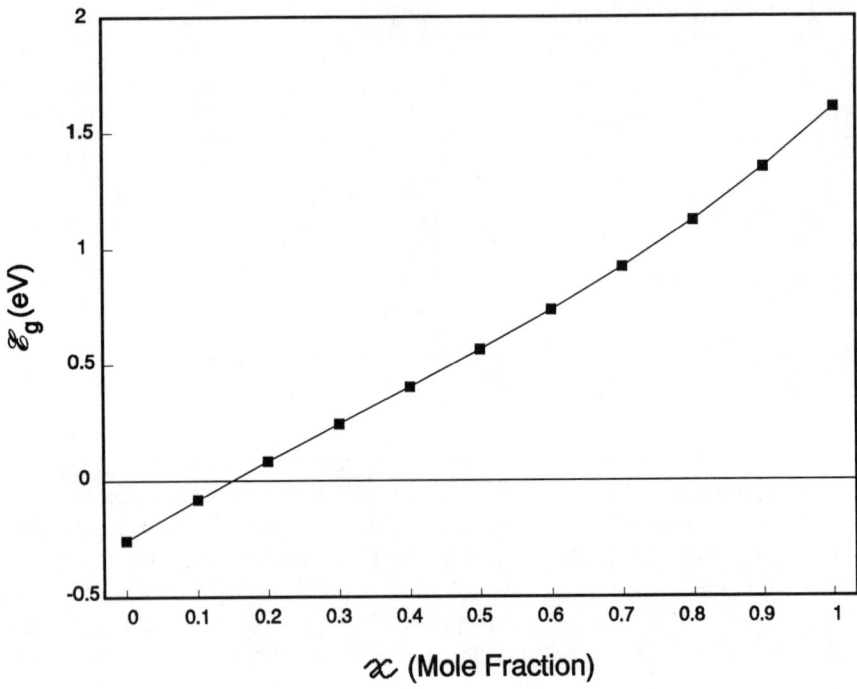

Figure 8.7. $\mathcal{E}_g$ vs. x for $T = 80$ K.

photons are absorbed within 10 μm of the surface. Plots of a versus photon energy for various x (Dornhaus and Nimtz, 1976) are shown in Fig. 8.8. The steepness in these curves leads to the rapid decline in detectivity for wavelengths larger than λ_c (Finkman and Nemerovsky, 1979).

The negative side of the variability of the energy gap is that the mole fraction is not well controlled in fabrication so the material must be analyzed by x-ray crystallography or Bragg diffraction to determine x. This uncontrollability in x leads to a nonuniform x throughout the material; thus the responsivity varies as a function of position. For large values of x, $\lambda_c < 5$ μm, the effect is negligible, but for $\lambda_c > 5$ μm, the effect is quite significant. This becomes a major problem when making an array with photoconductive $Hg_{1-x}Cd_xTe$ because of responsivity variations across the array, and detected background variation across the array attributable to differences in the cutoff wavelengths.

Because the effective mass of the electrons is much smaller than the effective mass of the holes, the mobility of the electrons is much greater than the mobility of the holes. Typical values at 77 K for the electron mobility and the hole mobility are $\mu_e = 3 \cdot 10^5$ cm^2/V-s and $\mu_h = 10^3$ cm/V-s, respectively; hence, one can neglect the electrical effects of the holes.

$Hg_{1-x}Cd_xTe$ photoconductors are typically fabricated from n-type material with an excess concentration of ionized donors of about 10^{14} cm^{-3}. This large

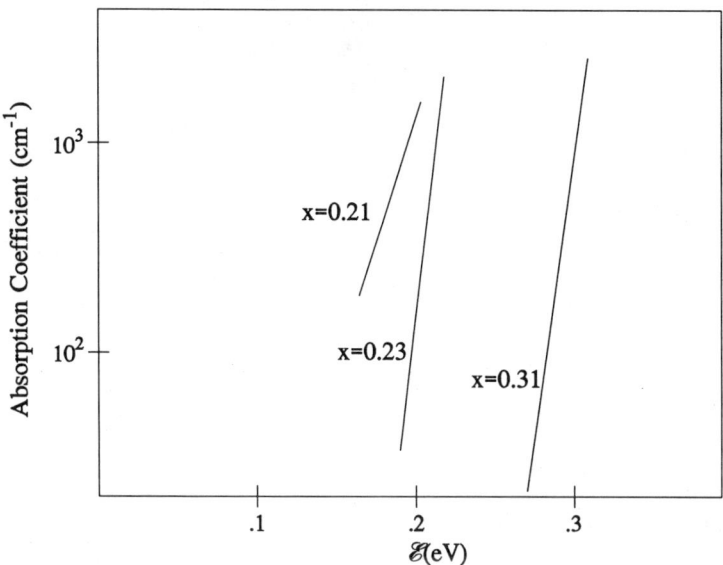

Figure 8.8. Approximate absorption coefficient for HgCdTe vs. photon energy for various values of x at $T = 300$ K.

intrinsic carrier concentration is the result of excess mercury atoms that exceed the stoichiometry and reside interstitially in the lattice. A plot of the intrinsic carrier concentration, n_i versus T for various values of x (Kruse, 1981; Hansen and Schmit, 1983), is shown in Fig. 8.9. The relationship is

$$n_i(x, T, \mathcal{E}_g) = [5.585 - 3.820x + 1.753 \times 10^{-3}\, T - 1.364 \times 10^{-3}\, Tx]$$
$$\cdot [10^{14}\, \mathcal{E}_g^{0.75}\, T^{1.5}\, e^{-\mathcal{E}_g/2kT}] \qquad (8.55)$$

The conductivity, σ_c, can be written as:

$$\sigma_c = \sigma_{c,i} + \sigma_{c,\mathrm{bkg}} + \Delta\sigma_c = n_i \mu_e q + n_b \mu_e q + \Delta n \mu_e q \qquad (8.56)$$

where

$\sigma_{c,i}$ = conductivity due to intrinsic carrier concentration
$\Delta\sigma_c$ = conductivity due to signal-photon-generated carriers
n_i = intrinsic-carrier concentration
n_{bkg} = concentration of background-photon-generated carriers
Δn = concentration of signal-photon-generated carriers
q = charge of an electron = 1.6×10^{-19} coulombs
μ_e = electron mobility

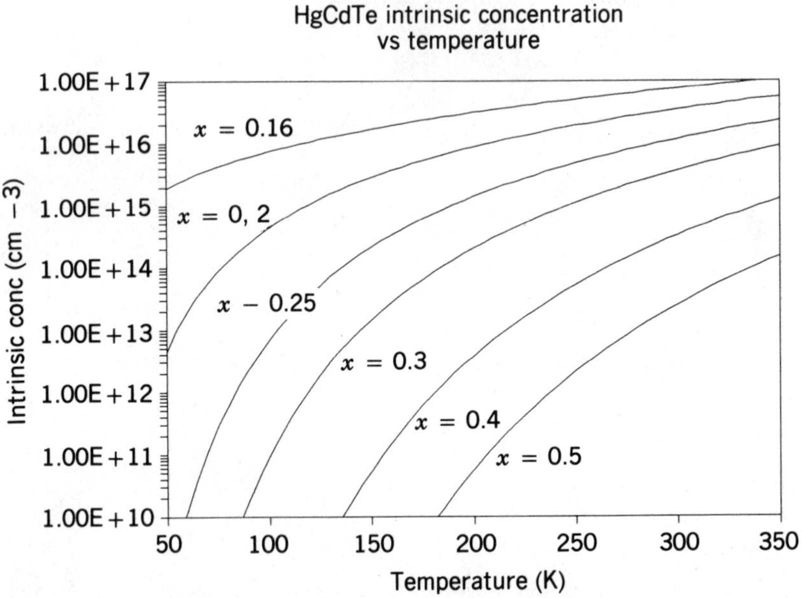

HgCdTe intrinsic concentration
vs temperature

Figure 8.9. Intrinsic carrier concentration vs. temperature for various values of x.

The intrinsic-carrier concentration is so high that $\sigma_{c,i} \gg \sigma_{c,\text{bkg}} + \Delta\sigma_c$. The magnitude of the conductivity is generally about 4.8 $(\Omega\text{-cm})^{-1}$. With the detector dimensions shown in Fig. 8.1, the detector equilibrium resistance is given by

$$R_d = \frac{\ell_z}{\sigma_c \ell_x \ell_y} \tag{8.57}$$

Using typical detector dimensions of $\ell_z = \ell_y = 0.05$ cm (or any square optically active area detector) and $\ell_x = 0.001$ cm, the resistance is about 200 Ω. The large intrinsic-carrier concentration of photoconductive $Hg_{1-x}Cd_xTe$ makes a very-low-resistance detector.

One disadvantage of such a low-resistance detector is that high currents can flow with low voltage. HgCdTe photoconductors can typically dissipate about 1 mW of power before burning out. Because $P_{av} = v_d^2/R_d$ (where P_{av} = average power dissipated, v_d = voltage across the detector, and R_d = detector resistance), a voltage of about 0.456 V (or an electric field of about 9 V/cm) can be applied across the detector. This requires that a very low bias voltage be applied.

8.3.1. Responsivity

The derivation of responsivity uses Eq. (8.56), the conductivity equation. In HgCdTe n-type photoconductors, $\mu_e \gg \mu_h$, $n > p$, and $\Delta n \cong \Delta p$, so the conductivity can be evaluated for electrons only:

$$\sigma_c = \sigma_{c,i} + \sigma_{c,bkg} + \Delta\sigma_c \cong (n_i + n_{bkg})\mu_{eq} + \Delta n\mu_{eq} \qquad (8.58)$$

Taking the derivative of the detector resistance, Eq. (8.57), we find

$$\frac{dR_d}{R_d} = -\frac{d\sigma_c}{\sigma_c} = -\frac{\Delta n\mu_e q}{\sigma_c} \qquad (8.59)$$

The excess carrier concentration resulting from signal-photon radiation was used (Eq. (8.12)). Substituting into Eqs. (8.19) and (8.21) yields

$$\frac{dR_d}{R_d} = -\frac{\mu_e q}{\sigma_c}\left[\frac{\eta\lambda\tau\ell_z}{hcA_d\ell_x}\right]\Delta\phi_{e,sig} \qquad (8.60)$$

The biasing circuit for an HgCdTe is the classic photoconductor circuit shown in Figure 8.10. The dc voltage across the detector from Fig. 8.10 is

$$v_d = \frac{v_B R_d}{R_L + R_d} \qquad (8.61)$$

To get the signal voltage resulting from an incremental charge in detector resistance caused by photon flux, we can take the differential of Eq. (8.61):

$$dv_d = \frac{-v_B R_L dR_d}{(R_L + R_d)^2} \qquad (8.62)$$

Substituting Eqs. (8.60) into Eq. (8.62) gives the signal voltage out of the

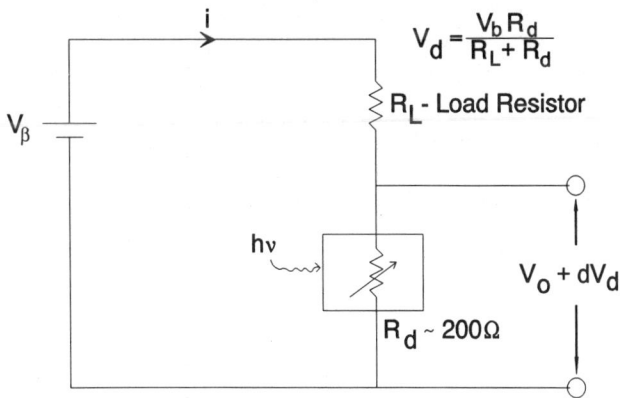

Figure 8.10. Typical photoconductor biasing configuration.

detector circuit:

$$dv_d = \left(\frac{q\lambda\eta}{hc}\right)\left(\frac{\mu_e \tau \ell_z i}{\sigma_c A_d \ell_x}\right)\Delta\phi_{e,\text{sig}}R_{\text{eq}} = \left(\frac{q\lambda\eta}{hc}\right)G\Delta\phi_{e,\text{sig}}R_{\text{eq}} \qquad (8.63)$$

By definition, the voltage responsivity of a photoconductive $Hg_{1-x}Cd_xTe$ detector (Eq. (8.63)) becomes

$$\Re_v(\lambda, f) = \frac{dv_d}{\Delta\phi_{e,\text{sig}}} = \left(\frac{q\lambda\eta}{hc}\right)GR_{\text{eq}} \qquad (8.64)$$

Recall that the photoconductive gain, G, is related to the carrier lifetime (τ) and transit time (τ_t). Typical values of τ and τ_t in $Hg_{1-x}Cd_xTe$ photoconductive detectors are 0.5 μs and 18 ns, respectively; thus, a typical value for G would be about 27. The gain is greater than one because of induced discontinuities producing localized electric fields, which leads to higher carrier velocity and secondary currents that far exceed the photo-generated currents.

This expression for the gain is true only for frequencies below the cutoff frequency of the detector. The gain is a function of frequency (Reine and Broudy, 1977):

$$G(f) = \frac{\tau}{\tau_t}\frac{1}{\sqrt{1 + (2\pi f\tau)^2}} \qquad (8.65)$$

and therefore the responsivity is also a function of chopping frequency, as was shown in Eq. (8.31). The cutoff frequency is given by the 3-dB point that is $f_c = 1/(2\pi\tau)$. A typical value of f_c is 0.32 MHz; thus, the signal should be chopped at frequencies lower than 0.32 MHz.

8.3.2. NEP and D^*

The total voltage noise in the detection circuit is the current-noise expression times the equivalent resistance. Assuming that the detector is operated such that the $1/f$ noise is negligible, the total-noise voltage equation becomes

$$v_n^2 = (4\,kTR_{\text{eq}}\Delta f) + (4q^2\eta E_{q,\text{bkg}}A_d G^2(f)R_{\text{eq}}^2\Delta f) + (v_{\text{pa}}^2\Delta f) \qquad (8.66)$$

Therefore, the NEP for a photoconductive $Hg_{1-x}Cd_xTe$ detector can be written as

$$\text{NEP}(\lambda, f)$$

$$= \frac{hc}{q\lambda\eta G(f)R_{\text{eq}}}\sqrt{(4\,kTR_{\text{eq}}\Delta f) + (4q^2\eta E_{q,\text{bkg}}A_d G^2(f)R_{\text{eq}}^2\Delta f) + (v_{\text{pa}}^2\Delta f)}$$

$$(8.67)$$

The lower the NEP (λ, f), the more sensitive the detector. A typical value of NEP for a HgCdTe detector is 2.5×10^{-12} W. The normalized detectivity of a HgCdTe photoconductive detector is

$$D^*(\lambda, f) = \frac{\left(\dfrac{\eta\lambda}{hc}\right)}{\sqrt{(4\eta E_{q,\text{bkg}}) + \left(\dfrac{4kT}{q^2 G^2(f) R_{\text{eq}} A_d}\right) + \left(\dfrac{v_{\text{pa}}^2}{q^2 G^2(f) R_{\text{eq}}^2 A_d}\right)}} \tag{8.68}$$

Notice that because of the $G(f)$ term, $D^*(\lambda, f)$ varies with chopping frequency:

$$D^*(\lambda, f) = \frac{D^*(\lambda)}{\sqrt{1 + (2\pi f\tau)^2}} \tag{8.69}$$

where $D^*(\lambda)$ is the normalized detectivity at low frequencies. The reason for the rolloff at high frequencies is that the Johnson and preamplifier noises become the dominant noises. A typical $D^*(\lambda, f)$ value for alloy compositions used for 10-μm detectors is 2×10^{10} for a 60° field of view (FOV) at 300-K background.

From Eq. (8.68) one can see the importance of photoconductive gain. We saw earlier that the $R_{\text{eq}} A_d$ product was quite low; however, a large $D^*(\lambda, f)$ comes about because of the large gain factor, G. Equation (8.68) seems to imply that $D^*(\lambda, f)$ will increase infinitely with decreasing temperature; however, it actually flattens out at lower temperatures (Dornhaus and Nimtz, 1976). The reason for constant $D^*(\lambda, f)$ at low T is that the minority carriers are swept out and the decrease in $D^*(\lambda, f)$ at larger T $(T > 77$ K$)$ is attributable to the temperature dependence of the carrier concentration. The normal operating temperature is 77 K. Under this operating temperature, the Johnson-noise limit is the ultimate noise limit.

8.3.3. Operational Constraints

A typical liquid-nitrogen Dewar houses the detector. A low-thermal-conductivity fiberglass sleeve isolates the liquid-nitrogen reservoir from the Dewar exterior to reduce heat leakage. By evacuating a chamber between the liquid-nitrogen reservoir and the room environment, a thermal insulator is created. In this manner the detector is cooled by conduction through the Dewar copper heat sink to its operating temperature of about 77 K. A preamplifier circuit for this low-impedance detector is shown in Fig. 8.11. Bipolar transistors (low input impedance) can be used because the detectors have such a low resistance. Cleaning of HgCdTe can be performed with a 20% solution of bromine and methyl alcohol, followed immediately by a H_2O wash. Typically, the electrodes are evaporated indium.

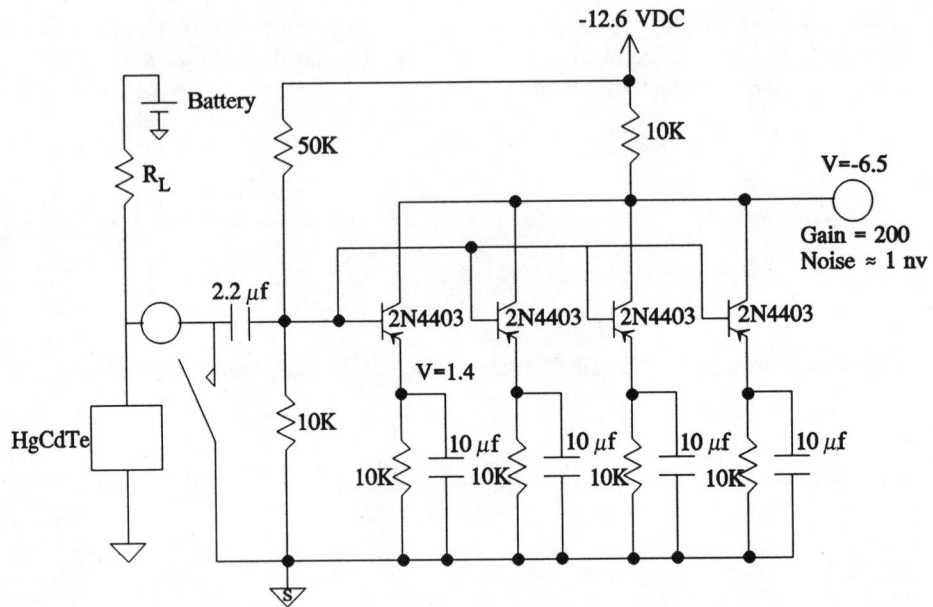

Figure 8.11. Photoconductive HgCdTe preamplifier circuit.

The steep slope of the optical-absorption-versus-mole-fraction curve in Fig. 8.5 shows that small changes in the mole fraction lead to large changes in absorption and, therefore, large changes in the cutoff wavelength. Also, fabricating compositionally uniform detectors, as discussed earlier, is difficult. Producing two identical detectors is extremely difficult. These drawbacks alone make it difficult to produce detector arrays. Detector arrays have large current densities (low resistance) and the potential wells fill up too quickly because of background currents. For all of these reasons, photoconductive HgCdTe is not used for making arrays in a charge-transfer device. An interesting effect occurs in some manufactured detectors at low background irradiances. The carrier lifetime increases; in fact, the background irradiance-carrier lifetime is a constant:

$$\tau E_q = \text{constant} \tag{8.70}$$

This causes the photoconductive gain to increase at low background: a somewhat surprising effect.

8.3.4. Summary

The photoconductive $Hg_{1-x}Cd_xTe$ detectors are excellent single-element detectors covering the spectral range of 1 to 30 μm, and typical specificatons have been given. They have filled the 8- to 14-μm atmospheric window that previously lacked a good detector. The drawbacks are of major concern, and at the present time research is being done on the growth of superlattices in the HgTe/HgCdTe system in hopes of reducing the steep compositional dependence of absorption, producing a better absorption coefficient and lower $1/f$ noise. At the present time, HgCdTe is still the best detector for the 8- to 14-μm spectral region.

8.4. SPRITE DETECTORS

SPRITE (Signal PRocessing In The Element) detectors are used to produce imaging systems in the 8- to 14-μm region using HgCdTe photoconductive material. This technique was invented by C. T. Elliott of the Royal Signals and Radar Establishment, Great Britain. Although 3- to 5-μm systems exist, they are less important because lower-cost focal-plane arrays are available to compete with the SPRITEs in the 3- to 5-μm band.

To increase the signal-to-noise ratio (SNR) in a background-noise-limited or Johnson-noise-limited imaging system, a technique called *time delay and integration* (TDI), is often used. Figure 8.12 illustrates TDI. The concept is to add the signal from several detectors while scanning an image across them. This improves the signal by the number of detectors (n_d). The noise increases by, and therefore the signal-to-noise ratio increases by, $n_d^{1/2}$. This is accomplished with no loss of optical resolution, because the detector geometrical size is unchanged. However, a mechanical scanner is needed to sweep the image across the TDI array of detectors. One drawback of the TDI technique is that at least one additional lead is required for each detector. In addition, the requirement of multiple delays make the electronics complicated (Leftwich and Ward, 1988).

The SPRITE detectors realize the equivalent of the TDI technique right in the element, without complex electronics or multiple leads. Basically, a SPRITE detector is a stretched HgCdTe n-type photoconductor, often referred to as a filament. A SPRITE is a three-terminal device, with two bias leads and

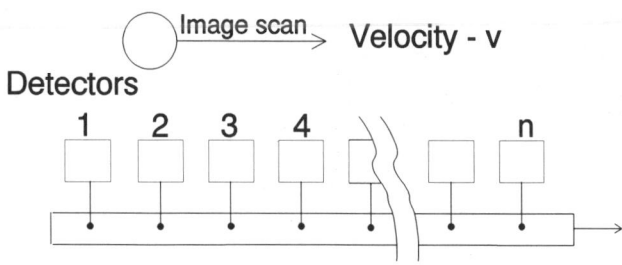

Figure 8.12. TDI using discrete detector and a scanned image.

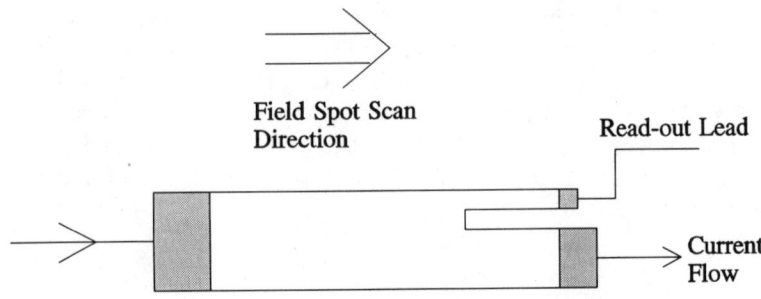

Figure 8.13. HgCdTe (PC) SPRITE.

a readout lead. An actual configuration is depicted in Fig. 8.13. Some typical values are: length $\ell_z = 1$ mm; width $\ell_y = 50$ μm; thickness $\ell_x = 10$ μm; and readout length $\ell_{z,\,\text{readout}} = 50$ μm. A constant-current source provides a virtually constant electric field in the filament. The change of excess minority-carrier concentration is detected by measuring the change of conductance in the read-out region (Elliott, 1981).

8.4.1. Theory of Operation

The current density is related to conductance by

$$J = \sigma_c E$$

For a semiconductor, there are always two kinds of carriers coexisting, that is, electrons and holes. The current inside the material is carried by both of them:

$$J = q(n v_{d,e} + p v_{d,h}) \tag{8.71}$$

Here $v_{d,e}$ and $v_{d,h}$ are the absolute values of the drift velocities of the electrons and holes, respectively. Although the electrons and holes travel in exactly opposite directions on the influence of an external electric field, it can be shown easily that both electrons and holes contribute to the electric current in a constructive manner. The magnitudes of both v_e and v_h are proportional to the applied electric field, with the proportionality constants their mobilities:

$$v_{d,e} = \mu_e E \tag{8.72a}$$

$$v_{d,h} = \mu_h E \tag{8.72b}$$

By substituting these equations into that of the current density, we can get the conductivity of a semiconductor (recall Eq. (8.1)):

$$\sigma_c = q(\mu_e n + \mu_h p) \tag{8.73}$$

For an n-type photoconductor, the material used for SPRITE detectors:

$$n \gg p \tag{8.74}$$

which means,

$$\sigma_c = q\mu_e n \tag{8.75}$$

When photons are absorbed by the HgCdTe material, electron–hole pairs are generated. These excess carriers will increase the photoconductor conductance:

$$\Delta n = \Delta p \propto E_q \tag{8.76}$$

where E_q is the effective photon irradiance onto the detector. If the excess carrier concentration is relatively small compared to the doping concentration, that is

$$\Delta n \ll n \tag{8.77}$$

the photon-generated carriers are not going to change the local conductance significantly, and the field inside the filament can be regarded as approximately constant.

Furthermore, the electrons generated, being majority carriers, are forced by the applied electric field toward the current injection end of the filament, while the holes are forced toward the current-collection end of the filament. Because we are to monitor excess charge concentration at the current-collection end of the filament, only the excess concentration of the minority carriers, that is, the holes, are important to the current.

The electric field inside the SPRITE detector is constant (Elliott et al., 1982). By controlling the value of a constant-current supply, we can choose any value for the electric field inside of the filament. Assume the image spot corresponding to a target region, d_{spot} in diameter, is scanned across the element at a constant velocity, v_{scan}. If we choose the electric field to produce a velocity of carriers in the detector equal to the scanned spot velocity on the detector (Fig. 8.14):

$$v_{\text{scan}} = v_{d,h} \tag{8.78}$$

where $v_{d,h}$ is the drift velocity of holes, then the continued effect of the scanning illumination spot is time integrated in terms of excess minority carrier concentration. The actual value of the current desired can be derived simply:

$$v_{d,h} = \mu_h E \tag{8.79}$$

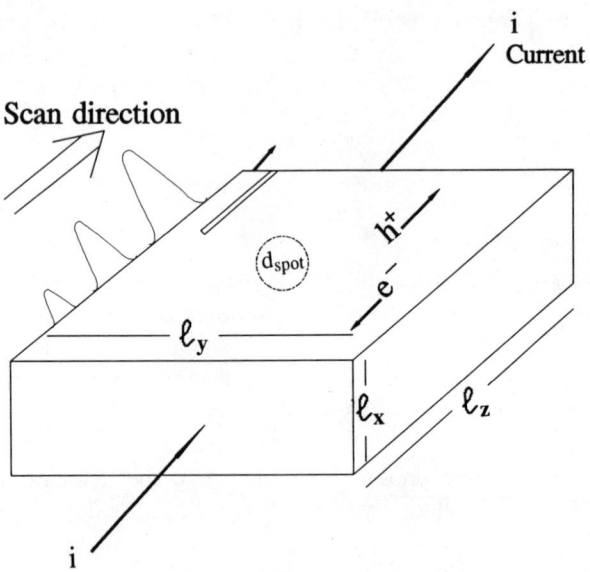

Figure 8.14.   Target spot is scanned at the same velocity as the holes are swept in the HgCdTe.

$$J = \sigma_c E = q n \mu_e \frac{v_{d,h}}{\mu_h} \tag{8.80}$$

$$i = \ell_y \ell_x J = \ell_y \ell_x q n \mu_e \frac{v_{d,h}}{\mu_h} \tag{8.81}$$

or

$$i = \frac{q n \mu_e \ell_y \ell_x}{\mu_h} v_{d,h} \tag{8.82}$$

The total integration time is

$$\tau_{\text{int}} = \frac{\ell_z}{v_{d,h}} \tag{8.83}$$

Two characteristics of the holes in HgCdTe make it an ideal choice for SPRITE detectors: long lifetime and small diffusion length. As explained earlier, the SPRITE detector works by virtue of integrating the excess hole concentration. The lifetime of the holes must be longer than the integration time

$$\tau_h > \tau_{\text{int}} \tag{8.84}$$

or the excess holes generated early in the integration period will have been largely lost by the time the integration is complete and the excess hole concentration is measured at the other end of the filament. Thus, the longer the lifetime of holes, the longer one can integrate the signal.

The other characteristic of the holes that makes HgCdTe the most feasible material for SPRITE detectors is the small diffusion length, D_h. The diffusion effect of the holes makes the signal spot smear out along the integration process, and deteriorate the detector impulse-response function. Although holes in HgCdTe have long lifetime, they have very small diffusion length because of limited mobility. Typical characteristics of the holes in HgCdTe include:

Band (μm)	8–14
Operating temperature (K)	80
τ_h (μs)	2
μ_h (cm^2/V s)	480
D_h (μm)	25

From the continuity equation of the holes (Elliott et al., 1982), we can solve for the excess hole concentration in the readout region of a SPRITE detector, assuming a constant signal-photon irradiance, $E_{q,\text{sig}}$:

$$\Delta p = \frac{\eta E_{q,\text{sig}} \tau_h}{\ell_x} \left(1 - \exp \left\{ -\frac{\ell_z}{\mu_h E \tau_h} \right\} \right) \tag{8.85}$$

The output voltage generated in the readout is

$$v_o = \frac{\Delta p E \ell_{z,\text{readout}}}{n} \tag{8.86}$$

and voltage responsivity for photon irradiance is

$$\Re = \frac{v_o}{E_{q,\text{sig}}(hc/\lambda)\ell_{z,\text{readout}}\ell_y} = \frac{E \eta \tau_h}{(hc/\lambda)\ell_x \ell_y n} \left(1 - \exp \left\{ -\frac{\ell_z}{\mu_h E \tau_h} \right\} \right) \tag{8.87}$$

Assuming a background-noise-limited situation, which is the case when generation–recombination noise caused by the background illumination is the dominant noise, we obtain the equation for D^*:

$$D^* = \frac{\lambda}{2 hc} \sqrt{\frac{\eta \ell_{z,\text{readout}}}{E_{q,\text{bkg}}\ell_y}} \frac{1}{\sqrt{1 - \frac{\tau_h}{\tau_{t,\text{readout}}} \left(1 - \exp \left\{ -\frac{\tau_{t,\text{readout}}}{\tau_h} \right\} \right)}} \tag{8.88}$$

Where $\tau_{t,\text{readout}}$ is the transit time of the holes through the readout: $\tau_{t,\text{readout}} = \ell_{z,\text{readout}}/v_{d,h}$. When the scan speed is sufficiently high, that is,

$$\tau_{t,\text{readout}} \ll \tau_h \tag{8.89}$$

we have

$$1 - \frac{\tau_h}{\tau_{t,\,\text{readout}}} \left(1 - \exp\left\{ -\frac{\tau_{t,\,\text{readout}}}{\tau_h} \right\} \right) \cong \frac{1}{2} \frac{\tau_{t,\,\text{readout}}}{\tau_h} \qquad (8.90)$$

and therefore

$$D^* = \frac{\lambda}{2hc} \sqrt{\frac{\eta \ell_{z,\,\text{readout}}}{E_{q,\,\text{bkg}} \ell_y}} \sqrt{2 \frac{\tau_h}{\tau_{t,\,\text{readout}}}} \qquad (8.91)$$

By identifying the term $(\lambda/2hc)\,(\eta/E_{q,\,\text{bkg}})^{1/2}$ as the D^*_{BLIP} for a conventional photoconductor detector, we get

$$D^* = D^*_{\text{BLIP}} \sqrt{\frac{\ell_{z,\,\text{readout}}}{\ell_y \tau_{t,\,\text{readout}}} \frac{2\tau_h}{}} \qquad (8.92)$$

We can identify the term $\ell_{z,\,\text{readout}}/\ell_y \tau_{\text{readout}} = v_s/\ell_y$ as the number of pixels (of dimension ℓ_y) per second that are scanned, that is, a scan rate. We see that SPRITE detectors provide an advantage over conventional photoconductive detectors when the product of this scan rate and the hole lifetime is greater than one-half:

$$\frac{v_s \tau_h}{\ell_y} > 1/2 \qquad (8.93)$$

This condition provides an estimate of the required scan rates of about 5×10^5 Hz for the 8- to 14-μm region and about 10^5 Hz for the 3- to 5-μm region, assuming the use of optimum temperatures of operation in each band.

Comparing to discrete detectors, a filament of SPRITE detector is equivalent to n_d detectors of size ℓ_y working in series (TDI mode). From the earlier D^* discussion, we identify the equivalent number of detectors as

$$n_d = \frac{\ell_{z,\,\text{readout}}}{\ell_y} \frac{1}{1 - \dfrac{\tau_h}{\tau_{t,\,\text{readout}}} \left(1 - \exp\left\{ -\dfrac{\tau_{t,\,\text{readout}}}{\tau_h} \right\} \right)} \qquad (8.94)$$

Again, in the limit of long hole lifetime comparing to the transition time in the readout region, we get (Bleicher, 1981)

$$n_d = 2 \frac{v_s}{\ell_y} \tau_h \qquad (8.95)$$

The modulation transfer function (MTF) of a SPRITE detector is determined by both the diffusion of the holes and the averaging effect of the readout region. Considering first the charge-carrier diffusion, we find that the MTF depends on the length of the integration region, ℓ_z. The longer the carriers reside in the SPRITE filament, the more they diffuse; they create a wider impulse response and contribute to poorer MTF. There is a tradeoff between the $D*$ of the SPRITE detector and the MTF. Longer filaments lead to both better $D*$s (because of longer integration times) and poorer MTFs (because of more charge-carrier diffusion). Identifying the diffusion length of the holes $L_h = (D_h \tau_h)^{1/2}$, we can write for the carrier-diffusion MTF (Boreman and Plogstedt, 1988)

$$\text{MTF}(\xi) = \frac{\dfrac{1}{(L_h^2 \xi^2 + 1)} \left\{ 1 - \exp - \dfrac{(L_h^2\xi^2 + 1)\ell_z}{\mu_h E \tau_h} \right\}}{1 - \exp \left\{ -\dfrac{\ell_z}{\mu_h E \tau_h} \right\}} \tag{8.96}$$

where ξ represents the along-scan spatial frequency. We can observe that a shorter bar length leads to a higher MTF. For the long-bar condition $\ell_z \gg \mu_h E \tau_h$, the carriers have achieved a steady-state diffusion spot size, and Eq. (8.96) reduces this to an expression independent of bar length ℓ_z:

$$\text{MTF}_{\text{long-bar}}(\xi) = \frac{1}{(L_h^2 \xi^2 + 1)} \tag{8.97}$$

The readout MTF is caused by spatial averaging of the image that falls on the readout zone. For a rectangular readout, the readout MTF is the same as for a rectangular discrete detector of dimension $\ell_{z,\text{readout}}$:

$$\text{MTF}_{\text{rectangular readout}}(\xi) = \text{sinc}(\ell_{z,\text{readout}}\xi) \tag{8.98}$$

Often the readout zone is tapered to enhance the readout MTF. For the case of a readout exponentially tapered by α_{taper} per unit length, the MTF of the readout becomes the convolution (Boreman and Plogstedt, 1989)

$$\text{MTF}_{\text{tapered readout}}(\xi) = \text{sinc}(\ell_{z,\text{readout}}\xi) * \mathcal{F}(\exp(-\alpha_{\text{taper}}x))$$

$$= \left| \left(\frac{\alpha_{\text{taper}}}{1 - \exp(-\alpha_{\text{taper}}x)} \right) \cdot \left(\frac{1 - \{\cos(\ell_{z,\text{readout}}\xi) \exp(-\alpha_{\text{taper}}\ell_{z,\text{readout}})\} + j \{\sin(\ell_{z,\text{readout}}\xi) \exp(-\alpha_{\text{taper}}\ell_{z,\text{readout}})\}}{(1 - \exp(-\alpha_{\text{taper}}\ell_{z,\text{readout}}))(-\alpha_{\text{taper}} + j\xi)} \right) \right|$$

$$\tag{8.99}$$

which is wider than the nontapered-readout MTF of Eq. (8.98). The MTF of the SPRITE is well approximated by the product of the carrier-diffusion MTF and the readout MTF. This determines, along with the other optics and system MTFs, the final image resolution.

8.4.2. Detector Description

An early successful use of the SPRITE is reported by Blackburn et al. (1982). In this case, a linear array of eight elements was fabricated. The detectors have lengths of 700 μm and widths of 62.5 μm. It is reported that a D^* (with background from an $F/1$ optical system) can be higher than 10^{10} cm · Hz$^{1/2}$/W. Some experimental results are:

Band (μm)	8–14
Operating Temperature (K)	77
Cooling method	Joule–Thomson
Bias field (V/cm)	30
f-number	$F/2.5$
μ_h (cm^2/V s)	390
Scan rate (s^{-1})	1.8×10^6
Resistance (Ω)	500
Power dissipation (mW)	< 80
D^* (500 K, 20 kHz, 1)	> 1.1×10^{11}
$\mathcal{R}$ (V/W)	6×10^4

The performance of a real detector can deviate from the simple theory as described earlier in at least two ways. One is that the detectors yield lower-than-predicted detectivity and responsivity at large-field angles. As shown earlier, the D^* of a SPRITE detector is predicted to be proportional to the D^*_{BLIP} of a photoconductor, which in turn is well known to be proportional to the square root of the projected solid angle of the FOV. It has been suggested without proof that the degradation at large fields of view is principally the result of the shorter lifetime of the holes. With large-field angles, background radiation increases, causing the carrier concentrations of both electrons and holes to increase, thereby increasing the recombination rate.

The behavior of SPRITE detectors deviates from the theoretical prediction when the scan rate is increased (Day and Shepherd, 1982). The decrease in performance in this case is associated with the high bias field required that causes Joule heating in the element. The mobility drops when the temperature of the element increases, decreasing the drift length of holes. Moreover, the thermal generation of excess carriers is no longer negligible and causes the detectivity to decrease. These heating effects could be reduced by increasing the thermal conductance between the detector and the heat sink of the cooler.

Another optimization technique changes the filament geometry to implement a serpentine, so that the filament is designed to zigzag about the direction of scanning. In this way, carriers are forced to travel in a longer path, and a higher drifting speed is achieved by a larger current supply. The power dissipation increases by the square of the increase in path length. In practice this meandering-path technique makes the fabrication process much more difficult, and some degradation of D^* arises from the nonuniformity of the field where the filament turns.

The original rectangular design of the SPRITE (Fig. 8.13) suffers from two shortcomings. One is related to the shape of the readout, and the other is related to charge-carrier integration effects.

The rectangular readout causes the equipotential contours (hence the current flow) to be badly distorted near the current collection end. The nonuniformity in the electric field in the readout leads to a spread of transit times for the photo-generated holes. Carriers drifting along the edge of the filament take more time to reach the readout contact than those drifting down the center, because they travel a longer distance and the electric field they experience is weaker. A tapered readout geometry in the shape of a horn, as shown in Fig. 8.15, is usually used in practice. This geometry yields a more uniform current flow with the electric field nearly symmetric about the center line, which leads to a more uniform hole transit time across the readout region. The MTF of this readout is described by Eq. (8.99). An added advantage of the tapered configuration is that the electric field near the electrode is increased. This increases the effective recombination velocity at the contact. A higher recombination velocity gives a shorter accumulation time at the contact and tends to increase the MTF.

The impact of the shape of the SPRITE filament on charge-carrier integration is considered next. A SPRITE detector responds to background as well as signal radiation, and one of the effects of the background is that the total carrier concentration along the filament increases toward the current-collection end. In a simplified analysis, this effect is neglected by assuming the photo-generated carrier density is negligible compared to the doping concentration. However, improvements in materials-fabrication techniques have led to longer carrier lifetime. To enable a longer integration time to result from the longer carrier lifetime, the filaments of the SPRITE have been lengthened to the point

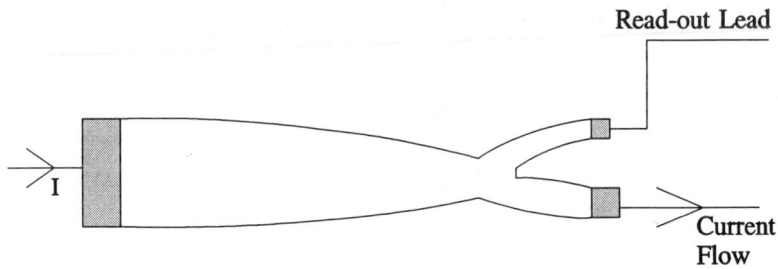

Figure 8.15. Tapered-readout configuration of a SPRITE detector.

where the increase in background-generated carrier concentration cannot be ignored. As the excess carrier concentration increases, however, the probability of carrier recombination increases, which results in a decrease in the effective carrier lifetime. The usual approach is to slightly taper the width ℓ_y of the filament so that it narrows toward the current-collection end. This decreases the effects of background integration, because the increased electric field in the tapered region causes the carriers to move faster. Experimental results show a better performance when the filaments are tapered.

As the length ℓ_z of the SPRITE increases to values comparable to $v_{d,h} \times \tau_h$, the D^* will eventually level out or even decrease because of background-charge integration, indicating that an optimum device length exists. Even without the effects of background integration, we know that an optimum device length does exist, because the signal integration will effectively stop when the device length exceeds the drift length of the photogenerated carriers ($\ell_z > v_{d,h} \times \tau_h$).

Another optimization approach suggests that the focal length of the optics be increased. The material-dependent parameter, D_h, is not changed in this process. If the image projected onto the SPRITE is made larger through a higher magnification, then the (fixed) size of the carrier-diffusion blur spot will subtend a smaller angle, and will be less of a limiting factor in the resolution. Provided that the detector remains background-noise limited, the system thermal sensitivity is not reduced. In fact, the system sensitivity is improved, as the excess carrier concentration is decreased, increasing the carrier lifetime (because the longer focal length (higher $F/\#$) reduces the background flux). However, the increase in the image size in the direction perpendicular to the filament long axis is unnecessary, leading to an anamorphic optical system that increases the image size only in the scanning direction.

8.5. EXTRINSIC PHOTOCONDUCTORS

Intrinsic materials become photoconductive when incident radiation creates electron–hole pairs by promoting valence electrons into the conduction band. Both holes and electrons contribute to the photocurrent and the relative contribution of each carrier type is given by the ratio of the carrier mobilities. In contrast to this, extrinsic photoconduction is solely the result of majority carriers created by the photoionization of the dopant atoms.

Two substrate materials used for extrinsic photoconductors are germanium and silicon. The present emphasis is on extrinsic silicon because of the advanced state of knowledge and material quality control compared to germanium. Although doped germanium was the IR detector of choice in the 1960s, doped silicon replaced germanium for most applications during the 1980s. Silicon dopants are limited to the high 30-μm range, whereas indium-doped germanium can reach 118 μm. Extrinsic germanium photoconductors offer spectral response to longer wavelengths than correspondingly doped silicon because the higher electrical permittivity of germanium means that the outer electrons

Table 8.3 Extrinsic Germanium Photoconductors

Material	$\mathcal{E}_g$ (eV)	λ_c (μm)	Highest Operating Temperature (K)
Ge : Au	0.138	9	77
Ge : Hg	0.09	14	28
Ge : Cd	0.06	21	4
Ge : Cu	0.041	28	4
Ge : In	0.010	118	2
Ge : Zn	0.031	40	4
Ge : Be	0.023	53	3

of the impurity band are less tightly bound, resulting in a lower ionization energy. The higher permittivity also means that the donor orbital (acceptor orbital in a p-type material) has a larger radius that limits the impurity concentration possible before the donor orbitals overlap, causing a broadening of energy bands and destroying the semiconductor properties of the crystal. The lower dopant concentration decreases the optical absorption of doped germanium and requires a thicker device to achieve quantum efficiencies similar to extrinsic silicon. Table 8.3 lists some dopants used in germanium with the corresponding cutoff wavelength.

Our emphasis will remain on extrinsic silicon because of its vast importance and because little interest exists currently in extrinsic germanium. Silicon dopants are available with ionization energies ranging from 0.034 to 0.55 eV, extending the spectral response to 37 μm. Extrinsic silicon detectors may be grouped into two broad classes, depending on the ionization energy of the dopant, shallow-level impurities and deep-level impurities. Shallow-level impurities produce the longest wavelength response. These dopants include easily ionized group III and V elements such as B, Al, Ga, Sb, P, As, and Bi. The ionization energies and cutoff wavelengths of these dopants are summarized in Table 8.4. Spectral D^* plots are shown in Figure 8.16 for some common extrinsic photoconductors.

Measured data have shown that the actual cutoff wavelength for n-type materials is somewhat greater than the cutoff wavelength for p-type materials with the same ionization energy. At present, no theoretical explanation for this phenomenon exists.

Deep-level impurities such as Zn, Hg, Cu, Ni, and Se have larger ionization energies and correspondingly shorter cutoff wavelengths. Many of these materials have multiple ionization states and the deep levels correspond to the second and third ionization energies. The shallower levels of these multileveled materials may be deactivated or compensated for by the introduction of shallow-level dopants of the opposite type. These deep-level detectors are sur-

Table 8.4   Extrinsic Silicon Photoconductors

Material	$\mathcal{E}_g$ (eV)	λ_c (μm)	Highest Operating Temperature
Si : In	0.155	8.0	45
Si : Mg	0.103	12	27
Si : Ga	0.074	16.8	15
Si : Bi	0.0706	17.6	18
Si : Al	0.0685	18.1	15
Si : As	0.054	23.0	12
Si : B	0.045	27.6	10
Si : P	0.045	27.6	4
Si : Sb	0.043	28.8	4
Si : Li	0.033	37.5	4

passed by intrinsic materials, such as HgCdTe, offering the same spectral response but with less-stringent cooling requirements.

For the longer wavelength detectors, compensation for boron is accomplished by neutron-transmutation doping (NTD). In silicon, sufficient silicon isotopes (about 2%) convert to phosphorous when irradiated with neutrons. This phosphorous compensates for the boron impurities because the two im-

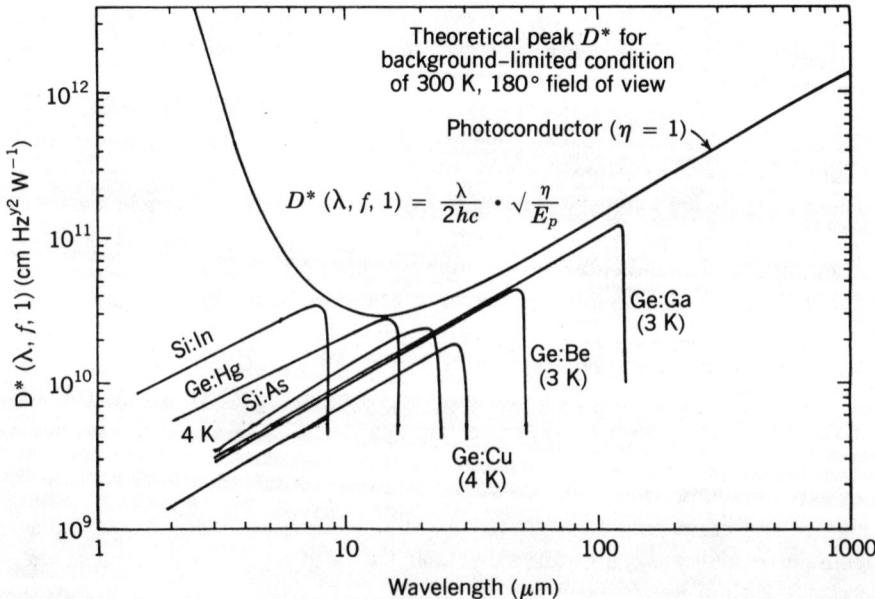

Figure 8.16. Spectral D^* for extrinsic photoconductive detectors, compared to theoretical limit for photoconductors.

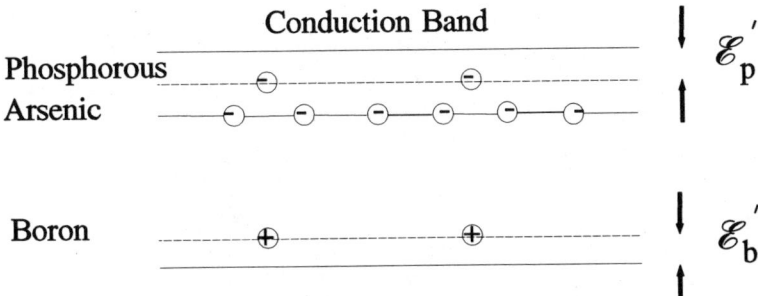

Figure 8.17. Compensation doping to balance boron.

purities have equal but opposite ionization energies. Figure 8.17 illustrates this schematically for Si : As material. The advantage of NTD is the ability to control the phosphorous doping level accurately. This results in a means of controlling the carrier lifetime, which in turn controls the photoconductive gain.

8.5.1. Response Time

The response time of the detector can be limited by several factors. The carrier lifetime for an n-type material, τ, is given by

$$\tau = \frac{1}{\mathcal{K}_r(n + n_a)} \tag{8.100}$$

where $\mathcal{K}_r$ is the recombination rate constant, n is the total carrier concentration, and n_a is the concentration of compensating acceptor atoms. For silicon, $\mathcal{K}_r$ is approximately 10^{-6} cm^3/s and n_a is usually between 10^{13} and 10^{15} cm^3. Because the thermally generated carriers have been frozen out at the standard operating temperature, except for special cases of intense illumination, $n \ll n_a$, while τ is calculated to be from 10 to 100 ns because of compensating atom concentration.

Referring back to Eq. (8.56), the intrinsic carrier conductance, σ_c, is very low, therefore, the total conductance is determined by the photon irradiance on the detector as in Eq. (8.3)

$$\sigma_c = \mu q \eta E_{q,\text{bkg}} A_d \tag{8.101}$$

By using Eq. (8.57), we can show that the product of the extrinsic-silicon detector resistance and the background irradiance is a constant.

$$R_d E_{q,\text{bkg}} = 5 \times 10^{-22} \left[\frac{\text{ohm}}{\text{cm}^2 \text{ s}} \right] \tag{8.102}$$

For low-background irradiance ($E_{q,\text{bkg}} \leq 10^{12}$ photon/s-cm^2), the detector response time may be limited by carrier sweep out and dielectric relaxation. From Eq. (8.102), the resistance then becomes very large (Bratt, 1977). The dielectric relaxation time τ_ρ from these effects is given by

$$\tau_\rho = \frac{\epsilon_r \epsilon_0}{n_{\text{bkg}} q \mu} \qquad (8.103)$$

where n_{bkg} is the carrier concentration due to background radiation and is given by

$$n_{\text{bkg}} = \frac{\eta \tau E_{q,\text{bkg}}}{\ell_x} \qquad (8.104)$$

For silicon, ϵ_r is 11.7, μ is 3×10^4 cm^2/V-s, the detector thickness ℓ_x is usually around 0.1 cm, and η is about 0.3. Thus, if the background is 10^{10} photons/s-cm^2 and the carrier lifetime τ is 50 ns, n_{bkg} is 1500/cm^3, giving a relaxation time of $\tau_\rho = 0.14$ s. This is a very slow response, and in fact almost unacceptable for a system application. We will see later how impurity-band-conduction (IBC) detectors solved this problem.

More often the response time is limited by the RC time constant of the detector/amplifier combination. A typical detector capacitance is 5 pF. Extrinsic silicon detectors usually operate with large load resistors to maximize responsivity. If R_{eq} is 1 MΩ, τ_{RC} is 5 μs. Higher resistances mean even longer response times.

8.5.2. Responsivity and Noise

The current responsivity of an extrinsic silicon detector follows Eq. (8.23) because only a single carrier (majority dopant) causes the conductance change. Therefore, in the strictest sense, Eq. (8.60) holds; however, the mobility must be chosen for the carrier type. When the frequency response is limited by sweep-out effects, the limitation appears as a frequency rolloff of G. The responsivity also shows the usual 3-dB-per-octave rolloff above a cutoff frequency given by $1/2\pi\tau$ used in Eq. (8.31).

The important noise sources are Johnson noise, $1/f$ noise, generation–recombination (G–R) noise, and photon noise. It is assumed that a low-noise preamplifier is used that does not limit overall performance. Furthermore, $1/f$ noise is usually neglected because it can often be controlled by careful manufacturing and by chopping the incident radiation at a frequency where this noise is insignificant. The total mean square noise current is therefore given by

$$i_n^2 = 4q^2 \eta E_{q,\text{bkg}} A_d G^2 \Delta f + \frac{4kT\Delta f}{R_{\text{eq}}} \qquad (8.105)$$

Of these noise sources, the G–R noise spectrum has the same frequency rolloff as the detector responsivity (see Eq. (8.32)). This is not shown in the preceding expression because it is usually the only noise when the detector is operated below its cutoff frequency.

8.5.3. Cooling Requirements to Achieve BLIP

Detector performance is optimized when the dominant noise source is the background radiation. The detectivity is given by Eq. (8.46), provided that the constraint condition of Eq. (8.43) is not violated. For typical detectors, Johnson noise is not a limiting factor except for low background levels (i.e., $E_{q,\text{bkg}} < 10^{12}$). The large G^2RA product reduces the background level at which the detector becomes Johnson-noise limited.

A much more stringent cooling requirement is imposed on extrinsic photoconductors by the condition that the thermal G–R noise term be negligible when compared to the background G–R noise. The condition $kT \ll \mathcal{E}_g$ is not very helpful, because kT at room temperature (~ 0.0025 eV) is only one-half the ionization energy of a detector with a cutoff wavelength of 25 μm. Also, an extrinsic silicon detector must be cooled to around 10 K where $kT \ll \mathcal{E}_g$ achieves BLIP.

It is sometimes thought that the cooling requirement is determined by the temperature above which most of the donor states are thermally ionized, leaving few donors for photo-generated carriers. A calculation of carrier concentrations for typical materials shows that this does not occur until the detector temperature is above 100 K, which is also well above the usually reported cooling requirement.

The cooling requirement can best be determined by looking at the expression for the G–R noise current. From Eq. (8.37), we see that background noise is dominant when

$$g_{\text{th}} \ll \eta E_{q,\text{bkg}} A_d \qquad (8.106)$$

This occurs when the thermal carrier-generation rate is much less than the photo-generation rate. For intrinsic materials, the thermal generation rate is usually lower at a given temperature for a given bandgap or ionization energy, which is why intrinsic materials do not need to be cooled as much as extrinsic materials. The mean-square thermal G–R noise current is

$$i_{\text{gr}}^2(T) = 4G^2q^2\left[\frac{2\ell_x \mathcal{K}_r A_d}{\delta_d}\, |n_d - n_a|\left(\frac{2\pi m^* kT}{h^2}\right)^{3/2} e^{-\mathcal{E}_g/kT}\right]\Delta f \qquad (8.107)$$

where

$\quad n_d$ = donor impurity concentration
$\quad n_a$ = acceptor impurity concentration

ℓ_x = crystal thickness
m^* = effective mass of majority carrier
δ_d = degeneracy factor (= 2 for donors in silicon)
h = Planck's constant
$\mathcal{K}_r$ = recombination coefficient.

This gives the detectivity as a function of temperature when other noise sources are negligible:

$$D^*(\lambda, T) = \frac{\eta\lambda}{2\,hc}\left[\eta E_{q,\text{bkg}} + \frac{2\ell_x\mathcal{K}_r}{\delta_d}\,|n_d - n_a|\left(\frac{2\pi m^*kT}{h^2}\right)^{3/2} e^{-\mathcal{E}_g/kT}\right]^{-1/2}$$

(8.108)

D^* for a given background ($E_{q,\text{bkg}}$) can be plotted versus detector temperature for a given spectral response. This would determine the minimum necessary cooling temperature for optimum detector operation. In practice, the detectors are usually operated somewhat below this temperature (Long, 1980).

8.5.4. Operational Constraints of Extrinsic Photoconductors

Extrinsic photoconductors are typically large in volume, approximately 1 mm × 1 mm × 3 mm. Figure 8.5a showed a technique to obtain a larger path length by total internal reflections. The electrodes are typically copper or indium. These devices are unique in that they can be tested with standard volt-meters.

In the case of germanium, the oxide is water soluble. Therefore, if a Ge oxide is exposed to water, the surface will deteriorate. Because the detector is the coldest surface in a Dewar, water will condense there first.

CP-4 can be used to clean the detector. CP-4 is made up of the following reagents: 3 mL nitric acid, 2 mL HF acid, 2 mL acetic acid, and bromine (1/2 g).

Using plastic tweezers, submerge the detector for a few seconds in CP-4 and then rinse with 10–18-MΩ deionized water.

Extrinsic arsenic-doped silicon (Si:As) is used to illustrate the calculations and evaluate the expected performance for extrinsic silicon detectors. Figure 8.18 shows a cross section of a cryoflask in which this detector is mounted and cooled to 4 K. The load resistor is mounted to the detector heat sink, as well as the preamplifier (PA).

By calculating the background irradiance on the Si:As detector, several important detector characteristics can be determined. Recall

$$E_{q,\text{bkg}} = \left[\int_0^{28\ \mu m} M_{q,\text{bkg}}(\lambda, 300\ K)\,d\lambda\right]\sin^2(\theta_{max}/2)$$ (8.109)

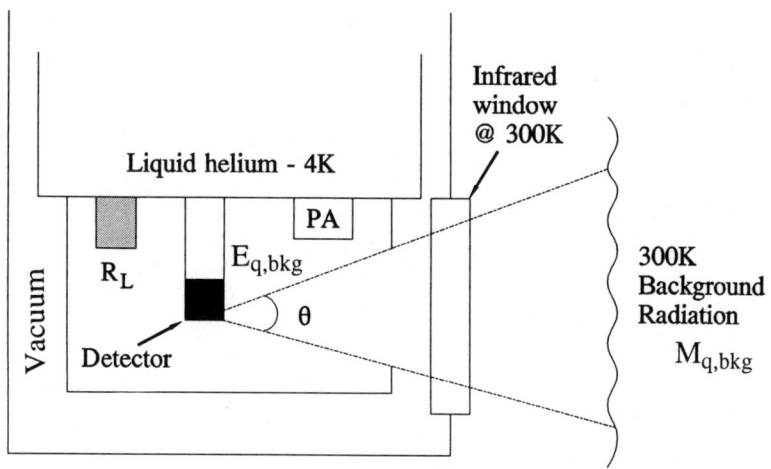

Figure 8.18. Si: As photoconductor mounted in cryoflask.

Once $E_{q,bkg}$ is known from the radiometry, the detector resistance is obtained from

$$R_d E_{q,bkg} \approx 5 \times 10^{22} \left[\frac{\Omega \ photon}{s \ cm^2} \right] \tag{8.110}$$

In addition, the $D^*(\lambda, f)$ is

$$D^*(\lambda, f) \times \sqrt{E_{q,bkg}} \approx 4.3 \times 10^{19} \ J \tag{8.111}$$

if the constraints expressed in Eq. (8.43) hold. Also Garcia and Dereniak (1991) showed that for Si: Ga,

$$D^* f^* \approx 3 \times 10^{16} \tag{8.112}$$

Figure 8.19 shows Eqs. (8.110) and (8.111) for resistance and D^* plotted against background irradiance. Note that large values of resistance correspond to low values of background irradiance. Similar operating characteristics for a Ge: Cu detector include:

Spectral response = 2–28 μm
Refractive index = 4
Optimum electric field = 200–600 V/cm
Cooling requirements $\leq$ 18 K
$R_d E_{q,bkg} \cong 10^{22} \ \Omega\text{-photon/s-cm}^2$
$D^*(\lambda_c, f) (E_{q,bkg})^{1/2} \cong 3.5 \times 10^{19} \ J$

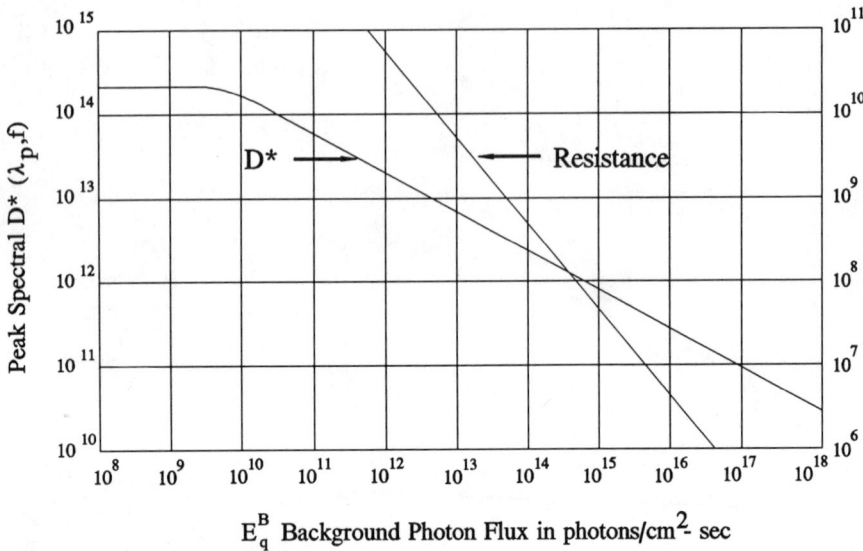

Figure 8.19. Resistance and D^* of an extrinsic silicon photoconductor as a function of background photon irradiance.

8.6. IMPURITY BAND CONDUCTION

A unique discovery in the Si:As photoconductor was the development of the IBC device by Petroff and Stapelbroek (1986). The IBC detector has a heavily doped donor band in silicon with a low compensating impurity concentration. Historically, arsenic was used in silicon, but other materials can be used. The heavily doped region causes the photons to be absorbed within a much thinner layer with greater quantum efficiency (Stetson et al., 1986).

Some additional benefits are lower optical cross talk between detectors, better spatial uniformity of absorption, lower applied bias, and faster response. Currently, the IBC is defining the state of the art in extrinsic photoconductors. In addition, the spectral response of the Si:As IBC has increased to a range of 2 to 28 μm (Petroff et al., 1987).

A cross section of the IBC photodetector is shown in Figure 8.20. Features unique to this detector are (a) blocking layer, (b) IR absorption region heavily doped by arsenic, (c) bias direction, and (d) transparent electrode.

Figure 8.21 shows the cross-sectional view of the IBC with the corresponding electric-field and energy-band diagram (Szmulowicz et al. 1987, 1988). From the energy-band diagram, one can visualize that the effect of high dopant concentration is to spread the width of the donor level by approximately 1 meV. This is sufficient, with some secondary effects, to cause an increase in cutoff wavelength to 28 μm, as opposed to 24 μm for Si:As conventional detectors.

If a conventional photoconductive detector was made by degenerative dop-

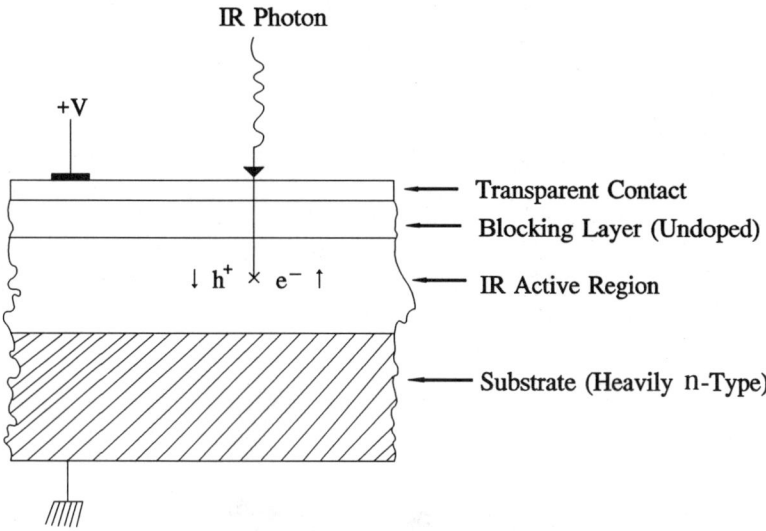

Figure 8.20. IBC detector layout.

ing, the dark current could never be eliminated by cryogenic cooling (Bratt, 1977). Without the blocking layer, the very thin highly doped region would produce excessive impurity conduction-band dark current (and hence excess noise). Even if the Si:As photoconductor operated at 4 K, the donor levels would still not be completely frozen out. Therefore, the blocking region is used to prevent dark current from dominating the carriers. This extra layer prevents the dark current from flowing out of the detector. This causes a preferred bias-voltage direction (+ on the blocking side).

Recall that we must cryogenically cool the detector to a temperature where thermally generated free electrons are negligible (Szmulowicz et al., 1988). The positive charges associated with the ionized donors are mobile in the heavily doped region. They migrate by hopping in the donor level in the opposite direction of the electron movement. Figure 8.21 shows that the photo-generated electrons are collected through the blocking layer once they are in the conduction band. The holes that migrate by hopping cannot cross from the depletion region into the blocking layer, preventing dark-current flow. The important idea is that this detector is a spinoff of photoconductive Si:As, but strictly speaking it is not a photoconductor. These detectors are now being used in two-dimensional arrays (Noel, 1992).

A spinoff of the IBC detector is the solid-state photomultiplier (SSPM) (Laviolette, 1989). This detector produces its gain with less noise than the APD, meeting the McIntyre (1966) requirements for low-noise gain. Extrinsic silicon causes gain by means of low impact-ionization electrons (n-type; Si:As). The applied bias level is around 20 V, and the ionization energy required to cause electron multiplication is around 54 meV. A cross section of

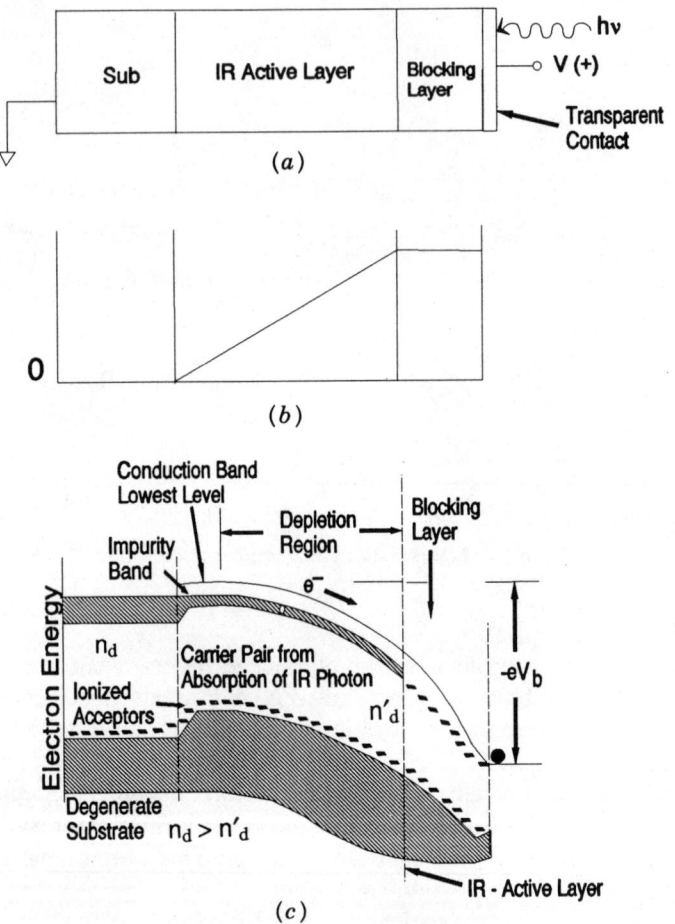

Figure 8.21. Schematic of IBC: (*a*) structure, (*b*) electric field, (*c*) energy-band diagram.

the SSPM is shown in Fig. 8.22. The SSPM is capable of doing single-photon detection out to 20 μm. In fact, concern exists that it is too sensitive for applications that have a 300-K background.

8.7. LEAD SALTS

The development of photoconductive lead-salt IR detectors (PbS, PbSe, PbTe) started in the 1930s. In 1933, at the University of Berlin, Edger W. Kutzscher discovered that lead sulfide exhibited photoconductivity and responded to infrared radiation to 3 μm. By 1943, the Germans had reached the manufacturing

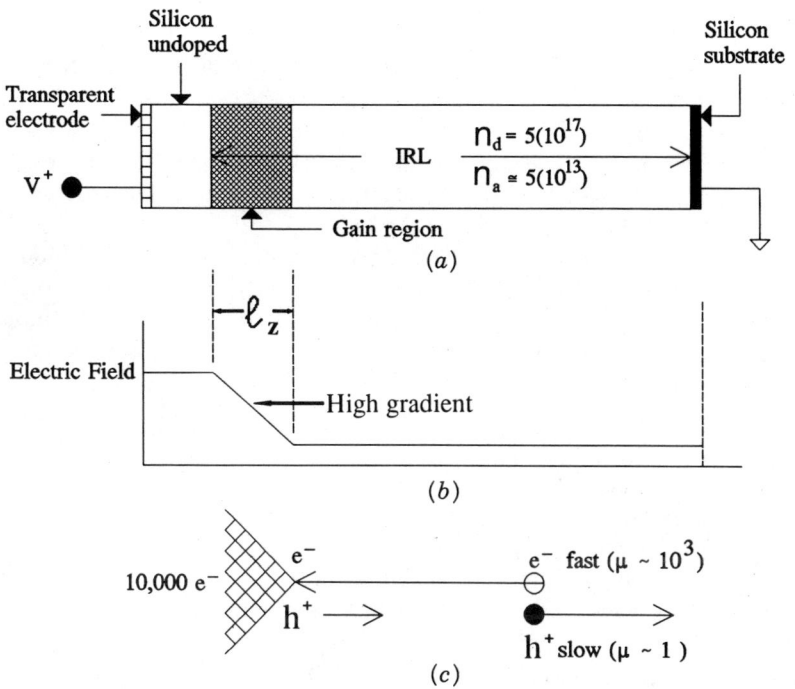

Figure 8.22. Solid-state photomultiplier (SSPM). (*a*) Structure of SSPM detector. (*b*) Electric-field configuration. (*c*) Multiplication mechanism.

state of development (Cashman, 1946). In 1944, a cell was produced at Northwestern University, Chicago. Sixty years later, PbS is still widely used because of its versatile and excellent performance, even when operated at room temperature. By controlling parameters such as dopants, deposition temperature, passivation process coating, and film thickness, the performance characteristics (resistance, responsivity, response time constant, spectral response, noise, and SNR) can be tailored over a wide range (Harris, 1976). Two forms of lead sulfide can be used for photoconductors. One is a polycrystalline thin film that exhibits superior performance and is used almost exclusively today; the other is a single-crystal material that can be cut into wafers.

The most common PbS detectors are used in the spectral region of 1 to 3 μm. The well-developed technology behind these detectors has resulted in their broad application as the sensing elements in a great number of passive and active systems. As examples, they have been used in automotive-pollution-monitoring devices, missile seekers, and missile-guidance systems. Current development with PbS is in focal-plane array configurations and PbS-Si heterojunctions. Such multielements are applicable toward the problem of spectrally selective target detection.

8.7.1. Lead Sulfide

Lead sulfide (PbS) is found naturally as the mineral galena. The direct energy gap at room temperature is about 0.37 eV, corresponding to a cutoff wavelength of about 3.3 μm (Humphrey, 1965), and at 4 K the energy gap is about 0.29 eV with a cutoff wavelength of 4.3 μm. It forms large face-centered cubic crystals in clumps (Kittel, 1976). Lead sulfide is a simple cubic structure, with different ions at alternate corners of the unit cell.

The lattice constant is 5.92 Å. The electron mobility is about 550 cm^2/V-s, and hole mobility is 700 cm^2/V-s at room temperature (Table 7.2). Lead sulfide is an extremely rugged material classified as a IV-VI compound with a molecular weight of 239.25. The melting point of PbS is 1114°C, hence the temperature of operation is not normally a concern, provided that it is operated below 80°C. At temperatures above 80°C, an oxidation process occurs that tends to decrease the resistance to decrease with time, and also results in an increased noise.

Pure PbS is an intrinsic semiconductor at 600°C and above. Purity is difficult to obtain below that temperature, but it is still believed that photoconductivity is an intrinsic property of the material and not the result of thermal vibrations, impurities, or crystal defects. Photoconductivity is, however, greatly enhanced by introduction of impurity ions into the lattice. The most sensitive PbS is usually in polycrystalline thin-film form.

Photoconductivity is the mechanism by which PbS detectors operate. Photons are absorbed in a PbS layer; the energy creating an electron–hole pair by exciting an electron from the valence to the conduction band. The electron becomes trapped at a site such as an oxide ion, so the hole is free to move through the lattice until the electron is freed and they recombine. This temporary increase in conductivity (decrease in resistance) can be attributed to a p-type semiconductor (Venkoba et al., 1963). The time between excitation of the electron and recombination of the electron and hole is the response time (Humphrey, 1965).

The material used in PbS detectors has a granular, nonstandard solid-state structure that lends itself to a wide range of characteristics, depending on the amount of dopants, deposition temperature, and coatings (Harris, 1976). The material parameters of PbS are summarized as follows:

Energy gap	
@ T = 300 K	0.42 eV
@ T = 77 K	0.34 eV
@ T = 4 K	0.29 eV
Spectral range	1–3 μm
Operating temperature	77–300 K
Index of refraction	2.3
Crystal structure	Cubic
Lattice constant	5.935 angstroms
μ_e (electron mobility)	550 cm^2/V-s

μ_h (hole mobility)	700 cm^2/V-s
Dielectric constant, ϵ	15
Detector resistance	0.2–20 MΩ/square
Time constant, τ	30–6000 μs
Density	7.7 gm/cm^3

Parameters for PbS detectors are given in Table 8.5.

8.7.2 Fabrication

Lead sulfide detectors are fabricated using two primary methods: vacuum sublimation and chemical deposition. In both methods, PbS is prepared as a thin polycrystalline film deposited on a substrate between two gold electrodes as shown in Fig. 8.23.

The thin film is deposited on a glass or quartz substrate, either over or under plated gold electrodes, to which fine leads are soldered for electrical connection. The entire device is overcoated for environmental protection and sensitivity improvement. Standard packaging consists of electrode deposition and placement in a metal case with a window over the top. Wire leads are placed in a groove in the substrate and gold electrodes are usually vacuum evaporated onto the film using a mask (over the leads to ensure good electrical contact). Typical PbS-detector structural components are shown in Figure 8.24.

Active areas may range from 0.025 to 10 mm^2 on quartz substrates. Each detector may be optimized for different temperature ranges and corresponding wavelength ranges: ambient-temperature operation at 295 K for 1- to 3.50-μm band; intermediate-temperature operation at 193 K (dry-ice temperature) yields a response from 1 to 4.0 μm; or low-temperature operation at 77 K (liquid nitrogen) yields a response from 1 to 4.5 μm. However, 4.5 μm is not the wavelength for optimum D^* for PbS detectors (W. Horter, 1992 private commun.). Sapphire windows are used for the longer wavelengths and quartz for the shorter wavelengths.

Various configurations are available beyond the standard square detectors. PbS may be used in multielement arrays to reduce scan time for sensor sys-

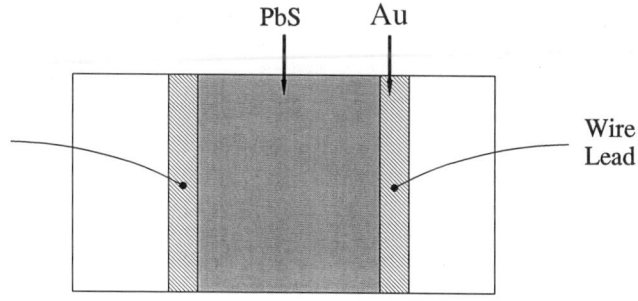

Figure 8.23. PbS detector layout.

Table 8.5 Typical Parameters for Lead-Salt Photoconductive Detectors

	D^* cm Hz$^{1/2}$W^{-1}	R_d (Ω/sq)	$\mathcal{R}_i$ (A_d dependent)	Time Constant (ms)	λ_c (μm)
PbS at 295 (ATO)	$>8 \times 10^{10}$	$<2 \times 10^6$	0.5–4 A/W	0.1–0.5	3
PbS at 195 (ITO)	$>4 \times 10^{11}$	$<10 \times 10^6$	1–10 A/W	2–4	4
PbS at 77 (LTO)	$>1 \times 10^{11}$	$<20 \times 10^6$	1–10 A/W	4–6	4.5
PbSe at 295	$>6 \times 10^9$	$0.5–10 \times 10^6$	$2–8 \times 10^3$ V/W	0.001–0.005	4.6
PbSe at 195	$>2 \times 10^{10}$	$1.5–30 \times 10^6$	$20–60 \times 10^3$ V/W	0.01–0.05	5.4
PbSe at 77	$>1.5 \times 10^{10}$	$1.0–60 \times 10^6$	$20–60 \times 10^3$ V/W	0.02–0.08	7.5

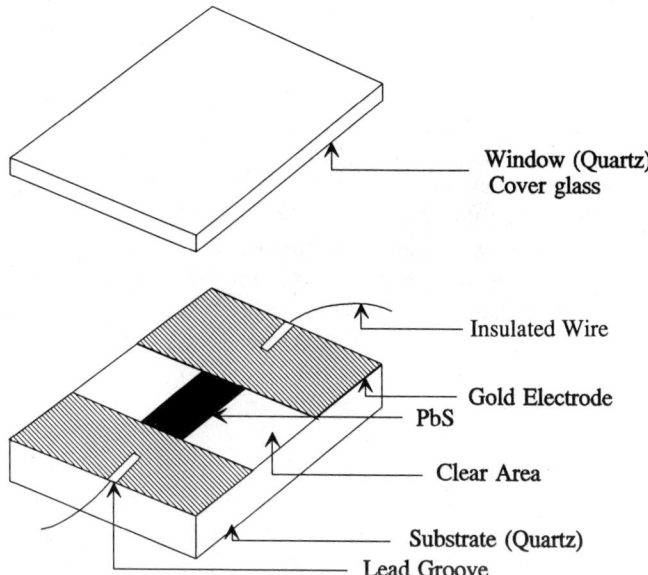

Window (Quartz)
Cover glass

Insulated Wire

Gold Electrode
PbS

Clear Area

Substrate (Quartz)
Lead Groove

Figure 8.24.   Construction of a PbS detector (exploded view).

tems. Until about 1980, these detectors were unchallenged for array applications. Detector elements can be directly deposited on immersion optics to reduce the required detector area, and hence reduce the noise. A useful material for this is strontium titanate, which has good IR transmission, a good thermal match to the PbS film, and a suitably high refractive index. Numerous packaging options are available, including units with filters, reticles, a second detector material for two-color operation, thermoelectric coolers, cryostats, refrigeration systems, temperature sensors, valves for liquid coolants, or preamplifiers. The detector elements may also be deposited on curved surfaces without major problems.

8.7.2.1.   Vacuum Evaporation of PbS.   In the evaporation method of fabrication, PbS is evaporated onto a substrate in a vacuum chamber under a small pressure of oxygen. A film is evaporated on a substrate between plated gold electrodes by heating a crucible containing PbS. Stringent controls on the deposition parameters of time, temperature, film thickness, composition, and homogeneity are required to obtain the proper film structure and performance characteristics. Often two to ten layers are required for good performance. The substrate is baked, and subsequently on antireflection coating is deposited to achieve optimum responsivity.

The resulting film is comparable to chemically deposited films, but control over the electrical and mechanical properties is an order of magnitude lower than for chemically deposited layers. Photoconductive devices have also been made from epitaxial films without baking that resulted in devices with uniform sensitivity, uniform response times, and no aging effects (Schoolar, 1970). These single-crystal devices are homogeneous over large areas. However, these enhancements do not offset the increased difficulty and cost of fabrication.

8.7.2.2. Chemical Deposition of PbS. The more common fabrication technique for PbS is chemical deposition. The fabrication of PbS using chemical deposition requires that the quartz detector substrate be placed in an aqueous (1 normal) solution of lead acetate and thiourea, a sulfur-bearing compound. Sodium hydroxide (NaOH) is added to the solution to complete the precipitation of PbS onto the substrate. A basic solution is needed to obtain a good layer of PbS semiconductor that has optimum mechanical and electrical properties.

The film is baked in oxygen to further enhance the material. The baking process changes the initial n-type film to a p-type film, and optimizes performance through manipulation of resistance. The best material is obtained using a specific level of oxygen and a specific bake time. It is this small percentage (3–9%) of oxygen that influences the absorption properties and response of the detector. PbS is considered an intrinsic photoconductor material, because the amount of doping and the control over the process are not sufficient to consider it an extrinsic. Also the oxygen is thought to provide traps, rather than electron–hole pairs.

Other impurities added to the chemical-deposition solution for PbS have a considerable effect on the photosensitivity characteristics of the films. $AgNO_3$, $HgNO_3$, and $CuSO_4$ shorten the induction period and reduce the photosensitivity. $SnCl_2$, SbC_{13}, and As_2O_3 prolong the induction period and increase the photosensitivity by up to ten times that of films prepared without these impurities. The increase is thought to be caused by the increased absorption of CO_2 during the prolonged induction period. This increases $PbCO_3$ formation and thus photosensitivity. Arsine sulfide also changes the oxidation states on the surface.

Finally, a passivation coating is deposited on the PbS to optimize transmission and surface passivation. In the case of the PbS detector, this passivation increases the sensitivity of the device by decreasing the noise. In the case of PbSe, the passivation increases the responsivity. A cover window of substrate material with an antireflection coating is cemented on and protects the detector surface by creating a sandwich structure where the sensitive material is placed between two substrates. This construction protects the lead from moisture that would otherwise harm the surface.

8.7.3. Lead-Salt Detection Mechanism

Lead-salt detectors are highly sensitive intrinsic photoconductive devices that exhibit a change in conductance (resistance) when incident radiation imparts enough energy for electrons to be ejected from covalent bonds, leaving behind vacancies or "holes." As an intrinsic device, the conductance increases as more electrons and holes become mobile. A bias voltage sweeps out the carriers, thereby producing a current as shown in Fig. 8.4 for the classic photoconductor. The responsivity is given as

$$\mathcal{R}_i = \frac{\eta \lambda q}{hc} G \qquad (8.113)$$

Figures 8.25 and 8.26 show the D^* versus wavelength for PbS and PbSe, respectively. The reasons for having different spectral D^* curves at different operating temperatures are discussed later.

Oxygen plays a major role in producing a lead-salt photoconductive film. x-Ray diffraction studies have shown that the films are composed of microscopic crystallites or grains in an approximately 1-μm-thick film. The grains are on the order of 0.1 μm and are separated by a 5-Å-thick barrier of oxidation products such as PbO and $PbSO_4$.

Because in a thin film of lead salt the photoconduction process is a direct-

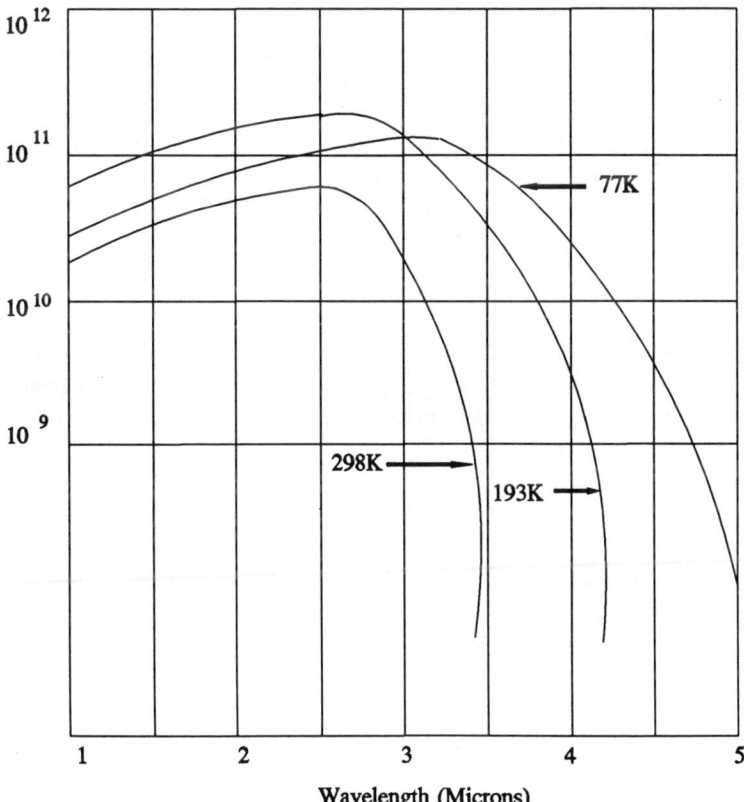

Figure 8.25. Spectral D^* for PbS photoconductor at 298 K, 193 K, and 77 K.

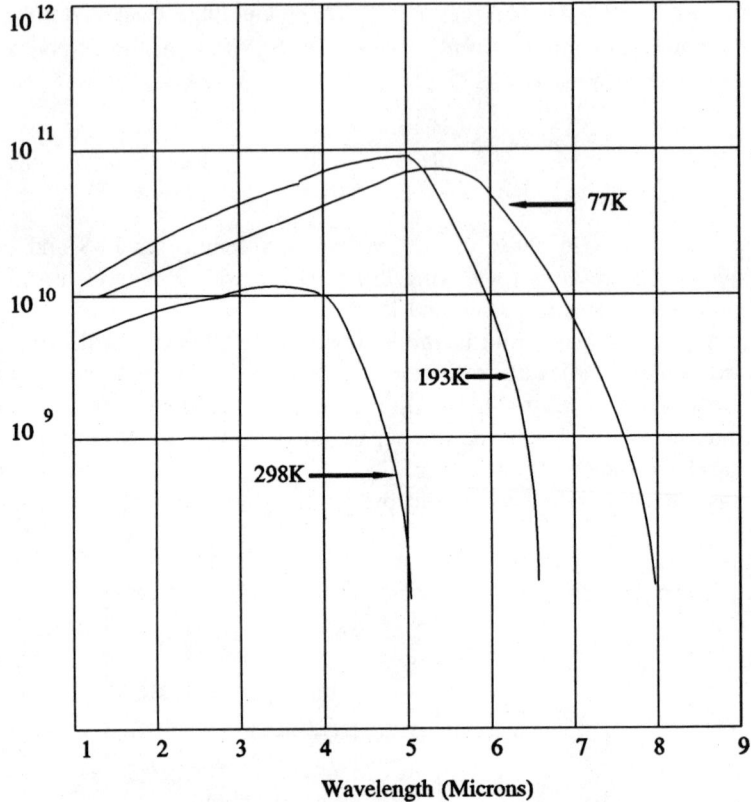

Figure 8.26. Spectral D^* for PbSe photoconductor at 298 K, 193 K, and 77 K.

band transition, the role of oxygen is in the recombination process. Three models are used in the discussion of the carrier-lifetime aspect of recombination: the intrinsic-carrier model, minority-carrier model, and majority-carrier model (Cashman, 1946).

In the intrinsic-carrier model, oxygen acts as a p-type impurity to compensate for the n-type impurities in the film. In this model, both the lifetime and concentrations are equal for both carrier types. Recombination can be either direct or through recombination centers. Maximum response is obtained when the lifetimes are long and the concentrations are low.

The minority-carrier model treats the PbS crystallites as an n-type material and the oxygen layers as p-type, forming a p-n junction. Because this is a diffusion process caused by minority carriers, the carriers cause a decrease in the space-charge region and thus permit more carriers to cross, giving a secondary amplification. The same conditions for carrier lifetime and concentration apply for this model as for the intrinsic model.

The majority-carrier model utilizes the oxygen as a minority-carrier trap. A minority/majority-carrier pair are created by the incoming photon. The minority carrier is trapped by the oxygen while the majority carrier continues to move through the material. No minority carrier is available for recombination, thus the majority carrier will have an increased lifetime. This provides a photoconductive gain as described by Long (1980).

8.7.4. Flash Effects/Precautions

When a lead-salt detector is operated below $-20°C$ and exposed to ultraviolet (UV) radiation, semipermanent changes in responsivity, resistance, and $D*$ occur, which are called *flash effects*. The amount of change and the degree of permanence depend on the intensity and length of UV exposure. Lead-salt detectors should be protected from fluorescent lighting. They are usually stored in a dark enclosure or overcoated with an appropriate UV-opaque material.

Humidity and corrosion will ruin the long-term stability of PbS detectors. They therefore should not be exposed to high humidity ($> 50\%$), and the combination of high temperature and high humidity should especially be avoided. Hermetically sealed detectors are usually used.

Large currents and voltages produce considerable heat, which is damaging to the detector. Therefore, overbiasing by more than 50% of the recommended voltage could be detrimental. Detectors can be cleaned with denatured alcohol only, and then only the substrate cover is really cleaned. Operating temperatures are between $-196°C$ and $100°C$; it is possible to operate at higher than the recommended temperature, but $150°C$ should never be exceeded.

The effects of radiation on PbS detectors have been observed by Billups and Gardner (1961). When exposed to protons from 7.5 to 450 MeV at a flux of 10^{10} protons/cm^2, lead salts had recovery times of years. They can also tolerate 10^{15} thermal neutrons/cm^2 without serious effects for a similar amount of time. In fact, nuclear radiation, in the form of protons, gamma rays, or neutrons, causes changes in resistivity and response with a recovery time of about one month.

Lead sulfide has a highly nonuniform response over its surface. A small spot can be moved a couple of hundred micrometers and the signal can change by an order of magnitude. The reason for these anomalies is the graininess and spatial nonuniformity of the inactive grains that do not have the correct oxygen content.

8.7.5. Noise Processes in Lead-Salt Detectors

Lead-salt detectors have three main noise sources: G–R, $1/f$, and Johnson noise. The general noise expression combining these three noise sources would be

$$i_n^2 = 4G^2 q(q\eta E_q A_d + qg_{th})\Delta f + \frac{B_{1/f}\overline{i}^2}{f}\Delta f + \frac{4kT}{R_d}\Delta f \qquad (8.114)$$

Generation–recombination noise results from the fluctuations in the generation, recombination, and/or trapping rates of carriers in the semiconductor, causing undesired modulations in the free-carrier current. Typically the photoconductive gain, G, is found to be on the order of 0.5 to 1. However, for PbS, G can be an order of magnitude larger, because of natural or induced discontinuities in the polycrystalline structure. Localized regions of high electric field are produced in these discontinuities, which increase the carrier velocity. The transit time of the carrier across the detector is reduced, and G becomes large. The second term in the G–R noise-current expression accounts for the thermally generated carriers.

Because PbS must be biased like any photoconductor, a dc current (and therefore $1/f$ noise) will always be present. This is true for all photoconductors. The $1/f$ noise becomes dominant when the detector is operated at low frequencies and is a major noise source in lead salts. Johnson noise, although present, is not a major factor in the noise contribution. It is insignificant compared to G–R noise and $1/f$ noise.

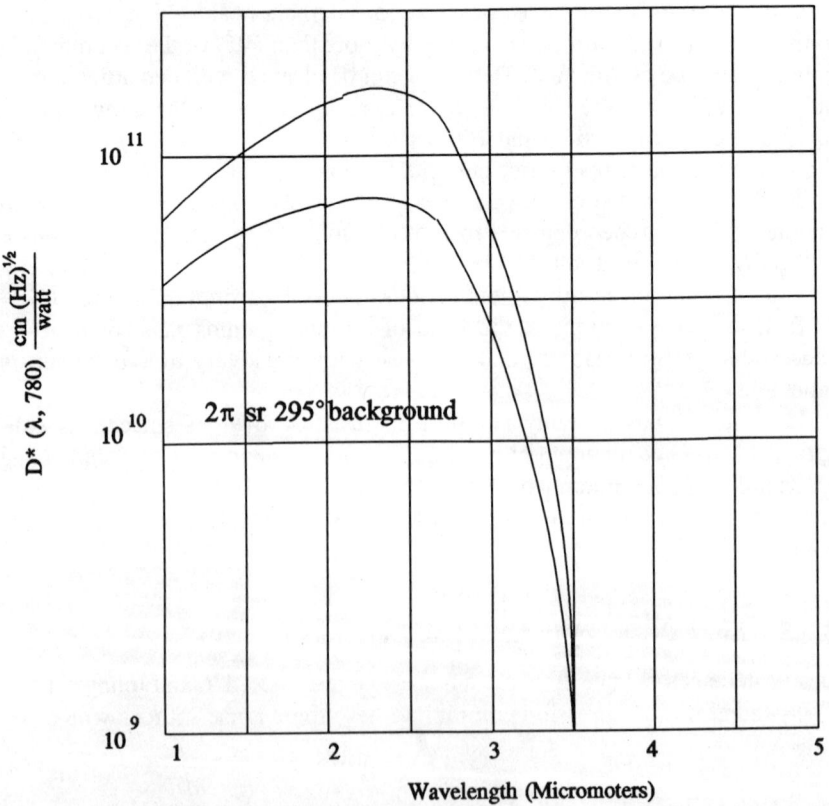

Figure 8.27.   Range of expected spectral D^* for PbS operating at ambient temperature (295 K).

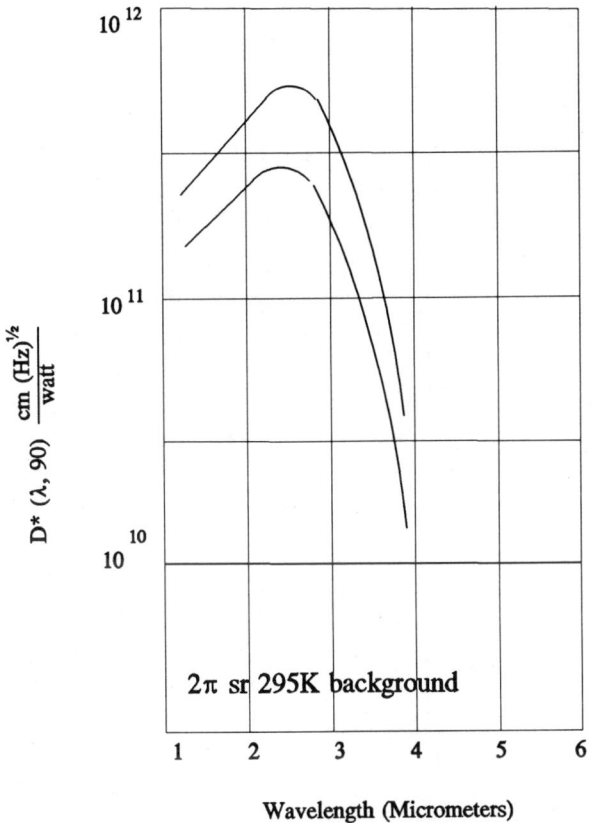

Figure 8.28. Range of expected spectral D^* for PbS operating at intermediate temperature (193 K).

The effects of cooling lead-salt detectors are significant (Figs. 8.27, 8.28, and 8.29), as can be seen in the spectral-D^* curves for ambient-, intermediate-, and low-temperature operation. At room temperature, the dominant noise is G–R from thermal generation (in addition to $1/f$). As the detector operating temperature is lowered, the thermal G–R and Johnson noise decrease because of their strong temperature dependence. Therefore, the D^* increases and reaches a maximum at 195 K. Upon further cooling to liquid nitrogen (77 K), the detector D^* decreases. Because the energy gap has decreased, the detector will respond to more longer-wavelength background photons that were previously too low in energy. This increased background response causes more noise, lowering D^*. The spectral response has been broadened, but the detectivity has decreased. The detector is now, however, operating in BLIP conditions. The cutoff wavelength shift is approximately 0.1 μm for every 25-K change in temperature.

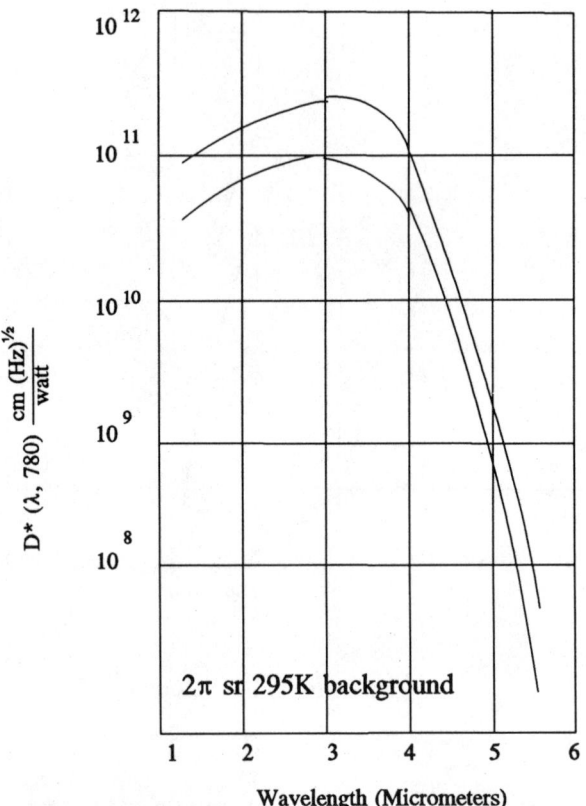

Figure 8.29.   Range of expected spectral D^* for PbS operating at low temperature (77 K).

8.7.6. Performance Characteristics

The performance characteristics of PbS detectors are best described by the plots of spectral D^* as shown in Figs. 8.27, 8.28, and 8.29. For any given temperature of operation, the D^* values can be expected to have a range of values as shown because of materials-processing variations, especially oxygen content.

The typical frequency response of a PbS detector is shown in Figure 8.30 by plotting the peak spectral D^* as a function of chopping frequency. There is an optimum operating frequency arising from the combined effects of the $1/f$ noise at low frequencies and the high-frequency rolloff in responsivity.

8.7.7. Preamplifier Configuration

Lead-salt detectors are often used in conjunction with a preamplifier. Preamplifiers are designed to boost the output signal without adding noise to the

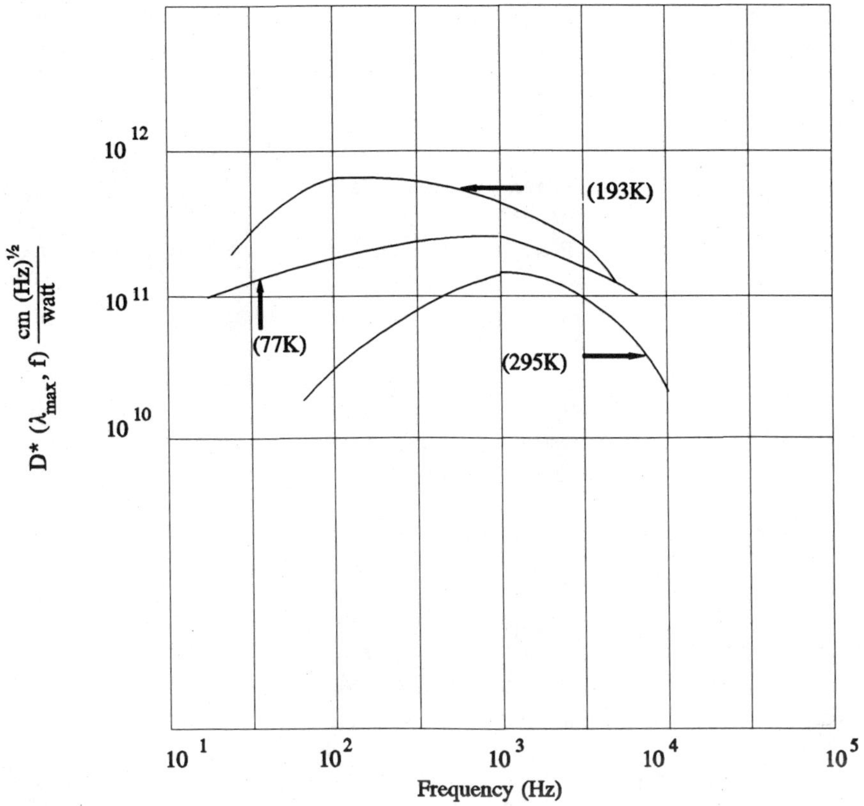

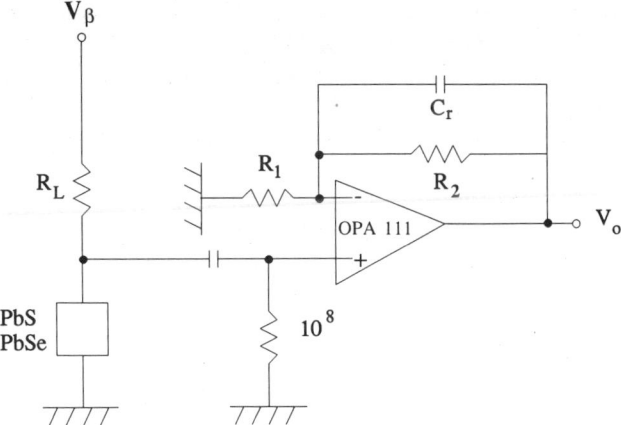

Figure 8.30. Peak spectral D^* as a function of chopping frequency for PbS photoconductive detector, at 295 K, 193 K, and 77 K operating temperatures.

Figure 8.31. Sample of a preamplifier circuit for PbS and PbSe photoconductor detectors.

detector noise. A typical circuit for a high-impedance detector such as photo-conductive PbS or PbSe is given in Fig. 8.31.

REFERENCES

Billups, R. R., and W. L. Gardner, "Radiation Damage Experiments on PBS Infrared Detectors," *Infrared Physics* **1**(3), 335 (1961).

Blackburn, A., et al., "The Practical Realisation and Performance of SPRITE Detectors," *Infrared Physics* **22,** 57 (1982).

Bleicher, I., "Theory and Experimental Results of a Charge Transport Detector," *SPIE Proc.* **304,** 108 (1981).

Boreman, G. D., and A. Plogstedt, "Modulation Transfer Function and Number of Equivalent Elements for SPRITE Detectors," *Appl. Opt.* **27,** 4331–4335 (1988).

Boreman, G. D., and A. Plogstedt, "Spatial Filtering by a Line-Scanned Nonrectangular Detector—Application to SPRITE Readout MTF," *Appl. Opt.* **28,** 1165–1168 (1989).

Bratt, P. R., "Impurity Germanium and Silicon Infrared Detectors," in *Semiconductors and Semimetals*, vol. 12 (R. K. Willardson and A. C. Beer, eds.), Academic Press, New York, 1977, pp. 29–141.

Broudy, R. M., and V. J. Mazurczyk, "(HgCd)Te Photoconductive Detectors," in *Semiconductors and Semimetals*, vol. 18 (R. K. Willardson and A. C. Beer, eds.), Academic Press, New York, 1981, pp. 157–199.

Cashman, R. J., New Photoconductive Cells," (abstract) *J. Opt. Soc. Am.* **36,** 336 (1946).

Day, D. J., and T. J. Shepherd, "Transport in Photo-Conductors 1: Focal Plane Processing," *Solid State Electron.* **8,** 707 (1982).

Dornhaus, R., and G. Nimtz, *Solid State Physics*, Springer Tracts in Modern Physics, vol. 78, Springer-Verlag, Berlin/New York, 1976.

Elliott, C. T., "New Detector for Thermal Imaging Systems," *Electron. Lett.* **17,** 312 (1981).

Elliott, C. T., D. Day, and D. J. Wilson, "An Integrating Detector for Serial Scan Thermal Imaging," *Infrared Physics* **22,** 31 (1982).

Finkman, E., and Y. J. Nemerovsky, "Infrared Absorption of $Hg_{1-x}Cd_xTe$," *J. Appl. Phys.* **50**(6), 4356 (1979).

Garcia, J., and E. L. Dereniak, "Extrinsic Silicon Photodetector Characterization," *Appl. Opt.* **29,** 559–569 and erratum 2838 (1991).

Hansen, G. L., and J. L. Schmit, "Calculation of Intrinsic Concentration in HgCdTe," *J. Appl. Phys.* **54**(3), 1638 (1983).

Hansen, G. L., J. L. Schmidt, and T. N. Casselman, "Energy Gap Versus Alloy Composition and Temperature in $Hg_{1-x}Cd_xTe$," *J. Appl. Phys.* **53,** 7099 (1982).

Harris, R., "PbS, Mr. Versatility of the Detector World," *Electro-Optical Systems Design*, p. 47 (Dec. 1976).

Humphrey, J. N., "Optimum Utilization of Lead Sulfide Detectors under Diverse Operating Conditions," *Appl. Opt.* **4**(6), 665 (1965).

Kittel, C., *Introduction to Solid State Physics*, 5th ed., Wiley, New York, 1976.

Kruse, P. W., "The Emergence of $(Hg_{1-x}Cd_x)Te$ as a Modern Infrared Sensitive

Material, in *Semiconductors and Semimetals* (R. K. Willardson and A. C. Beer, eds.), Academic Press, New York, 1981, pp. 1–20.

Laviolette, R. A., "A Non-Markovian Model of Avalanche Gain Statistics for a Solid-State Photomultiplier," *J. Appl. Phys.* **65,** 830 (1989).

Leftwich, R. F., and R. Ward, "Latest Developments in SPRITE Detector Technology," *SPIE Proc.* **930,** 76 (1988).

Long, D., "Photovoltaic and Photoconductive Infrared Detectors," in *Optical and Infrared Detectors*, Topics in Applied Physics, vol. 19 (R. J. Keyes, ed.), Springer-Verlag, Berlin/New York, 1980.

Long, D., and J. L. Schmit, "Mercury-Cadmium Telluride and Closely Related Alloys," in *Semiconductors and Semimetals* (R. K. Willardson and A. C. Beer, eds.), Academic Press, New York, 1970, pp. 175–255.

McIntyre, R. J., "Multiplication Noise in Uniform Avalanche Diodes," *IEEE Trans. Electron Devices* **ED13,** 164–168 (1966).

Noel, R. A., "Large Area Blocked Impurity Band Focal Plane Array Development," *SPIE Proc.* **1685,** 250 (1992).

Petroff, M. D., and M. G. Stapelbroek, "Blocked Impurity Band Detectors," U.S. Patent 4,568,960, Feb. 4, 1986.

Petroff, M. D., M. G. Stapelbroek, and W. A. Kleinhans, "Detection of Individual 0.4–28 μm Wavelength Photons via Impurity-Impact Ionization in a Solid-State Photomultiplier," *Appl. Phys. Lett.* **51,** 406 (1987).

Reine, M. B., and R. M. Broudy, "A Review of HgCdTe Infrared Detector Technology," *SPIE Proc.* **124,** 80 (1977).

Schoolar, R. B., "Epitaxial Lead Sulfide Photovoltaic Cells and Photoconductive Films," *Appl. Phys. Lett.* **16,** 446 (1970).

Stetson, S. B., D. B. Reynolds, M. G. Stapelbroek, and R. L. Stermer, "Design and Performance of Blocked Impurity Band Detector Focal Plane Arrays," *SPIE Proc.* **686,** 48 (1986).

Szmulowicz, F., F. L. Madarasz, and J. Diller, "Photoconductive Gain of a Longitudinal Detector with an Arbitrary Absorption Profile," *J. Appl. Phys.* **62,** 310 (1987).

Szmulowicz, F., F. L. Madarasz, and J. Diller, "Temperature Dependence of the Figures of Merit for Blocked Impurity Band Detectors," *J. Appl. Phys.* **63,** 5583 (1988).

Van der Ziel, A., *Noise in Measurements*, Wiley, New York, 1976.

Venkoba Rao, H. N., J. Kuppuswamy, and R. Lakshminarayan, "Some Preliminary Studies and Observations on the Chemically Deposited Photosensitive Lead Sulphide Layers," *Pure Appl. Phys.* **1**(9), 335 (1963).

Wolfe, W. L., and G. J. Zissis, *The Infrared Handbook*, Optical Engineering Press, Bellingham, WA, 1990.

BIBLIOGRAPHY

Hudson, R. D., Jr., and Z. B. Marinkovic, "Influence of Impurities on Photosensitivity of Chemically Deposited Lead Sulphide Layers," in *Infrared Detectors* (R. D. Hudson and J. W. Hudson, eds.),

Dowden, Hutchinson & Ross, Stroudsburg, PA, 1975, p. 189.

Kruse, Paul W., Laurence D. McGlauchlin, and Richmond B. McQuistan, *Elements of Infrared Technology*, Wiley, New York, p. 344.

Sclar, N., "Extrinsic Silicon Detectors for 3–5 and 8–14 μm," *Infrared Physics* **16**, 435 (1976).

Simic, V. M., and Z. B. Mirinkovic, "Influence of Impurities on Photosensitivity of Chemically Deposited Lead Sulphide Layers," *Infrared Physics* **8**, 189 (1968).

Szmulowicz, F., and F. L. Madarasz, "Blocked Impurity Band Detectors—An Analytical Model: Figure of Merit," *J. Appl. Phys.* **62**, 2533 (1987).

PROBLEMS

8.1.   Plot the intrinsic carrier concentration (n_i) for silicon vs. temperature in the range of 200–300 K. Assume an effective mass for electrons $m_e^* = 1.1m_0$ and an effective mass for holes of $m_h^* = 0.56m_0$, where m_0 is the free-electron rest mass.

   (a) What temperature change (ΔT) from room temperature produces a factor of 2 change in the carrier concentration?

   (b) Repeat for germanium. In this case, $m_e^* = 0.55m_0$ and $m_h^* = 0.37m_0$.

8.2.   Why is the dc resistance of HgCdTe not a function of background irradiance $E_{p,bkg}$?

8.3.   Suppose a photoconductive detector is to be built with a given photoactive area. Discuss the tradeoffs that should be considered in selecting its thickness (i.e., its dimension in the direction of the incident light). Suggest improvements that could be made at the detector surfaces to improve quantum efficiency.

8.4.   What is photoconductive gain? (See figure.) Derive the expression

$$G = \tau/\tau_t$$

where τ = carrier life time and τ_t = carrier transit time. Also show that

$$G = (i\tau\mu)/(\sigma_c A_d \ell_x)$$

where

   i = current

   μ = carrier mobility

   σ_c = conductivity of material

   A_d = area of detector

   ℓ_x = thickness of detector

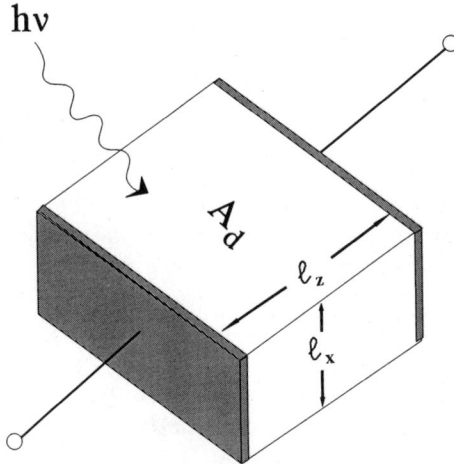

Figure P8.4

8.5. Derive the theoretical relation between resistance–irradiance product $(R_d E_q)$ for extrinsic silicon.

8.6. If the blackbody responsivity $\mathcal{R}_i(T, f)$ of a photoconductor is 5 A/W and its resistance is 10^6 Ω, what is the NEP and D^*? Assume room temperature, Johnson-noise-limited operation, active area $(1 \text{ mm})^2$, and $\Delta f = 10$ Hz.

8.7. Why does peak spectral D^* vs. detector temperature for PbS and PbSe photoconductors have a maximum?

8.8. The intrinsic HgCdTe photoconductor shown in the figure should be just BLIP when operated at 80 K. What constraint is placed on n_i, the

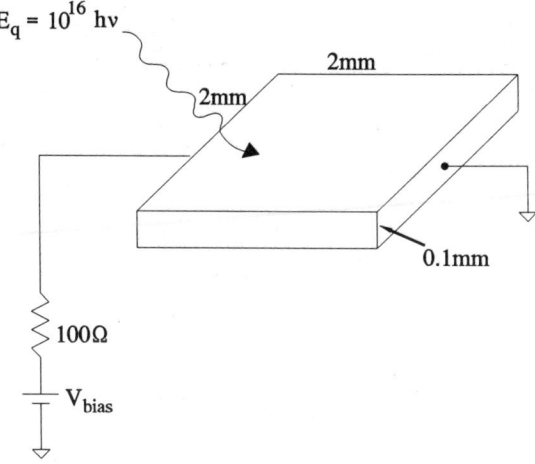

Figure P8.8

free-carrier concentration of the material? The photoconductive gain G = 5.0 and η = 0.3.

8.9.   An arsenic-doped silicon (Si:As) photoconductor has an energy gap from donor to conductor band of 0.052 eV. What is the cutoff wavelength? What is the cutoff wavelength for 50% absorption?

8.10.   Find the value for the resistance-background-photon-irradiance product $(R_d E_{q,\text{bkg}})$ for Si:Ga. Make reasonable assumptions about mobility, quantum efficiency, and carrier lifetime.

8.11.   Explain the operation of impurity-band-conduction (IBC) detectors.

8.12.   (a) Derive a general expression for the D^* of an intrinsic photoconductive detector, taking into account only the G–R noise and the Johnson noise of the detector itself.

         (b) Make a plot of the expression of part (a) vs. $E_{q,\text{bkg}}$ that illustrates the BLIP and Johnson-noise-limited regions. Assume a wavelength of 10 μm, an operating temperature of 77 K, and a photoconductive gain and quantum efficiency of 1.0. The RA product is 1 Ω-cm^2.

         (c) Again using part (a), solve for the background flux level at which the G–R and Johnson noise are identical. For a given temperature, bias voltage, active area, and thickness, explain which material parameters must be improved to lower this background flux level.

8.13.   Calculate the photoconductive gain (G) for a 1 mm $\times$ 1 mm $\times$ 1 mm Si:As photoconductor with optimum bias of 20 V across the detector. The detector is operated with a 60° full field of view and 300 K background temperature. The detector temperature (T_d) is 4 K.

8.14.   Explain neutron transmutation doping. How does it affect photoconductive gain (G)? What effect would be noticed in G if overcompensation takes place?

8.15.   Why does the solid-state photomultiplier provide low noise amplification, whereas the avalanche photodiode does not?

8.16.   The resistance-area product $(R_d A_d)$ of a detector material is often used as a figure of merit. Discuss (by means of mathematical relationships) the appropriateness of this figure of merit as applied to the following two photoconductor bias conditions:

         (a) Illumination and photoconductor bias current in the same direction

         (b) Illumination and photoconductor bias current in transverse directions

         (*Hint:* Consider this figure of merit for a constant thickness in the direction of the illumination, and variable detector photoactive area.) Also, is D^* of a photoconductor an appropriate figure of merit for the orientations of (a) and (b)? Why, or why not?

9

Thermal Detectors

9.1. INTRODUCTION

A thermal detector responds thermally to optical radiation, which means that its temperature depends on incident optical radiation. The word thermal comes from the Greek word $\tau'\varepsilon\rho\mu o\sigma$, meaning heat. A confusion that often arises is that infrared radiation is called *heat radiation*, and although it is often detected by thermal detectors, it can, as we have seen, be detected by photon detectors as well.

The thermal detector absorbs optical radiation, which produces a change in its temperature. This temperature change produces a corresponding change in another material parameter. For example, the temperature change could produce a resistance change (bolometer or thermistor). The resistance could be monitored by an ohmmeter to determine the presence of optical radiation. Unlike photodetectors, where the optical radiation gets absorbed and causes changes within the crystal lattice, thermal detection is based on a surface phenomenon. The outer surface determines whether or not the optical radiation will be absorbed. The absorptance is related to emissivity by

$$\alpha = \varepsilon \tag{9.1}$$

The spectral response of a thermal detector is completely determined by its outer-surface emissivity. If the radiation is absorbed, it is detected; if not, it is reflected and lost. The higher the emissivity, the more efficient the thermal detector; $\varepsilon = 1$ is ideal. Because surface properties such as emissivity do not change rapidly with wavelength, thermal detectors typically have very wide spectral responses and respond from the visible far into the infrared region.

Because thermal detection involves heating phenomena, and the temperature-change process is slow, thermal detectors do not respond quickly. By far, the most common thermal detector is a mother's hand. She determines the temperature of her child by conduction to her hand. Examples in everyday living of mechanical-response thermal detectors are:

1. Glass thermometers with alcohol
2. Bimetal springs used to set temperature in a house thermostat
3. Crooke's radiometer
4. Golay cell

Common thermal detectors providing an electrical output are:

1. Bolometer, where temperature change produces a resistance change
2. Thermocouple, where temperature change produces a voltage
3. Pyroelectric, where temperature change causes a material polarization change

All thermal detectors have at least two essential parts: (1) absorber of optical radiation, and (2) temperature sensor or transducer. The most suitable materials for these detectors produce the largest temperature change for a given amount of incident radiant power, which reduces the thermal mass of the detector.

Some general comparisons of thermal detectors and photon detectors are listed in Table 9.1. Table 9.2 lists some examples of thermal detectors in common use today and their figures of merit.

Thermal detectors are unique because radiation can be detected without the detector itself being cooled to a temperature below the surrounding environment. This unique capability makes them the best choice for many applications in the infrared. Thermal detectors occur in nature as well as being man-made devices. For example, certain snakes have eyes for night-time vision and a second set of eyes for daylight. Rattlesnakes have infrared eyes that are well-developed pinhole cameras, which gives them a marked advantage over man in the desert at night. Reptiles are cold blooded and thus their body temperature must be about the same as the ambient, yet they can strike with amazing accuracy just above the boot line in the calf of the leg. This accuracy indicates that high sensitivity is not necessarily of paramount concern in a system application.

The operation of thermal detectors depends on the flow of heat energy. Elementary texts discuss the rate at which energy is transferred as heat (Reynolds and Perkins, 1977). The rate of energy transfer is clearly vital to the speed of the thermal-detection process. Optical radiation, incident on the detector surface, heats the detector. The heat-loss mechanisms for the detector are conduction and convection. Fundamental relationships for heat transfer relate the rate of energy transfer (heat) between two systems as a function of the thermodynamic properties of the components. The heat-transfer equations, which involve an energy balance, boundary conditions, and material proper-

Table 9.1   Comparison of Thermal Detectors and Photodetectors

Property	Thermal	Photodetector
Speed/frequency of response	Slow	Fast
Spectral response	Wide	Limited
Operating temperature	Room	Cooled
Cost	Cheap	Expensive

Table 9.2 Thermal Detector Examples

Type	Operating Temperature (K)	D^* (cmHz$^{1/2}$W^{-1})	NEP (W Hz^{-1})	Time Constant (ms)	Size (mm^2)
Thermistor Bolometer	300	$1-6 \times 10^8$		1–8	0.01–10
Germanium Bolometer	2–4	9×10^{11}	0.005	0.4	1.5
Carbon Bolometer	2–4		0.03	10	20
Superconducting Bolometer	16–70		5	0.2–2	5×0.25
Thermocouples	300	$7-10 \times 10^8$	2–3	10–20	0.1×1 to 0.3×3
Thermopiles	300	$1.5-9 \times 10^8$		4–30	1–100
Pyroelectrics	300	1×10^9	5×10^5	$10^{-4}-0.01$	
Golay cell	300	10^9	0.6	10–30	10

ties, will yield temperature as a function of time for any component of the system.

Temperature change is the primary indicator of heat transfer. Temperature is the driving potential for energy transfer, and the heat-transfer rate is proportional to temperature gradients. This is the only type of heat transfer of interest in thermal detectors. It should be noted, however, that heat transfer can take place without producing a temperature change, for example, in the cases of heat of vaporization and heat of fusion. These examples of heat flow relate to phase changes from solid to liquid or liquid to gas; these processes are not significant in terms of optical detection.

The two fundamental mechanisms for energy transfer as heat are conduction and radiation. A third mechanism is convection (energy transfer between solids and moving fluids). In fact, convection is one of the most important considerations in fabricating thermal detectors. For instance, the gas encapsulation surrounding the detector requires detailed analysis and engineering tradeoffs to optimize the detector response. Convection is a form of energy transfer, but because it depends on the conduction within the moving fluid or conduction within the solid counterpart, it is not considered as fundamental a mechanism as conduction or radiation.

Conduction is the process of energy transfer through a medium such as copper, water, or air. The energy transfer is caused by atoms at a higher temperature vibrating more than the adjacent atoms, thus inducing vibrations at the microscopic level. In solids, electrons are the carriers of heat. In liquids and gases, mobile molecules cause heat transfer by collisions, which is why heat transfer by conduction is difficult in a vacuum, and radiation becomes the dominant heat-transfer mechanism.

Heat transfer can also occur as a consequence of emitted radiation. Any body at a temperature above zero emits thermal radiation as photons (see Chapter 2). This process causes the temperature of the source to continuously drop, until it reaches equilibrium with the surroundings.

9.2. THEORETICAL PERFORMANCE OF THERMAL DETECTORS

The best possible detector performance occurs when the incoming radiation fluctuation determines the noise of the detector. In the case of thermal detectors, which depend on temperature change, the incoming optical-radiation fluctuation causes a corresponding equivalent temperature-fluctuation noise that is the lowest noise situation that one can obtain.

This is analogous to photon noise manifesting itself as shot noise in a photodiode and as generation/recombination noise in photoconductors. Fluctuations in the incident radiation cause a temperature-fluctuation noise in the thermal detector. If this noise dominates, it represents the best possible performance for the thermal detector.

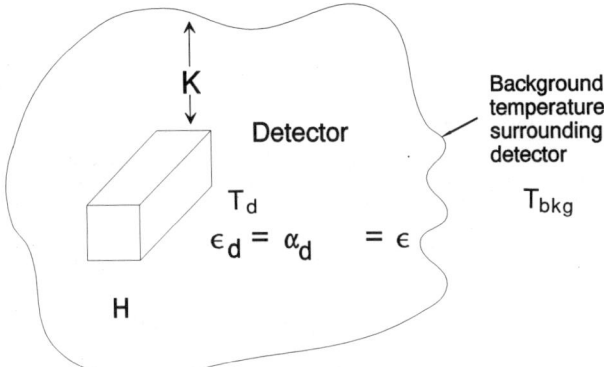

Figure 9.1.   Ideal thermal detector in a radiative-exchange environment.

In this section, expressions for noise-equivalent power (NEP) and $D*$ are derived for a background-limited case. For the derivation, we assume the following idealizing conditions (Putley, 1980): the detector (Fig. 9.1) has only radiative heat transfer with its surroundings. The detector is at temperature T_d, and the background is at temperature T_{bkg}. The assumptions for this thermodynamic situation are:

1. Detector is in a perfect vacuum
2. No leads to the outside from the detector
3. Detector emissivity and absorptivity are independent of wavelength
4. Detector temperature, T_d
5. Background $(2\pi$ sr$)$, temperature T_{bkg}

Any energy transfer within the system must be heat. We develop an equation for the rate of heat transfer through the vacuum, at first assuming that the temperature variable T, the detector temperature T_d, and the background temperature T_{bkg} are all very close. The mean-square radiative power that causes a temperature variance can be expressed as

$$\overline{\Delta T^2} = \frac{4kT^2}{K} \Delta f \tag{9.2}$$

where K is the thermal conductance. Recall that the radiation emitted by the detector into the surround is

$$\phi_e = \sigma_e T_d^4 \varepsilon A_d \tag{9.3}$$

where σ_e is the Stefan–Boltzmann constant for radiant exitance. The total differential is

$$d\phi_e = 4\sigma_e T_d^3 \varepsilon A_d dT \tag{9.4}$$

The thermal conductance (K) is the heat-transfer property between the detector and heat sink, exactly the relationship shown in Eq. (9.4). Therefore

$$K \equiv \frac{d\phi_e}{dT} = 4\sigma_e T_d^3 \varepsilon A_d \tag{9.5}$$

For the special case of a thermal detector in a vacuum, the radiative conductance of Eq. (9.5) is the only conductance between the detector and the surroundings.

To express the amount of optical radiation required to produce a temperature rise of ΔT, we set the input radiation equal to the energy conducted away in equilibrium:

$$\varepsilon\phi_e = K\Delta T \tag{9.6}$$

Using Eqs. (9.2) and (9.6), we find the radiation necessary for a signal-to-noise ratio (SNR) of one, the NEP:

$$\Delta T = \sqrt{\overline{\Delta T^2}} \tag{9.7}$$

$$\frac{\varepsilon\phi_e}{K} = \sqrt{\frac{4kT^2\,\Delta f}{K}} \tag{9.8}$$

or

$$NEP = \frac{\sqrt{4kT^2\,K\,\Delta f}}{\varepsilon} \tag{9.9}$$

Note that the smaller the temperature and the thermal conductance, the better the NEP. Substituting for the thermal conductance using Eq. (9.5), the NEP becomes

$$NEP = \sqrt{\frac{16kA_d\sigma_e T^5 \Delta f}{\varepsilon}} \tag{9.10}$$

The corresponding $D*$ is

$$D* \equiv \frac{\sqrt{A_d \Delta f}}{NEP} = \sqrt{\frac{\varepsilon}{16\,k\sigma_e T^5}} \tag{9.11}$$

Evaluating for the constants in Eq. (9.11), and allowing the emissivity to be a function of wavelength, we get

$$D* = 2.8 \times 10^{16} \sqrt{\frac{\varepsilon(\lambda)}{T^5}} \quad (\text{cm } \sqrt{\text{Hz}} \text{ W}^{-1}) \quad (9.12)$$

If we assume an ambient temperature of 300 K, the theoretical limit of the best possible $D*$ is 1.8×10^{10}. This is a blackbody $D*$; however, the peak spectral $D*$ has the same value because the responsivity of thermal detectors is flat. This derivation (in particular, Eq. (9.4)) assumed the detector temperature was very near the ambient background temperature. In real cases, the detector temperature (T_d) may be, and probably is, appreciably different from the surround temperature (T_{bkg}). In this case, the NEP is

$$\text{NEP} = \sqrt{\frac{8 \, k A_d \sigma_e \Delta f \, (T_d^5 + T_{bkg}^5)}{\varepsilon}} \quad (9.13)$$

where we simply add the variances of the two temperatures of interest. The corresponding $D*$ is

$$D*(\lambda, f) = D*(T_1, T_2, f) = \sqrt{\frac{\varepsilon}{8 \, k\sigma_e (T_d^5 + T_{bkg}^5)}} \quad (9.14)$$

which for the case of a 300-K detector and background temperature is again $D* = 1.8 \times 10^{10}$.

$D*$ is an appropriate figure of merit for this radiation-noise-limited case where the noise is proportional to the detector area. Typically this is not the case for thermal detectors, but it does provide a means of comparison with the optimum performance of other detectors.

Figure 9.2 is a plot of $D*$ for various background temperatures (T_{bkg}) and some operating detector temperatures (T_d). The behavior depicted in Fig. 9.2 is idealized, in that the thermal conductance results from the radiative exchange between detector and background. In practice, electrical leads are attached to the detector to monitor parameter changes. The $D*$ curves shift down appreciably lower from those shown in Fig. 9.2 for the theoretical limits. The maximum advantage to be gained by cooling the detector to 4 K, given a 300-K background temperature, is about $\sqrt{2}$, not a significant increase in performance, which is why most thermal detectors used with a room-temperature background are not cooled.

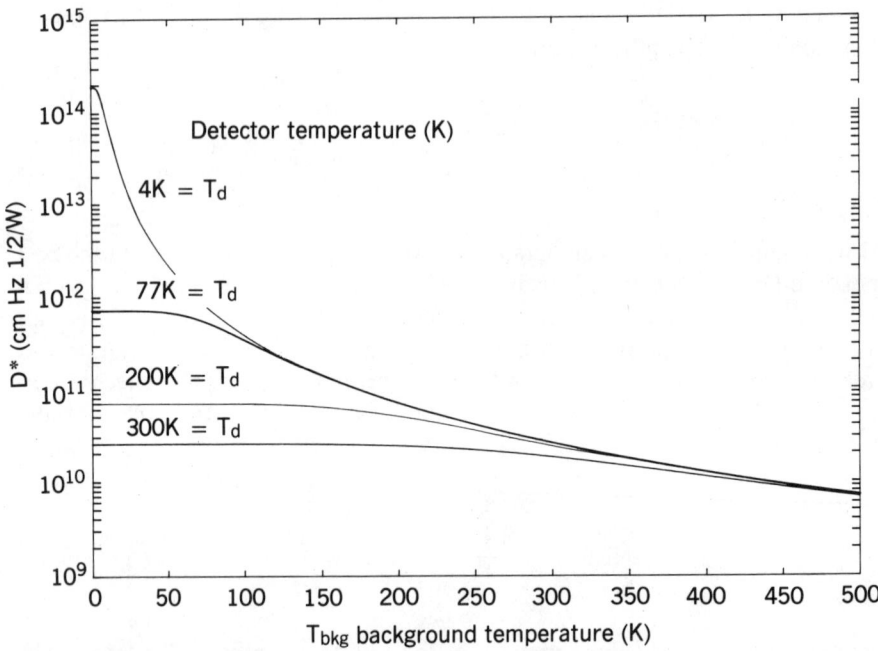

Figure 9.2. Thermal-fluctuation-noise-limited D^* for ideal thermal detector, given detector temperatures of 4 K, 77 K, 200 K, and 300 K, as a function of background temperature (background assumed to subtend 2π steradians).

9.3. SPECTRAL-RESPONSE DETERMINATION

The spectral response is determined by the absorber of the thermal detector. To be a good thermal detector, the absorptance must be high. Recall that absorptance equals emissivity, and is also a function of wavelength. The degree to which an absorber converts the incident radiation to heat is a function of wavelength of incident radiation, with higher absorptance being better. Surface characteristics affect the spectral response, or the "blackness" at different wavelengths. A highly absorbing black detector surface is produced by three primary methods: (1) paints, (2) polymers, and (3) metal sintering. Paints seem to be the most obvious material, but are not the best for optical detectors because they typically increase the thermal mass (and hence the heat capacity) too much. Paints that appear to be black in the visible region are often highly reflective beyond 8 μm.

Carbon black is by far the oldest technique. In this case a smoking candle is used to deposit carbon black on the detector surface. The emissivity is about 0.5 from the visible to 15 μm. More recently, gold black and bismuth black have proved to have a much wider spectral response and higher emissivity. Table 9.3 lists various high-emissivity paints that were measured by Henninger (1984).

Table 9.3 Emissivities of Various Black Paints

Paint	ε	λ_{short}	λ_{long}	T (K)
MH2200	0.86	0.35	2.15	300
Normal at 10.6 μm	0.95	10.6	10.6	300
Nextel 2010	0.95	0.3	2.5	300
Aeroglaze Z series				300
L300	0.84	0.30	1.0	300
Z004	0.84	0.30	1.0	300
Z302	0.86	0.30	1.0	300
Z306	0.86	0.30	1.0	300
Z306 w/microspheres	0.91	0.30	1.0	300
Z307	0.86	0.30	1.0	300
Z313	0.86	0.30	1.0	300
Cat-a-Lac black	0.88	0.30	2.5	300
DeSoto black	0.96	0.2		300
Carbon black (NS-7)	0.88	0.3		300
Chemglaze black	0.86	0.3	1.0	300
Aquadag, 4 coats on copper	0.49	0.3	10	300
Microband, 4 coats on magnesium	0.85	0.3	10	300
TiO_2, gray	0.87	0.3	10	300
TiO_2, white	0.94	0.3	10	300
Tar, asphalt, matte black paint	0.90–0.97	0.3	10	300

Various black surfaces can be formed by metals directly or by causing small varigations on metal surfaces. These metals are deposited in a low vacuum, allowing the evaporation of the metal to be sputtered. Instead of a shiny surface, the surface is black because of all the cavities that trap light at the surface. The coated surface is made up of micro-cavities that have very high absorptivity. Table 9.4 lists some metal surfaces with high emissivities.

The anodizing process, which consists of etching the surface using an oxide-

Table 9.4 Sintered Metal Surfaces to Produce a Black Surface

Surface Preparation	ε
Carbon black	0.5
Anodized aluminum	
Black	0.87
Blue	0.82
Brown	0.86
Chromic	0.56
Green	0.88
Gold	0.82
Red	0.88

forming chemical reaction, will also produce high-emissivity surfaces. Aluminum is used as the anode in an electrolytic process. The electrolyte typically used is sulfuric acid, which produces a high current flow and a high oxidation rate. This process leaves the surface very porous and pitted. The aluminum part is then submerged in a color dye (see Table 9.4) that fills the voids with a chosen color. The surface is then sealed with boiling water, forming a layer of aluminum oxide.

Polymers are also used to produce black surfaces for detectors. This approach is very viable and provides a means for mass production of detectors. The polymers used frequently have conjugate double bonds of carbon/nitrogen that are stable and that absorb IR radiation. Examples of this are pyrozole and pyrodium, which have carbon/nitrogen double bonds in the classical carbon ring. This molecule has a wide range of energy levels for absorption, making the emissivity high over a wide spectral range. In addition, these polymers do not break down from either IR radiation or aging, so they are very stable. Also, the absorption taking place in the molecule is very fast, so these detectors have a good time response.

9.4. RESPONSIVITY ANALYSIS FOR THERMAL DETECTORS

A less idealized situation is a thermal detector mounted as shown in Fig. 9.3, with leads and a heat sink. The wires not only get the signal from the detector but also are a path for thermal conduction to mechanical mounts. The optical radiation incident on the detector goes into heating the chip, but conduction losses along the wires and radiation must be accounted for. A heat-balance equation for the equilibrium conditions of heat in equal to heat out can be calculated. The signal radiant power input to the detector is

$$\frac{dQ_{e,\,\text{input}}}{dt} = \varepsilon\phi_e \tag{9.15}$$

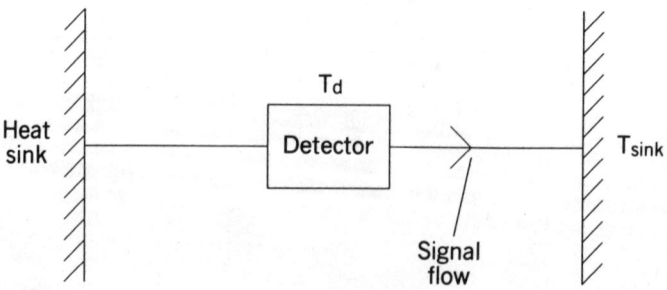

Figure 9.3. Thermal detector suspended on two wires to a heat sink.

The amount of heat to warm the detector to an elevated temperature is

$$Q_{e,\,input} = H\frac{dT_d}{dt} \tag{9.16}$$

where H is the heat capacity of the detector (Joule/Kelvin), the product of the specific heat and the detector mass. The heat loss results from radiative and conductive processes along wires:

$$Q_{e,\,output} = A_d(\sigma_e T_d^4 - \sigma_e T_{bkg}^4) + KdT_d \tag{9.17}$$

where K is the thermal conductance of the wires and the first term is the radiative loss between the detector at temperature T_d and the background temperature T_{bkg}. The heat-balance equation relating incident radiation and heat-loss terms is the following differential equation

$$H\frac{d\Delta T_d}{dt} + K\Delta T_d = \varepsilon\phi_e \tag{9.18}$$

Typically the radiation is modulated or chopped so that it can be expressed as an exponential, because any square wave can be decomposed into a Fourier series of sinusoidal components as

$$\phi_e = \phi_{e,0}e^{j\omega t} \tag{9.19}$$

If we substitute Eq. (9.19) into (9.18) we get

$$H\frac{d\Delta T_d}{dt} + K\Delta T_d = \varepsilon\phi_{e,0}e^{j\omega t} \tag{9.20}$$

To solve Eq. (9.20), we can assume a solution of the form

$$\Delta T_d = A_{\Delta T_d}e^{j\omega t} \tag{9.21}$$

where $A_{\Delta T_d}$ is the amplitude of the solution. Substituting into Eq. (9.20),

$$HA_{\Delta T_d}(j\omega)e^{j\omega t} + KA_{\Delta T_d}e^{j\omega t} = \varepsilon\phi_{e,0}e^{j\omega t} \tag{9.22}$$

Solving for $A_{\Delta T_d}$

$$A_{\Delta T_d} = \frac{\varepsilon\phi_{e,0}}{K + j\omega H} \tag{9.23}$$

This gives us the expression for the (small) ΔT_d, the temperature change in the sensor that is induced by the optical radiation. Solving for $|\Delta T_d|$, we get

$$|\Delta T_d| = |A_{\Delta T_d} e^{j\omega t}| = \left|\left(\frac{\varepsilon\phi_{e,0}}{K + j\omega H}\right)e^{j\omega t}\right| = \frac{\varepsilon\phi_{e,0}}{\sqrt{K^2 + \omega^2 H^2}} \qquad (9.24)$$

where the magnitude is the important quantity in terms of the temperature rise of the detector. The homogeneous part of the solution of the differential equation goes to zero with time, so it has been suppressed in Eq. (9.24). If the terms are rearranged, the ratio of H/K is the thermal time constant, τ_{th}:

$$\tau_{\text{th}} \equiv \frac{H}{K} \qquad (9.25)$$

$$|\Delta T_d| = \frac{\varepsilon\phi_{e,0}}{K\sqrt{1 + \omega^2 \tau_{\text{th}}^2}} \qquad (9.26)$$

To obtain a large change in temperature (ΔT_d) from the optical radiation, this analysis indicates that a high emissivity (absorptivity) is desirable, along with a small thermal conductance to the heat sink, and a small heat capacity.

The frequency response of the temperature change has a 3-dB cutoff frequency f_c, determined by the thermal time constant, similar to the action of an RC time constant in an electrical circuit. The cutoff frequency is

$$f_c = \frac{K}{2\pi H} \qquad (9.27)$$

For a fast temporal response, we want a large thermal conductance and a small heat capacity (H). For thermal detectors, a tradeoff is required between a large magnitude of response and fast response.

This heat balance equation provides an expression for the temperature responsivity (temperature rise per unit radiant-power input) of the thermal detector. This temperature-induced change results in the classic responsivity (volts or amps per unit radiant-power input), but because of the thermal time constant, the rise in temperature, and then the electrical response, the process is inherently slow. If, for example, a thermocouple is used as a detector, the output voltage produced for a temperature rise is

$$v_0 = \alpha_s dT_d \qquad (9.28)$$

where α_s is the Seebeck coefficient. What we now have is an output-voltage-to-radiant-power incident ratio (a voltage responsivity) which, using Eq. (9.26), equals

$$\mathcal{R}_v = \frac{v_0}{\phi_e} = \frac{\alpha_s \varepsilon}{K \sqrt{1 + \omega^2 \tau^2}} \tag{9.29}$$

From the example of this simple thermocouple, the thermal detector can be thought of as a two-part device: an absorber and a temperature sensor. A generic thermal detector is illustrated in Fig. 9.4. The figure shows the absorber, which must have a high emissivity to absorb the radiant power, and the temperature sensor, which has a temperature-dependent electrical property that responds to the temperature of the structure. An important observation one can immediately make is that the detector collecting area is not the temperature sensor's area.

An additional observation must be made about the properties of the wires or mechanical structure supporting the detector chip in Fig. 9.4. The thermal conductivity should be low so that the sensor has a thermal path for heat conduction such that a good detector response can be obtained. As was mentioned earlier; however, these same electrode wires are used to get the electrical signal off of the detector, and we do not want to impede the electrical signal flow. Material properties of thermal conductance and electrical conductance are related. The electrical conductivity (σ_c) and thermal conductance (K) are related through the Lorentz number ($\mathcal{L}$) at a given temperature by

$$\mathcal{L} = \frac{K}{\sigma_c T} \tag{9.30}$$

where $\mathcal{L}$ = Lorentz number $\sim 2.45 \times 10^{-8}$ V/K.

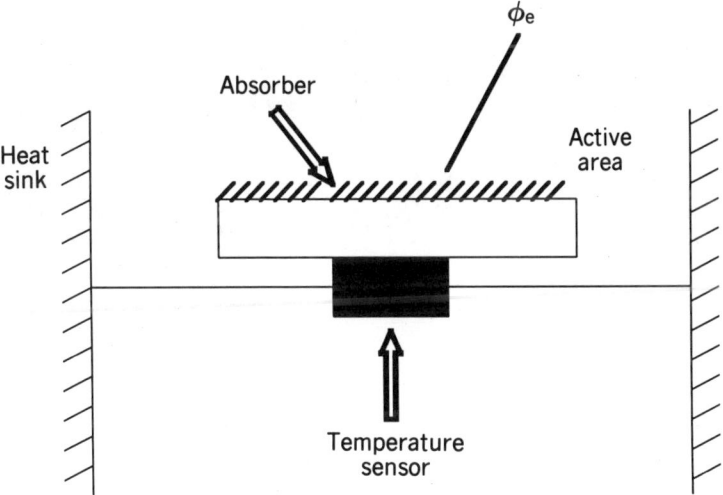

Figure 9.4. Schematic of a thermal detector, including the absorber and the temperature sensor.

The thermal conductance determines the heat flow from the detector, reducing its temperature. The rate of heat flow can be expressed as a one-dimensional model:

$$\frac{dQ}{dt} = K_t A_{\text{wire}} \frac{dT}{dx} \tag{9.31}$$

where dQ/dt is the rate of heat conduction, K_t is the thermal conductivity (W/m-degree), which is not the same as the thermal conductance (see below), and A_{wire} is the cross-sectional area of wire.

If we consider the simple case where the thermal conductivity is independent of temperature and the cross-sectional area does not change, we can find the expression for heat loss down a length of wire ℓ_x connecting the detector temperature T_d and the sink temperature T_{sink}:

$$\frac{dQ}{dt} = K_t A_{\text{wire}} \frac{(T_d - T_{\text{sink}})}{\ell_x} \tag{9.32}$$

where the relation between the thermal conductance (K) and the thermal conductivity (K_t) is

$$K = \frac{K_t A_{\text{wire}}}{\ell_x} \tag{9.33}$$

However, because the thermal conductivity is actually a function of temperature, the expression must be integrated between the two temperatures. Assuming that the diameter of the wire does not change, the expression for heat flow is

$$\frac{dQ}{dt} = \frac{A_{\text{wire}}}{\ell_x} \int_{T_{\text{sink}}}^{T_d} K_t \, dT = K_0 \, (T_d - T_{\text{sink}}) \tag{9.34}$$

where K_0 is the integrated thermal conductance for finite temperature differences. Table 9.5 evaluates the integrated thermal conductivity for various wire materials.

9.5. BOLOMETERS

Bolometers operate on the principle that the temperature change produced by the absorbed radiation produces a change in the resistance of the material. From an external observation, a bolometer acts as a photoconductor; however, the fundamental process is quite different. The temperature coefficient of a bolometer depends on the material and the amount of resistance change produced for a temperature change. The three types of bolometers are (1) metal,

Table 9.5 Integrated Thermal Conductivity for Common Wire Materials

Material	T_d	T_{sink}	$\int K_t dT$
Copper	300	76	934
	300	4	1620
	76	4	686
Constantan	300	76	42.9
	300	4	51.6
	76	4	8.8
	300	250	11
Stainless steel	300	4	30.6
	76	4	3.2
	300	250	6.3

(2) semiconductor, and (3) superconductor. Metal bolometers are used in the microwave region and have a linear dependence on temperature. Typically the temperature coefficient of resistance (α) is 0.5%/C, which is quite low. The linear equation relating resistance and temperature change is

$$R = R_0(1 + \alpha \Delta T_d) \qquad (9.35)$$

where α is the temperature coefficient of the detector material and R_0 is the resistance at the base temperature.

The temperature coefficient for a semiconductor is negative, and has an exponential dependence on temperature. Because of this exponential dependence, the relative resistance change is larger, approximately 3%/C. However, this large change brings with it the added possibility of burnout or thermal runaway because, when α is negative, current causes heating, which decreases the resistance and increases the current, which increases the heating, until the detector burns out.

The superconducting bolometer operates on the transition edge of being a superconductor, that is, having no resistance. Figure 9.5 is an expected plot of resistance versus temperature. The steep slope of the transition region produces a very large temperature coefficient. Optical detectors are produced using only the semiconductor and superconductor types of bolometers. The semiconductor type is discussed here. The most common semiconductors are the thermistor, cryogenic Ge, and the composite bolometer.

The electrical circuits used to interface the bolometer into the system are either the photoconductor type or the Wheatstone bridge. These circuits are shown in Fig. 9.6. The photoconductor circuit was evaluated in Chapter 8 and need not be discussed further. The Wheatstone-bridge circuit shown in Fig. 9.6 is frequently used, and deserving of discussion. With the Wheatstone bridge circuit, two bolometer chips are placed together with one detector optically covered to keep it in the dark, or at ambient levels. A second (uncovered) detector is used to detect optical radiation. With no radiation incident on the

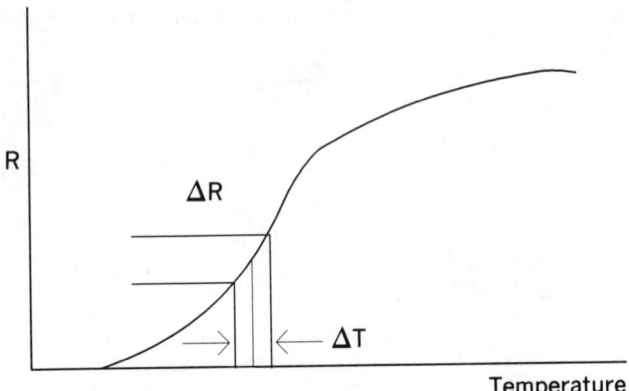

Figure 9.5. Resistance vs. temperature for a superconducting bolometer.

second detector, the bridge is balanced, and no current flows through resistor R_2 (resistors R_1 and R_3 are properly chosen to obtain this condition). Incident infrared radiation on the detector will cause its temperature to rise and its electrical resistance to decrease. This unbalances the bridge, causing a current to flow through resistor R_2, and voltage can be detected as a signal. The change in electrical resistance in the bolometer resulting from an increase in temperature can be expressed in terms of the temperature coefficient of resistance, α:

$$\alpha = \frac{1}{R_d} \frac{dR_d}{dT_d} \tag{9.36}$$

where R_d is the detector resistance and T_d is the detector temperature.

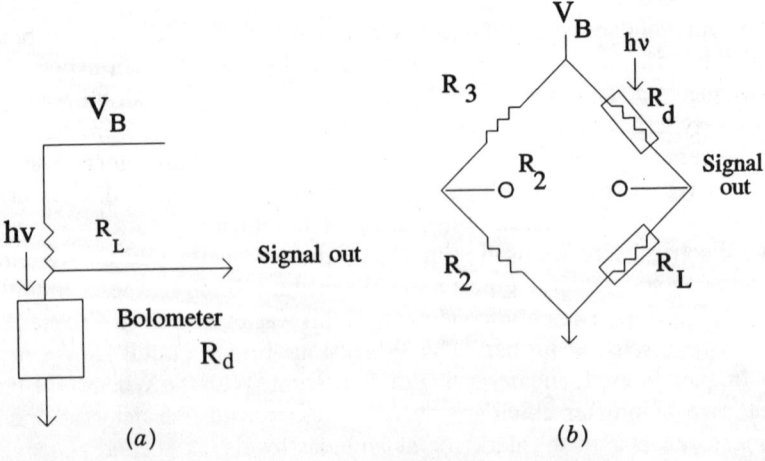

Figure 9.6. Electrical biasing circuits for the bolometer: (a) single-load-resistor circuit; (b) Wheatstone-bridge circuit.

The signal output for the circuit shown in Fig. 9.6 is

$$v_0 = \frac{iR_2\Delta R_d}{(2R_2 + R_1 + R_3)}$$

(9.37)

where

$$\Delta R_d = \frac{dR_d}{dT_d}\Delta T_d$$

(9.38)

i = steady-state current
ΔR_d = change in detector resistance caused by optical radiation
ΔT = change in detector temperature
v_o = change in output voltage

The detector temperature change (ΔT_d) is more complicated than the expression derived earlier (Eq. 9.26), because now we have a current flowing through the detector, the wires, and the load, which introduces some Joule heating (i^2R). Taking into account this added source of heat, the increased bolometer temperature must be offset by higher thermal conductance to the heat sink by way of the wire leads:

$$H\frac{d\Delta T_d}{dt} + K_0(T_d - T_{\text{sink}}) = i^2R$$

(9.39)

In the steady-state case (time derivative is zero),

$$i^2R = K_0(T_d - T_{\text{sink}})$$

(9.40)

From Fig. 9.6b,

$$K_0(T_d - T_{\text{bkg}}) = \frac{v_B^2 R_d}{(R_d + R_L)^2}$$

(9.41)

As the bolometer resistance changes, a second-order effect occurs on the Joule heating; the resistance goes down so that the i^2R heating is lower. The mathematical representation of this differential heating is

$$d(i^2R) = \frac{d(i^2R)}{dT_d}\Delta T_d + \frac{d^2(i^2R)}{dT_d^2}(\Delta T_d)^2 + \frac{d^3(i^3R)}{dT_d^3}(\Delta T_d)^3 + \cdots$$

(9.42)

We will only consider significant the first term in this heat-balance differential equation. Again setting up an expression equating heat input and heat outflow,

$$H\frac{d\Delta T_d}{dt} + K\Delta T_d = \frac{d(i^2R)}{dT_d}\Delta T_d + \varepsilon\phi_e$$

(9.43)

Redefining the differential Joule heating and using circuit analysis on Fig. 9.6b:

$$\frac{d(i^2R)}{dT_d} = \frac{d}{dT_d}\left[\frac{v_B^2 R_d}{(R_L + R_d)^2}\right] = \frac{v_B^2(R_L - R_d)}{(R_L + R_d)^3}\frac{dR_d}{dT_d} \qquad (9.44)$$

Substituting into Eq. (9.43) also using Eq. (9.36),

$$H\frac{d\Delta T_d}{dt} + K\Delta T_d = \frac{v_B^2(R_L - R_d)}{(R_L + R_d)^3}(\alpha R_d \Delta T_d) + \varepsilon\phi_e \qquad (9.45)$$

From Eqs. (9.40) through (9.42),

$$i^2R = \frac{v_B^2 R_d}{(R_L + R_d)^2} = K_0(T_d - T_{\text{sink}}) \qquad (9.46)$$

$$H\frac{d\Delta T_d}{dt} + K\Delta T_d = K_0(T_d - T_{\text{bkg}})\frac{(R_L - R_d)}{(R_L + R_d)}(\alpha\Delta T_d) + \varepsilon\phi_e \quad (9.47)$$

$$H\frac{d\Delta T_d}{dt} + \left\{K - K_0(T_d - T_{\text{bkg}})\frac{(R_L - R_d)}{(R_L + R_d)}\alpha\right\}\Delta T_d = \varepsilon\phi_e \quad (9.48)$$

We define the term in the curly brackets as K_{eff}, the effective conductance, including the Joule heating contribution. We rewrite the heat balance equation with the effective conductance (K_{eff}),

$$H\frac{d\Delta T_d}{dt} + K_{\text{eff}}\Delta T_d = \varepsilon\phi_e \qquad (9.49)$$

and assume a periodic radiant power incident on the detector;

$$\phi_e = \phi_{e,0}e^{j\omega t} \qquad (9.50)$$

This provides a solution (ΔT_d) to the heat-balance differential equation:

$$\Delta T_d = A_{\Delta T_d}e^{-(K_{\text{eff}}/H)t} + \frac{\varepsilon\phi_{e,0}}{[K_{\text{eff}}^2 + \omega^2 H^2]^{1/2}} \qquad (9.51)$$

If K_{eff} is positive, then the transient part of the solution goes to zero and the second term of Eq. (9.51) dominates. If K_{eff} becomes negative, the exponential term goes to infinity, the temperature increases without limit, and the bolometer burns up.

Assuming normal operation (no burnout), the relative change in resistance from Eq. (9.36) is

$$\frac{dR_d}{R_d} = \alpha\Delta T_d = \frac{\alpha\varepsilon\phi_{e,0}}{[K_{\text{eff}}^2 + \omega^2 H^2]^{1/2}} \qquad (9.52)$$

Recalling from the biasing circuit analysis for a photoconductor, the voltage signal is

$$dv = \frac{v_B R_L}{(R_L + R_d)^2} dR_d \tag{9.53}$$

For the case where $R_L = R_d$, as in a Wheatstone-bridge circuit,

$$dv = \frac{v_B}{4} \frac{dR_d}{R_d} = \frac{v_B}{4} \frac{\alpha \varepsilon \phi_{e,0}}{[K_{\text{eff}}^2 + \omega^2 H^2]^{1/2}} \tag{9.54}$$

Therefore, the voltage responsivity (V/W) is

$$\mathfrak{R}_v = \frac{\alpha \varepsilon v_B}{4[K_{\text{eff}}^2 + \omega^2 H^2]^{1/2}} \tag{9.55}$$

Rewriting in terms of the thermal time constant $\tau_{\text{th}} = H/K_{\text{eff}}$:

$$\mathfrak{R}_v = \frac{\alpha \varepsilon v_B}{4K_{\text{eff}}[1 + \omega^2 \tau_{\text{th}}^2]^{1/2}} \tag{9.56}$$

This again shows that, in a bolometer, if the thermal conductance increases (more heat flow through the wires), the responsivity decreases, the time constant decreases, and frequency response is higher. Conversely, if the thermal conductance decreases, the responsivity increases, the time constant increases, and frequency response is lower. Basically, we cannot achieve both high responsivity and fast response in a bolometer.

9.5.1. Thermistor Bolometer

The temperature-variable resistor (thermistor) operates as an optical detector at room temperature. The spectral response is essentially flat with an upper wavelength determined by the transmission of the window that encapsulates the chip. Typically, thalium-bromoiodide windows (KRS-5) are used, which give a response out to 38 μm. The frequency response is limited to 20 Hz, because of the mounting of the chip. Often, an immersion lens is placed on the chip to increase its effective area. Because the noise is not proportional to the area (but radiant-signal collection is), an increase of sensitivity occurs. If a hemispheric lens of high-index material (Ge; $n \cong 4$) is used, the improvement in D^* is a factor of 4 (Sec. 9.5.3). A hyperhemispheric lens improves the performance further (a factor of n^2), but narrows the acceptance angle more than the hemispheric immersion lens (Jones, 1962).

The resistance of these devices are in low megohms per square. The temperature coefficient is about 0.05/°C for sintered manganese, cobalt, and nickel oxides. The dominant noises are Johnson noise and preamplifier noise. Because neither is proportional to the optically active area of the detector, $D*$ is an inappropriate figure of merit to use. However, NEP (noise divided by responsivity) is

$$\text{NEP} = \frac{1}{\alpha\varepsilon} \sqrt{4kTR_d\Delta f(K_{\text{eff}}^2 + \omega^2 H^2)} \qquad (9.57)$$

For a thermistor bolometer operated at room temperature, the spectral response is typically from the visible to about 40 μm, but depends on window transmission and emissivity of the absorber on the detector. For a 1-mm^2-area bolometer, the detector resistance R_d is in the range from 250 kΩ to 2 MΩ. The NEP is in the range of 2×10^{-10} W, and the time constant is typically 1 to 5 ms.

9.5.2. Cryogenic Bolometer

The bolometer performance is improved by cryogenic cooling. Bolometric properties that ensure a low NEP are (1) a high value for the temperature coefficient of resistance α; (2) a resistance compatible with a low-noise preamplifier; (3) a high absorptance (ε); and (4) a low value of thermal capacitance. By cooling the bolometer and its surroundings to a very low temperature, the ultimate detector sensitivity can be orders of magnitude higher than at room temperature. Low and Hoffman (1963) were the first to develop the Ge bolometer operating in the liquid-helium temperature range. Temperature-fluctuation noise-limited performance has been obtained at these low temperatures using a bolometer made by blackening a chip fabricated from single-crystal germanium doped with gallium. The blackening was typically done using black paint; however, this increases the thermal mass of the bolometer. The theoretical limit was achieved by the proper tradeoff of the Johnson-noise and the background-radiation-noise components. The acceptance angle ($F/\#$) viewing the background radiation is chosen for the optical system, with the detector thermal conductance sufficient to conduct all the heat resulting from radiation. If this condition is met, then the bolometer is Johnson-noise limited. If the background aperture is increased, the detector becomes photon-noise (temperature-fluctuation-noise) limited. If the aperture is made smaller (larger $F/\#$), the bolometer operates in the Johnson-noise limit. These devices are background-noise limited at $F/10$ viewing a 300-K background. In normal operation, with a small aperture, Johnson-limited performance is more common. The performance is enhanced by cooling the bolometer below 4 K. This is achieved by creating a vacuum above the helium reservoir, thus reducing the temperature of operation. Figure 9.7 shows an open-cycle cryogenic-bolometer layout, along with a readout circuit. The germanium bolometers are made by lapping and etching (CP4) to the desired dimensions. Gold is used to attach the leads that allow the bolometer to be suspended in air (and vacuum) in a strain-free manner. These leads must also have a low value of thermal conductance. During vacuum operation, these two wires provide the only conductance to the

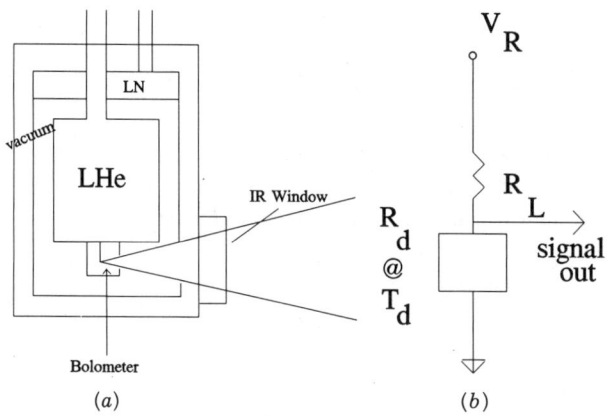

Figure 9.7. Cryogenic bolometer layout and readout circuit. (*a*) Mechanical layout; (*b*) Electrical circuit.

liquid-helium heat sink. These wires are typically made out of constantan or stainless steel. The thermal conductance can be varied over a very large range, by controlling length, diameter, and material. The wires also provide an electrical contact to a load resistor (Fig. 9.7) mounted on the helium heat sink (also reducing the Johnson noise of the load resistor).

To analyze the cryogenic bolometer, we must have an additional term in the heat-balance equation to account for the effects of the background-radiation heating:

$$H\frac{d\Delta T_d}{dt} + K_0(T_d - T_{\text{sink}}) = i^2R + \varepsilon\phi_{e,\text{bkg}} + \varepsilon\phi_{e,\text{sig}} \qquad (9.58)$$

where $\phi_{e,\text{bkg}}$ is the radiant power absorbed from background flux and $\phi_{e,\text{sig}}$ is the radiant power absorbed from the signal flux.

The background radiation power will normally satisfy the condition

$$\phi_{e,\text{bkg}} \leq i^2R \qquad (9.59)$$

If $\phi_{e,\text{bkg}} \ll i^2R$ and steady-state conditions apply, and $\phi_{e,\text{sig}} = 0$, we can determine the steady-temperature T_d of the bolometer

$$K_0(T_d - T_{\text{sink}}) = i^2R \qquad (9.60)$$

If a detector has good thermal conductance to the heat sink, the thermal time constant increases. Recall that K_0 is integrated thermal conductance, which depends on the length, diameter, and composition (typically constantan) of the leads. K_0 can be controlled by a factor of about 10,000. A rule of thumb in the design is

$$K_0 T_{\text{sink}} \approx 10\,\phi_e \qquad (9.61)$$

which says that higher background radiation is compensated for by higher K_0 values. Therefore, in practice, the detector baffling, Dewar baffling, and spectral filters are tailored to the particular application, with an estimate of the expected radiant power on the detector.

Suppose the temperature of the bolometer is $T_d + \Delta T_d$ when the signal input $\phi_{e,\text{sig}}$ is 0. The Joule heating power will also change by $\Delta (i^2 R) = (di^2 R/dT_d) \Delta T_d$. We then have from Eq. (9.58) that if ΔT is small,

$$H \frac{d\Delta T_d}{dt} + K \, \Delta T_d = \frac{d(i^2 R)}{dT_d} \, \Delta T_d + \varepsilon \phi_{e,\text{sig}} \tag{9.62}$$

where K is the usual thermal conductance defined for small temperature changes. Using the biasing circuit shown in Fig. 9.7 and similar analysis as was shown earlier in Eqs. (9.36), (9.44), and (9.46),

$$\frac{d(i^2 R)}{dT_d} = v_B^2 \frac{d}{dT} \left\{ \frac{R_d}{(R_L + R_d)^2} \right\} \tag{9.63a}$$

$$\frac{d(i^2 R)}{dT_d} = i^2 \left(\frac{R}{R_d} \right) \frac{(R_L - R_d)}{(R_L + R_d)} \frac{dR_d}{dT} \tag{9.63b}$$

$$\frac{d(i^2 R)}{dT_d} = \alpha i^2 R \frac{(R_L - R_d)}{(R_L + R_d)} \tag{9.63c}$$

$$\frac{d(i^2 R)}{dT_d} = \alpha K_0 (T_d - T_{\text{sink}}) \frac{(R_L - R_d)}{(R_L + R_d)} \tag{9.63d}$$

where α is the temperature coefficient of resistance defined by Eq. (9.36):

$$\alpha = \frac{1}{R_d} \frac{dR_d}{dT_d} \tag{9.64}$$

We may write Eq. (9.62) in the form

$$H \frac{d\Delta T_d}{dt} + K_{\text{eff}} \Delta T_d = \varepsilon \phi_{e,\text{sig}} \tag{9.65}$$

Here we use the effective thermal conductance K_{eff} given by Eqs. (9.48) and (9.49):

$$K_{\text{eff}} = \left\{ K - K_0 (T_d - T_{\text{bkg}}) \frac{(R_L - R_d)}{(R_L + R_d)} \alpha \right\} \Delta T_d \tag{9.66}$$

If $K_{\text{eff}} < 0$, Eq. (9.65) has an exponentially increasing solution with no signal radiant power input. The bolometer is thus unstable and will continue to heat until the material burns out. For semiconductor bolometers, $\alpha < 0$. If we use a load resistor larger than the detector resistance $R_L \geq R_d$, then K_{eff} will always be positive and no burnout occurs. Now let us assume ΔT is sinusoidally varying as $\exp(j\omega t)$. Referring to Eq. (9.24), the solution to Eq. (9.65) is

$$|\Delta T_d| = \frac{\varepsilon \phi_{e,\text{sig}}}{K_{\text{eff}} \sqrt{1 + \omega^2 \tau_{\text{th}}^2}} \tag{9.67}$$

where $\tau_{\text{th}} = H/K_{\text{eff}}$ is the thermal time constant. The output voltage v_o appearing across the load resistor in the ac arrangement of Fig. 9.7 is given by

$$v_o = \frac{iR_L \, \Delta R_d}{R_L + R_d} = \frac{iR_L \, R_d \, \alpha \Delta T_d}{R_L + R_d} \tag{9.68a}$$

$$v_o = \frac{R_L}{R_L + R_d} \frac{\varepsilon \phi_{e,\text{sig}} \alpha}{K_{\text{eff}} \sqrt{1 + \omega^2 \tau_{\text{th}}^2}} v_d \tag{9.68b}$$

where $v_d = iR_d$ is the voltage across the bolometer element. We therefore have the responsivity given by

$$\mathcal{R}_v = \frac{v_o}{\phi_{e,\text{sig}}} = \frac{R_L}{R_L + R_d} \frac{\varepsilon}{K_{\text{eff}} \sqrt{1 + \omega^2 \tau_{\text{th}}^2}} v_d \alpha \tag{9.69a}$$

Assuming that $\varepsilon = 1$, $R \approx R_L \gg R_d$, and evaluating at low frequencies ($\omega^2 \tau_{\text{th}}^2 \ll 1$):

$$\mathcal{R}_v = \frac{1}{K_{\text{eff}}} v_d \alpha = \frac{v_d \alpha}{K_{\text{eff}}} = \frac{v_d \alpha}{K - \alpha K_0(T_d - T_{\text{sink}})} = \frac{v_d \alpha}{K - (\alpha i^2 R)} \tag{9.69b}$$

The time constant is

$$\tau_{\text{th}} = \frac{H}{K_{\text{eff}}} = \frac{H}{K - (\alpha i^2 R)} \tag{9.70}$$

The parametric equations for the load curve are

$$v_d(T_d) = \sqrt{KR(T_d - T_{\text{sink}})} \tag{9.71}$$

and

$$i(T_d) = \sqrt{\frac{K(T_d - T_{\text{sink}})}{R}} \tag{9.72}$$

Substituting for v_d and using Eq. (9.69b), the responsivity ($\mathcal{R}_v$) can be written as an explicit function of T:

$$\mathcal{R}_v(T_d, T_{\text{sink}}) = \frac{v_d \alpha}{K - (\alpha i^2 R)} = \frac{\sqrt{KR(T_d - T_{\text{sink}})}\ \alpha}{K - \left(\alpha \dfrac{K(T_d - T_{\text{sink}})}{R} R\right)} \tag{9.73a}$$

$$\mathcal{R}_v(T_d, T_{\text{sink}}) = \frac{\sqrt{KR(T_d - T_{\text{sink}})}\ \alpha}{K - (\alpha K(T_d - T_{\text{sink}}))} = \sqrt{\frac{R}{K}}\ \frac{\alpha\ \sqrt{(T_d - T_{\text{sink}})}}{1 - \alpha\ (T_d - T_{\text{sink}})} \tag{9.73b}$$

The overall time constant becomes

$$\tau = \frac{\tau_{\text{th}}}{1 - \alpha\ (T_d - T_{\text{sink}})} \tag{9.74}$$

where $\tau_{\text{th}} = H/K$ is the thermal time constant. An empirical relationship between the resistance and temperature of the bolometric material has been deduced from measurements between 1.1 K and 4.2 K (Low, 1961):

$$R_d(T_d) = R_{d@T_{\text{sink}}}(T_{\text{sink}}/T_d)^{\mathcal{Q}} \tag{9.75}$$

where $R_{d@T_{\text{sink}}}$ is the bolometer resistance at temperature T_{sink}. The constant $\mathcal{Q}$ is typically around 4, but it is uniquely defined for any particular bolometer. The temperature coefficient of resistance α is thus

$$\alpha(T_d) = \frac{1}{R_d}\frac{dR_d}{dT_d} = -\frac{\mathcal{Q}}{T_d} \tag{9.76}$$

Introducing this expression into Eq. (9.73b) we have, for the temperature-dependent responsivity,

$$\mathcal{R}_v(T_d, T_{\text{sink}}) = \sqrt{\frac{R_d(T_d)}{K}}\ \frac{(-\mathcal{Q}/T_d)\ \sqrt{(T_d - T_{\text{sink}})}}{1 - (-\mathcal{Q}/T_d)\ (T_d - T_{\text{sink}})} \tag{9.77a}$$

With the identification of $\mathcal{B} \equiv T_d/T_{\text{sink}}$, we have for the responsivity (Low, 1961):

$$\mathcal{R}_v(\mathcal{B}) = -\sqrt{\frac{\mathcal{Q}^2(\mathcal{B} - 1)}{[(\mathcal{Q} + 1)\mathcal{B} - \mathcal{Q}]^2\ \mathcal{B}^{\mathcal{Q}}}}\ \sqrt{\frac{R_{d@T_{\text{sink}}}}{T_{\text{sink}} K}} \tag{9.77b}$$

For a given bath temperature T_{sink}, the responsivity has a maximum value that depends only on the constant $\mathcal{Q}$. In Fig. 9.8, Eq. (9.77b) is plotted for three different values of $\mathcal{Q}$. The position of the maximum for other values of $\mathcal{Q}$ is shown by the dashed line. For $\mathcal{Q} = 4$, the maximum value of the respon-

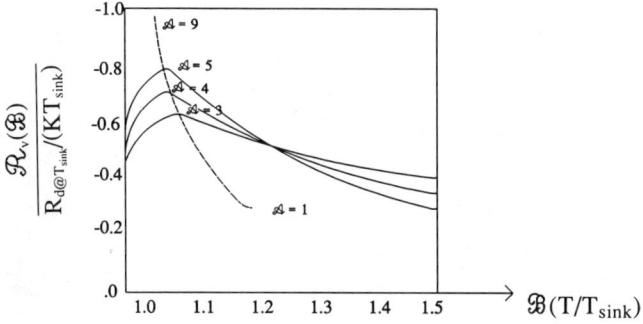

Figure 9.8. Normalized responsivity vs. normalized temperature.

sivity is

$$|\mathfrak{R}_{v,\max}| = 0.7 \sqrt{\frac{R_{d@T_{\text{sink}}}}{T_{\text{sink}} K}} \tag{9.78}$$

which assumes, as stated earlier, that

$$T_d = 1.1 T_{\text{sink}} \tag{9.79}$$

From Eq. (9.60), the optimum value for Joule-heating power is $0.1(T_{\text{sink}}) \times K$. The Johnson-noise-limited and the photon-noise-limited NEPs can be computed (Low, 1961) by using Eq. (9.78).

Assuming that Johnson-noise power limits the performance and $T \approx T_{\text{sink}} \cong T_d$, the NEP is

$$\text{NEP} = \frac{\sqrt{4\ kT R_d \Delta f}}{\mathfrak{R}_{v,\max}} = \frac{\sqrt{4\ kT^2 K \Delta f}}{0.7} \tag{9.80}$$

The radiation-noise-limited NEP, in the absence of phonon noise, is

$$\text{NEP} \approx 4 T_{\text{sink}} \sqrt{kK\Delta f} \tag{9.81}$$

The NEP depends on the temperature of the bath T_{sink} and on the conductance K. The NEP is thus independent of the detector-element dimensions (K is the thermal conductance of the leads). It is important to note that the usual square-root-of-area dependence of NEP is not followed for the cryogenic bolometer. This invalidates the use of D^* for describing the performance of these detectors, because an assumption inherent in the normalization of D^* is that the root-mean-square (rms) noise is proportional to the square root of the area.

Figure 9.9 is a plot of NEP as a function of thermal conductance for a Ge

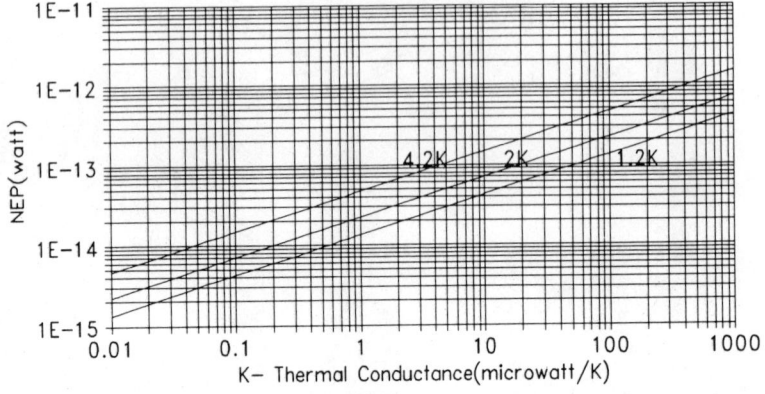

Figure 9.9. NEP as a function of thermal conductance for various temperatures.

bolometer at various operating temperatures. Other materials are often used as bolometers because their heat capacity is much less than germanium; for example, sapphire has a heat capacity of about 1/60 that of germanium and diamond is 1/600. The characteristics of a typical Ge bolometer are summarized in Table 9.6 (Low, 1961). Similar data are included for the carbon (C) bolometer of Boyle and Rodgers (1959). Values of D^* are included but have limited significance.

Pure Ge does not absorb strongly in the far infrared. If it is lightly doped

Table 9.6 Cryogenic Bolometer Characteristics

	Ge Bolometer	C Bolometer
T_{sink} (K)	2.15	2.1
A_d (cm^2)	0.15	0.20
Thickness (cm)	0.012	0.0076
R_L (Ω)	5.0×10^5	3.2×10^6
R_d (Ω)	1.2×10^4	1.2×10^6
$\mathcal{R}_v$ (V/W)	4.5×10^3	2.1×10^4
τ (μs)	400	1.0×10^4
f (Hz)	200	13
K (μW/K)	183	36
NEP (W)	5×10^{-13}	1×10^{-11}
D^* (cm Hz$^{1/2}$W^{-1})	8×10^{11}	4.5×10^{10}

with suitable impurities (gallium), it has extrinsic photoconductivity to just beyond the 100-μm wavelength. If it is more heavily doped but compensated to maintain a high resistivity (and also a high resistance temperature coefficient), the impurities enhance the absorption, especially in the submillimeter region. The energy absorbed is transferred to the lattice, raising the temperature of the sample instead of the number of free carriers as in a photoconductor, making the material useful for a bolometer.

Equation (9.76) shows that α increases as T is reduced. Because the responsivity is proportional to α and it is desirable to make this as large as possible, operation below 4 K is preferred. Lowering the temperature causes the specific heat to fall and increases the resistance R. If R is not too large, this may also be advantageous. However, if R rises to a large value, impedance matching of the amplifier becomes difficult. Thus, there is often an optimum value for R, from the point of view of operating temperature and frequency response.

The most important characteristic of bolometer detectors is the load curves (i–v curves), which are generated for each bolometer/Dewar/filter configuration. Figure 9.10a is a typical curve. Once this curve is established, we can solve for NEP and responsivity by the simple construction bias lines (Fig. 9.10b). To determine the optimum bias, we plot the SNR versus bias voltage, choosing a maximum as the operating bias point. For a particular operating point (v_d', i') that optimizes the SNR, a construction line is drawn tangent to the load curve. The thermal conductance is

$$K = \frac{\phi_e}{0.1 \times T_{\text{sink}}} = \frac{i' \, v_d'}{0.1 \times T_{\text{sink}}} \tag{9.82}$$

Now, from Eq. (9.81), NEP can be found because we know the electrical bandwidth. In addition, the responsivity can be shown to equal

$$\mathfrak{R}_v = \frac{v_d''}{2 \, T_{\text{sink}} v_d'} \tag{9.83}$$

where v_d'' is the tangent-line intercept seen in Fig. 9.10.

9.5.3. Immersion Optics

The effective D^* of the bolometer can be increased by means of immersion optics (Jones, 1962). The lens is usually made from a high-index-of-refraction material such as germanium ($n = 4$) for the infrared. The lens increases the detector area by a factor of n^2 for the hemispherical lens, and a factor of n^4 larger for the hyperhemispherical lens (when the detector is located at one of the aplanatic points of the lens). The lens effectively minifies the image by those amounts compared to the image that would be formed without the lens,

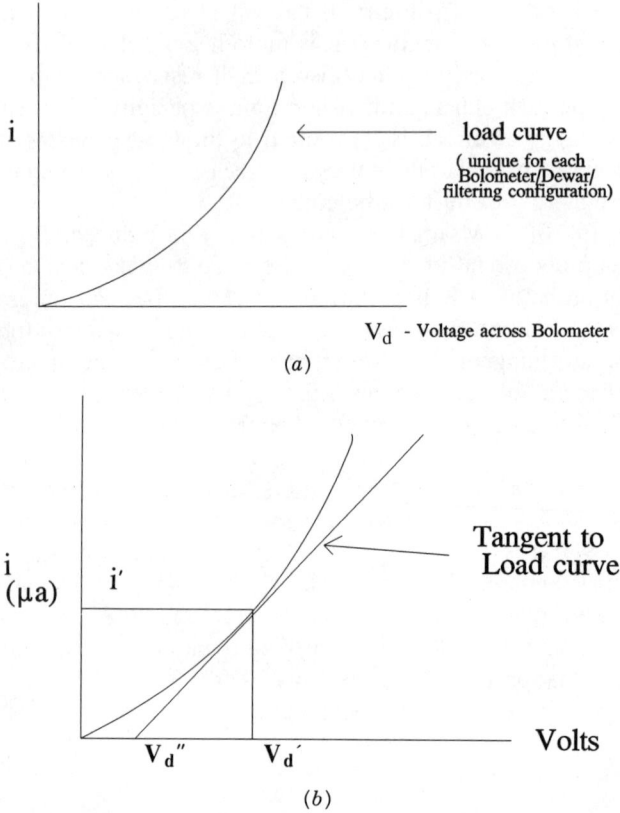

Figure 9.10. (a) Typical i-v curve for bolometer; (b) construction lines on i-v curve for bolometer.

resulting in an improvement of D^* by a factor of n or n^2, respectively (remember that D^* varies as the square root of the area).

In the case of thermal detectors at room temperature, the detector sensitivity almost always varies as the inverse of the detector-area square root. With an immersion lens, we compress the radiation on the detector. The limit to this reduction is governed by the Lagrange invariant ($A\Omega$ product) and the sine condition for an aplanatic system:

$$n_{lens}^2 A_d \sin \theta' = n_{air}^2 A_{enp} \sin \theta \qquad (9.84)$$

where

$$
\begin{aligned}
n_{lens} &= \text{refractive index of the lens} \\
\theta' &= \text{marginal ray angle at the image} \\
A_{enp} &= \text{area of entrance pupil} \\
\theta &= \text{marginal ray angle before refraction at the lens} \\
n_{air} &= \text{refractive index in object space}
\end{aligned}
$$

By increasing n_{lens}, the image area (and hence the detector area) can be decreased, because the size of the entrance pupil and the ray angle in object space will not change. This is analogous to increasing the numerical aperture in an oil-immersion microscope. The hemispherical lens as shown in Fig. 9.11a has the detector mounted at the center of curvature. This situation provides no induced spherical aberration or coma (aplanatic imaging). In fact, the position of the image plane is not changed with or without the hemispherical lens.

A hyperhemisphere can also be used as an aplanatic lens (Fig. 9.11b). Theoretically an apparent increase in area of n^4 can be achieved as stated earlier; however, one must be careful in practice. The location of the image plane is shifted toward the object (Fig. 9.11b). Also note that the hyperhemispherical lens adds optical power to the system. This will only produce the desired smaller image if the ray crossing into the detector does not experience total internal reflection. Typically the acceptance angle (θ') is determined by the index match (i.e., between germanium, where $n = 4$, and the detector, where $n = 2.5$), or $\theta' = 39°$ by Snell's law. Unlike the hemispherical immersion lens, in the case of the hyperhemisphere, there is also an effect on the solid angle. The solid angle is multiplied by a factor of n_{det}^2/n_{Ge}^4, so a decrease in area of only n^2 is seen. Therefore, the hyperhemisphere is no better than the hemisphere.

High-index immersion lenses with antireflection coatings are best, but several other factors must be taken into account if a thermal detector is in contact with an immersion lens. How do we make good contact without losing heat or insulation from surroundings? Typically a glue must hold these two components together that in addition to its mechanical properties, must act as an electrical and thermal isolator, have good index matching (high index), and high spectral transmission in the infrared. The electrical insulation is necessary so that it won't break down with an applied electric field and it must be thermally isolated so it won't affect the sensitivity of the detector. In the past, arsenic-doped amorphous selenium or Mylar have been used as this adhesive. Various practical limitations prevent the theoretical increase in $D*$ of n to be realized. In fact, if Ge is used as a hemispherical lens, the realizable improvement in $D*$ is around 3.5 instead of the 4 predicted.

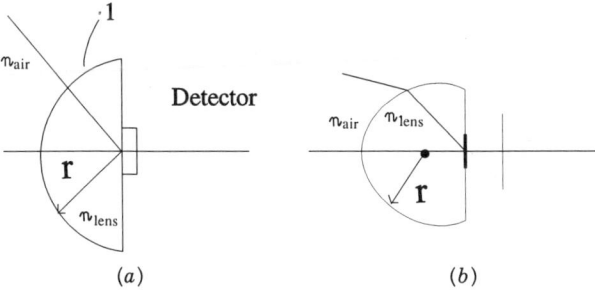

Figure 9.11. Immersion lenses for thermal detector: (a) hemispherical; (b) hyperhemispherical.

9.6. PYROELECTRIC DETECTORS

Pyroelectric detectors were developed in response to the need for sensitive uncooled thermal-radiation detectors. These detectors outperform the thermistor bolometer at room temperature. Pyroelectrics have the advantage of both a wide spectral response and dynamic range, and additionally possess a very good frequency response unlike the typical thermal detector. Lithium tantaliate is probably the most common detector material and triglycerine sulfate is the most sensitive crystal.

A pyroelectric detector is made from a single monoclinic crystal of a ferroelectric material. These materials are inherently polarized, even without an electric field, because of the asymmetry in the basic cell structure (i.e., like the water molecule). This spontaneous polarization exists for temperatures less than the Curie temperature of the material. The polarization (P) is strongly affected by the material temperature, as illustrated in Fig. 9.12. Incident radiation heats the sample, expanding the crystal lattice and changing the dielectric constant, which alters the polarization of the material. As the crystal absorbs heat, there is a motion of bound positive ions, producing a change in the charge near the surface. The polarization change affects the charge on the electrodes mounted on the detector, causing current to flow in an external circuit if electrodes are placed on the crystal face perpendicular to the polarization axis. The detector can be modeled as a capacitor filled with a ferroelectric material. This change in polarization (realignment of electric dipole concentration) results in a proportional change in the dielectric constant of a capacitor. A voltage is produced by the charge on the capacitor. Ferroelectric crystals cause a change in spontaneous polarization with temperature change. In the

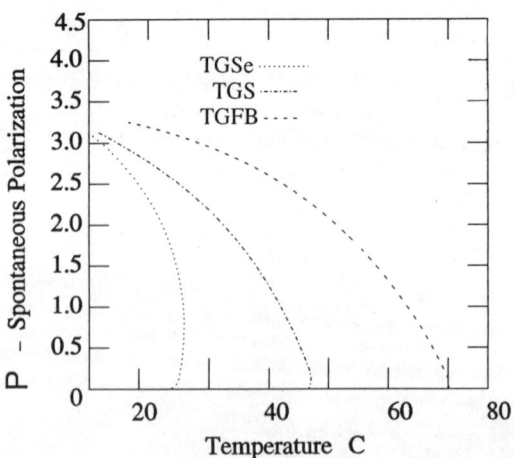

Figure 9.12. Spontaneous polarization vs. temperature for previously poled crystals of triglycerine sulfate (TGS), triglycerine fluoroberyllate (TGFB), and triglycerine selenate (TGSe).

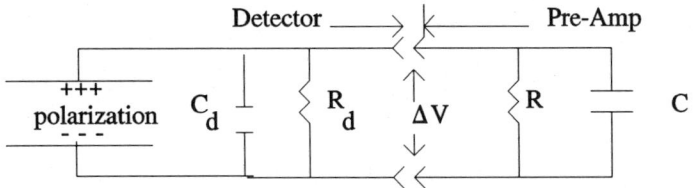

Figure 9.13. Equivalent circuit for a pyroelectric detector.

static case (constant temperature) the material aligns internally to display no net charge. Therefore, the pyroelectric is a derivative-responding detector.

Figure 9.13 is a schematic of a pyroelectric detector, modeled as a lossy capacitor. The capacitor is filled with a ferroelectric that remains polarized as long as the temperature is below the Curie temperature. The Curie temperature is the temperature at which the polarization goes to zero, as shown in Fig. 9.14. This limits the temperature range in which the detector can operate. If the temperature of the material is changed, the electric polarization will be changed. The degree to which this polarization changes with temperature defines a parameter called the pyroelectric coefficient (p).

The pyroelectric coefficient p is the measure of the rate of change of electric polarization with respect to the temperature change defined as

$$p \equiv \frac{\partial P}{\partial T} \tag{9.85}$$

Table 9.7 gives the magnitude of the pyroelectric coefficient for some typical materials at 300 K.

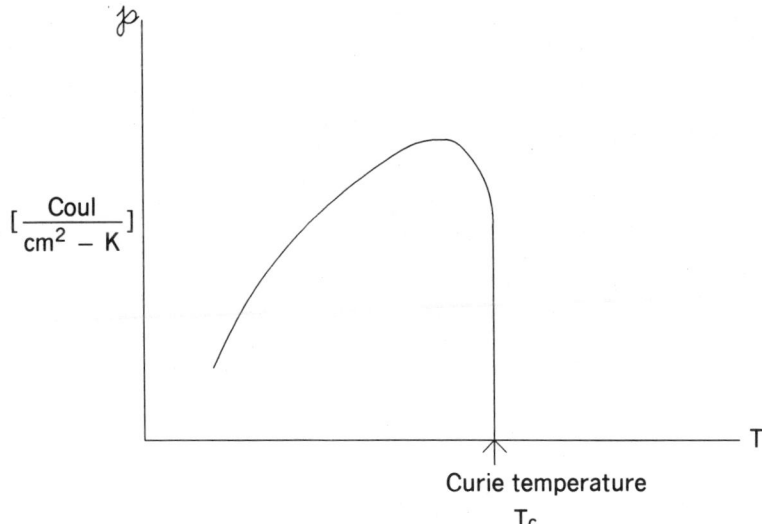

Figure 9.14. Pyroelectric coefficient vs. temperature.

Table 9.7 Ferroelectric Material Characteristics

Material	Pyroelectric Coefficient (p) (C cm^{-2} K^{-1})	Dielectric Constant (ε)	Specific Heat (J g^{-1} K^{-1})	Thermal Conductivity K_t (W cm^{-1} K^{-1})	Density (gm cm^{-1})
Tourmaline	4×10^{-10}	160 ($\parallel$ to axis)		9×10^{-3}	6.0
BaTiO$_3$	2×10^{-8}	4100 ($\perp$ to axis)	0.5	6.8×10^{-3}	
TGS	$(2\text{-}3.5) \times 10^{-8}$	$25 \sim 50$	0.97	17×10^{-3}	1.69
Li$_2$SO$_4 \cdot$ H$_2$O	1.0×10^{-8}	10	0.4		2.05
LiNbO$_3$	4.0×10^{-9}	30 ($\parallel$) 75 $\perp$			4.64
LiTaO$_3$	6×10^{-9}	58			
SbSI	2.6×10^{-7}	10^4	0.29		8.2
NaNO$_3$	1.2×10^{-8}	8.0	0.96		2.1

The pyroelectric coefficient has units of coulomb $cm^{-2} K^{-1}$ and is a function of temperature. A larger pyroelectric coefficient generally indicates better detector performance. Figure 9.12 illustrates typical polarization-versus-temperature plots. The pyroelectric coefficient is the slope of that curve.

9.6.1. Poling

If a Curie-temperature phase transition occurs, the material is no longer polarized or "poled." To make the material operate as a detector, repoling is necessary. A detector can be poled by heating it above the Curie temperature, applying a bias of about 1000 V/cm, and holding it for several minutes, and then slowly cooling the detector, with the bias applied, back to room temperature. The depoling can also occur from shock or simply aging.

9.6.2. Theory of Operation

To investigate how a pyroelectric detector produces current, let's start with Maxwell's equation:

$$D = \epsilon E + P \approx P \qquad (9.86)$$

where ϵ is the permittivity of free space of the detector material and we have assumed small or zero applied electric fields in the detector to drop the first term. Taking the divergence of both sides gives the charge density

$$\nabla \cdot D = \nabla \cdot P = \rho_c \qquad (9.87)$$

Integrating over a cylindrical-shaped volume that just covers the area of one side of the detector that has an electrical lead, and applying Stokes' theorem, we arrive at

$$\int_v \rho_c \, dV = \int_v \nabla \cdot P \, dV = \int_s P \cdot dA_d = PA_d = Q \qquad (9.88)$$

The polarization is offset by surface charge density that is now a function of temperature because polarization is a function of temperature.

When the detector is aligned so that its intrinsic polarization P is perpendicular to the surface through which current flows, the current flowing through the detector is given by

$$i = \frac{dQ}{dt} = \frac{dP}{dt} A_d = \frac{dP}{dT} \frac{dT}{dt} A_d = p \frac{dT}{dt} A_d \qquad (9.89)$$

where $p = dP/dT$ is the pyroelectric coefficient of the material. The detector area is assumed to be constant in time, and the only way P can be changed is by heating the detector. Therefore, it is clear that current only flows through the detector when a temperature change occurs (dT/dt), resulting in a zero response to dc (an ac detector).

The heat balance equation gives the classic solution from Eq. (9.26) to be

$$\Delta T = \frac{\varepsilon\phi_e(t)R_{th}}{\sqrt{1 + \omega^2 R_{th}^2 C_{th}^2}} \tag{9.90}$$

assuming a graybody detector with emissivity ε. A thermal time constant $\tau_{th} = R_{th}C_{th}$ can also be defined, where R_{th} is the thermal resistance equal to the inverse of the thermal conductance K, and C_{th} is the thermal capacitance. Recognizing that $d\Delta T/dt = dT/dt$, and assuming a sinusoidal input radiant power $\phi_e(t) = \phi_{e,0} \exp\{j\omega t\}$, we have

$$\frac{dT}{dt} = \frac{d\Delta T}{dt} = \frac{\varepsilon j\omega\phi_e(t)\ R_{th}}{\sqrt{1 + \omega^2\tau_{th}^2}} \tag{9.91}$$

$$\left|\frac{dT}{dt}\right| = \frac{\varepsilon\omega\phi_{e,0}R_{th}}{\sqrt{1 + \omega^2\tau_{th}^2}} \tag{9.92}$$

By plugging Eq. (9.92) into Eq. (9.89) we get an expression for the current flow through the detector when sinusoidal radiation is incident on it.

$$i = \frac{p\varepsilon\omega\phi_{e,0}R_{th}A_d}{\sqrt{1 + \omega^2\tau_{th}^2}} \tag{9.93}$$

By dividing Eq. (9.93) by the incident power $\phi_{e,0}$, we have the current responsivity of the detector

$$\mathfrak{R}_i = \frac{p\varepsilon\omega R_{th}A_d}{\sqrt{1 + \omega^2\tau_{th}^2}} \tag{9.94}$$

This equation reveals three important properties of the current response of a pyroelectric detector. First, the responsivity is zero for dc illumination; second, $\mathfrak{R}_i \approx$ constant for chopping frequencies $f > 1/(2\pi\tau_{th})$, and third, there is no upper cut-off frequency for $\mathfrak{R}_i$.

To calculate the voltage responsivity, the voltage on the detector is needed. This voltage is simply the current times the parallel electrical impedance of the detector (from evaluation of the circuit in Fig. 9.13) and the amplifier. The voltage is thus given by

$$v = \frac{iR_d}{\sqrt{1 + \omega^2 \tau_{RC}^2}} \tag{9.95}$$

where $\tau_{RC} = R_d C_d$ the electrical time constant of the circuit, and R_d and C_d are the parallel electrical resistance and capacitance of the detector/amplifier combination. The voltage responsivity is simply

$$\Re_v = \Re_i \left(\frac{v}{i} \right) \tag{9.96a}$$

$$\Re_v = \frac{p\varepsilon\omega R_{th} R_d A_d}{\sqrt{1 + \omega^2 \tau_{th}^2} \sqrt{1 + \omega^2 \tau_{RC}^2}} \tag{9.96b}$$

An important difference between Eq. (9.94) for the current response and Eq. (9.96b) for the voltage response is that $\Re_v$ has an upper frequency cutoff given by $f_{RC} = 1/(2\pi\tau_{RC})$. As a result, there are three frequency regions of operation in terms of $\Re_v$. The first is for $f < 1/(2\pi\tau_{th})$, where $\Re_v$ increases linearly with f. For $1/(2\pi\tau_{th}) < f < 1/(2\pi\tau_{RC})$, the electrical and thermal time constants offset each other, and $\Re_v$ is constant. Finally, for $f > 1/(2\pi\tau_{RC})$, $\Re_v$ falls off linearly with f. Figure 9.15 shows a typical plot of voltage responsivity versus frequency.

9.6.3. Figures of Merit

Pyroelectric detectors can suffer from four different noises. Temperature noise is rarely a problem because these detectors are seldom used in background-radiation noise-limited situations. The high sensitivity of pyroelectrics requires that careful attention be paid to minimize both preamplifier noise and microphonic noise and (hopefully) reduce them below the Johnson noise of the detector. For the most part, pyroelectric detectors are operated in this Johnson-

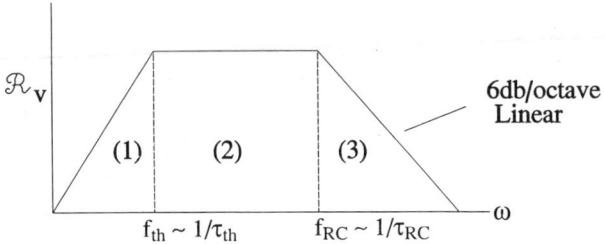

Figure 9.15. Responsivity vs. frequency for a pyroelectric detector.

noise-limited (JOLI) region, for which

$$v_j = \sqrt{4kTR_d\Delta f} \qquad (9.97)$$

Given Eq. (9.96b) and (9.97) for $\mathfrak{R}_v$ and the noise, the NEP may be found:

$$\text{NEP} = \frac{v_j}{\mathfrak{R}_v} = \frac{\sqrt{4kT\Delta f}\,\sqrt{1 + \omega^2\tau_{\text{th}}^2}\,\sqrt{1 + \omega^2\tau_{RC}^2}}{p\varepsilon\omega R_{\text{th}}\,\sqrt{R_d}\,A_d} \qquad (9.98)$$

Given the NEP, D^* follows as

$$D^* = \frac{\sqrt{A_d\Delta f}}{\text{NEP}} = \frac{p\varepsilon\omega R_{\text{th}}A_d\,\sqrt{R_dA_d}}{\sqrt{4\,kT}\,\sqrt{1 + \omega^2\tau_{\text{th}}^2}\,\sqrt{1 + \omega^2\tau_{RC}^2}} \qquad (9.99)$$

Equation (9.99) provides for some important insight into the pyroelectric detector. The detectivity is still strongly frequency dependent, in the same way as $\mathfrak{R}_v$ in Eq. (9.96b). The resistance-area RA product also plays an important role in the pyroelectric detector. There are, in fact, two RA products: the customary electrical one and a thermal RA product. Thus we have D^* dependencies:

$$D^* \propto \sqrt{R_dA_d} \quad \text{and} \quad D^* \propto R_{\text{th}}A_d \qquad (9.100)$$

The D^* depends linearly on the thermal RA product, whereas the dependence of D^* on the electrical RA product is as a square root.

Based on Eq. (9.100), the desired characteristics for a pyroelectric material include a large RA product. By reducing the heat capacity and the dielectric constant of the material, a faster detector can be created. Detectors with a response time as fast as 170 ps have been constructed (Hudson and Hudson, 1975). A large pyroelectric coefficient is also desirable, and can be achieved by cooling the detector to a point where dP/dT is largest. This temperature point is material specific, and lower is not necessarily better. In most applications, pyroelectric detectors can be operated effectively at room temperature. For a barium titanate pyroelectric detector operated near 381 K, the minimum detectable power has been calculated as 1.4×10^{-11} W, comparable to superconducting bolometers (Putley, 1980).

Practically speaking, detectors with a large Curie temperature are desirable. However, detectors can be repoled if the need should arise. Microphonics are a critical consideration in the design of pyroelectric detectors, and sources of such noise should be avoided.

9.7. DETECTOR TEMPERATURE CONSTRAINTS

At this point, the reader should realize that there is an ultimate sensitivity of an optical detector limited by the thermodynamics of the detector itself. Simply stated, the temperature of the detector limits its sensitivity. The photo-generated carrier energy must be above the temperature-noise-generated carrier energy to be detected. There have been two approaches to establish this range of operating temperatures for infrared detectors:

1. The shot noise associated with the self-radiation of the detector material being at a temperature above absolute zero. Several authors (Kinch and Yariv, 1989; Blouke et al., 1973) have analyzed particular detector types.
2. The temperature noise associated with the carrier generation rate commensurate with a detector being at a finite temperature.

We would like to establish the ultimate theoretical sensitivity limit that can be expected for a detector operating at a given temperature. This establishes the highest temperature at which a detector can operate and yet be photon-noise limited. In other words, the detector must be at this temperature or below to be a background-limited infrared photodetector (BLIP), if it is above this temperature, the other detector noises are dominant.

To obtain this operating-temperature limit, we start with the concept that the self-radiation of the detector itself at a finite temperature is the dominant limiting noise. The derivation will be valid for both photodetectors and thermal detectors. The radiation inside a detector is assumed to follow Planck's radiation law (noting that, inside the detector, the speed of light and the wavelength of the radiation are both reduced because of division by the index of refraction (n). In addition, because of total internal reflection, we assume that no radiation escapes. Thus, all of the radiation with photon energy greater than the energy gap of the material is assumed to produce carriers within the detector. The combination of these factors results in an increase of allowable blackbody-radiation cavity modes, causing an increase in flux density inside the detector material.

The carrier generation rate within the detector is (Jensen, 1990)

$$g = 8\pi c n^2 \int_0^{\lambda_c} \frac{d\lambda}{\lambda^4 (e^{hc/\lambda kT} - 1)} \quad \text{[carriers/s-cm}^2\text{]} \quad (9.101)$$

For BLIP operation, the background radiation incident of the detector should just equal the radiation from self-emission. Actually, at that point, the detector D^* will be $\sqrt{2}$ below BLIP performance. However, this is a common benchmark to establish the required operating temperature for BLIP operation. The power-spectral density of the detector noise can be expressed as the sum of the background radiation and the self-emission, assuming that both act as shot-noise terms:

$$\frac{di_n^2}{df} = 2q^2 \ell_z a A_d g + 2q^2 A_d \eta E_{q,\text{bkg}} \quad (9.102)$$

where

q = charge of an electron
ℓ_z = detector thickness or absorption distance
a = absorption coefficient
η = quantum efficiency
A_d = optical active area
$E_{q,\text{bkg}}$ = background photon irradiance within the detectors spectral response

Assuming that the incoming photons incident on the detector produce the same amount of noise as the detector self-emission, we equate the two noise components. Solving for the background photon irradiance incident on the detector yields

$$E_{q,\text{bkg}} = g\left(\frac{\ell_z a}{\eta}\right) \tag{9.103}$$

This the equivalent photon irradiance of the detector that gives equal noise contribution from the background photon noise and the self-radiation noise. Actually the noise is $\sqrt{2}$ above BLIP at this point, as mentioned earlier. This gives a means of determining the operating temperature of the detector for this level of background flux. If the detector is operated at a higher temperature, then the system cannot be BLIP (Jensen, 1993). The background irradiance can be calculated from the radiometry.

Figure 9.16 shows a background-irradiance-versus-detector-temperature plot needed to ensure BLIP operation, where we assume that the detector refractive index $n = 3$. We also assumed that the detector is large enough to absorb all the radiation and thus have a unit quantum efficiency. Therefore,

$$\left(\frac{\ell_z a}{\eta}\right) \approx 1 \tag{9.104}$$

The user must know the background irradiance incident on the detector to quantify the necessary detector operating temperature. The curves in Fig. 9.16 determine the minimum operating temperature for various detector cutoff wavelengths. If the detector rises above the temperature, it cannot possibly be BLIP for this environment.

If a 5-μm-cutoff detector is used, the background irradiance on it is 10^{12} photon/s-cm^2. Then the temperature of operation to assure BLIP conditions must be 110 K or lower, as shown in Fig. 9.16. Once the radiometry is done (background irradiance on the detector determined), the detector operating temperature must be below the temperature shown in Fig. 9.16, if there is any chance of BLIP operation for a photodetector.

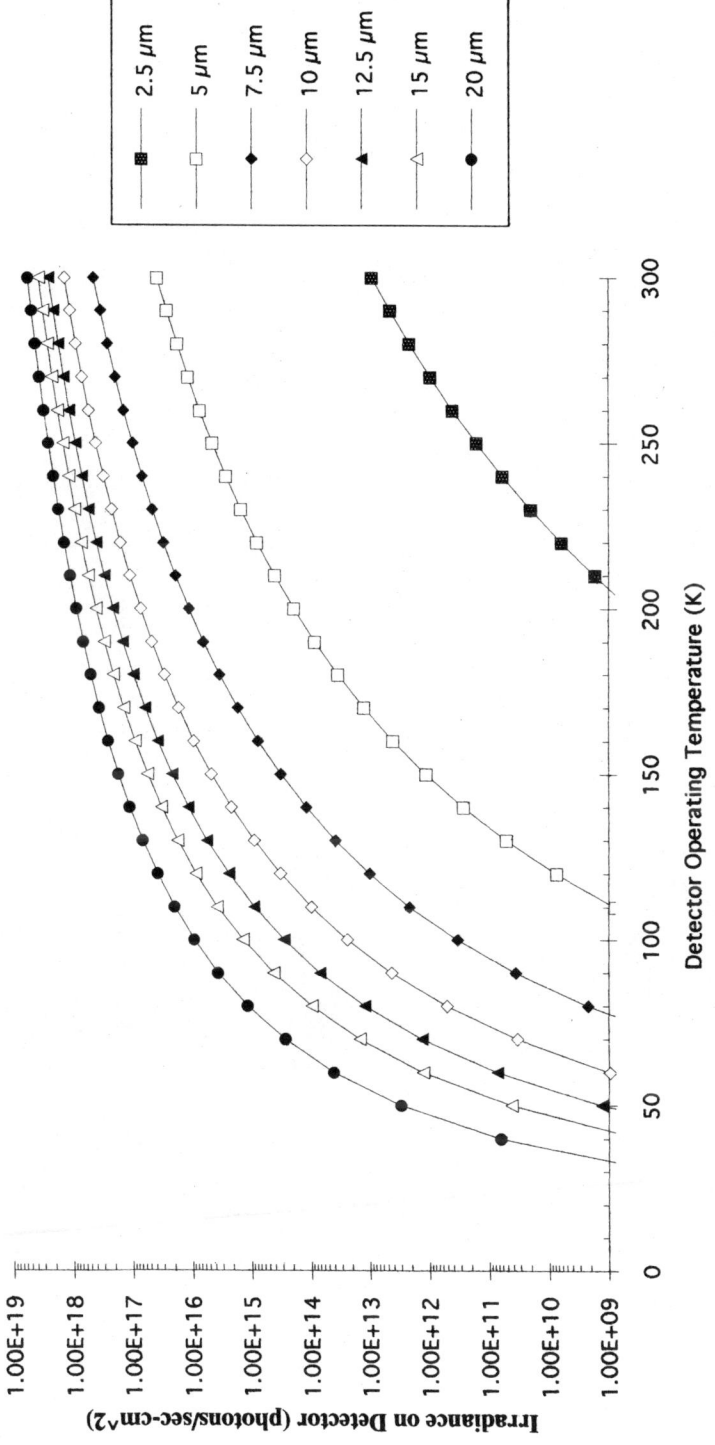

Figure 9.16. Background irradiance vs. required temperature for BLIP operation, as a function of cutoff wavelength.

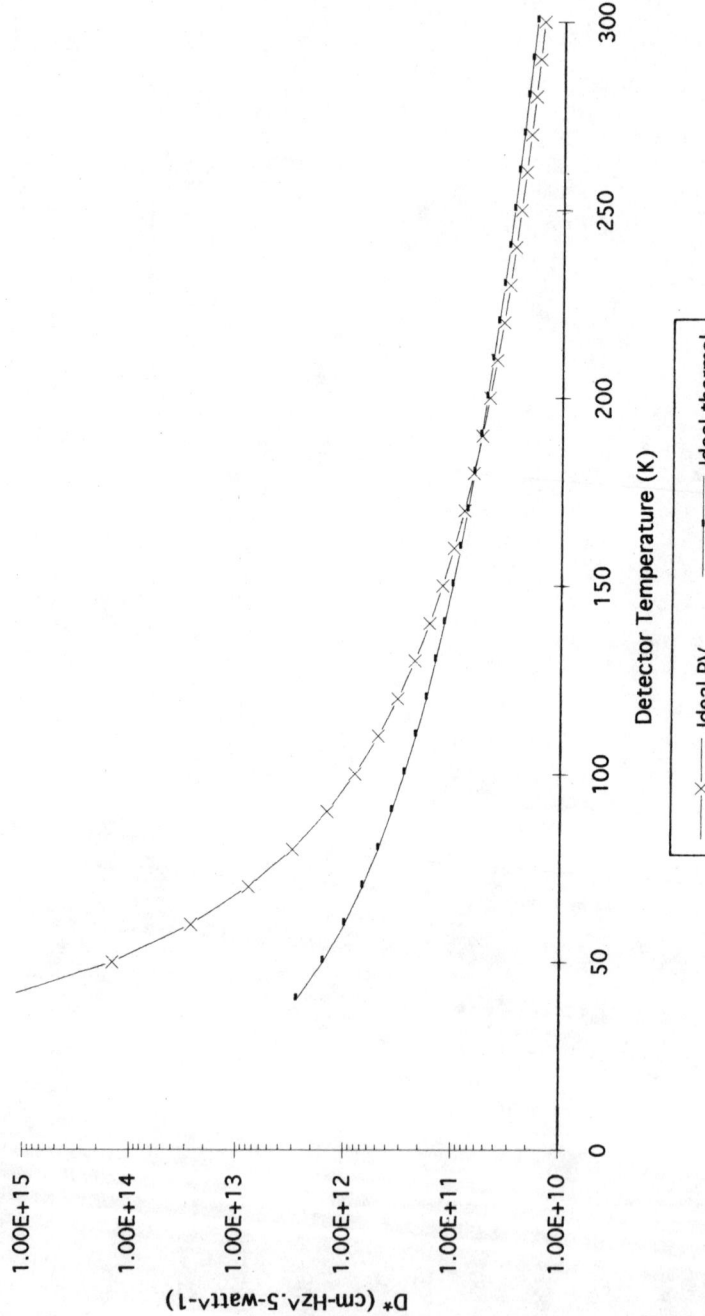

Figure 9.17. BLIP D^* vs. detector temperature for cutoff wavelength of 15 μm, for an ideal photovoltaic and an ideal thermal detector.

For an ideal photovoltaic detector, BLIP $D*$ versus detector temperature can now be plotted as shown in Fig. 9.17 for a cutoff wavelength of 15 μm. This figure is plotted from Fig. 9.16, curve labeled 15 μm, and assuming $\eta = 1$. The ideal thermal detector $D*$ versus temperature, assuming temperature-noise-limited operation, from Eq. (9.14), is also plotted in Fig. 9.17. At 190 K, these curves cross (Low and Hoffman, 1963; Jensen, 1993). So, for room-temperature operation, thermal detectors are better than photodetectors, a somewhat surprising result.

For long-wavelength infrared (LWIR) detectors, sensitive in the 8- to 14-μm atmospheric window, we can see a big improvement in $D*$ for cooling the detector (in fact, several orders of magnitude). However, at detector operating temperatures above 190 K, it is better to choose a thermal detector because its $D*$ is higher. At room temperature, the thermal detector is better than the photodetector.

REFERENCES

Blouke, M. M., C. B. Burgett, and R. L. Williams, "Sensitivity Limits for Extrinsic Infrared Detectors," *Infrared Physics* **13**, 61 (1973).

Boyle, W. S., and K. F. Rodgers, "Performance Characteristics of a New Low-Temperature Bolometer," *J. Opt. Soc. Am.* **49**, 66 (1959).

Henninger, J. H., *Solar Absorptance and Thermal Emittance of Some Common Spacecraft Thermal-Control Coatings* (NASA Reference Publication 1121), National Aeronautics and Space Administration, Huntsville, AL, 1984.

Hudson, R. D., and J. W. Hudson (eds.), *Infrared Detectors*, Dowden, Hutschinson & Ross, Stroudsburg, PA, 1975.

Kinch, M. A., and A. Yariv, "Performance Limitations of GaAs/AlGaAs Infrared Superlattices," *Appl. Phys. Lett.* **55**, 2093 (1989).

Jensen, A. S., "Temperature Limitations to Infrared Detectors," *SPIE Proc.* **1308**, 284 (1990).

Jensen, A. S., "Limitations to Room-Temperature IR Imaging Systems" *SPIE Proc.* **2020**, 284 (1993).

Jones, R. C., "Immersed Radiation Detectors," *Appl. Opt.* **1**, 607 (1962).

Low, F. J., "Low-Temperature Germanium Bolometer," *Opt. Soc. Am.* **51**, 1300–1304 (1961).

Low, F. J., and A. R. Hoffman, "The Detectivity of Cryogenic Bolometers," *Appl. Opt.* **2**, 649 (1963).

Putley, E. H., "Thermal Detectors," in *Optical and Infrared Detectors* (R. J. Keyes, ed.), Springer-Verlag, Berlin/New York, 1980, p. 71.

Reynolds, W. C., and H. C. Perkins, *Engineering Thermodynamics*, McGraw-Hill, New York, 1977.

BIBLIOGRAPHY

Andrews, D. H., R. M. Milton, and W. deSorbo, "A Fast Superconducting Bolometer," *J. Opt. Soc. Am.* **36**, 518 (1946).

Cooper, J., "Minimum Detectable Power of a Pyroelectric Thermal Receiver," *Rev. Sci. Instr.* **33**(1), 92 (1962).

Coron, N., "Infrared Helium Cooled Bolometers in the Presence of Background Radiation: Optical Parameters and Ultimate Performances," *Infrared Physics* **16**, 411 (1976).

Coron, N., G. Dambier, J. Le Blanc, and J. P. Moliac, "High-Performance Far Infrared Bolometer Working Directing in a Helium Bath and Note," *Rev. Sci. Instr.* **46**, 492 (1975).

Dereniak, E. L., and D. G. Crowe, "Thermal Detectors and Thermopiles," *Optical Radiation Detectors*, Wiley, New York, 1984, pp. 133–150.

Dewaard, R., and E. Wormser, "Description and Properties of Various Thermal Detectors," *Proc. IRE* **47**, 1508 (1959).

Fuson, N. J., "The Infrared Sensitivity of Superconducting Bolometers," *Opt. Soc. Am.* **38**, 845 (1948).

Gallinaro, G., and R. Varone, "Construction and Calibration of a Fast Superconducting Bolometer and Note," *Cryogenics* **15**, 292 (1975).

Golay, M. J. E., "A Pneumatic Infrared Detector," *Rev. Sci. Instr.* **18**, 357 (1947).

Golay, M. J. E., "Theoretical Consideration in Heat and Infrared Detection, with Particular Reference to the Pneumatic Detector," *Rev. Sci. Instr.* **18**, 347 (1947).

Golay, M. J. E., "The Theoretical and Practical Sensitivity of the Pneumatic Infrared Detector," *Rev. Sci. Instr.* **20**, 816 (1949).

Hickey, J. R., and D. B. Daniels, "Modified Optical System for Golay Detector," *Rev. Sci. Instr.* **40**, 732 (1969).

Hornig, D. F., and B. J. O'Keefe, "Design of Fast Thermopiles and the Ultimate Sensitivity of Thermal Detectors," *Rev. Sci. Instr.* **18**, 474 (1947).

Jones, R. C., "The Ultimate Sensitivity of Radiation Detectors," *J. Opt. Soc. Am.* **37**, 879 (1947); errata, **39**, 343 (1949).

Jones, R. C., "The General Theory of Bolometer Performance," *J. Opt. Soc. Am.* **43**, 1 (1953).

Milton, R. M., "A Superconducting Bolometer for Infrared Measurements," *Chem. Rev.* **39**, 419 (1946).

Nudelman, S., "The Detectivity of Infrared Detectors," *Appl. Opt.* **1**, 627 (1962).

Putley, E. H., "The Pyroelectric Detector," in *Semiconductors and Semimetals*, (R. K. Willardson and A. C. Beer, eds.), Academic Press, New York, 1970, pp. 259–285.

Putley, E. H., Chapter 7, in "The Pyroelectric Detector—An Update" in *Semiconductors and Semimetals* (R. K., Willardson and A. C. Beer, eds.), Academic Press, New York, 1977, pp. 441–449.

Roundy, C. B., R. L. Byer, D. W. Phillion, and D. J. Kuizenga, "A 170 ps Pyroelectric Detector," *Optic Comm.* **10**(4), 374–377 (1974).

Smith, R. A., F. E. Jones, and R. P. Chasmar, *The Detection and Measurement of Infrared Radiation*, Oxford University Press, Oxford, UK, 1968.

Zwerdling, S., R. A., Smith, and J. P. Thierault, "A Fast High-Responsivity Bolometer Detector for Very-Far Infrared," *Infrared Physics* **8**, 271 (1968).

PROBLEMS

9.1.   For a radiation-noise-limited thermal detector at 18 K (liquid hydrogen temperature), what is the best D^* possible with a 100-K 2π-sr background for a 1-mm^2 detector with $\Delta f = 1$ Hz?

9.2.   Compare and contrast bolometers and photoconductors

9.3.   Why should we chop infrared radiation? What detectors demand chopping of radiation?

9.4.   The RA product has significance in a photovoltaic detector; however, it is also in the theoretical D^* expression for a pyroelectric detector. Explain its significance in the case of the pyroelectric.

9.5.   What material characteristics are needed to make a good pyroelectric detector?

9.6.   Sketch a typical D^*-versus-electrical-frequency plot for a pyroelectric detector. (See the figure.) Explain each region.

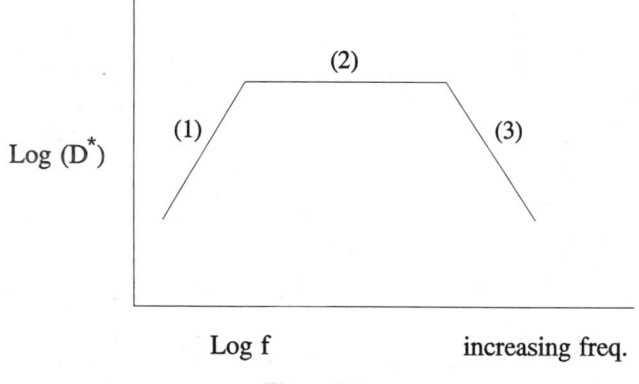

Figure 9.6

9.7.   A thermal detector ($\varepsilon = 1$ for all λ) is mounted on two posts to a heat sink. The total steady-state thermal conductance to the heat sink via the posts is 1.0 μW/K. If the detector is a 1-mm^3 TGS device, what is the thermal time constant?

9.8.   The plot of log D^* versus electrical frequency (f) can be approximated by three straight-line portions for a (a) photoconductive and (b) pyroelectric detector.

9.9.   A thermal detector is mounted on two posts to a heat sink. The steady-state heat flow to the heat sink via the posts is 1 μW/K. Assuming equal thermal conduction in each post, if one post breaks, then (numerically) what will happen to the responsivity and the time constant?

9.10. From curves of responsivity versus frequency for a pyroelectric detector, estimate τ_{thermal} and C_d.

9.11. What is the best $D*$ achievable for a thermal detector at a temperature of 300 K, looking into a 1000-K background? Assume $\varepsilon = 0.90$. Is this best $D*$ a function of wavelength? Why or why not?

9.12. Show that the following relationships regarding the detector irradiance when the detector is put in contact with the given optical elements of refractive index n. (See the figure.)

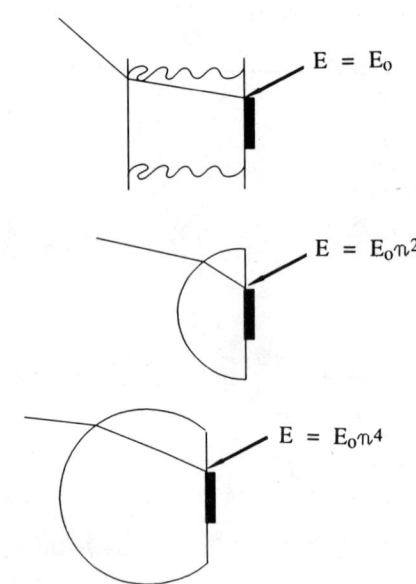

Figure 9.12

(a) Plane-parallel slab: Show that E is unchanged, with or without.
(b) Hemispherical lens: Show that E increases by a factor of n^2.
(c) Hyperhemispherical lens: Show that E increases by a factor of n^4.

10

Schottky-Barrier Photodiodes

10.1. INTRODUCTION

Schottky-barrier photodiodes, first suggested by Peters (1967), consist of a metal film on a silicon substrate, which is back-illuminated through the silicon (usually p-type). This metal–silicide junction produces a potential-energy barrier over which photo-generated holes can be excited to produce internal photoemission into the semiconductor. A variety of metals can be used to make the Schottky barrier. The most important is platinum (Pt), which produces a response to approximately 5.6 μm; other possibilities include nickel (Ni) responding out to about 1.8 μm, palladium (Pd) out to about 2.6 μm, and iridium (Ir) out to about 9.5 μm. Illumination of the junction through the silicon substrate produces a natural long-wave pass filter that blocks radiation at wavelengths less than 1 μm. Typical dimensions of a PtSi detector in the photo-propagation direction are about 500 μm for the thickness of the Si and from about 2 to 10 nm for the thickness of the PtSi layer. An additional optical cavity is often employed to enhance the quantum efficiency by the creation of a standing wave inside the PtSi layer, which is composed of an additional dielectric layer (typically SiO_2) and an aluminum reflector, as seen in Fig. 10.1.

Schottky photodiodes are of interest primarily because they are suitable for integration into focal-plane arrays. Two characteristics are important in this regard. The detectors are fabricated on a silicon substrate and are well suited to standard silicon VLSI processing, which facilitates coupling of the detector structure to the readout–multiplexer circuitry and results in monolithic focal planes made of a single material system, rather than hybrid focal planes that require a different material system for the detector array and the change-transfer electronics. Monolithic focal planes are cheaper to manufacture, have a lower defect rate, and are mechanically and thermally more rugged than hybrid focal planes. In addition, because the photons are absorbed near the interface, the responsivity of the Schottky photodiodes does not critically depend on the thickness of the silicide layer. This solves one of the major problems that affects other detector-array technologies, the detector-to-detector uniformity. Residual nonuniformity (after digital correction) remains one of the fundamental limitations to the performance of detector arrays made of HgCdTe and InSb. Schottky photodiodes can be fabricated that have a response nonuniformity of

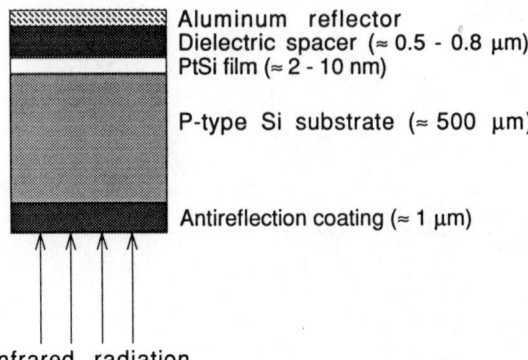

Aluminum reflector
Dielectric spacer ($\approx$ 0.5 - 0.8 µm)
PtSi film ($\approx$ 2 - 10 nm)

P-type Si substrate ($\approx$ 500 µm)

Antireflection coating ($\approx$ 1 µm)

Infrared radiation

Figure 10.1. Typical structure of Schottky-barrier photodiode, with optical cavity.

less than 1% in large arrays (greater than 512×512 elements). This advantage often outweighs the relatively low quantum efficiency of these detectors (typically 0.1% to 1%) for use in array applications where the uniformity is crucial. They are rarely seen as individual detectors, because other detector materials have a higher quantum efficiency ($\eta > 50\%$) in most of the wavelength ranges covered by the metal–silicide detectors. One exception to this is NiSi, which is under development as a detector for fiber optics, where its response speed and ease of silicon integration are important factors.

10.2. FABRICATION OF PtSi

Given the importance of the PtSi-material system, we first consider some aspects of the fabrication processes for PtSi photodiodes. The solid–solid chemical reaction between the Pt and Si substrate is well defined and controllable, leading to stable and reproducible Schottky barriers. Another attractive feature of PtSi is that it forms at a relatively low annealing temperature of around 300°C. However, care must be taken to avoid contamination with O_2 or Al, as these elements can affect the reaction products and thus the electrical properties of the contact. The Schottky-barrier height can be adjusted by the predeposition and postdeposition heat treatments.

The main steps in the formation of a detector structure such as that illustrated in Fig. 10.1 are given in Fig. 10.2. The initial PtSi interface is formed by evaporation of a thin Pt film on the substrate. The initial phase of Pt_2Si begins to form at a temperature of 300°C. The thickness of the Pt_2Si layer grows with the square root of the anneal time, until all of the Pt is consumed, yielding a Pt_2Si–Si interface. The absence of a further source of elemental Pt leads to a change in the composition of the interface in favor of formation of a more Si-rich phase, namely PtSi. With additional time at the annealing temperature, the conversion of Pt_2Si to PtSi proceeds, with the thickness of the PtSi layer also proportional to the square root of the annealing time.

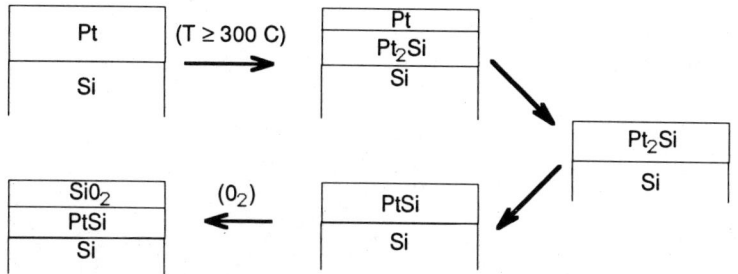

Figure 10.2. Formation of PtSi. Arrows indicate increasing time or temperature.

Ultimately, all of the Pt_2Si is consumed, and the final junction interface is PtSi–Si.

With postannealing in an oxygen atmosphere, an SiO_2 layer forms on the outer surface of the PtSi film, suitable for formation of the resonant cavity seen in Fig. 10.1. The final step is deposition of the Al reflector on top of the SiO_2.

10.3. INTERNAL PHOTOEMISSION

Schottky-barrier infrared photodetectors act on the process of internal pho-toemission of hot holes across the Schottky barrier between the silicide layer and the *p*-type Si substrate. Three fundamental processes are involved in pho-ton detection: photoexcitation, transport to the barrier, and emission across the barrier. We discuss each process separately. Figure 10.3 shows the energy-band diagram and carrier flow for the detector, where the potential Ψ (eV) is the Schottky-barrier height developed at the contact.

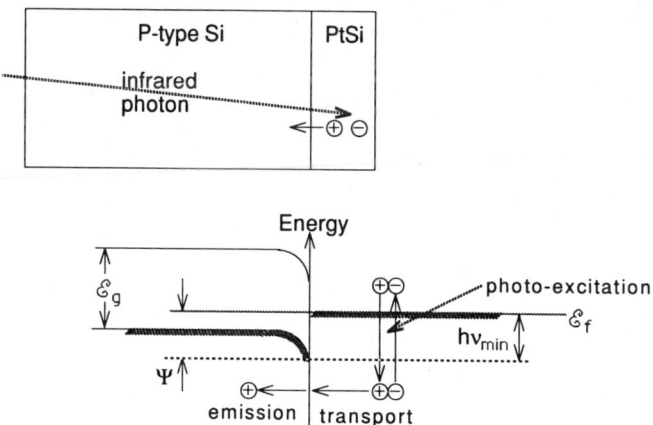

Figure 10.3. Carrier flow and band diagram for PtSi Schottky photodiode.

10.3.1. Photoexcitation

The photon absorption takes place in the metal–silicide layer. During absorption, photons with energy $h\nu$ such that

$$\Psi < h\nu < \mathcal{E}_g \tag{10.1}$$

will pass through the semiconductor substrate to excite holes in the silicide from states above the Fermi energy $\mathcal{E}_f$ to states below the emission barrier. $\mathcal{E}_g$ is the difference between the valence- and conduction-band energies for the semiconductor through which the junction is illuminated. The shaded area in Fig. 10.3 below the Fermi energy in the silicide and below the valence-band edge in the Si indicate that states below the shading are occupied by electrons. Photoexcited hot holes can thus move into those regions.

Schottky-barrier heights for the common metal–silicides are NiSi, $\Psi = 0.69$ eV; Pd_2Si, $\Psi = 0.48$ eV; PtSi, $\Psi = 0.22$ eV; IrSi, $\Psi = 0.13$ eV. The usual relationship applies for cutoff wavelength and barrier height:

$$\lambda_c \, [\mu m] = 1.24/\Psi \, [eV] \tag{10.2}$$

consistent with the cutoff wavelengths quoted earlier. As with any detector where an energy barrier is involved in photon detection, the use of smaller barrier heights necessitates a lower operating temperature. PtSi is generally operated at 77 K, with IrSi operated around 35 K, Pd_2Si operated around 135 K, and NiSi operated around 200 K.

10.3.2. Transport

Photoexcited holes reach the PtSi–Si interface by either a ballistic path or by a sequence of scattered paths. Most of the photoexcited holes are lost from the point of view of the emission process because they either are moving away from the interface or lose enough energy to collisional processes that they no longer have sufficient kinetic energy to cross the Schottky barrier. The addition of the Al reflector and optical cavity seen in Fig. 10.1 provide scattering mechanisms that favor the emission process.

10.3.3. Emission

The Si provides the medium into which photoexcited holes from the silicide are injected if they have the correct direction and momentum (see Sec. 10.4.3). The photo-generated negative charges remain behind and accumulate on the silicide electrode. The signal readout is accomplished by periodically transferring this negative charge into a charge-coupled device (CCD) by means of a voltage-reset pulse.

10.4. THEORETICAL PERFORMANCE

10.4.1 Quantum Efficiency and Responsivity

The quantum efficiency $[e^-/\text{photon}]$ of the Schottky-barrier photodiode is given by (Vickers, 1971)

$$\eta = C_1 \frac{(h\nu - \Psi)^2}{h\nu} \tag{10.3}$$

where both $h\nu$ and Ψ are in eV, and C_1 is the quantum yield $[\text{eV}^{-1}]$ that depends on the properties of the metal (Dalal, 1971), which is in the range of 0.2 to 0.4 for PtSi photodiodes. We will derive Eq. (10.3) in Section 10.3.4, including an approximate evaluation of C_1.

We would like to write Eq. (10.3) in the form of a current responsivity $\mathcal{R}_i$ (amp/W) to facilitate comparison with other detector technologies. The usual expression (from Chap. 3) that relates quantum efficiency to current responsivity is

$$\mathcal{R}_i = \frac{\eta q}{h\nu} \tag{10.4}$$

but it should be noted that Eq. (10.4) assumes that $h\nu$ is in units of joules. For $h\nu$ in units of eV, we write Eq. (10.4) as

$$\mathcal{R}_i = \frac{\eta}{h\nu} \tag{10.5}$$

because the Planck constant h (eV-s) implicitly contains the factor of q in its definition. We substitute Eq. (10.3) into Eq. (10.5) to obtain

$$\mathcal{R}_i = \frac{C_1((h\nu - \Psi)^2/h\nu)}{h\nu} = C_1\left(\frac{h\nu - \Psi}{h\nu}\right) = C_1\left(1 - \frac{\Psi}{h\nu}\right)^2 \tag{10.6}$$

where it should be noted that $\mathcal{R}_i$ is in A/W, C_1 is in eV^{-1}, and both Ψ and $h\nu$ are in eV. We can write Eq. (10.6) in terms of wavelength

$$\mathcal{R}_i = C_1\left(1 - \frac{\Psi\lambda}{hc}\right)^2 \tag{10.7}$$

We can write the term λ/hc as $\lambda/1.24$, when λ is in μm, h is in eV-s, and c is in m/s. We can write Eq. (10.7) as

$$\mathcal{R}_i = C_1\left(1 - \frac{\Psi\lambda}{1.24}\right)^2 \tag{10.8}$$

and substitute Eq. (10.2) to yield

$$\mathcal{R}_i = C_1 \left(1 - \frac{\lambda}{\lambda_c} \right)^2 \tag{10.9}$$

where $\mathcal{R}_i$ is in A/W and C_1 is in eV^{-1}. A typical spectral-responsivity curve (Ewing et al., 1985) is shown in Fig. 10.4 for PtSi, along with lines of constant quantum efficiency. For instance, PtSi has a quantum efficiency of approximately 1% at a wavelength of 4 μm.

If we compare relative responsivities, normalizing Eq. (10.9) to unity at λ = 1 μm (the cut-on wavelength for a back-illuminated device), we obtain the curves seen in Fig. 10.5. It should be noted that, in contrast to other detectors that exhibit an energy gap such as photovoltaics and photoconductors, the spectral responsivity does not rise with increasing wavelength, but rather exhibits a decreasing responsivity as the cutoff wavelength is approached.

10.4.2. Fowler Plot

To facilitate data analysis to yield numerical values for C_1 and Ψ, Eq. (10.3) is typically plotted in the "Fowler-plot" form (after Fowler, 1931) as

$$\sqrt{\eta h\nu} = \sqrt{C_1}(h\nu - \Psi) \tag{10.10}$$

When $(\eta h\nu)^{1/2}$ is plotted versus $h\nu$, the slope of the straight line is $(C_1)^{1/2}$ and the x-intercept value is Ψ, as seen in Fig. 10.6. The Schottky-barrier height is a slow function of the reverse-bias voltage v_B applied to the diode. The change in Ψ is (Cabanski and Schulz, 1991) proportional to the fourth root of v_B, with larger reverse biases producing smaller values of Ψ by around 10 meV, and hence longer cutoff wavelengths.

10.4.3. Fermi Sphere

Once the incident photon has been absorbed and the hot hole has moved to the junction, its probability of emission into the silicon can be calculated by requiring that the momentum of the hole normal to the semiconductor–silicide interface correspond to a kinetic energy in excess of the barrier height

$$\frac{((h/2\pi)k_n)^2}{2m_h^*} + h\nu > \mathcal{E}_f + \Psi \tag{10.11}$$

where k_n is the normal component of the momentum and m_h^* is the effective mass of the hole. This relation defines a cap of an escape cone of angle θ, typically between 1 and 10 degrees. The rather small escape angle is one reason why the quantum efficiency of metal–silicide photodetectors is low. Figure

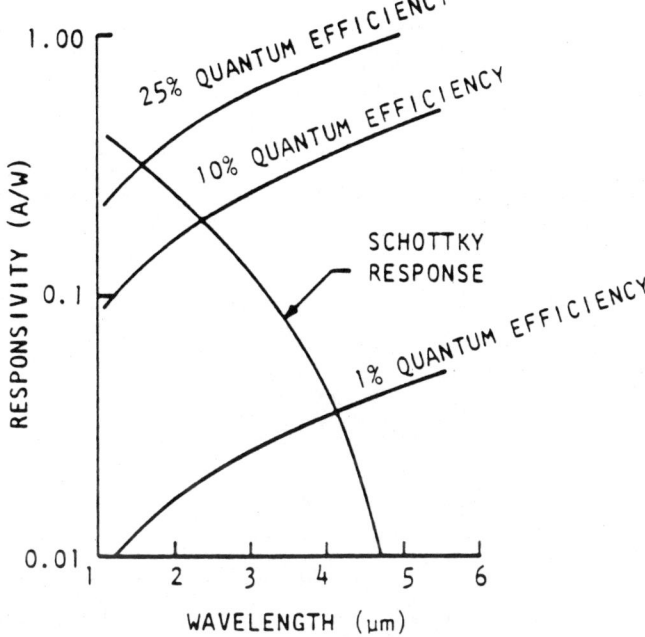

Figure 10.4. Spectral responsivity of PtSi, showing lines of constant quantum efficiency. (Originally appeared in W. S. Ewing et al., "Applications of an Infrared Charge-Coupled-Device Schottky Diode Array in Astronomical Instrumentation," *SPIE Proc.* **331,** 1985. Reprinted by permission of SPIE.)

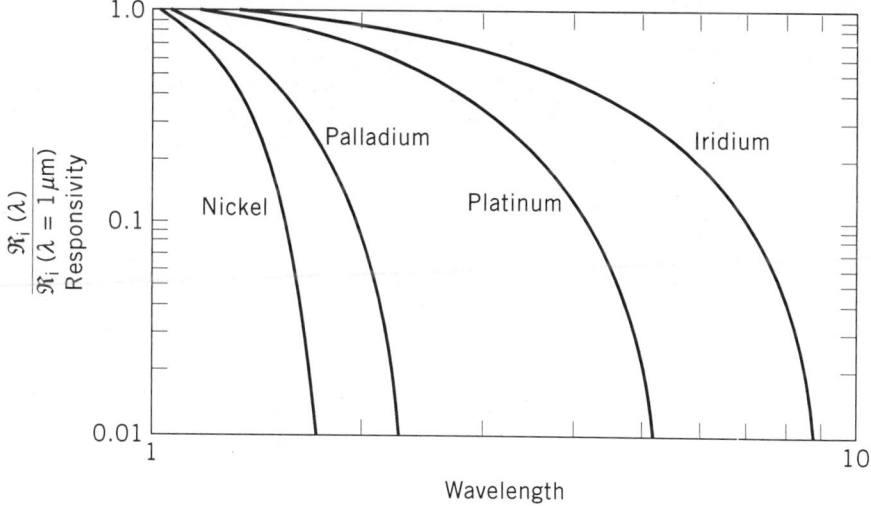

Figure 10.5. Normalized responsivities for various metal silicides.

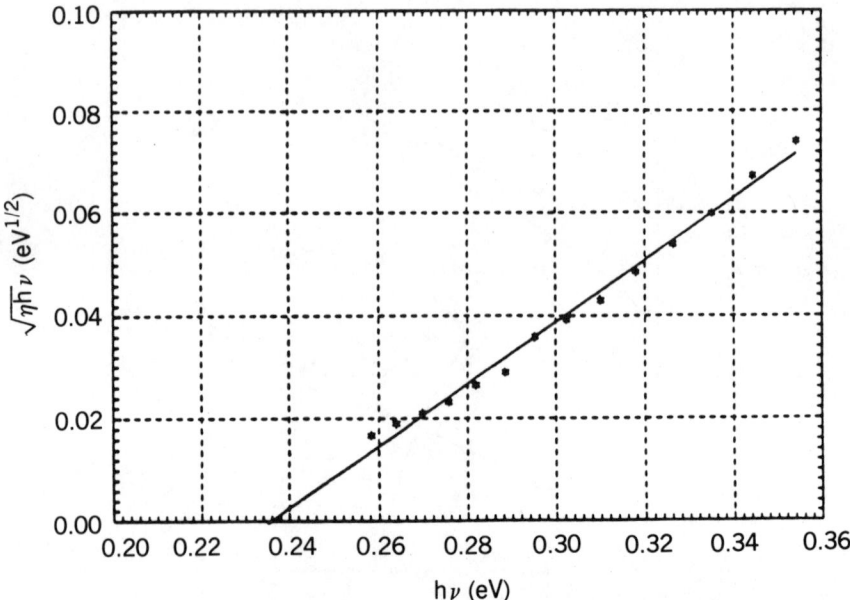

Figure 10.6. Fowler plot for PtSi-photodiode data, yielding $C_1 = 0.37$ eV^{-1} and $\Psi = 0.236$ eV. (After Fowler, 1931.)

10.7 shows a k-space model of the emission process, where the emission probability of a hot hole can be expressed as the ratio of the number of emitted electrons to the number of excited electrons.

The hole-excitation condition occurs when the energy of the hole is less than $h\nu$ below the Fermi level. The probability of hole excitation is the entire

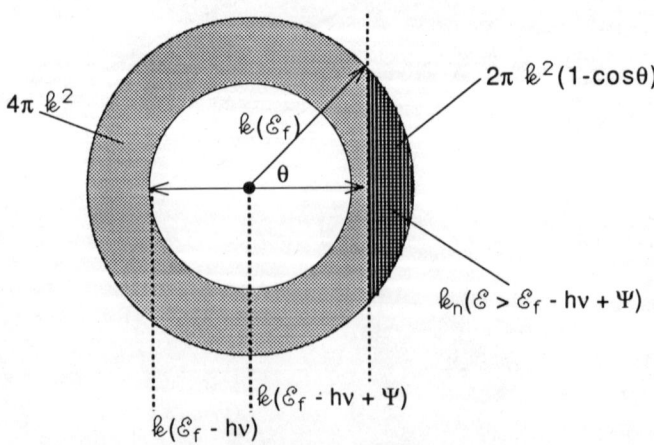

Figure 10.7. Emission-probability model in k space.

shaded region in Fig. 10.7, and defines a spherical shell in k space bounded by $k(\mathcal{E}_f - h\nu)$ and $k(\mathcal{E}_f)$, where

$$k(\mathcal{E}) = \frac{(2m_h^* \mathcal{E})^{1/2}}{h/(2\pi)} \tag{10.12}$$

with differential volume of $4\pi k^2$. The probability of hole emission, given excitation, is calculated as the volume of the dark-shaded spherical cap in Fig. 10.7, $2\pi k^2(1 - \cos\theta)$. Making use of conditional probabilities (Chap. 4) we find

$$P_{\text{emission}} = \frac{P(\text{emission}|\text{excitation})}{P(\text{excitation})} = \frac{2\pi k^2(1 - \cos\theta)}{4\pi k^2} = \frac{(1 - \cos\theta)}{2}$$

$$\tag{10.13}$$

where

$$\cos\theta = \sqrt{\Psi/\mathcal{E}_f} \tag{10.14}$$

assuming that the density of states is uniform in k space.

The limiting angle for a photon of wavelength λ, with energy $h\nu$, is given by

$$\theta_{\text{max}} = \cos^{-1}\sqrt{\Psi/h\nu} = \cos^{-1}\left(\sqrt{\frac{\Psi\lambda}{1.24}}\right) \tag{10.15}$$

typically between 1 and 10 degrees. Equation (10.15) can be substituted into Eq. (10.13) to yield

$$P_{\text{emission}} = \frac{1}{2}(1 - \sqrt{\Psi/h\nu}) = \frac{1}{2}\left(1 - \sqrt{\frac{\Psi\lambda}{1.24}}\right) \tag{10.16}$$

shown in Fig. 10.8, the emission probability as a function of wavelength (Elabd and Kosonocky, 1982).

10.4.4. Quantum Yield C_1

We now obtain an expression for C_1, the quantum-yield coefficient introduced in Eq. (10.3). We basically follow the development of Elabd and Kosonocky (1982). The number of states from which hole emission can occur (n) is

$$n = \int_{\Psi}^{h\nu} \text{DOS} \times P_{\text{emission}}\, d\mathcal{E} = \int_{\Psi}^{h\nu} \left(\frac{dn}{d\mathcal{E}}\right) P_{\text{emission}}\, d\mathcal{E} \tag{10.17}$$

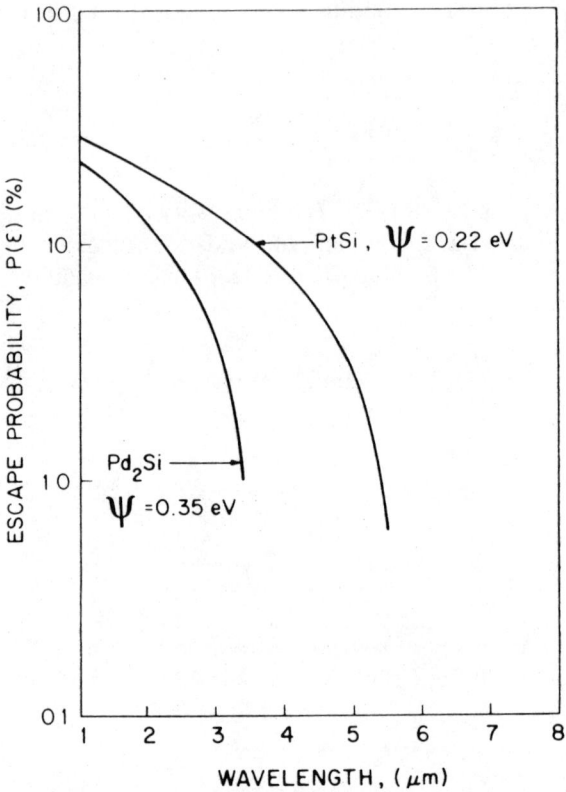

Figure 10.8. Hot-hole emission probability as a function of wavelength. (Originally appeared in H. Elabd and W. Kosonocky, ''Theory and Measurements of Photoresponse for Thin Film Pd$_2$Si and PtSi Infrared Schottky-Barrier Detectors with Optical Cavity,'' *RCA Review* **43,** 1982. Reprinted by permission of General Electric Company.)

where DOS is the density of states, also written as $dn/d\mathcal{E}$,

$$n = \int_{\Psi}^{h\nu} \left(\frac{dn}{d\mathcal{E}}\right) \frac{1}{2} (1 - \sqrt{\Psi/\mathcal{E}})\, d\mathcal{E} = \frac{1}{2} \left(\frac{dn}{d\mathcal{E}}\right) \int_{\Psi}^{h\nu} (1 - \sqrt{\Psi/\mathcal{E}})\, d\mathcal{E} \quad (10.18)$$

$$n = \frac{1}{2} \left(\frac{dn}{d\mathcal{E}}\right) (h\nu - 2\sqrt{\Psi h\nu} + \Psi) \quad (10.19)$$

and

$$n = \frac{1}{2} \left(\frac{dn}{d\mathcal{E}}\right) h\nu \left(1 - \sqrt{\frac{\Psi}{h\nu}}\right)^2 \quad (10.20)$$

With the identification of the total number of possible excited hole states as

$$n_T = \int_0^{hv} \left(\frac{dn}{d\mathcal{E}}\right) d\mathcal{E} = \frac{dn}{d\mathcal{E}} \int_0^{hv} d\mathcal{E} = \frac{dn}{d\mathcal{E}} hv \qquad (10.21)$$

we express the quantum efficiency as

$$\eta = a(\lambda) \frac{n}{n_T} = a(\lambda) \frac{\frac{1}{2}(dn/d\mathcal{E})hv(1 - \sqrt{\Psi/hv})^2}{(dn/d\mathcal{E})hv} = \frac{a(\lambda)}{2}\left(1 - \sqrt{\frac{\Psi}{hv}}\right)^2$$

$$(10.22)$$

where $a(\lambda)$ is the spectral absorptance of the detector material, which is in the range of 20–60%, given the thicknesses and materials commonly used in Schottky photodiodes. For the situation where the photon energy is close to that of the barrier height, we can write Eq. (10.18) in the approximate form

$$n = \frac{1}{2}\left(\frac{dn}{d\mathcal{E}}\right)\int_\Psi^{hv}(1 - \sqrt{\Psi/\mathcal{E}})\, d\mathcal{E} \cong \frac{1}{2}\left(\frac{dn}{d\mathcal{E}}\right)\int_\Psi^{hv}\frac{1}{2}\left(1 - \frac{\Psi}{\mathcal{E}}\right) d\mathcal{E} \qquad (10.23)$$

leading to

$$n = \frac{1}{4}\left(\frac{dn}{d\mathcal{E}}\right)\int_\Psi^{hv}\left(1 - \frac{\Psi}{\mathcal{E}}\right) d\mathcal{E} = \frac{1}{4}\left(\frac{dn}{d\mathcal{E}}\right)\left\{(hv - \Psi) - \Psi \ln\left(\frac{hv}{\Psi}\right)\right\}$$

$$(10.24)$$

Using the approximation (consistent with the case under consideration of $hv \approx \Psi$)

$$\ln\left(\frac{hv}{\Psi}\right) \approx \left(\frac{hv}{\Psi} - 1\right) - \frac{1}{2}\left(\frac{hv}{\Psi} - 1\right)^2 \qquad (10.25)$$

we write quantum efficiency as

$$\eta = a(\lambda) \frac{n}{n_T} = a(\lambda) \frac{\frac{1}{4}\left\{(hv - \Psi) - \Psi \ln(hv/\Psi)\right\}}{hv}$$

$$\approx \frac{a(\lambda)}{8\Psi} \frac{(hv - \Psi)^2}{hv} \qquad (10.26)$$

which we compare with Eq. (10.3) to evaluate the quantum-yield coefficient C_1

$$C_1 = \frac{\alpha(\lambda)}{8\Psi} \qquad (10.27)$$

Taking $\alpha(\lambda) \approx 50\%$ and $\Psi = 0.22$ eV for PtSi yields an approximate value of 0.35 eV^{-1} for C_1, consistent with the range quoted earlier.

It should be noted that Eq. (10.26) was derived under the assumption of a thick (compared to the mean free path of the hot holes) metal–silicide layer. There is an additional gain factor in C_1 if the layer is thin, because well scattering redirects photoexcited holes that are moving from the interface and toward the barrier, enhancing the effective-emission probability. This gain factor can be as high as 5 at 1 μm and as high as 100 at 5 μm. Because the maximum escape angle θ_{max} is so small, these additional scattering processes tend to favor emission, rather than discriminate against it.

10.5. DARK CURRENT

The current–voltage relationship for thermionic emission over a Schottky-barrier photodiode may be written (Sze, 1969) as

$$J = J_s(\exp\{qv/kT\} - 1) \qquad (10.28)$$

The reverse-saturation (dark) current density J_s (amp/cm^2) is assumed to be caused entirely by the diffusion of holes over the Schottky barrier and is given by

$$J_s = A^{**}T^2 \exp\{-q\Psi/kT\} \qquad (10.29)$$

where A^{**} is the effective Richardson constant (Sze, 1969). For p-type Si at 77 K A^{**} is approximately 32 amp cm^{-2} K^{-2}. Consistent with Eq. (10.29), the dark current for Schottky photodiodes increases rapidly with increasing temperature, and decreases exponentially with increasing barrier height. Values of J_s for PtSi at 77 K are in the range of 20 nA/cm^2. The barrier height can be adjusted for any metal–silicide material system by controlling the metal thickness and the anneal time, but it should be noted that increasing the barrier height to decrease the dark current will result in a reduced quantum efficiency (Kosonocky et al., 1985).

If Eq. (10.29) is written in the form of a semi-log Arrhenius plot

$$\ln(J_s/T^2) = \ln(A^{**}) - \left(\frac{q\Psi}{k}\right)\left(\frac{1}{T}\right) \qquad (10.30)$$

with $\ln(J_s/T^2)$ plotted versus $1/T$ as in Fig. 10.9, we can determine Ψ from the slope of the line and can also solve for the value of A^{**}.

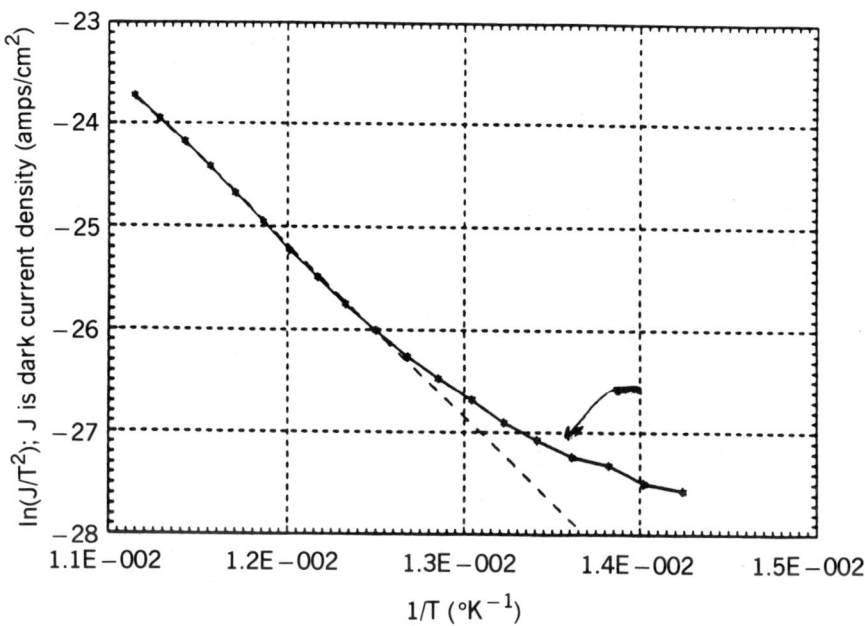

Figure 10.9.   Plot of Eq. (10.30) and dark-current data.

One issue of dark current is that the Schottky diodes are used in a reverse-bias mode (in the range of 2- to 10-V reverse bias) and then electrically isolated, enabling the detector to float. This allows the silicide electrode to accumulate negative charge left behind by the injection of the photo-generated hot holes into the Si. The barrier height is lowered by the reverse bias, leading to better quantum efficiencies, but also causing edge-leakage currents in addition to the hole-diffusion currents noted in Eq. (10.29). These surface-leakage currents become severe because the thin silicide layer causes the creation of large local electric fields, even for modest reverse biases. Guard rings are used to control the edge-leakage current (Kosonocky et al., 1985), but there is a penalty in the active area of the detector that is important when closely packed detector arrays are to be fabricated. An alternative approach (McNutt, 1988) is to place a compensating voltage on the aluminum reflector of Fig. 10.1. This field-plate method reduces the edge-leakage current without the active-area penalty of a guard ring.

10.6. SUMMARY

Metal–silicide Schottky-barrier photodiodes are widely used in thermal-imaging applications. PtSi detector arrays with pixel counts as high as 1024 × 1024 are becoming commercially available, while cameras based on 512 × 512 arrays are commonplace. These arrays can achieve background-limited operation (Kosonocky et al., 1985) at 30 frames/s, with a room-temperature background, an operating temperature of 77 K, and an $F/2.5$ cold-shield angle. Other metal–silicide detectors are under development for applications in other portions of the infrared spectrum.

REFERENCES

Cabanski, W. A., and M. J. Schulz, "Electronic and IR-Optical Properties of Silicide/ Silicon Interfaces," *Infrared Physics* **32**, 29–44, (1991).

Dalal, V. L., "Simple Model for Internal Photoemission," *J. Appl. Phys.* **42**, 2274–2279 (1971).

Elabd, H., and W. F. Kosonocky, "Theory and Measurements of Photoresponse for Thin Film Pd_2Si and PtSi Infrared Schottky-Barrier Detectors with Optical Cavity," *RCA Review* **43**, 569–589 (1982).

Ewing, W. S., F. D. Shepherd, R. W. Capps, and E. L. Dereniak, "Applications of an Infrared Charge-Coupled-Device Schottky Diode Array in Astronomical Instrumentation," *SPIE Proc.* **331** (1985).

Fowler, R. H., "The Analysis of Photoelectric Sensitivity Curves for Clean Metals at Various Temperatures," *Phys. Rev.* **38**, 45–56 (1931).

Kosonocky, W. F., F. V. Shallcross, T. S. Villani, and J. V. Groppe, "160-by-244 Element PtSi Schottky-Barrier IR-CCD Image Sensor," *IEEE Trans. Electron Devices* **ED 32**, 1564–1573 (1985).

McNutt, M. J., "Edge Leakage Control in Platinum-Silicide Schottky-Barrier Diodes Used for Infrared Detection," *IEEE Electron Device Lett.* **9**, 394–396 (1988).

Peters, D. W., "An Infrared Detector Utilizing Internal Photoemission," *Proc. IEEE* **55**, 704–705 (1967).

Sze, S. M., "Metal-Semiconductor Devices," in *Physics of Semiconductor Devices*, Wiley, New York, 1969, pp. 363–417.

Vickers, V. E., "Model of Schottky Barrier Hot-Electron-Mode Photodetection," *Appl. Opt.* **10**, 2190–2192 (1971).

PROBLEMS

10.1. Verify the construction of the lines of constant quantum efficiency in Fig. 10.4.

10.2. Plot Eq. (10.9) for spectral responsivity on a linear–linear graph.

10.3. Given the responsivity curve in Fig. 10.4, calculate the blackbody responsivity (A/W) for a blackbody of (a) 300 K; (b) 500 K; and (c) 1000 K.

10.4. Verify the values for C_1 and Ψ given in the caption of Fig. 10.6.

10.5. Verify Eq. (10.13) for the probability of emission, from the construction in Fig. 10.7.

10.6. What is the maximum angle of escape for a PtSi detector, given a photon wavelength of (a) 4 μm and (b) 5 μm?

10.7. Find values of Ψ and A^{**} consistent with the data in Fig. 10.9.

10.8. Perform the approximation inherent in Eq. (10.23) to the next highest order in Ψ/ε. Carry this approximation through to Eq. (10.26).

11

Bandgap-Engineered Photodetectors: Multiple Quantum Wells and Superlattices

11.1. INTRODUCTION

Within the last ten years, a new approach to material development was established that included fabrication techniques that allowed researchers to create their own semiconductor materials. These semiconductor materials have optical and electrical properties and characteristics that are desirable for optical-radiation detectors. This approach to making new materials is known as *bandgap engineering* (Capasso et al., 1983; Ferry 1985). Critical to bandgap engineering has been the technology development of molecular-beam epitaxy (MBE) and metal-organic chemical-vapor deposition (MOCVD). These processes make materials one atomic monolayer at a time. Using these techniques, the growth of a material thickness can be controlled to less than 10 Å, and layer interfaces can ideally be controlled to the discrete atomic layer (approximately one atom thickness—3 Å). Spatial composition and doping profiles can be varied over thicknesses of 3 Å to a few nanometers, with excellent control of the constituents. The primary applications driver for MBE and MOCVD was the laser diode, which needed to be designed with numerous output wavelengths from visible to infrared.

Bandgap-engineered photodetectors were developed in response to two requirements: (1) photodetectors in the 1.4-μm region for fiber-optic communications, and (2) infrared detectors in the 8- to 14-μm band that have excellent spatial uniformity, D^*, and relaxed cryogenic requirements.

The use of bandgap-engineered semiconductors for infrared (IR) detectors was an obvious application (Levine, 1993), particularly in the long-wave infrared (LWIR) detection region. The detector materials were developed by alternating different semiconductor layers and exercising precise control over the thickness at the atomic level. These layers were so thin that energy levels showed confinement. The continuous energy levels found in bulk material became discrete energy levels within these thin layers. Multiple quantum wells and superlattices are two types of bandgap-engineered semiconductor materials. These can be further classified as to whether or not the atomic lattice induces strain within the layers (i.e., strained-layer superlattices).

Some examples of detectors made using these techniques are listed in Table 11.1. The majority of the infrared detectors listed in Table 11.1 are detectors

Table 11.1 Detector Properties for Various Engineered-Bandgap Materials

Material Type	Composition	$\lambda_1-\lambda_2$ (μm)	λ_c (μm)	Temp. (T_d) (K)	$\mathcal{R}_i$ (A/W)	D^* (cm Hz$^{1/2}$W^{-1})	Reference
Quantum wells	GaAs/AlGaAs	5.8–7.5	6.7	77	0.67	3 (10^{10})	Paska et al. (1991)
	GaAs/AlGaAs	8.4–11.7	10.2	20	1.5	1 (10^{10})	Levine et al. (1989)
	GaAs/AlGaAs		8	77	0.42	1 (10^{10})	Levine et al. (1989)
	GaAs/AlGaAs	9.0–11.0	10.7	50	1.2	1 (10^{10})	Levine et al. (1990)
	GaAs/AlGaAs	–14	11.7	10		6 (10^{12})	Lacoe et al. (1992)
Strained	InGaAs/InAlAs		8.1	77		5.9 (10^{10})	Li et al. (1994)
quantum wells	InGaAs/InAlAs	3.7–4.2	4	120		2.3 (10^{10})	Hasnain et al. (1990)
Superlattices	GaAs/AlGaAs	10.2–11.3	10.8	15	0.52		Levine et al. (1987)
	InSb		10.0	77		8.1 (10^{10})	
	InP/InGaAs	1.0–1.7					Ackley et al. (1990)
Strained layer	InAsSb		10.0	77	2.0	1.0 (10^{10})	Kurtz et al. (1988a)
superlattices (SLS)	GaAlAs/GaInAs						
	GeSi/Si						

of multiple-quantum-well construction. Note that quantum-well detectors have a relatively narrow-passband spectral response, unlike a normal photodetector. This is a result of the limited selection of energy-band transitions that can be photoexcited for bandgap-engineered materials.

Multiple quantum wells are very thin layers grown either on or between thick barrier layers so that electron wavefunctions do not overlap between adjacent thin layers. Typically, quantum wells are created by sandwiching a thin, narrow-bandgap material between wider layers of a wide-bandgap material (e.g., GaAs between AlGaAs). The electron-in-a-box energy states are quantized by the presence of potential-energy barriers (the wide-gap material) on both sides of the narrow-gap material. The quantity of discrete energy levels is controlled by the thickness of the narrow layer and the relative energy gap of the barrier material. Thicker quantum-wells will contain more energy states. The optical-detection process (absorption of a photon) takes place within the conduction band (intersubband), not across the energy gap as with a typical photoconductor.

Because the thin layer is a material with a different bandgap than the substrate, the Fermi levels will line up, but the conduction and valence band edges will not be continuous. For example, if a narrow-bandgap material is grown on a wide-bandgap material, followed by a layer of wide-bandgap material, the offsets in the conduction and valence bands will produce a "potential well," as shown in Fig. 11.1a. Electrons are confined within the well. Recall that the quantum-well layers are so thin that the energy bands are not a continuum as in a bulk material but form subbands as indicated in Fig. 11.1a. This results in a potential well that has only discrete energy levels.

Figure 11.1b shows the construction of the quantum well. What well width (ℓ_W) is sufficiently small to produce quantum-confinement effects? If the width of the well approaches the de Broglie wavelength of the electron (Λ), then quantum confinement forms a potential well. The expression for de Broglie wavelength is

$$\Lambda = \frac{h}{2\pi p} \tag{11.1}$$

where h = Planck's constant and p = momentum of the electron in the material.

For GaAs, the de Broglie wavelengths lie between 10 Å and 500 Å. Because these potential wells are physically so small, the electrons can only occupy discrete energy levels. The condition for a bound energy state is that $\Lambda/2$ be an integral multiple of the well width, and that the energy be smaller than the energy-band offset. The bound energy levels are calculated (West and Eglash, 1985) from

$$\mathcal{E}_m = \frac{h^2 m^2}{8 m_e^* \ell_W} \tag{11.2}$$

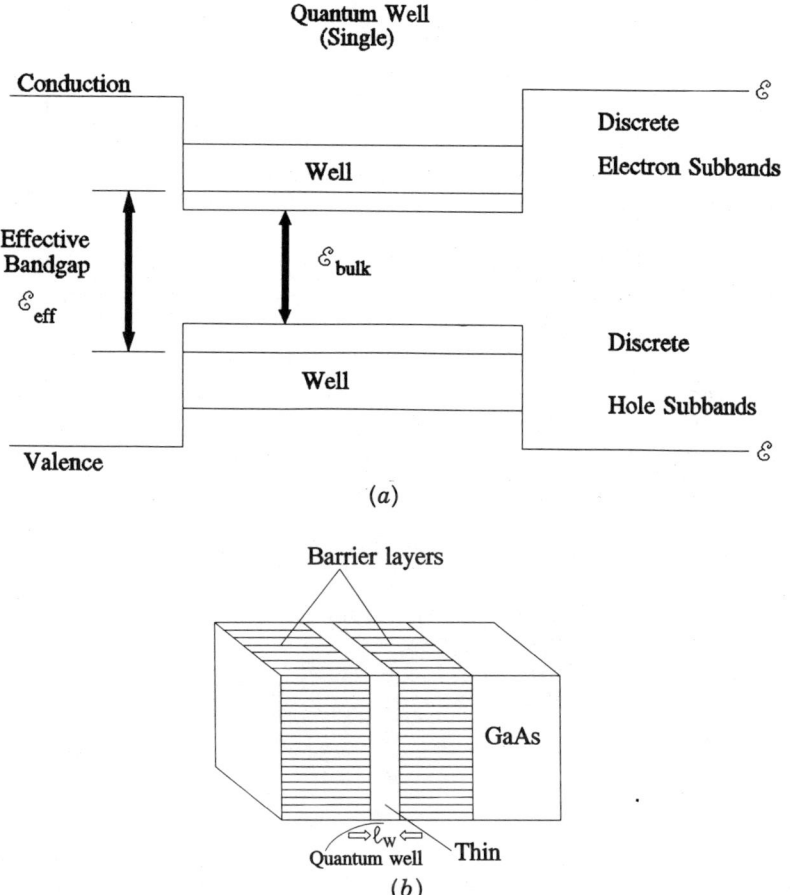

Figure 11.1. (a) Band structure of a single quantum well. The effective bandgap is the difference between the highest valence subband and the lowest conduction subband. (b) Construction of a single quantum well.

where

   m_e^* = effective mass of electron
   ℓ_W = well width
   m = indexing parameter.

For an effective mass of 10% of the true electron mass, and a 50-Å well width, the ground state is 149 meV, the first excited state is 598 meV, and the second excited state is 1341 meV (Saleh and Teich, 1991).

   A photon with sufficient energy to create an intersubband transition (598 meV to 149 meV) can be detected with this material. Because the width of the barriers (ℓ_B) are not infinite, these energies are only approximate (Peyghambarian et al., 1993). The infinite potential-well solution gives us slightly higher energies, but it is a good approximation to the more complex problem of determining the energy levels of a finite potential well.

A superlattice is a periodic structure consisting of thin alternating layers of two or more semiconductors, creating a new semiconductor material with optical and electrical properties controlled by the layer thickness, spacing, and composition. These properties are radically different from any of the bulk constituent-material properties.

Barrier layers in a superlattice (unlike those of quantum wells) are sufficiently thin that the electronics or holes have a substantial probability of tunneling through the barriers between the wells. The overlap of the carrier's wavefunctions lifts the degeneracies of the discrete energy levels to create minibands at those energies. Figure 11.2a shows the quantum-well construction where the wells are separated by a thick barrier. Figure 11.2b shows a superlattice structure, where the barrier is thin enough that the wavefunctions

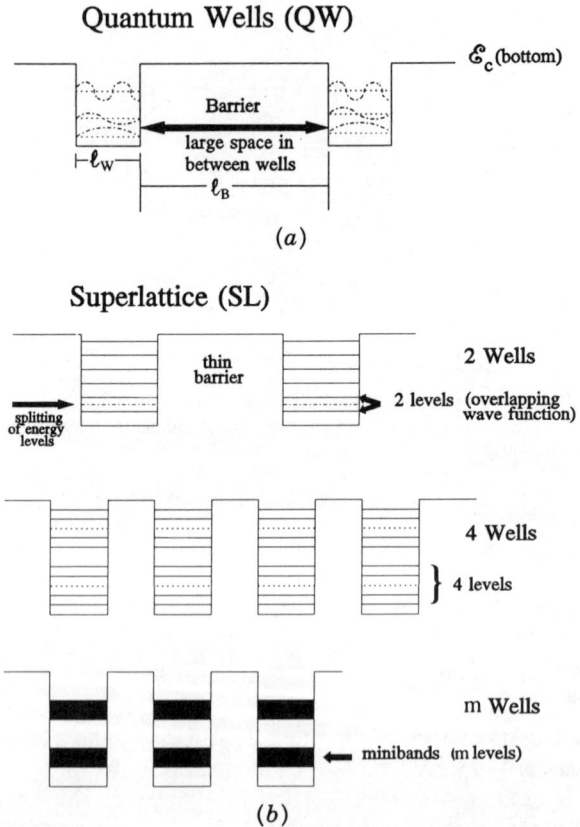

Figure 11.2. (a) Construction and energy structure of a multiple quantum well. (b) Superlattice formed by the overlap of wavefunctions of 2, 4, and m quantum wells.

of the electrons overlap, producing minibands instead of discrete levels. The first example in Fig. 11.2*b* shows two wells splitting each discrete level into two energy levels. If there are four wells with overlapping wavefronts, the energy levels near the discrete level split into four energy levels. Finally, for *m* interacting wells, minibands are formed that are made of *m* energy levels. The structure of superlattices is shown in Fig. 11.3, where layers of GaAs are sandwiched between GaAlAs to form a superlattice material.

If we were restricted to choosing materials with similar lattice constants to form a superlattice, our range of options would be quite limited. Figure 11.4 shows the energy-gap properties of bulk semiconductors and alloys. The black rectangular regions shown in the figure indicate where composite materials could be made without introducing strain, because there are materials with different bandgap but similar lattice constants. The plots show that very few bulk materials can be used in the LWIR detector region (10 μm or 0.124 eV). In fact, HgCdTe and InAsSb are the only intrinsic material choices, having an electronic transition from the valence to conduction band around 0.1 eV.

A way to generate more choices of effective bandgap of the superlattice is to form a material structure with strained layers. In this case, the materials used in alternating layers do not have the same lattice constant; that is, the atomic spacing is different for the well and the barrier material. This introduces strains in the crystalline structure, which changes the optical and electronic

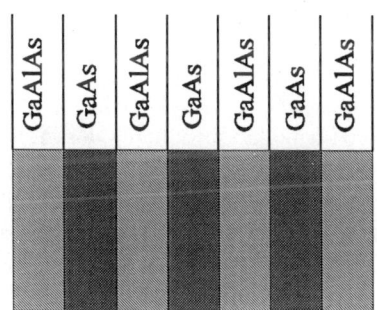

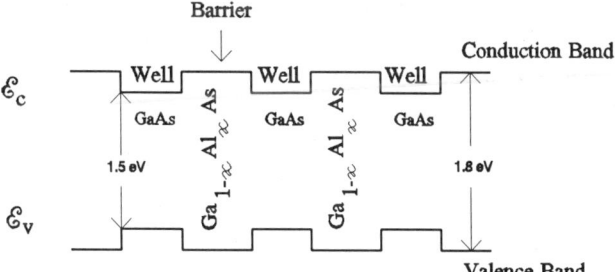

Figure 11.3. Construction of a superlattice, consisting of alternating epitaxial layers with different bandgaps.

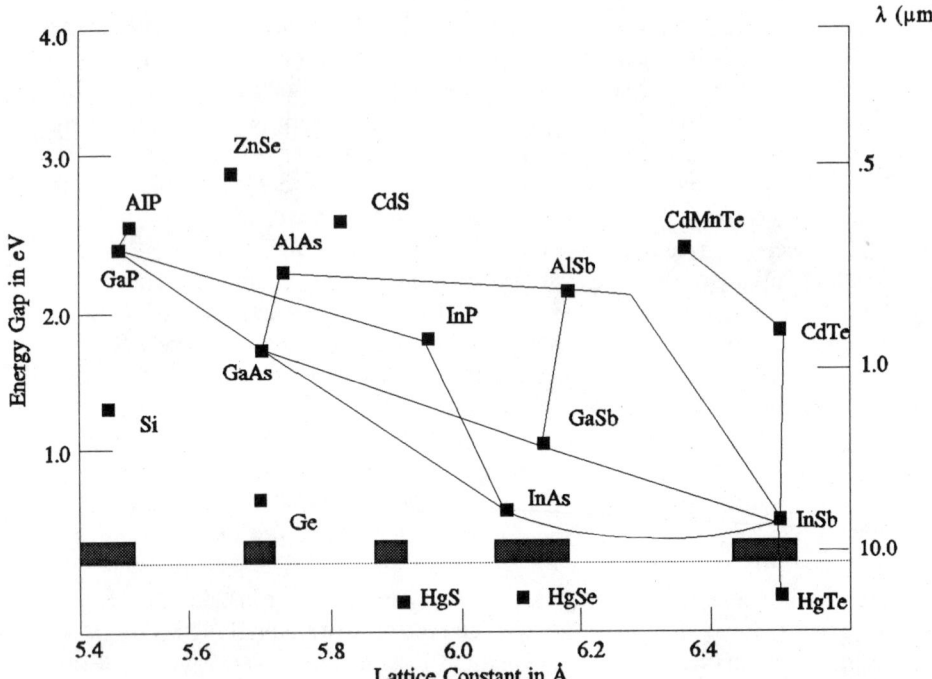

Figure 11.4. Plot of bulk energy gaps and lattice constants for common semiconductor materials.

properties of the superlattice. We will see later in this chapter that a strained lattice can result in a reduced bandgap, producing detectors with a longer cutoff wavelength. The bandgap is further reduced by decreasing the thickness of the layers. Thus the spectral response is tailored by controlling the layer thickness, the alloy composition, and the amount of strain. The use of strained layer superlattices for construction of photodetector materials is generally restricted to the long-wave portion of the spectrum (8 to 12 μm and beyond). The reason for this development is that the bulk semiconductors addressing that spectral region have a number of disadvantages in terms of detector fabrication, including high cost, nonuniformity, and low sensitivity.

An example of a strain layer can be illustrated by putting InAs between InSb, where the lattice constant from Fig. 11.4 is about 6.05 Å to 6.5 Å, respectively. Table 11.2 lists the lattice constants for the various alloys.

11.2. MULTIPLE-QUANTUM-WELL PHOTODETECTORS

A multiple quantum well is made up of many layers of a narrow-bandgap material alternating with a larger-bandgap material. The large number of layers

Table 11.2 Lattice Constants and Energy Gaps for Some III-V Materials

Material	Lattice Constant (Å)	Energy Gap (eV)
AlP	5.451	2.45
AlAs	5.662	2.16
GaP	5.451	2.26
GaAs	5.653	1.44
InP	5.869	1.35
AlSb	6.136	1.50
GaSb	6.096	0.72
InAs	6.058	0.35
InSb	6.479	0.18

increases the probability of photon capture over that for one layer alone. A structural example and energy-band diagram are shown in Fig. 11.5. There is an excellent review article on quantum-well photodetectors by Levine (1993).

The quantum-well barrier thicknesses (ℓ_B) are sufficiently large in a multiple-quantum-well detector that the individual wavefunctions do not overlap. The energy bands are thus discrete levels. The optical absorption (along the photon-propagation direction) of any individual quantum well is multiplied by m, the number of periods that make up the whole detector structure. The energy levels of each individual well exhibit some variation because of the fabrication tolerances on ℓ_W and ℓ_B. There is a $\pm \ell_{W,B}$ associated with the fabrication. The energy levels thus combine into minibands rather than keeping an exactly discrete nature, even when the individual wells do not interact by tunneling. If we consider these levels to be combined at the location of a single well for conceptual purposes, we obtain the energy diagram shown in Fig. 11.6a.

The effect of the alternating layers is to cause the absorption to be spectrally selective. Only select transitions can take place from the lower to the upper subband levels. This results in spectral responsivity and D^* that are substantially different from most photodetectors. As seen in Fig. 11.6b, the normalized spectral D^* (and hence spectral responsivity) shows a peaked response rather than a linear slope to the cutoff wavelength.

Three types of quantum-well structures are important for optical-detector applications. We designate the three types as I, II, III (Peyghambarian et al., 1993), as shown in Fig. 11.7. The lowest energy of the conduction band is denoted as $\mathcal{E}_c$, and the highest energy of the valence band is $\mathcal{E}_v$. The corresponding quantities for the barrier and the well, respectively, are written as $\mathcal{E}_c^B$, $\mathcal{E}_c^W$, $\mathcal{E}_v^B$, and $\mathcal{E}_c^W$. The energy gap of the barrier material is written as

$$\mathcal{E}_g^B \equiv \mathcal{E}_c^B - \mathcal{E}_v^B \qquad (11.3)$$

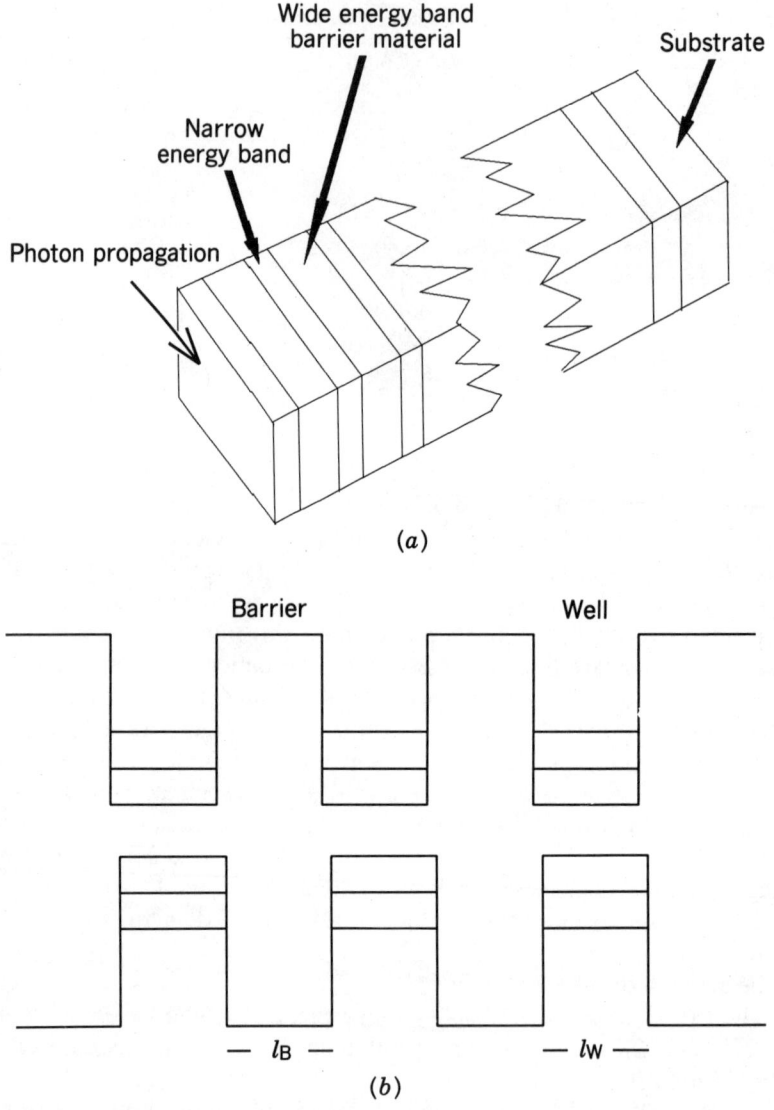

Figure 11.5. (a) Structure of a multiple-quantum-well detector. (b) Energy-band diagram (upper-conduction, lower-valence) for a multiple-quantum-well detector. The barrier widths are sufficient to make the individual wavefunctions discrete (nonoverlapping).

For a type-I quantum well, the energy difference between the valence bands $\Delta\mathcal{E}_v$ (the convention is well minus barrier) is a positive quantity

$$\Delta\mathcal{E}_v \equiv \mathcal{E}_v^W - \mathcal{E}_v^B > 0 \tag{11.4a}$$

and the energy difference between the conduction bands of the two materials

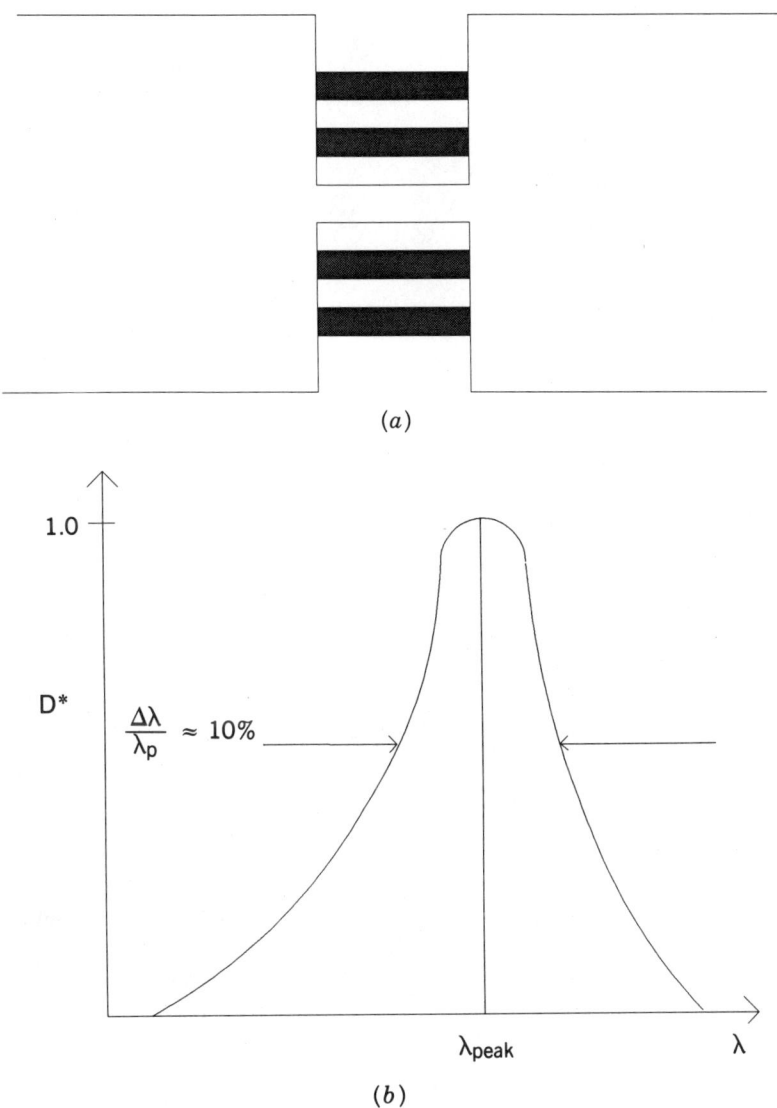

(a)

(b)

Figure 11.6. (a) Energy-band diagram for a multiple-quantum-well structure. The shaded regions show the conduction (upper) minibands and valence (lower) minibands. These minibands are the combined result of a large number of noninteracting discrete levels of slightly different energy. (b) Normalized plot of spectral D^* for a multiple-quantum-well photodetector.

is negative

$$\Delta \mathcal{E}_c \equiv \mathcal{E}_c^W - \mathcal{E}_c^B < 0 \qquad (11.4b)$$

As shown in Fig. 11.7, for a type-I structure, the valence- and conduction-band offsets ($\Delta \mathcal{E}_v$ and $\Delta \mathcal{E}_c$) are less than the differences of the bulk energy

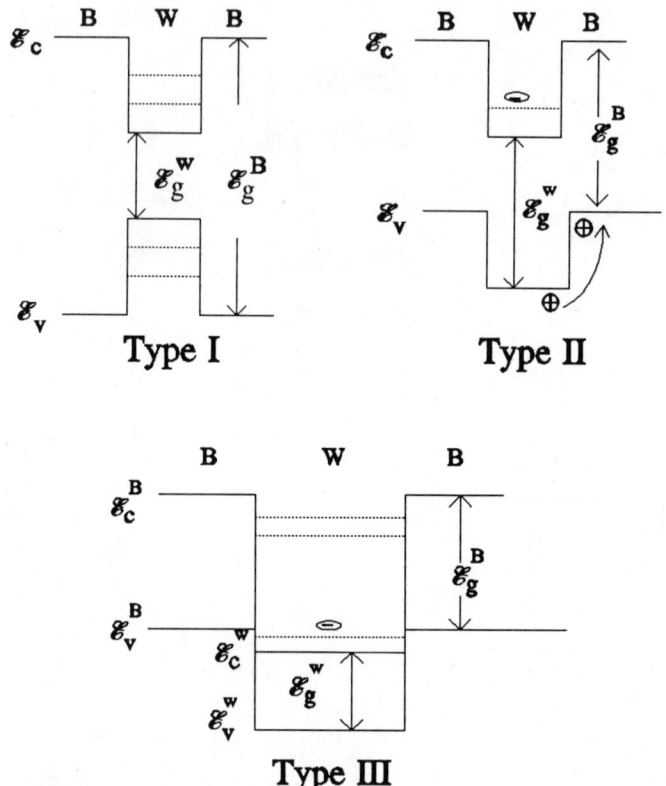

Figure 11.7. Energy diagrams for type I, II, and III quantum-well structures.

bands of the materials. The result is that the valence and conduction bands of the type-I well structure are formed within the same quantum-well layer.

The energy-band structure for type-II structures is shown in Fig. 11.7, with both energy differences

$$\Delta \mathcal{E}_c \equiv \mathcal{E}_c^W - \mathcal{E}_c^B < 0 \qquad (11.5)$$

and

$$\Delta \mathcal{E}_v \equiv \mathcal{E}_v^W - \mathcal{E}_v^B < 0 \qquad (11.6)$$

being negative quantities. Also, we see from Fig. 11.7 that

$$\Delta \mathcal{E}_v - \Delta \mathcal{E}_c < \mathcal{E}_g^B \qquad (11.7)$$

Type-II quantum wells have different optical and electrical properties from the type-I structures because the conduction band in the well and the valence band in the barrier have spatial overlap of the carrier wavefunctions in the two layers. For instance, a hole, shown at the top of the valence band in the well, can relax to its lowest energy state at the top of the barrier valence band. Recombination of a hole–electron pair results, with the electron residing in the conduction band of the well. This process can be caused by photon absorption. Thus a type-II quantum well will absorb (detect) photons having a lower energy than the corresponding type-I structure by an amount equal to $\Delta \mathcal{E}_v$.

A type-III structure is also shown in Fig. 11.7, and illustrates a quantum-well formed when the conduction-band shift of the well is greater than the energy gap of the barrier material

$$\Delta \mathcal{E}_c \equiv \mathcal{E}_c^W - \mathcal{E}_c^B > \mathcal{E}_g^B \qquad (11.8)$$

The lowest conduction subband in the well lies below the energy of the valence-band edge in the barrier material. The top of the valence band in the barrier is above the bottom of the conduction band of the well. Thus, electrons in the valence band emit into the conduction subband, which is at lower energy state for an electron and the electrons and holes are confined to different separate layers. An electron can be optically excited to an energy state outside the well. This produces a photoconductive response, a photoinduced change in conductivity.

As photons are absorbed in a type-III multiple-quantum-well structure, the electrons are first excited from the ground state $\mathcal{E}_1$ to the excited state $\mathcal{E}_2$, as shown in Fig. 11.8a. However, even with the electron at $\mathcal{E}_2$, it still cannot escape the potential well (neglecting tunneling processes). To escape into the continuous conduction band, a bias voltage (external electric field) must be applied, as shown in Fig. 11.8b. The application of a bias reduces the barrier seen by electrons at the excited level $\mathcal{E}_2$ to a level where tunneling processes allow them to escape into the conduction band.

We now briefly consider the performance concerns common to all quantum-well photodetectors. Relatively high values of thermally generated dark current are caused by electron tunneling between wells, resulting in dark-current noise. The detectors are quite selective spectrally, exhibiting a passband type of spectral response because there is only a limited energy range where photons can be absorbed by the minibands. There is also the issue of polarization sensitivity because the detector structure itself is anisotropic. The quantum wells absorb the two polarization states preferentially.

11.2.1. GaAs/GaAlAs

Quantum-well detectors can be made to respond in the 8- to 14-μm band and can be developed into large-area two-dimensional arrays (Beck, 1993) with good spatial uniformity. It should be noted that the quantum-well photoconductive detector is not theoretically as good as a HgCdTe photovoltaic detector operating under similar conditions (Kinch and Yariv, 1989). However, this

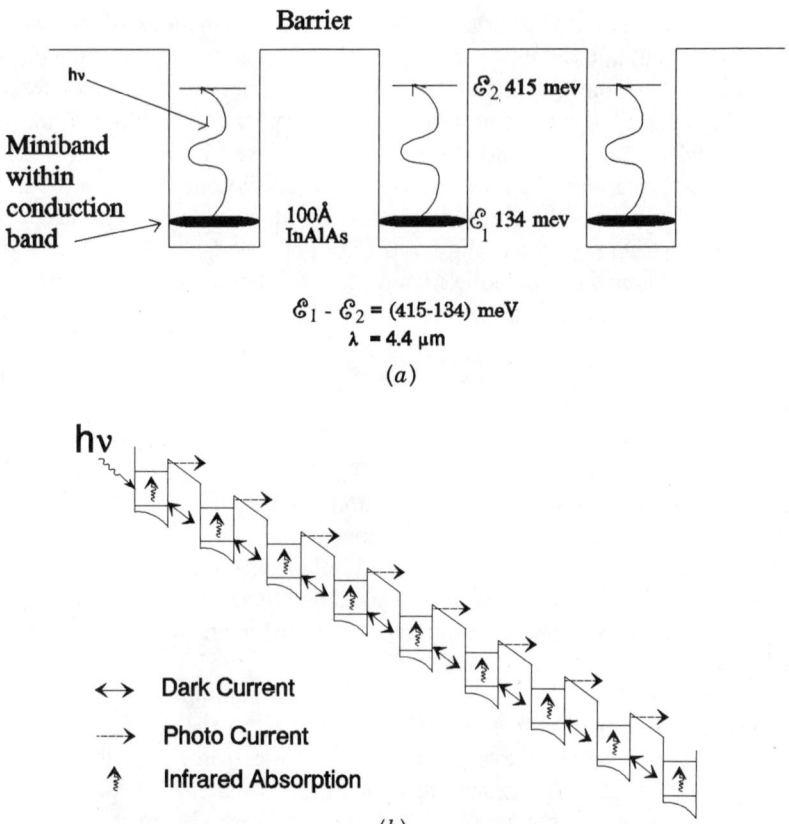

Figure 11.8. Photon absorption in a type-III multiple-quantum-well structure. (*b*) Application of an external electric field lowers the barrier height enough so that tunneling processes allow electrons to escape the well.

analysis does not consider spatial uniformity, which is a limiting performance factor for focal-plane arrays.

One multiple quantum-well structure that has been the subject of much study is the GaAs/Al$_x$Ga$_{1-x}$As multiple quantum well (Levine, 1993), which consists of alternating layers of GaAs and Al$_x$Ga$_{1-x}$As. Normally, the thicknesses are on the order of 100 Å. The bandgap energy of the GaAs ($\mathcal{E}_g = 1.4$ eV) is less than that of the Al$_x$Ga$_{1-x}$As ($\mathcal{E}_g = 1.83$ eV).

The transition between the ground state and the first excited state of the conduction-band well, Fig. 11.8*a*, is used in the GaAs/GaAlAs system to detect radiation at wavelengths around 10 μm. The two processes critical to the operation of this device as a detector are thus intersubband absorption and tunneling (Levin et al., 1987). The typical configuration consists of the device with a voltage applied across it, changing the square-wave potential to a staircase-type potential, as shown in Fig. 11.8*b*. The interaction of a photon of energy $\mathcal{E}_2 - \mathcal{E}_1$ and an electron in the ground state causes a transition to the first excited state. Unless the photon is energetic enough to excite carriers from the ground state directly to the continuum, as in Fig. 11.9a, the next process required is tunneling. This electron in the first excited state has an energy that is very close to the energy of the potential barrier. This condition, coupled with the fact that the potential barrier is narrowed by the application of the external voltage, allows it to tunnel across the barrier. The tunneling mechanism works in biased multiple-quantum-well devices because of the steep electric-field gradient, even without the very thin barriers of the superlattice. Once the electron has escaped the confines of the well, it accelerates toward the side with the applied positive voltage until it is captured by collision with a ground-state electron. The distance that the electron travels before it collides with the other electron is called the *transport length L_t*. If the kinetic energy of the first electron is large enough, it can cause the transition of a second electron to its first excited state. This process repeated many times down the chain of wells will cause avalanche multiplication as illustrated in Fig. 11.9b.

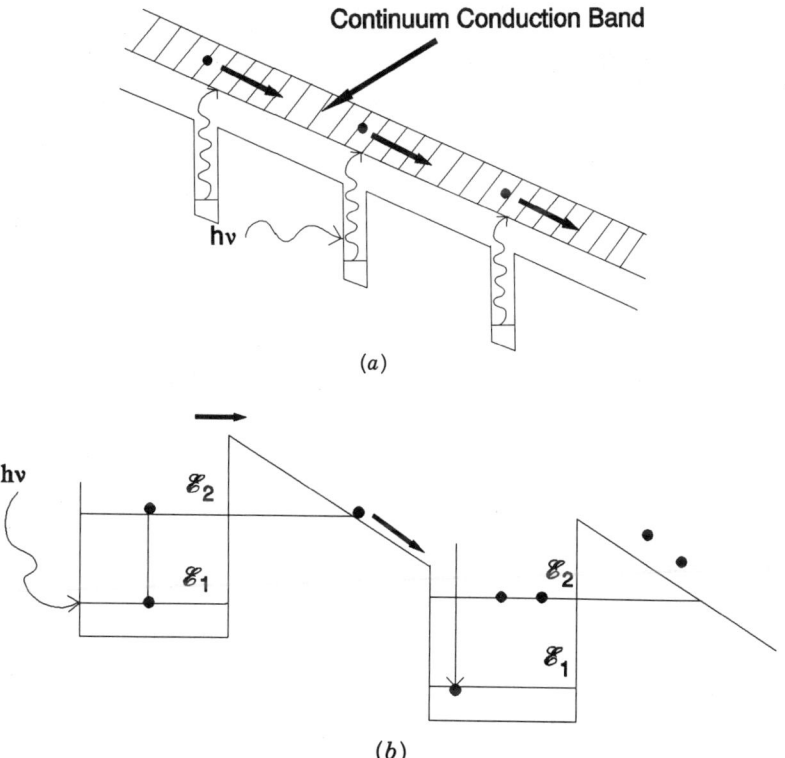

Figure 11.9. (a) Excitation of an electron from the ground state of the well directly to the conduction-band continuum in a biased multiple-quantum-well detector. (b) Avalanche multiplication in a biased multiple-quantum-well detector.

The expressions for detector responsivity and quantum efficiency need further development for us to understand device operation. Following Levine et al. (1987), the photocurrent at any particular location, z, of the sample can be expressed as:

$$i(z) = \frac{q}{h\nu} aP_{\text{tunneling}} L_t(z)\phi(z) \tag{11.9a}$$

where q is the charge on an electron, ν is the optical-radiation frequency, a is the absorption coefficient, and $P_{\text{tunneling}}$ is the probability of an electron in the first excited state tunneling through the potential barrier. The transport length $L_t(z)$ depends on the mean free path L_{mfp} as

$$L_t(z) = L_{\text{mfp}}(1 - e^{-z/L_{\text{mfp}}}) \tag{11.9b}$$

and the optical power (in watts) at a location z is given by

$$\phi(z) = \phi_0 e^{-az} \tag{11.9c}$$

where ϕ_0 is the incident power at $z = 0$.

Integration over a detector of length ℓ_z, the average photocurrent is given by

$$\bar{i} = \frac{1}{\ell_z} \int_0^{\ell_z} i(z)\, dz \tag{11.10}$$

Substituting variables from Eq. (11.9a–c) gives

$$\bar{i} = \frac{qaP_{\text{tunneling}} L_{\text{mfp}} \phi_0}{h\nu\ell_z} \int_0^{\ell_z} (1 - e^{-z/L_{\text{mfp}}}) e^{-az}\, dz \tag{11.11}$$

Performing the integration yields

$$\bar{i} = \frac{qaP_{\text{tunneling}} L_{\text{mfp}} \phi_0}{h\nu\ell_z} \left\{ \frac{1}{a}(1 - e^{-a\ell_z}) + \left(\frac{\ell_z}{L_{\text{mfp}} + a}\right)(e^{(1/L_{\text{mfp}} + a)\ell_z} - 1) \right\} \tag{11.12}$$

The responsivity is defined as

$$\mathcal{R}_i = \bar{i}/\phi_0 = \frac{q\eta}{n\nu} \tag{11.13a}$$

Solving for quantum efficiency yields

$$\eta = \frac{aP_{\text{tunneling}}L_{\text{mfp}}}{\ell_z} \left\{ \frac{1}{a}(1 - e^{-a\ell_z}) + \left(\frac{\ell_z}{L_{\text{mfp}} + a}\right)(e^{(1/L_{\text{mfp}} + a)\ell_z} - 1) \right\}$$

$$(11.13b)$$

Levine et al. (1987) measured a maximum responsivity of 0.52 A/W with a bias of 2.6 V across an active region of 33 periods. By curve fitting the responsivity data, they determined that $L_{\text{mfp}} = 15$ periods, and $\eta = 10\%$. The probability of tunneling out of the first excited state was found to be $P_{\text{tunneling}} = 60\%$. In this experiment, the wells were 65 Å thick, the barriers were 95 Å thick, and L_{mfp} was determined to be 2400 Å.

Choi et al. (1987) demonstrated improved responsivity by using thicker (140 Å) barrier layers. They measured responsivity of 1.9 A/W and $L_{\text{mfp}} = 4500$ Å.

When an electric field is applied across the device, the conduction band tilts, which causes the original square well to have a triangular bottom (see Fig. 11.8b). The ground-state energy level slides into this triangular bottom, thus lowering the energy. The excited states see a generally unchanged overall energy. This results in an increase in energy differences between the ground state and the excited states. This difference in energy between the states results in a blue shift of the absorption wavelength with applied electric field (Harris and Harwit, 1987). This means that, in addition to their application as detectors, quantum-well devices hold potential as electrical, high-speed IR light modulators.

The polarization sensitivity exists in these detectors because the electric field only interacts in one confined direction (Fig. 11.10). Therefore, the quantum efficiency has a maximum theoretical value of 50%. Measured data indicate that $\eta \approx 19\%$ is a typical value. In the case of polarized light, the electric-field component of the incident radiation can be divided into two components, either perpendicular or parallel to the quantum-well layers.

The D^* values predicted for a given peak-response wavelength (λ_p) is given by (Rogalski, 1994):

$$D^* = 1.1 \times 10^6 \exp\left\{\frac{hc}{2\lambda_p kT}\right\} \qquad (11.14)$$

which is plotted in Fig. 11.11. The values of peak spectral D^* measured by Levine et al. (1992a), Levine et al. (1992b), Zussman et al. (1991), and Gunapala et al. (1991) can be superimposed on these curves to confirm that the model is in good agreement with published data. Equation (11.14) can be used to predict the D^* for other quantum-well detectors, given a spectral response and an operating temperature.

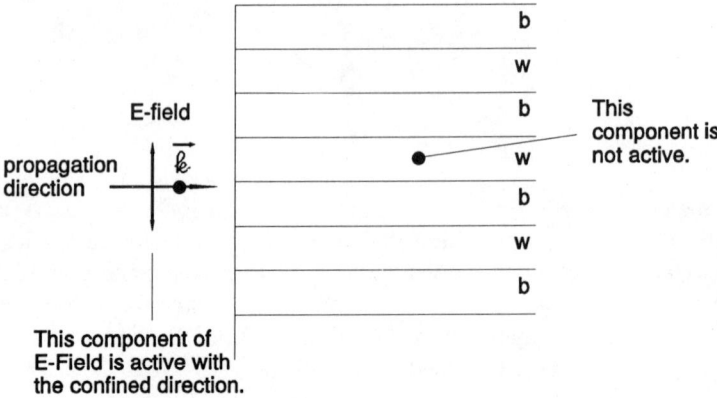

Figure 11.10. Polarization sensitivity of quantum-well materials.

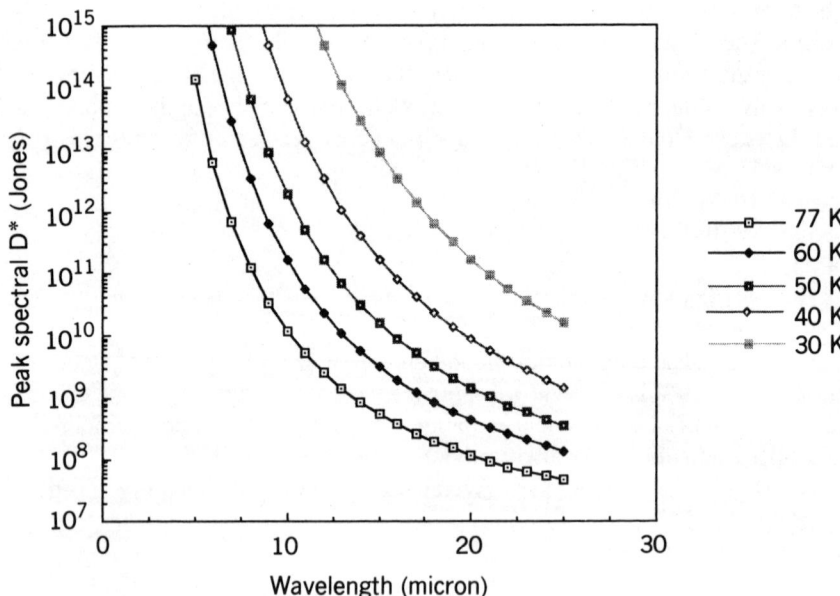

Figure 11.11. Predicted spectral D^* for quantum-well detectors for various operating temperatures.

11.3. SUPERLATTICE OPTICAL DETECTORS

Superlattice structures are simply multiple quantum wells with a thin barrier layer. The barriers are thin enough that the wavefunctions in neighboring wells overlap, causing energy-level perturbation and level shifting. The thin barrier allows tunneling of electrons between wells. This provides a new degree of freedom in the material for optical properties. The effects of thin barriers between wells is as shown in Fig. 11.2a where two wells were illustrated. The wavefunctions exponentially decay in the barrier but perturb the energy levels in the adjacent well because the barrier thickness is small. The perturbation causes the energy levels in the well to split by means of the selection rules into two levels ($\varepsilon \pm \Delta\varepsilon$), where $\Delta\varepsilon$ is the overlap energy resulting from the wavefunction interaction. (*Note*: For a system of three quantum wells, the energy level would be split into three levels; therefore, for m levels, we have m degenerate levels forming a miniband, as was illustrated in Fig. 11.2b).

The superlattice types are divided depending on the relaxation time constants of electrons and holes. Type I is fast, and type II is slower. The superlattice type I follows the definition of Fig. 11.7, where the quantum wells are all confined in the narrow-bandgap layer. In the type-II superlattice, the top of the valence band of the wide-bandgap barrier material is above the conduction band of the narrowband material. Thus valence-band quantum wells are formed in one layer, and conduction-band quantum wells are formed in an adjacent layer. Superlattices can be fabricated using materials with matched lattice constant and with unmatched lattice constants (strained layers).

11.3.1. Lattice-Matched Superlattice

A lattice-matched superlattice is a periodic array of alternating thin layers of well and barrier semiconductors where the lattice constants are matched. Examples of this situation (Fig. 11.4) include GaAs/AlAs and GaSb/AlSb. The most straightforward way to determine the new material's properties is to treat the superlattice as a series of semiconducting layers, with each layer contributing its own characteristic properties. The periodic layers give rise to a periodic variation of electronic potential. Inside each potential well, only certain energy states are available to conduction-band electrons (each state is split into a quasi-continuum miniband, following the Pauli exclusion principle). The situation is similar to that of a crystalline solid, where atoms are periodically spaced. Each atom creates a potential well, and in each atom, only certain energy states are available to electrons. There is an important difference, however, in the case of superlattices. The electronic properties of the atom are determined by nature, while the electronic properties of the superlattice are under the control of the designer of the material. The values of the energy levels available to electrons and holes can be tailored by the appropriate choice of semiconductors, and the widths of their layers ℓ_B and ℓ_W. In addition, the width of each miniband can be tailored by the strength of the interaction between neighboring potential wells. This interaction increases with decreasing width of the layers of the barrier semiconductor. Calculations for a superlattice consisting of 40- to 100-Å-thick layers of GaAs and AlGaAs show that the minibands are much narrower than the bands in any natural bulk semiconductor and also that the minibands are separated from each other in the conduction band by relatively large gap energies.

A special kind of lattice-matched superlattice is the doping superlattice. It is a single bulk semiconductor modulated only by periodic n doping and p doping, giving a large versatility to this type of material. Sometimes these n- and p-doped layers are separated by a layer of intrinsic material, and therefore it is called an n-i-p-i superlattice. In the doping superlattice, a certain fraction of the donors and acceptors are ionized, due to electron hole recombination. The resulting charge in the doped layers (positive charge in each n layer; negative change in each p layer) produces a periodic electrostatic potential, and this potential modulates the conduction and valence bands much as it does in a two-semiconductor superlattice. Any semiconductor can be the host material for a doping superlattice, provided only that both n and p doping are possible. Furthermore, the effective bandgap of the doped suerlattice can be given any value from zero to the bandgap of the undoped host material by choosing the appropriate combination of doping concentration and layer thickness.

Heterostructure superlattices can also be doped to increase the conductivity of the material. However, this n doping must be done selectively only to the large-bandgap layers. One concern is with the tunneling currents through the potential barriers that contribute to the dark-current noise of the detector.

11.3.2. Strained-Layer Superlattice

High-quality superlattices can also be grown from lattice-mismatched materials if the superlattice layers are kept sufficiently thin. For thin lattice-mismatched layers, the mismatch must be totally accommodated by biaxial lattice strain so that no misfit dislocations are generated at the interfaces or at boundaries. Misfit defects cause trapping and recombination sites, lowering material quality. Strained-layer superlattices (SLSs) with no misfit defect have crystalline quality equal to that of conventionally lattice-matched superlattices (i.e., GaAs/AlGaAs). A great deal of research has gone into the tailoring of the optical, electrical, and structural properties of such materials. Strained-layer superlattices can be grown on a wide variety of alloy substrates, with the allowable layer thicknesses in the SLS depending on the amount of lattice mismatch. These provide great flexibility in the selection of specific electronic and optical properties by the choice of layer materials and thicknesses. High crystalline quality can be obtained without requiring a good lattice match to the substrate by putting graded layers between the substrate and the optical-detector materials. This provides a wide choice of substrate properties. The SLS electronic and optical properties are determined by the quantum mechanical wavefunction overlap as in a typical superlattice. What is unique about SLS materials are the elastic stress–strain effects caused by the lattice mismatch. The energy gap is determined by the material composition, the periodicity, and the induced strain.

The idea of strained-layer superlattices is illustrated in Fig. 11.12a, where one crystal has a lattice constant smaller than the other crystal. As seen in Fig. 11.12b, the lattice constants adjust, resulting in periodic variations in strain,

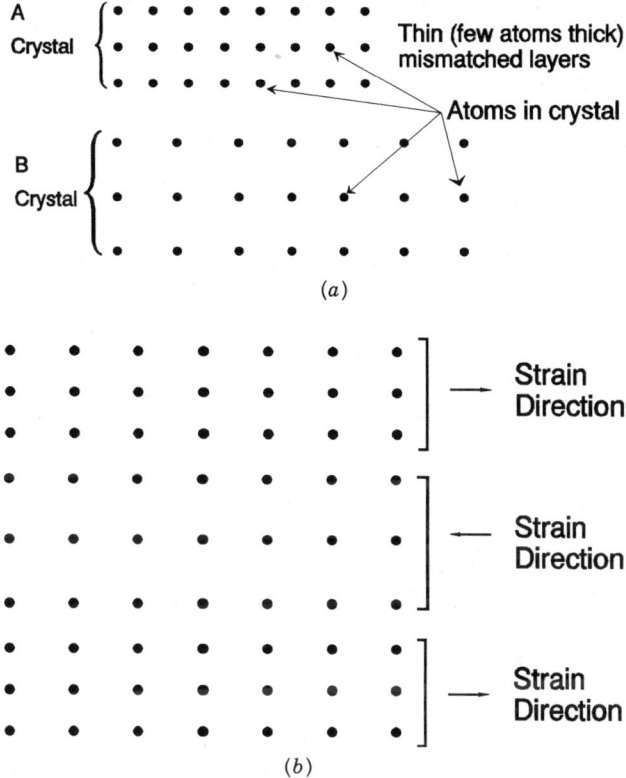

Figure 11.12. (a) Bulk materials to be used in an SLS have different lattice constants. (b) When grown together, the lattice constants adjust, leaving periodic strain variations.

modifying the optical, electrical, and magnetic properties from those observed in the constituent bulk components. The periodic strain variation of the new crystalline lattice affects the energy gap of the semiconductor, as shown in Fig. 11.13. Compressive strain tends to increase the energy gap, while tensile strain has the opposite effect. This is primarily caused by a hydrostatic shift of the conduction band. There is a bump in the $\mathcal{E}$-versus-k curve created by the strain near $k = 0$, which effectively reduces the energy gap, enabling longer-wavelength radiation to be detected. The strain also increases the mobility of holes in the valence band because the effective mass is reduced by band bending. Recall that the effective mass is related to energy-band curvature by

$$\frac{1}{m^*} = \frac{1}{h^2} \frac{\partial^2 \mathcal{E}}{\partial k^2} \tag{11.15}$$

The result of Eq. (11.15) is that the rate of change of the slope in Fig. 11.13 is larger in the region near $k = 0$ than in the rest of the band.

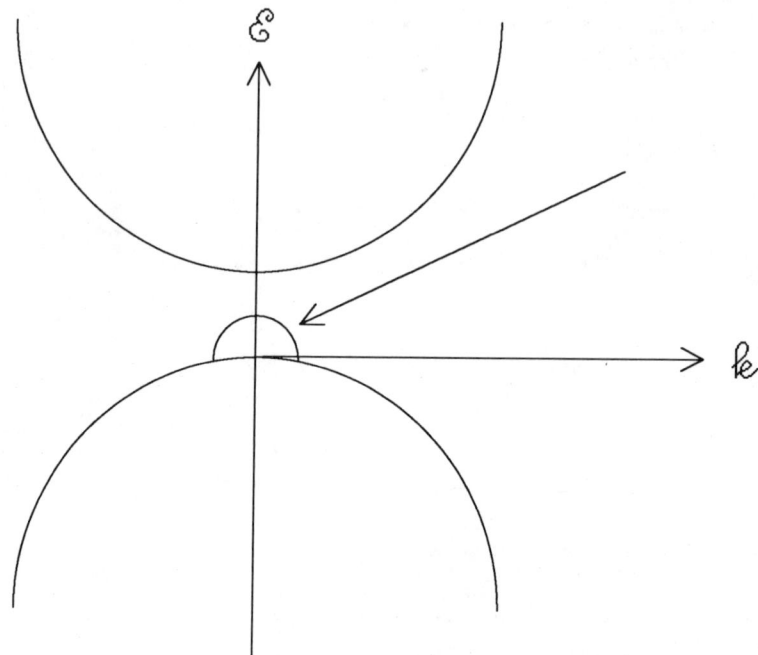

Figure 11.13.   Effect of periodic strain on the $\mathcal{E}$-vs.-k curve, reducing the energy gap.

A major research effort was required for the development of a SLS material to replace HgCdTe in the 8–14-μm region. One advantage of SLS materials based on the III-V material group is that the strong covalent bonds make the material better able to withstand higher growth temperatures, leading to a chemically stable and spatially uniform material, which is free of native defects. The second advantage is that compositional variation in SLS does not have a strong effect on the bandgap, leading to a spatially more uniform spectral response than HgCdTe.

Bulk-detector materials must be grown on a substrate that provides a good match of the detector-material lattice constant to prevent the growth of dislocations and defects from strain. Alternatively strained-layer superlattice detectors can be grown on nonlattice-matched substrates, and the lattice constant is set by the SLS itself. The SLS is typically made of as many as 50 alternating layers. The SLS lattice constant in the plane parallel to the superlattice interface determines how well it matches the substrate. The thicknesses of the layers must be less than some critical thickness where the size would start to induce defects in the lattice. Osbourn (1984) and Matthews and Blakeslee (1974, 1975, 1976) developed theoretical relationships between the relative lattice-constant mismatch and the critical thickness for typical materials such as GaAs and InAs. The fabrication process can cope with a lattice-constant mismatch of approximately 7% to 9% for the layer sizes required for an SLS

material. Similarly, the entire SLS stack has a critical thickness determined by
the substrate lattice and average SLS lattice mismatch, which is often the practical limit to the number of layers used.

If a large mismatch exists between the SLS and substrate, a buffer layer
must be placed between them, which is graded in terms of lattice constant from
the substrate to the SLS lattice values. This grading is accomplished by varying
the composition of the buffer material as it is grown. The buffer is grown very
thick, so that the dislocations are not important at the interface between the
SLS and buffer. The structure is shown in Fig. 11.14 (Hughes, 1987), where
each layer in the buffer is kept thin enough to maintain good crystalline quality,
and dislocations from the buffer layer of the SLS migrate to the edges. Thus,
only the first few layers contain defects.

By varying the thickness and composition of the SLS, the lattice constant
can be determined, affecting the energy gap and spectral response of the detector. The same lattice constant can be made with a variety of materials and
thicknesses. Theoretical studies of SLS structures using the Kronig–Penny
models have given reasonable predictions of bandgap energies (Osbourn,
1984). The elastic strains in the layers can be varied by changing the ratio of

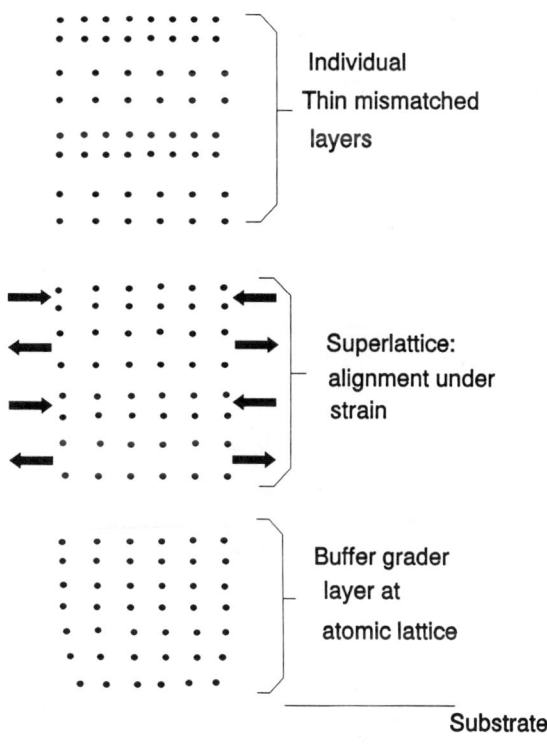

Figure 11.14. SLS structure on a mismatched substrate, showing the required buffer layer.

their thicknesses, also affecting the energy gap. The strain induces energy shifts of 10 to 100 meV (Osbourn, 1986) in the energy gap. The quantum-size effects generally counteract the strain effects and act to increase the energy gap, but it is usually possible to choose a layer thickness large enough to minimize these effects and yet thin enough to be below the critical thickness.

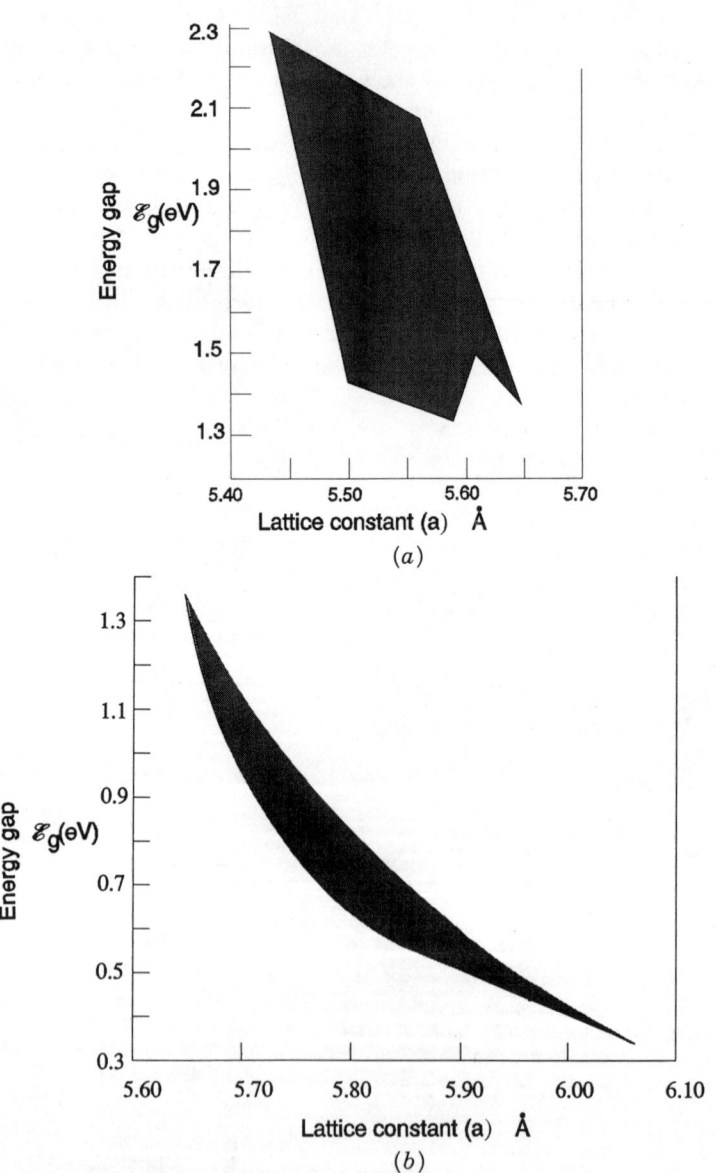

Figure 11.15. (a) Possible energy gaps using GaAsP material system, as a function of lattice constant. (b) Possible energy gaps using InGaAs material system, as a function of lattice constant.

A wide variety of energy gaps is achievable with the same material by changing composition and thickness. Figure 11.15a shows the available energy gaps by using $GaAs_xP_{1-x}/Ga_yAs_{1-y}P$, which can be controlled from about 1.31 to 2.28 eV. This is made possible by varying the compositional thickness of GaAsP. Figure 11.15b similarly shows the $In_xGa_{1-x}As/In_yGa_{1-y}As$ system, producing energy gaps from about 0.4 to 1.4 eV.

Figure 11.16 shows the variation of energy gap when the lattice constant is kept fixed but the thickness and compositions are varied. The ratio of the layer thicknesses was kept constant at two. Therefore, several different combinations of compositions and layer thicknesses can have not only the same lattice parameter, but also the same energy gap. This provides the possibility of changing other material properties, such as carrier mobility, independently of both of these quantities. The effective mass in the direction perpendicular to the SLS interface is determined by the curvature of the energy bands, which, in turn, determines the spectral bandwidths and the material response. Controlling the curvature of the conduction/valence band not only affects mass but also the spectral bandwidth.

As seen in Fig. 11.17, the spread in energy levels varies with layer thickness, with thinner layers having the wider bandwidths. Wider bandwidths have larger effective masses. This can be useful in the material development because the effective mass in the normal direction can be varied by changing the layer thickness. In addition, if the thickness ratio is held constant, as illustrated in Fig. 11.16, the effective mass can be controlled by the layer thicknesses while

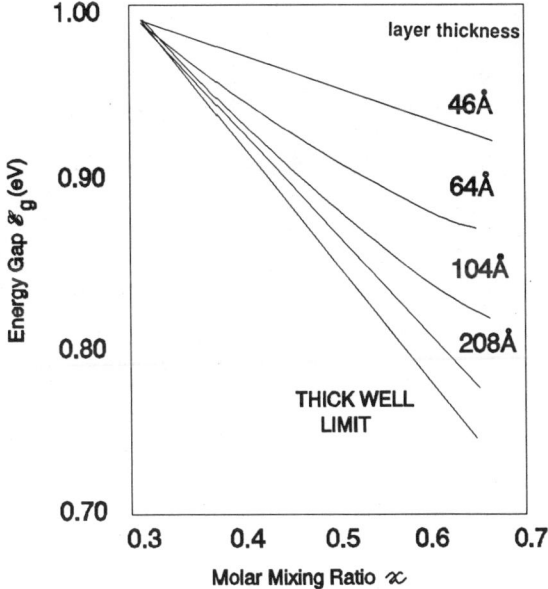

Figure 11.16. Energy gap as a function of mixing ratio x and (narrower) layer thickness. (After Osbourn, 1984.)

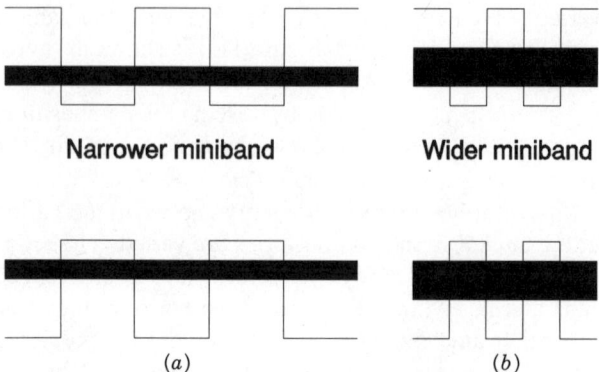

Figure 11.17. Effect of layer thickness on the energy-level spread in the miniband. (*a*) Thick layers; (*b*) thin layers.

the lattice parameter and energy gap remain fixed. This would result in control of the carrier mobility and diffusion length along the perpendicular direction. Thus, control of structure (lattice constant), optical property (energy gap), and transport property (carrier mobility) is possible in the direction perpendicular to the SLS direction.

11.3.2.1. InAsSb. InAsSb used in strained-layer superlattice is a novel III-V semiconductor material with the potential for extended-wavelength response (LWIR) intrinsic-detector applications. Osbourn (1984) pointed out that the InAsSb SLS material offers a number of potential advantages in the LWIR region. The first advantage is the use of stable III-V materials, which are typically the easiest materials to work with. Second, theoretical studies show that the energy gap varies slowly with composition so that small lateral compositional variations across the detector array will not result in spectral variations from detector to detector. Third, the tunneling current for the SLS is considerably reduced in a low band-gap material compared to similar-band-gap bulk materials, resulting in less dark-current noise.

Figure 11.4 summarizes the bandgap wavelength cutoffs and lattice constants available in III-V alloy materials, showing that the InAsSb system is the only III-V material that can have cutoff wavelengths in the LWIR region. However, bulk InAsSb only detects out to 9 μm for typical alloy compositions. The effects of alloy composition are shown in Fig. 11.18 for InAsSb, where energy gap is plotted against composition (Kurtz et al., 1988a). The nominal 77-K curve is shown for the bulk material alloy, where the cutoff wavelength is about 9 μm ($x \sim 0.75$). The SLS shifts the cutoff wavelength into the 12–14-μm range, increasing the usefulness of this material for LWIR applications, even for relatively low Sb concentrations ($x \geq 0.8$).

11.3.2.2. Strain-Caused Energy-Level Shifts. Since the SLS superlattices are grown from lattice-mismatched alloy layers with layer thicknesses thin

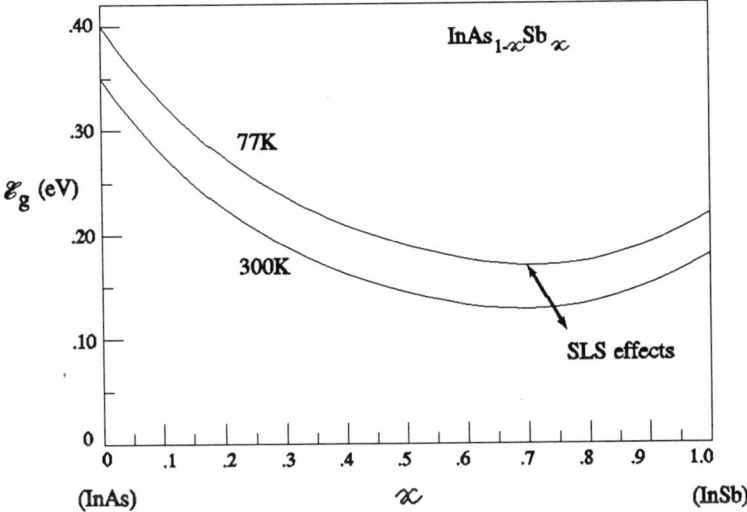

Figure 11.18. Energy gap of the InAsSb alloy system as a function of mixing ratio. SLS structures lower the energy gap compared to the bulk. Mixing ratio of 0.1 corresponds to a lattice constant of 6.1 Å, $x = 0.35$ corresponds to a lattice constant of 6.2 Å, $x = 0.55$ corresponds to a lattice constant of 6.3 Å, $x = 0.8$ corresponds to a lattice constant of 6.4 Å. (After Kurtz et al., 1988a.)

enough to allow complete elastic strain accommodation, the SLS layers with larger bulk lattice constants are under biaxial compression, while the layers with smaller bulk lattice constants are under biaxial tension. The typical SLS geometry is a layer of $As_{1-x}Sb_x$ between InSb layers. The lattice constant for InSb (as shown in Fig. 11.18) is 6.5 Å. Thus, the $InAs_{1-x}Sb$ layers are in tension, and the InSb layers are in compression. The layers under biaxial tension exhibit a reduction (compared to the bulk) conduction-band minimum energy and a splitting of the bulk light- and heavy-hole bands. The layers under biaxial compression exhibit an increased value of conduction-band minimum energy and a lowering of the valence-band level. In addition, a splitting of the bulk light- and heavy-hole bands occurs, with the heavy-hole band energy higher than the light-hole band (Kurtz et al., 1988b).

For the case of $InAs_{1-x}Sb_x/InSb$, the energy gap of the small-bandgap-material ($InAs_{1-x}Sb_x$) is decreased by tension strain. Therefore, the SLS material responds to a longer wavelength than the corresponding alloy material. The effects of pressure on the energy gap is on the order of a few (10^{-11}) eV per pascal (Newton/m^2), depending strongly on the crystal lattice orientation (Zallen and Paul, 1964). Typical energy-gap shifts are on the order of 50 meV, so the pressure is very high. This superlattice is a type II as shown in Fig. 11.7, where the optical transitions for the longest wavelength take place as an indirect transition, as shown in Fig. 11.19. Most of the bandgap reduction is caused by the type II energy-level offset.

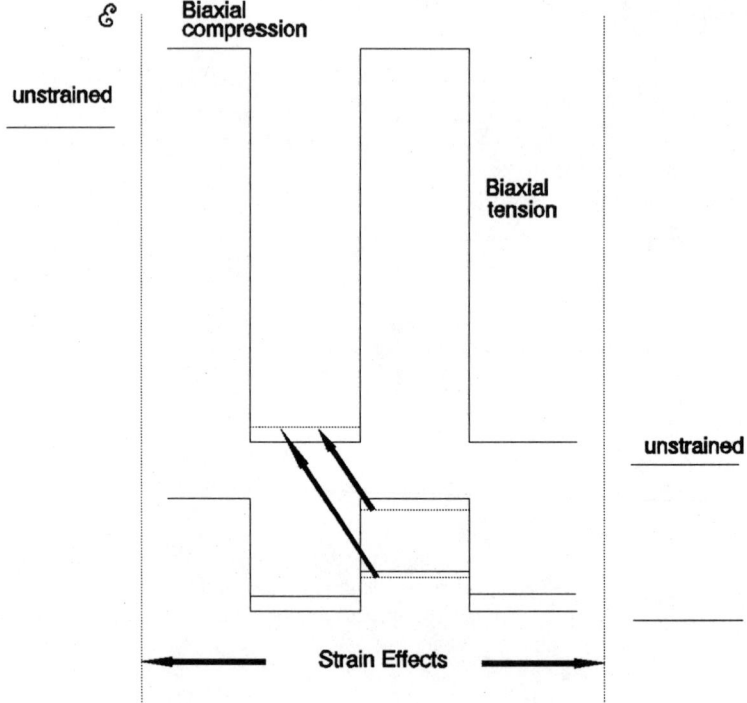

Figure 11.19.   Strain effects on the $InAs_{1-x}Sb/InSb$ indirect transition.

The absorption in the type II SLS is sensitive to the layer thicknesses, because the wave functions for the electrons and holes are separated, and the strength of the transition is influenced accordingly. The tails of the wave functions decrease with increased barrier thickness and increased effective mass. Barrier-layer (InSb) thicknesses are chosen so that good absorption can take place. A tradeoff exists between barrier thickness, long-wavelength cutoff, and good absorption.

The carrier-transport properties are anisotropic because of the barrier in the structure. The absorption length $(1/a)$ is on the order of micrometers in $InAs_{1-x}Sb_x$. However, the *PN* junction depletion region is only about 1 μm wide and, to obtain good quantum efficiency (photon absorption), the carrier-diffusion length must be long. Recall that the diffusion length was determined by mobility, temperature, and carrier lifetime, and because the infrared detectors must be cooled (at least 77 K), the only means of making a good photovoltaic detector is to have high mobility and a long lifetime. Prototype LWIR InAsSb SLS detectors have been demonstrated experimentally. The absorption is well beyond the cutoff wavelength of the alloy materials (i.e., 14 μm). However, these materials will require significant improvement in detectivity to compete effectively against the much better developed HgCdTe technologies.

11.4 APPLICATIONS

The development of avalanche photodiodes (solid-state photomultipliers) has been underway since 1985. The SLS avalanche photodiodes work on the basis that electrons gain a kinetic energy and are accelerated by a much greater electric field, thus causing multiplication. Figure 11.20 shows the energy-band diagram. The holes do not cause multiplication because they do not acquire sufficient energy. The material used is typically an InGaAs/GaAsAs SLS system with a 1000-Å buffer layer of InAlGaAs grown on the edge. The GaAs substrate is followed by 4000 Å of the SLS structure. Beryllium (Be) was used to fabricate a $P+$ region down from the buffer layer. Thus, carriers move in the plane of the surfaces, which provides higher mobility than in the perpendicular direction. An AlGaAs lattice-matched window was grown on the SLS structure to passivate the surface. At low reverse bias, the detector had uniform response over a large area and a quantum efficiency of 50% for a spectral region of 775 to 850 nm. Although we observed impact ionization and signal multiplication of the electrons with light bias, these devices have not proved

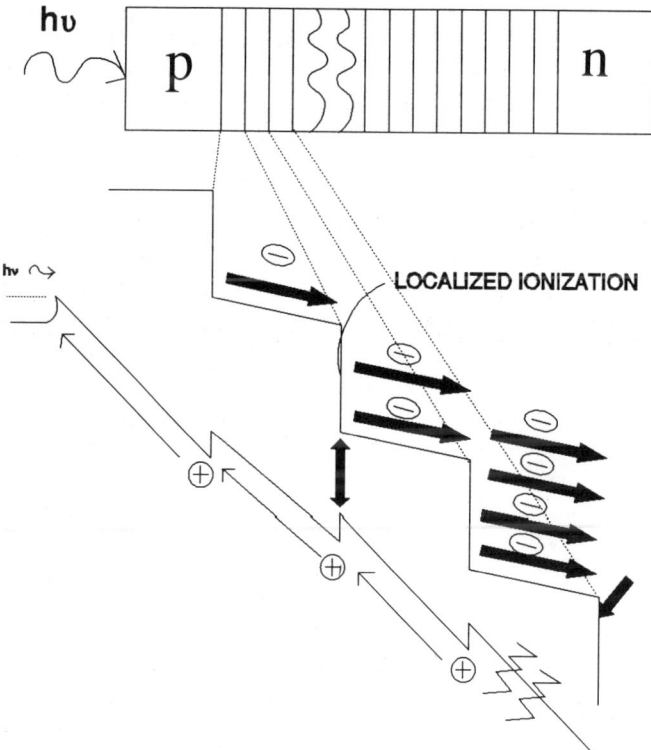

Figure 11.20. Energy-band diagram of a solid-state photomultiplier.

to be successful in laboratory demonstration as solid-state multipliers. They are, however, conceptually interesting.

GeSi/Si photodetectors have also been extensively studied at the Jet Propulsion Laboratory and elsewhere (Lin et al., 1990). This vertical SLS acts as a light-guiding absorbing layer that can be used in fiber-optics applications. The thin layers minimize the transport length in the direction perpendicular to the interfaces, facilitating high-speed operation. A p-n junction was fabricated in a $In_{0.1}Ga_{0.9}As$, grown on an n-type GaAs substrate. The n-type InGaAs acted as a buffer layer and was graded to match the lattice constant of the superlattice. This SLS structure was initially doped with silicon to obtain n-type material and then with beryllium (Be) to obtain p-type material. A p-type window layer covers the SLS. This structure is a double-heterojunction photodiode. Experimental results agreed with the theoretical values for the bandgap (about 1.2 eV).

11.5. FUTURE OUTLOOK

Recent developments have enhanced the potential of quantum-well-detector technology for long-wavelength focal-plane array applications. A 128×128 focal-plane-array camera has been demonstrated at 15 μm (Gunapala et al., 1995a) using GaAs/AlGaAs quantum wells. The noise-equivalent temperature difference is 30 mK against a 300-K background and the uniformity is 99.99%. Other efforts have resulted in a demonstration of a 256×256 focal-plane camera at 9.5 μm, suitable for integration into portable cameras, and the development of an IR detector in InGaAs/GaAs that responds out to 20.5 μm (Gunapala et al., 1995b).

Quantum-well detectors are strong competition for the established HgCdTe technology. Fabrication costs for quantum-well focal-plane arrays have been reduced two orders of magnitude compared to HgCdTe arrays (S. D. Gunapala, 1995, private commun.). Even state-of-the-art HgCdTe long-wave detectors suffer from high nonuniformities and high dark currents. Quantum-well arrays exhibit very low noise, especially at low operating temperatures, and provide an avenue for photon-counting applications at long wavelengths.

REFERENCES

Ackley, D. E., J. Hladky, M. J. Lange, S. Mason, G. Erickson, G. H. Olsen, V. S. Ban, S. R. Forrest, and C. Staller, "Linear Arrays of InGaAs/InP Avalanche Photodiodes for 1.10–1.7 μm," *SPIE Proc.* **1308**, 261–272 (1990).

Beck, W., "Image Enhancement and Moving Target Detection in IR Image Sequences," *SPIE Proc.* **2020**, 187–195 (1993).

Capasso, F., W. T. Tsang, and G. F. Williams, "Staircase Solid-State Photomultipliers and Avalanche Photo-Diodes with Enhanced Ionization Rates Ratio," *IEEE Trans. Electron Devices* **ED-30**, 381 (1983).

Choi, K. K., B. F. Levine, C. G. Bethea, J. Walker, and R. J. Malik, "Multiple Quantum-Well 10 μm GaAs/Al$_x$Ga$_{1-x}$As Infrared Detector with Improved Responsivity," *App. Phys. Lett.* **50**, 1814–1816 (1987).

Ferry, D. K., *Gallium Arsenide Technology*, H. W. Sams, Indianapolis, 1985.

Gunapala, S. D., B. F. Levine, L. Pfeiffer, and K. West, "Dependence of the Performance of GaAs/AlGaAs Quantum Well Infrared Photodetectors on Doping and Bias," *J. Appl. Phys.* **69**, 6517–6520 (1991).

Gunapala, S. D., J. S. Park, G. Sarusi, T. L. Lin, J. K. Liu, P. D. Maker, R. E. Muller, B. F. Levine, C. A. Shott, and T. Hoelter, "128 × 128 GaAs/AlGaAs Quantum Well Infrared Photodetector Focal Plane Array for Imaging at 15 μm," submitted to IEEE Electron. Device Letters (1995a).

Gunapala, S. D., J. S. Park, T. L. Lin, J. K. Liu, and W. T. Pike, "20.5 μm Cutoff Long Wavelength InGaAs/GaAs Quantum Well Infrared Photodetectors," in *Proceedings of the First International Symposium on Long Wavelength Infrared Detectors and Arrays*, F. Radpour and V. R. McCray (eds.), The Electrochemical Society, Pennington, NJ, 1995b, pp. 46–52.

Harris, J. S., Jr., and A. Harwit, "Observation of Stark Shifts in Quantum-Well Intersubband Transitions," *Appl. Phys. Lett.* **50**, 685–687 (1987).

Hasnain, G., B. F. Levine, D. L. Sivco, and A. Y. Cho, "Mid-Infrared Detectors 3–5 μm Band Using Bound to Continuum State Absorption in InGaAs/InAlAs Multiquantum-Well Structures," *Appl. Phys. Lett.* **56**, 770–772 (1990).

Hughes, R. C., "III-V Compound Semiconductor Superlattices for Infrared Photodetector Applications," *Opt. Eng.* **26**, 249–255 (1987).

Kinch, M. A., and A. Yariv, "Performance Limitations of GaAs/AlGaAs Infrared Superlattices," *Appl. Phys. Lett.* **55**, 2093–2095 (1989).

Kurtz, S. R., L. R. Dawson, R. M. Biefled, and G. C. Osbourn, "InAsSb Strained Layer Superlattices: A New Class of Far Infrared Materials," *SPIE Proc.* **930**, 101–113 (1988a).

Kurtz, S. R., G. C. Osbourn, R. M. Biefeld, L. R. Dawson, and H. J. Stein, "Extended Infrared Response of InAsSb Strained-Layer Superlattices," *Appl. Phys. Lett.* **52**, 831 (1988b).

Lacoe, R. C., M. J. O'Loughlin, D. A. Gutierrez, W. L. Bloss, R. C. Cole, P. A. Dafesh, and M. Isaac, "Modified Quantum-Well Infrared Photodetector Designs for High Temperature and Long Wavelength Operation," *SPIE Proc.* **1685**, 230–240 (1992).

Levine, B. F. "Quantum-Well Infrared Photodetectors," *J. Appl. Phys.* **74**, 81 (1993).

Levine, B. F., K. K. Choi, G. G. Bethea, J. Walker, and R. J. Malik, "New 10 μm Infrared Detector Using Intersubband Absorption in Resonant Tunneling GaAlAs Superlattices," *Appl. Phys. Lett.* **56**, 1092–1094 (1987).

Levine, B. F., G. Hasnain, G. G. Bethea, and N. Chand, "Broadband 8–12 μm High Sensitivity GaAs Quantum-Well Infrared Photodetectors," *Appl. Phys. Lett.* **54**, 2704–2706 (1989).

Levine, B. F., G. B. Bethea, G. Hasnain, V. O. Shen, E. Pelve, R. R. Abbott, and S. J. Hseih, "High Sensitivity Low Dark Current 10 μm GaAs Quantum-Well Infrared Photodetectors," *Appl. Phys. Lett.* **56**, 851–853 (1990).

Levine, B. F., A. Zussman, J. M. Kuo, and J. de Jong, "19 μm Cutoff Long-Wavelength GaAs/AlGaAs Quantum-Well Infrared Photodetectors," *J. Appl. Phys.* **71,** 5130–5135 (1992a).

Levine, B. F., A. Zussman, S. D. Gunapala, M. T. Asom, J. M. Kuo, and W. S. Hobson, "Photoexcited Escape Probability, Optical Gain, and Noise in Quantum Well Infrared Photodetectors," *J. Appl. Phys.* **72,** 4429–4443 (1992b).

Li, S., Y. Wang, J. Chu, and P. Ho, "A Normal Incidence P-Type Strained Layer 0.3 Ga 0.7 As/In 0.52 Al 0.48 As Quantum-Well Infrared Photodetectors with Background Limited Performance at 77 K," *Proc. SPIE* **2225,** 139–150 (1994).

Lin, R. L., T. Maserjian, N. Krabach, A. Ksendzov, M. L. Huberman, and R. Terhune, "Novel $Si_{1-x}Ge_x$/Si Heterojunction Internal Photoemission Long Wavelength Infrared Detectors," Infrared Detector Workshop, JPL 1990.

Matthews, J. W., and A. E. Blakeslee, "Defects in Epitaxial Multilayers, I. Misfit Dislocation," *J. Cryst. Growth* **27,** 118 (1974).

Matthews, J. W., and A. E. Blakeslee, "Defects in Epitaxial Multilayers, II. Dislocation Pile-Ups, Threading Dislocations, Slip Lines and Cracks," *J. Cryst. Growth* **29,** 273 (1975).

Matthews, J. W., and A. E. Blakeslee, "Defects in Epitaxial Multilayers, III. Preparation of Almost Perfect Multilayers," *J. Cryst. Growth* **32,** 265 (1976).

Osbourn, G. C., "InSb Strained-Layer Superlattices for Long Wavelength Detector Applications," *J. Vac. Sci. Tech.* **B2,** 176 (1984).

Osbourn, G. C., "Recent Trends in III-V Strained Layer Superlattices," *J. Vac. Sci. Tech.* **B4,** 1423 (1986).

Paska, Z. F., J. Y. Andersson, L. Lundquist, and C. Olsson, "Growth and Characterization of AlGaAs/GaAs Quantum Well Structures for the Fabrication of Long Wavelength Infrared Detectors," *J. Cryst. Growth* **107,** 845–849 (1991).

Peyghambarian, N., S. Koch, and A. Mysyrowicz, *Introduction to Semiconductor Optics,* Prentice Hall, Englewood Cliffs, NJ, 1993.

Rogalski, A., "GaAs/AlGaAs Quantum Well Infrared Photoconductors Versus HgCdTe Photodiodes for Long-Wavelength Infrared Applications," *SPIE Proc.* **2225,** 118–129 (1994).

Saleh, B. E. A., and M. C. Teich, *Fundamentals of Photonics,* Wiley, New York, 1991.

West, L. C., and S. J. Eglash, "First Observation of an Extremely Large-Dipole Infrared Transition Within the Conduction Band of a GaAs Quantum-Well," *Appl. Phys. Lett.* **46,** 1156–1158 (1985).

Zallen, R., and W. Paul, "Band Structure of Gallium Phosphide from Optical Experiments at High Pressure," *Phys. Rev.* **134,** A1628 (1964).

Zussman, A., B. F. Levine, J. M. Kuo, and J. de Jong, "Extended Long-wavelength $\lambda = 11$–15 μm $GaAs/Al_xGa_{1-x}As$ Quantum-Well Infrared Photodetectors," *J. Appl. Phys.* **70,** 5101–5107 (1991).

BIBLIOGRAPHY

Kurtz, S. R., L. R. Dawson, T. E. Zipperian, R. D. Whaley, "High Detectivity ($>71 \times 10^{10}$) InAsSb Strained-Layer Superlattice Photovoltaic Infrared Detectors," *IEEE Electron Device Letters* **7,** 54–56 (1990).

Manasreh, M. O., ed., *Semiconductor Wells and Superlattices for Long-Wavelength Infrared Detectors*, Artech, Boston, 1993.

Willardson, R. K., and A. C. Beer, "Principles and Applications of Semiconductors Strained-Layer Superlattices," in *Semiconductors and Semimetals* (R. K. Willardson and A. C. Beer, eds.), Academic, New York, 1987, pp. 459–503.

PROBLEMS

11.1. Calculate the de Broglie wavelength for GaAs.

11.2. For a well width of 39 angstroms, and an effective mass that is 20% of the rest mass, what are the various energy levels that can exist in a GaAs superlattice?

11.3. Explain the reason for minibands in quantum wells and alternatively in superlattices. Do they have the same reason for appearing?

11.4. How is the effective mass influenced by the well thickness?

11.5. You are a detector manufacturer, and your goal is to build a solid-state photomultiplier. You want to replace Ge ($\mathcal{E}_g = 0.7$ eV) as the material of choice because of excess noise in the avalanche-multiplication process. From the data of Fig. 11.4 of energy gap versus lattice constant for various compounds, list all of the choices of materials that could replace Ge. Which could be superlattices, and which could work as strained superlattices?

12

Infrared Search Systems

12.1. INTRODUCTION

This chapter introduces and explains the engineering tradeoffs inherent in the design of infrared search systems. We begin by comparing the functionalities of search systems and thermal-imaging systems. Indeed, the same infrared imager might perform both functions, depending on the range and distance to the target, but the design philosophies differ, depending on which task the system is optimized for. Thermal-imaging systems are considered separately in Chapter 14.

Search systems are designed to detect and locate a target that has a pre-scribed minimum intensity within a specified search volume. The search system operates on point sources, because the targets are first detected at maximum range. Any finite-sized source will act as a point target when it is first detected. At the range where the target is first detected, the angular size of the source is smaller than the angular resolution of the search system. The end product from a search system is a statistical decision about the existence of a target within the search volume. The design equations are concerned with maximizing the probability of correct detection and minimizing the false alarm rate and probability of missing the target. These concerns are met by maximizing the signal-to-noise ratio (SNR). The SNR will increase for larger collection apertures and for longer integration times. Thermal-imager systems, on the other hand, operate on extended targets, which are typical of targets at closer ranges than the search systems. The function of a thermal imager is to provide a spatially resolved flux variation related to temperature and emissivity differences across an extended target. The end product of a thermal-imager system is typically an image displayed on a video monitor or on a hard copy.

The discussion of performance and design issues for both search systems and thermal-imaging systems will be cast in terms of the detector D^*. We develop equations that describe the performance of these systems, initially making the assumption that the dominant noise is internally generated detector noise (e.g., Johnson noise). The crux of this assumption is that the D^* of the detector does not depend on the $F/\#$ of the optic that illuminates the detector. This assumption is consistent with operation in the internal-detector-noise-limited case. However, for background-noise-limited (BLIP) operation, the D^* achieved by the detector, and ultimately the SNR achieved by the system, depend on the $F/\#$ of the system that illuminates the detector. We will bring in the BLIP case separately for both system types, taking into account the dependence of D^* on $F/\#$, and noting the modifications that result in the design equations.

12.2. SCAN FORMATS

As we noted in Chapter 1, scanning is often used in infrared systems to cover a two-dimensional field of view (FOV) with fewer detectors than would be required for a staring system. We denote HFOV as the complete horizontal FOV and VFOV as the complete vertical FOV. The instantaneous field of view (IFOV) is that portion of object space being sensed by the detector(s) at any particular instant in time, the instantaneous footprint of the detector(s) in object space. We denote by HIFOV and VIFOV the horizontal and vertical IFOVs, respectively. For convenience, we usually consider the IFOVs to move in a (stationary) object space, when it is actually the image that is scanned across the stationary detectors.

A single detector, or an array that does not cover the whole vertical field of view (VFOV), is required to have two directions of scan motion, while a one-dimensional array of sufficient height to cover the VFOV need only scan in one direction.

12.2.1. Single-Detector Scan Formats

We begin our discussion with the single-detector scan format, as seen in Fig. 12.1. The IFOV scans in a raster format; horizontally across each line, then drops to the beginning of the next line until the entire scene is scanned. The entire scene is scanned in τ_f, the frame time. The scene to be scanned has

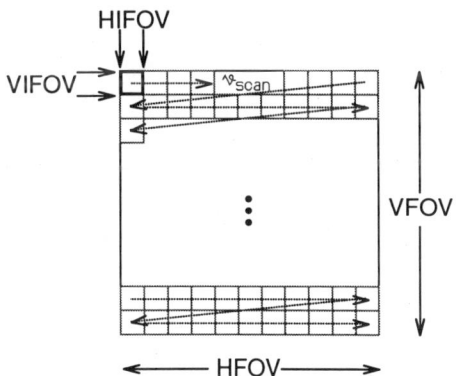

Figure 12.1. Single-detector scan format.

angular dimensions HFOV × VFOV. We assume in our development that no time is wasted in the scan of the scene (scan efficiency $\eta_{sc} = 100\%$). Scan inefficiencies are, however, unavoidable and arise in various places including overlap between scan lines; finite retrace time to move the detector to the start of the next line; and overscan of the IFOV beyond the region to be sensed. Scan inefficiencies cause the data (and noise) bandwidth to be higher than our equations will indicate. This can be compensated for by making the dwell time smaller by a multiplicative factor of η_{sc}, which leads to an increased noise bandwidth and hence a reduced SNR.

We can write an expression for the number of horizontal lines (n_L) that make up the scene:

$$n_L = \frac{\text{VFOV}}{\text{VIFOV}} \tag{12.1}$$

We can now express the time taken to scan one particular line, τ_L:

$$\tau_L = \frac{\tau_f}{n_L} = \tau_f \times \frac{\text{VIFOV}}{\text{VFOV}} \tag{12.2}$$

The dwell time τ_d is the line time divided by the number of horizontal IFOVs contained in that line:

$$\tau_d = \frac{\tau_L}{\text{HFOV/HIFOV}} \tag{12.3}$$

We can further interpret the dwell time in terms of v_{scan}, the along-scan (horizontal) angular velocity of the IFOV in mrad/s. One line, having an angular subtense HFOV, will be scanned in a time τ_L:

$$v_{scan} = \frac{1}{\tau_L} \times \text{HFOV} \tag{12.4}$$

The IFOV collects information about any point in the image for a time τ_d. Thus, as seen in Fig. 12.2, dwell time is the time that the HIFOV remains over any particular scene feature:

$$\tau_d = \text{HIFOV}/v_{scan} \tag{12.5}$$

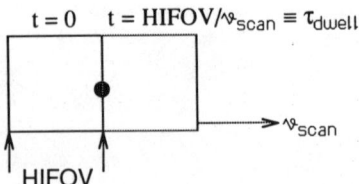

Figure 12.2.   Relation of dwell time to scan velocity and HIFOV.

One other useful relationship for dwell time is found by substituting Eq. (12.2) into Eq. (12.3):

$$\tau_d = \frac{\tau_f}{((\text{VFOV}/\text{VIFOV}) \times (\text{HFOV}/\text{HIFOV}))} \qquad (12.6)$$

The factor in the denominator of Eq. (12.6) is the number of pixels, n_p, contained in the scene:

$$n_p = \frac{\text{VFOV}}{\text{VIFOV}} \times \frac{\text{HFOV}}{\text{HIFOV}} \qquad (12.7)$$

A pixel corresponds to the smallest resolution element in the received picture.

We are interested in the dwell time, from a systems point of view, because the signal bandwidth Δf for the detector is

$$\Delta f \cong \frac{1}{2\tau_d} \qquad (12.8)$$

which is also the noise-equivalent bandwidth. The equality in Eq. (12.8) is approximate, depending on the exact form of the temporal impulse response. Assuming white noise, the root-mean-square (rms) noise is proportional to the square root of the bandwidth:

$$v_n \propto \sqrt{\Delta f} \cong \sqrt{\frac{1}{2\tau_d}} \qquad (12.9)$$

Any modifications to the system design that yield a longer dwell time will reduce the noise, but tradeoffs must be made. For example, given a fixed number of detectors, the frame time can be increased (thus lengthening the dwell time), which improves the SNR. However, typically the frame time is set by the overall task of the sensor system, which cannot afford to take an arbitrarily long time to scan through a scene. There is usually a minimum acceptable frame rate, based on the expected rate of change of scene elements. For a fixed frame time, having more detectors increases the dwell time on a scene element at the cost of a larger number of detectors. This is the most common approach for increasing the SNR. We investigate some options in the next section.

12.2.2. Multiple-Detector Scan Formats

For multiple-detector systems, the detectors can be arranged in various configurations. For scanning systems, the additional detectors are either added in parallel (in the cross-scan direction), or in series (in the along-scan direction).

For staring systems, the detectors are arranged in a two-dimensional matrix to cover the whole FOV simultaneously. Whether the detectors are arranged in series, in parallel, or as a staring matrix, we will find a common result that describes the impact of additional detectors on the SNR, as compared to the single-detector case:

$$\text{SNR} \propto \sqrt{n_d} \qquad (12.10)$$

12.2.3. Parallel Scan

The addition of detectors in the cross-scan direction, as in Fig. 12.3, will decrease the required signal (and thus also the noise) bandwidth, given a fixed frame time. If there are a sufficient number of detectors ($n_d = \{\text{VFOV}/\text{VIFOV}\}$) in the vertical direction to cover the entire vertical FOV, only a horizontal scan motion is required. If $n_d < \{\text{VFOV}/\text{VIFOV}\}$, a two-dimensional raster motion will be required, with the scan format such that any given detector will drop ($n_d \times \text{VIFOV}$) on the next scan. For a fixed frame time, the impact of having multiple detectors is to allow an increase in the dwell time, and hence the SNR. With additional detectors in the cross-scan direction, the scan velocity does not need to be as fast to complete the scan of the entire FOV in the same amount of time.

For a system with n_d detectors in parallel and a fixed frame time τ_f, we have for the line time, analogously to Eq. (12.2),

$$\tau_L = \frac{\tau_f}{n_L/n_d} = \tau_f \times \frac{1}{(\text{VFOV}/\text{VIFOV} \; n_d)} = \tau_f n_d \left(\frac{\text{VIFOV}}{\text{VFOV}}\right) \quad (12.11)$$

Referring to Eq. (12.3), the dwell time will also be increased by a factor of n_d

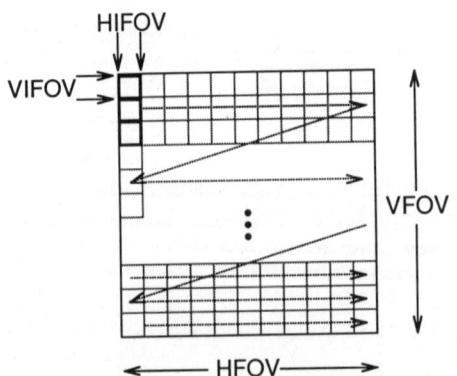

Figure 12.3. Parallel-scan format.

over the single-detector case:

$$\tau_d = \frac{\tau_L}{\dfrac{\text{HFOV}}{\text{HIFOV}}} = \frac{\tau_f n_d}{\left(\dfrac{\text{HFOV} \times \text{VFOV}}{\text{HIFOV} \times \text{VIFOV}}\right)} \tag{12.12}$$

By Eq. (12.8), the bandwidth Δf will decrease, being multiplied by a factor of $1/n_d$, compared to the single-detector case:

$$\Delta f \cong \frac{1}{2\tau_d} = \frac{\left(\dfrac{\text{HFOV} \times \text{VFOV}}{\text{HIFOV} \times \text{VIFOV}}\right)}{2\tau_f n_d} \tag{12.13}$$

and by Eq. (12.9), the rms noise will decrease the factor of $(1/n_d)^{1/2}$, compared to the single-detector case:

$$v_n \propto \sqrt{\Delta f} \cong \sqrt{\frac{\left(\dfrac{\text{HFOV} \times \text{VFOV}}{\text{HIFOV} \times \text{VIFOV}}\right)}{2\tau_f n_d}} \tag{12.14}$$

Thus for a parallel-scan system, the overall SNR will increase by the factor $(n_d)^{1/2}$ over the single-detector case. It should be noted that the increase in dwell time has only changed the rms noise level, and not the signal level. The signal is an instantaneous voltage or current that is proportional to the flux collected from the source (either watts or photon/s) that is brought to the detector. The speed at which the IFOV scans past some feature in the scene does not affect the signal level. However, because scan speed affects the temporal bandwidth required to pass the signal, the scan velocity does affect the noise and thus the SNR.

12.2.4. Serial Scan with Time Delay and Integration

A serial-scan system is a modification of the single-element serial scan, and arranges the n_d detectors in the along-scan direction, as shown in Fig. 12.4. We will show that this scan format has the same SNR advantage (compared to a single-element case) as the parallel scan, that of the square root of the number of detectors. The number of detectors used in practical systems is typically between 2 and 10. The scan velocity is unchanged from the single-detector case, so the bandwidth of any given detector channel is unchanged. The mechanism used to implement a serial scan is time delay and integration (TDI), as shown in Fig. 12.5. For clarity, we have taken the point of view that the image is scanned across the (stationary) detectors. In TDI, the output charge

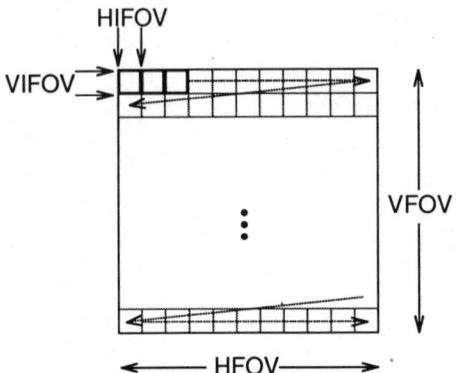

Figure 12.4.   Serial-scan format.

from each detector in the series is summed with the output from all subsequent detectors in the series as scan progresses. There is an integration of the flux on the detector for a dwell time τ_d, the accumulated charge is shifted one position along the scan, and the charge packet is added to the charge packet already in that cell. The transfers and additions occur in an analog delay line, with the velocity of the charge motion matched to the image-scan velocity. Over the course of the TDI operation, the magnitude of the charge packet increases by a factor of n_d, in that the charge packet grows with each addtion. We assume that the noise from each detector is uncorrelated, and hence will

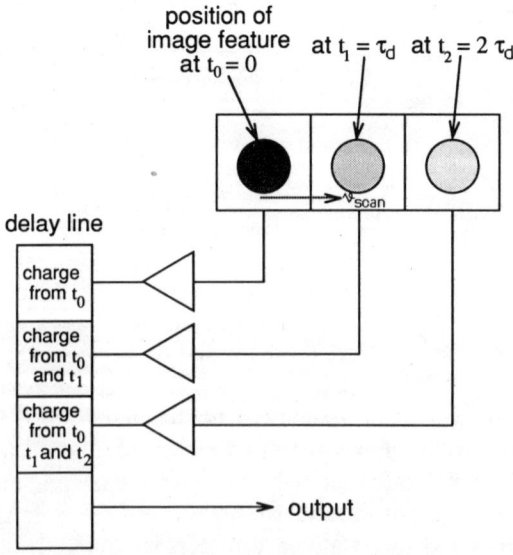

Figure 12.5.   Time delay and integration.

add in quadrature. Thus the rms noise will increase as square root of n_d. Thus the SNR will increase as square root of the number of detector elements in TDI.

TDI is useful in increasing the SNR obtained from a scanning system. The disadvantage of a TDI implementation is that there is a synchronized delay line required, which increases the weight and power of the electronics subsystem. It should be noted that the square-root-of-n_d advantage in SNR is obtained only when all of the detectors are identical in terms of responsivity and noise level, and the practical result typically falls around 10% short of that ideal.

Although our example has used TDI in the serial-scan format, TDI is also used to good advantage in conjunction with parallel-scan formats. An example of this TDI/parallel scan is commonly found in second-generation forward-looking infrared (FLIR) imagers, an intermediate technology between a simple parallel scan of a single vertical column of detectors and a full staring system. Several vertical columns of detectors are used, each column having sufficient detectors to cover the VFOV. These columns are scanned in a normal parallel-scan format, but with TDI being applied in the along-scan direction. The overall signal-to-noise (S/N) improvement, over a single-detector serial scan system of the same frame time, is once again proportional to the square root of the number of detectors used.

12.2.5. Staring Systems

The main SNR advantage of a staring system over any type of scanned system is that there are enough detectors to cover the whole FOV for the entire frame time. This modifies the dwell time from that of Eq. (12.6) and, because the number of detectors is equal to the number of pixels, the dwell time is equal to the frame time:

$$\tau_d = \tau_f \tag{12.15}$$

Each detector channel has a reduced bandwidth because of the increased dwell time, which reduces the noise bandwidth and thus the rms noise. The SNR is increased by a factor of the square root of n_d, as compared to a single-detector system. This SNR gain will not be completely obtained, because of the effect of spatial noise on staring systems (Milton et al., 1985) which, under certain conditions, constitutes the limiting noise mechanism, rather than the temporal noise considered in this section.

Example. We can use the square-root-of-n_d dependence of SNR to compare the potential performance of various system configurations. For example, given a system with the parameters of fixed frame time, fixed detector D^*, and a 128 $\times$ 128 pixel format, we compare the SNRs of a 16-element vertical array, operating in the parallel-raster format of Fig. 12.3, with a staring array of 128 $\times$ 128 detectors. The SNR of the staring system is higher than the scanner by a factor of

$$\frac{(\text{SNR})_{\text{staring}}}{(\text{SNR})_{\text{scanning}}} = \sqrt{\frac{128^2}{16}} = 32 \qquad (12.16)$$

12.3. RANGE EQUATION

The engineering tradeoffs for search-system design are cast in terms of the range equation, which expresses the distance at which a given point source can be detected. The target for a search system is a point source, specified by its radiant intensity (W/sr). The general design goal is to be able to detect the point source as far away as is possible. In the development of the range equation, we follow the general approach of Hudson (1969).

Within the context of Fig. 12.6, we can express the amount of flux reaching the detector (assuming that all of the radiation incident on the lens is transferred to the detector) as

$$\phi_d = \frac{I \times A_{\text{enp}}}{r^2} = I \times \Omega_{\text{enp}} \qquad (12.17)$$

where A_{enp} is the area of the collection aperture (entrance pupil) of the optics, and the range r is the distance from the point source to the system. We can write an expression for the signal voltage v_{sig} produced by the detector if we multiply Eq. (12.17) by the responsivity:

$$v_{\text{sig}} = \mathcal{R}_v \times \frac{I \times A_{\text{enp}}}{r^2} \qquad (12.18)$$

where it is understood that detector responsivity and target intensity are functions of wavelength and that their product $\mathcal{R}_v \times I$ actually denotes an integral

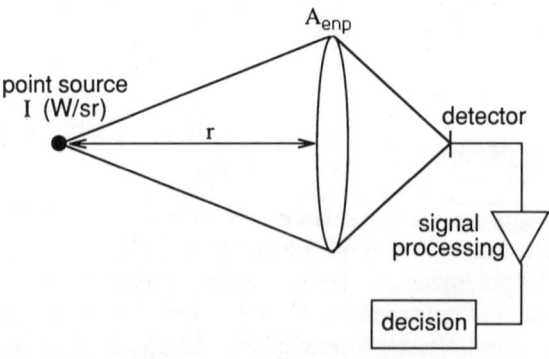

Figure 12.6.   Search-system schematic.

over wavelength of the product of the spectral functions, over the passband of the system ($\lambda_1 \leq \lambda \leq \lambda_2$):

$$\Re_v \times I \equiv \int_{\lambda_1}^{\lambda_2} \Re_v(\lambda)\, I(\lambda)\, d\lambda \tag{12.19}$$

We can formally divide each side of Eq. (12.18) by the rms noise v_n to yield the SNR:

$$\text{SNR} = \frac{v_{\text{sig}}}{v_n} = \frac{\Re_v}{v_n} \times \frac{I \times A_{\text{enp}}}{r^2} \tag{12.20}$$

For search systems, which look for a pulse signal representing a target, SNR is considered to be the ratio of peak signal to rms noise. Recalling the definition of noise equivalent power (NEP):

$$\text{NEP} = \frac{\phi_p}{\text{SNR}} = \frac{v_n}{\Re_v} \tag{12.21}$$

we can write for Eq. (12.20):

$$\text{SNR} = \frac{v_{\text{sig}}}{v_n} = \frac{1}{\text{NEP}} \times \frac{I \times A_{\text{enp}}}{r^2} \tag{12.22}$$

As an aside, it should be noted that NEP is the flux reaching the detector, divided by the SNR produced. Another useful quantity is the noise-equivalent irradiance (NEI), which is the required irradiance at the entrance pupil to produce an SNR = 1:

$$\text{NEI} = \frac{\text{NEP}}{A_{\text{enp}}} = \frac{E_{\text{enp}}}{\text{SNR}} \tag{12.23}$$

Continuing with Eq. (12.22), we want to express the NEP in terms of detector parameters D^* and A_d, as well as the bandwidth Δf:

$$\text{SNR} = \frac{v_{\text{sig}}}{v_n} = \frac{1}{\text{NEP}} \times \frac{I \times A_{\text{enp}}}{r^2} = \frac{D^*}{\sqrt{A_d \Delta f}} \times \frac{I \times A_{\text{enp}}}{r^2} \tag{12.24}$$

Let us check the dependences in Eq. (12.24). The SNR produced by the system increases with increasing collection area, target intensity, and detector D^*. The SNR produced by the system decreases as target distance, bandwidth, or detector area increase. At this point we recall our assumptions in the derivation of Eq. (12.24): that we are collecting radiation from a point source, not an

extended source, and that the limiting noise of the system is internally generated detector noise (in other words, the detector is not BLIP). Thus, it is consistent with our assumptions that making the detector larger does not increase
the signal, assuming that the detector is large enough to collect the focused
flux from the point source. Increasing the detector area merely increases the
noise without changing the signal.

We can now solve Eq. (12.24) formally for the range r:

$$r = \sqrt{\frac{D^*}{\sqrt{A_d \Delta f}} \times \frac{I \times A_{\text{enp}}}{\text{SNR}}} \qquad (12.25)$$

where it should be noted that r is now interpreted as the maximum range at
which the point source can be detected, and SNR, rather than being the SNR
produced by the system as in Eq. (12.24), is now interpreted as the minimum
SNR required to reach a detection decision with an acceptable degree of certainty. The decision certainty increases as SNR increases, as seen in Fig. 12.7.
Analysis of the costs involved in possible erroneous decisions sets the required
certainty of decision in the system specification. A common certainty of decision used in design work is 50%. If a higher SNR is required from the system
(a more certain decision) before a decision is made, the system will have a
shorter range.

We would like to recast the range equation, Eq. (12.25), in terms of parameters that can be derived simply from overall performance requirements, such
as the $F/\#$ of the optic, the diameter of the collection optic, and the solid-angle
subtense of the detector ($\Omega_d = \text{IFOV}^2$). The $F/\#$ of the optic will usually be as
low as possible for optimum flux collection, within the constraint that a lower-
$F/\#$ system is harder to correct for aberrations. The diameter of the collection
optic will generally be as large as the space, weight, and cost constraints will
allow to enhance the flux-collection capabilities of the system. The solid-angle
subtense of the detector is determined from the imaging-resolution requirements, given that the ultimate resolution of the system will surely be no better
than the IFOV. These parameters are a convenient place to begin a tradeoff
study, because they can be determined and fixed early in the design process.

Let us proceed to identify the design variables. Consistent with the objec-

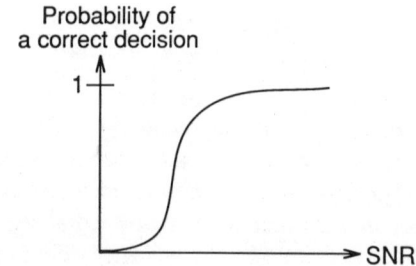

Figure 12.7. Decision certainty increases as
SNR increases.

tives of a search system, we let the range r be sufficiently long that the point-source image is formed at a distance F behind the optic. We further assume that the entrance pupil is located at the lens, and thus the entrance pupil diameter and the lens diameter are equal. With the substitutions

$$A_{\text{enp}} = \frac{\pi D_{\text{lens}}^2}{4} \qquad (12.26)$$

$$\Omega_d = \frac{A_d}{F^2} \qquad (12.27)$$

$$\sqrt{A_d} = \sqrt{\Omega_d} \times F \qquad (12.28)$$

Eq. (12.25) becomes

$$r = \left[\frac{I\, D^*}{\text{SNR}\, \sqrt{\Delta f}} \times \frac{A_{\text{enp}}}{\sqrt{A_d}} \right]^{1/2} \qquad (12.29)$$

$$r = \left[\frac{I\, D^*}{(\text{SNR})\, \sqrt{\Delta f}} \times \frac{\pi}{4} \frac{D_{\text{lens}}^2}{\sqrt{A_d}} \right]^{1/2} \qquad (12.30)$$

$$r = \left[\frac{I\, D^*}{\text{SNR}\, \sqrt{\Delta f}} \times \frac{\pi}{4} \frac{D_{\text{lens}}^2}{\sqrt{\Omega_d}\, F} \right]^{1/2} \qquad (12.31)$$

$$r = \left[\frac{I\, D^*}{\text{SNR}\, \sqrt{\Delta f}} \times \frac{\pi}{4} \frac{D_{\text{lens}}}{\sqrt{\Omega_d}\, F/\#} \right]^{1/2} \qquad (12.32)$$

We can see the potential tradeoffs more clearly if we regroup the terms in Eq. (12.32) according to the portion of the system in which they arise:

$$r = \left[\frac{\pi}{4} \frac{D_{\text{lens}}}{F/\#} \right]^{1/2} [I]^{1/2}[D^*]^{1/2} \left[\frac{1}{\text{SNR}\, \sqrt{\Delta f}\, \sqrt{\Omega_d}} \right]^{1/2}$$

$$\qquad\qquad \uparrow \qquad\qquad \uparrow \quad \uparrow \qquad\qquad \uparrow$$

$$\text{optics} \qquad \text{target detector} \quad \text{signal processing} \qquad (12.33)$$

Equation (12.33) is the range equation for a non-BLIP system. We now consider each subsystem separately.

12.3.1. Optics

A larger aperture collects more flux, which enables the system to see farther. Intuitively, the range is proportional to the square root of the flux collected, and hence the range should be proportional to the square root of the collecting area. This is seen to be the case when considering the term $[D_{\text{lens}}/(F/\#)]^{1/2}$, which is proportional to the square root of the collecting area. However, D_{lens} and $F/\#$ are written as separate terms to facilitate their independent selection in the design process. The range is inversely dependent on the square root of the focal length because decreasing the focal length (if aberration correction can be maintained) will result in a smaller diffraction-limited spot diameter, and a higher image irradiance. Thus, decreasing the focal length will allow smaller detectors (with less noise) to be used, and still capture the entire diffraction-limited target image.

12.3.2. Target

The term "target" also encompasses the impact of the background, the atmosphere, and the effects of any spectral filtering used in the system. The target intensity is a parameter over which the system designer usually has no direct control. The main influence that the designer has in this area is in the choice of system bandpass, seen in Fig. 12.8. The bandpass determines the contrast between the target and the background, and determines the amount of photon noise entering the system from the background. Given these constraints, the spectral bandpass can be chosen to maximize either the contrast between target and background or to maximize the SNR. If the flux being received is high, but the contrast of the scene is low, an advantage exists in choosing the passband to include only that spectral region where the target and background have the largest difference. An example of this situation is in terrestrial thermal-imaging systems that operate against a 300-K background, and are looking for targets that are only nominally above that temperature. The combination of absorption effects from the atmosphere and the spectral emissivity of the target may produce spectral regions of high contrast that are conducive to selective filtering. However, if there is relatively little flux on the detectors, such that they are in an internal-noise-limited situation, there is SNR benefit to be gained from a wider bandpass, that will allow more flux to pass, while still excluding those spectral regions that contribute primarily background radiation.

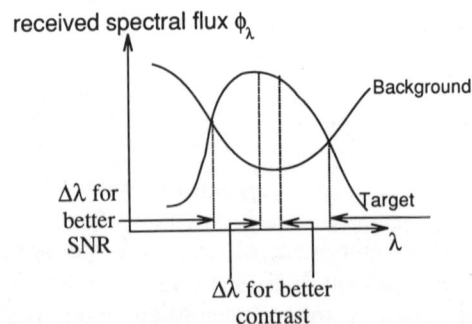

Figure 12.8.   Spectral bandpass affects contrast and SNR.

12.3.3. Detector

The range is proportional to the square root of the D^* of the detector. The options for effectively enhancing this D^* come from serial-scan techniques such as SPRITE and TDI approaches. The D^* in Eq. (12.33) accounts for temporal noise effects only. In situations where other noise sources are significant, such as spatial noise or row-to-row variations (D'Agostino and Webb, 1991), they can be added in quadrature to the temporal noise. Also, the same comments and conditions in relation to Eq. (12.19) apply to the integral of the product of $D^*(\lambda)$ and $I(\lambda)$ over the passband of the system:

$$D^* \times I \equiv \int_{\lambda_1}^{\lambda_2} D^*(\lambda)\, I(\lambda)\, d\lambda \qquad (12.34)$$

12.3.4. Signal Processing

The range is proportional to the inverse fourth root of the product of the angular subtense of the detector and the bandwidth. Thus decreasing either one will increase the range, but only slowly because of the fourth-root dependence. This is particularly useful in comparing the potential performance of a system at two different frame rates (such as 30 Hz and 60 Hz). The slower system will have a slightly longer range. The product $\Omega_d \Delta f$ represents an angular scan rate, and has units of sr/s. An increased range will result from a smaller angular-scan rate. This observation presents some interesting possibilities for tradeoffs. Decreasing the bandwidth (increasing the dwell time) can be accomplished in at least three ways: if the frame rate is decreased, the bandwidth is decreased directly, and the range will increase; if the angular coverage of the system (HFOV × VFOV) is decreased for a fixed frame time, the range will also increase because of the reduced bandwidth; if the number of detectors is increased, and the frame time is kept constant, the bandwidth will decrease because of the increased dwell time. The range is thus proportional to the fourth root of the number of detectors.

12.3.5. A Search System (non-BLIP) Preliminary Design

We illustrate the use of these techniques to calculate the expected range of a search system, under the assumption of non-BLIP operation. The parameters of the system are as follows. We use a single detector with a D^* of 2×10^{10} cm $Hz^{1/2}$ W^{-1}. The collection optic is $F/2$, with a 15-cm diameter. The scan format is two-dimensional raster, with unit scan efficiency. The detector size is 0.5 mm by 0.5 mm, and the SNR required to achieve the desired detection probability is 2.5. The target is point source at a distance of 5 km, with an intensity of 3 W/sr, and is to be found in a $10° \times 10°$ search field. We want to find the minimum frame time that is consistent with these parameters. How fast can the system operate?

We could use Eq. (12.32) directly, but it is also instructive to carry out the calculations from first principles: Starting from the equation for D^*:

$$D^* = 2 \times 10^{10} \text{ cm } \sqrt{\text{Hz}} \text{ W}^{-1} = \frac{\sqrt{A_d} \sqrt{\Delta f}}{\text{NEP}} = \frac{\sqrt{A_d} \sqrt{\Delta f}}{\phi_d} \times \text{SNR} \quad (12.35)$$

We can calculate ϕ_d from radiometry:

$$\phi_d = I \times \Omega_{\text{enp}} = 3 \text{ W/sr} \times \frac{\pi}{4} \left[\frac{15 \text{ cm}}{5 \text{ km}} \right]^2 = 2.12 \times 10^{-9} \text{ W} \quad (12.36)$$

Knowing the area of the detector and the required SNR, we can use Eqs. (12.35) and (12.36) to solve for the bandwidth $\Delta f = 115$ kHz and dwell time $\tau_d = 4.34$ μs. From knowing τ_d, IFOV, HFOV, and VFOV, we can determine the frame rate:

$$\text{HIFOV} = \text{VIFOV} = \frac{\sqrt{A_d}}{F} = \frac{\sqrt{A_d}}{F/\# \times D_{\text{lens}}} = \frac{0.5 \text{ mm}}{2 \times 15 \text{ cm}} = 1.67 \text{ mrad}$$

$$(12.37)$$

Using Eq. (12.5) to solve for v_{scan}:

$$v_{\text{scan}} = \frac{\text{HIFOV}}{\tau_d} = \frac{1.67 \text{ mrad}}{4.34 \text{ } \mu\text{s}} = 385 \times 10^3 \text{ mrad/s} \quad (12.38)$$

Using Eq. (12.4) to solve for the line time τ_L:

$$\tau_L = \frac{1}{v_{\text{scan}}} \times \text{HFOV} = \frac{1}{385 \times 10^3 \text{ mrad/s}} \times 175 \text{ mrad} = 455 \text{ } \mu\text{s} \quad (12.39)$$

Using Eq. (12.1) to find the number of lines n_L:

$$n_L = \frac{\text{VFOV}}{\text{VIFOV}} = \frac{10°}{1.67 \text{ mrad}} = \frac{175 \text{ mrad}}{1.67 \text{ mrad}} = 105 \text{ lines} \quad (12.40)$$

Using Eq. (12.2) to solve for τ_f:

$$\tau_f = \tau_L \times n_L = 47.8 \text{ ms} \quad (12.41)$$

or about 21 frames per second. Equation (12.41) expresses the minimum frame time consistent with the specifications. A longer frame time would allow for a longer dwell time and thus decrease the bandwidth, increasing the SNR obtained.

12.3.6. Range Equation for Search Systems: BLIP

We can obtain a range equation for the BLIP case by generalizing the range equation Eq. (12.32) for the non-BLIP case:

$$r = \left[\frac{\pi}{4} \frac{D_{lens}}{F/\#} \right]^{1/2} [I]^{1/2} [D*]^{1/2} \left[\frac{1}{SNR \sqrt{\Delta f} \sqrt{\Omega_d}} \right]^{1/2} \tag{12.42}$$

by substitution of the expression for background-limited $D*$ (the detector is assumed here to be a cooled photovoltaic):

$$D^*_{BLIP}(\lambda, f) = \frac{\lambda}{hc} \sqrt{\frac{\eta}{2E_{q,bkg}}} \tag{12.42a}$$

From radiometry, we can express the background photon irradiance on the detector as a function of the photon radiance of the background and the cold-shield angle θ (actually the marginal-ray angle), through which the detector views the warm background

$$E_{q,bkg} = \pi \int_0^{\lambda_p} L_q(\lambda, T_{bkg}) \, d\lambda \times \sin^2 \theta \tag{12.43}$$

Recall the definition of $F/\#$ for a system with object substantially at infinity:

$$(F/\#) = \frac{F}{D} \tag{12.44}$$

We can express the tangent of the marginal-ray angle as

$$\tan \theta = \frac{D/2}{F} \cong \sin \theta \tag{12.45}$$

where the last equality assumes small angles. We then have the relation between $F/\#$ and angle:

$$\sin^2 \theta \cong \frac{1}{4(F/\#)^2} \tag{12.46}$$

and thus finally for background-limited D^*:

$$D^*_{\text{BLIP}}(\lambda, f) = \frac{\lambda}{hc} \sqrt{\frac{\eta}{2\pi L_{q,\text{bkg}}(1/4(F/\#)^2)}}$$

$$= (F/\#) \frac{\lambda}{hc} \sqrt{\frac{2\eta}{\pi L_{q,\text{bkg}}}} \qquad (12.47)$$

Substituting Eq. (12.47) into Eq. (12.41):

$$r = \left[\frac{\pi}{4} \frac{D_{\text{lens}}}{F/\#}\right]^{1/2} [I]^{1/2} \left[(F/\#) \frac{\lambda}{hc} \sqrt{\frac{2\eta}{\pi L_{q,\text{bkg}}}}\right]^{1/2} \left[\frac{1}{\text{SNR} \sqrt{\Delta f} \sqrt{\Omega_d}}\right]^{1/2}$$

$$(12.48)$$

we find that for a BLIP system, the range is independent of the $F/\#$. The dependence on the diameter of the collecting optics remains. Other new dependences are that the range is proportional to the fourth root of the quantum efficiency, and inversely proportional to the fourth root of the in-band background flux.

REFERENCES

D'Agostino, J., and C. Webb, "3-D Analysis Framework and Measurement Methodology for Imaging System Noise," *SPIE Proc.* **1488**, 110–121 (1991).

Hudson, R. D., "The Analysis of Infrared Systems," in *Infrared System Engineering*, Wiley, New York, 1969, pp. 417–437.

Milton, A. F., F. R. Barone, and M. R. Kruer, "Influence of Nonuniformity on Infrared Focal Plane Array Performance," *Opt. Eng.* **24**(5), 885–862 (1985).

BIBLIOGRAPHY

Acetta, J. S., "Infrared Search and Track Systems," in *Passive Electro-Optical Systems* (S. B. Campana, ed.), SPIE, Bellingham, WA, 1993, pp. 209–344.

Johnson, R. B., and W. L. Wolfe, eds., *Proc. SPIE*, Volume 513, 1985.

Lloyd, J. M., *Thermal Imaging Systems*, Plenum, New York, 1975.

Seyrafi, E., *Electro-Optical Systems Analysis*, Electro-Optical Research Company, Los Angeles, 1985.

Spiro, I. J., and M. Schlessinger, *Infrared Technology Fundamentals*, Dekker, New York, 1989.

PROBLEMS

12.1. Given the following non-BLIP search-system parameters: $F/\# = 3$; diameter of optic $= 20$ cm; single detector scanned in 2-dimensional raster; frame rate $= 30$ s^{-1}; detector size 200 μm $\times$ 200 μm; average in-band $D* = 10^{11}$ cm Hz$^{1/2}$ W^{-1}; required SNR $= 2.5$; and search field is $2° \times 2°$, what temperature would a point-source blackbody need to be in order to be detected at a distance of 10 km?

   (a) assume flat passband of 3–5 μm

   (b) assume flat passband of 8–12 μm

   (c) repeat a and b for a system with a 10-cm diameter

12.2. Given the following non-BLIP search-system parameters: $F/\# = 2$; diameter of optic $= 15$ cm; single detector scanned in 2-dimensional raster; frame rate $= 60$ s^{-1}; detector size 200 μm $\times$ 200 μm; average in-band $D* = 10^{11}$ cm Hz$^{1/2}$ W^{-1}; required SNR $= 2.5$; and search field is $4° \times 10°$, assume that the target is a point source of in-band intensity $I = 2$ watt/sr.

   (a) Find the maximum distance at which the target can be detected.

   (b) Find the frame rate that would double the range.

   (c) Find the aperture diameter that would double the range.

   (d) Repeat (a) for a three-detector TDI serial scan.

   (e) Repeat (a) if the $F/\#$ is increased to $F/\# = 3$, while keeping the optic diameter the same.

   (f) Repeat (a) if the required SNR was increased to 10.

12.3. Find the ratio $(\text{SNR}_{\text{staring system}})/(\text{SNR}_{\text{scanning system}})$ for the following situation. The staring system is a 64-by-64 detector array, and the scanned system has a 1-by-64 vertical array of detectors, scanned in a 1-dimensional raster. Both systems have identical A_d, $D*$, optics, IFOV, and frame time.

12.4. Given an IR system with a 10-cm-aperture diameter, a 32-element vertical detector array, where each detector has a $D*$ of 10^{11} cm Hz$^{1/2}$ W^{-1}. Let the frame time be 1/30 s. This baseline system satisfies performance requirements.

   (a) Suppose we have available a 32-by-32 detector array; each detector has a $D* = 8 \times 10^{10}$ cm Hz$^{1/2}$ W^{-1}. If we keep the same aperture diameter, what frame time can we now achieve (keeping SNR a constant)?

   (b) If the frame time is fixed at 1/30 s, what aperture diameter could be used with the detector array (keeping SNR a constant)?

12.5. A 30-frames/s search system has a maximum range of 5 km. If the frame rate is increased to 60 frames/s, what will be the new maximum range? If the frame rate is decreased to 15 frames/s, what will be the new maximum range?

12.6. Why do the design equations (in terms of maximum range) for search systems specify targets in terms of point sources (intensity), rather than extended sources (radiance)?

12.7. Given the following parameters for a BLIP search system: $F/\# = 2$; diameter of optic = 15 cm; single detector scanned in 2-dimensional raster; frame rate = 60 s^{-1}; detector size 200 μm $\times$ 200 μm; average in-band $D^* = 10^{11}$ cm Hz$^{1/2}$ W^{-1}; required SNR = 2.5; and search field is $4° \times 10°$, assume the target is a point source of in-band intensity $I = 2$ W/sr.

(a) Find the maximum distance at which the target can be detected.

(b) Repeat (a) if the $F/\#$ is increased to $F/\# = 3$ by decreasing the optic diameter.

(c) Repeat (a) if the $F/\#$ is increased $F/\# = 3$ by increasing the focal length.

(d) If the background temperature changed from 350 K to 250 K, and the system remained BLIP, what would be the percent increase in the range? Assume flat passband of 3–5 μm.

(e) If the background temperature changed from 350 K to 250 K, and the system remained BLIP, what would be the percent increase in the range? Assume flat passband of 8–12 μm.

13

Modulation Transfer Function

13.1. INTRODUCTION

Transfer functions are a powerful tool for analyzing infrared-imaging systems. The interpretation of image quality in the frequency domain makes the entire range of linear-systems analysis techniques available, which facilitates insight, particularly when several subsystems are combined. Several subsystems are often combined in the analysis of infrared systems, where effects of the atmosphere, optics, detector, signal processing, display, and observer collectively produce the final image. Each subsystem has its own impulse response and transfer function. In the Fourier domain, the overall transfer function is the multiplication of the individual transfer functions. This is easier to visualize and calculate than the equivalent spatial-domain operation of convolution of the individual subsystem impulse responses.

13.2. DEFINITIONS

The image quality of an infrared system can be characterized by either the impulse response of the system or its Fourier transform, the transfer function. The impulse response $h(x, y)$ is the two-dimensional image formed in response to a point-source object (mathematically, a delta function). Because of the effects of diffraction and aberrations, the image quality depends on the wavelength passband, system $F/\#$, and field of view.

An object can be described by its irradiance as a function of two-dimensional position $f(x, y)$. This function can be decomposed into a regularly spaced set of point sources, each with a strength proportional to the brightness of the object at that location. This decomposition uses the sifting property of delta functions (Gaskill, 1978). Each point source in the object gives rise to an impulse-response irradiance $h(x, y)$ in the image plane. The impulse response depends on both diffraction and aberrations. The image radiance $g(x, y)$ is the sum of the individual impulse responses, each with weight proportional to the brightness of the corresponding object location, which is equivalent to the convolution of the object with the impulse response

$$f(x, y) ** h(x, y) = g(x, y) \tag{13.1}$$

where the double asterisk denotes a two-dimensional convolution.

The representation of the imaging process as a convolution in Eq. (13.1) requires shift invariance and linearity. The definition of a single-impulse response for the entire image requires shift invariance. The superposition of impulse responses requires linearity. These assumptions are often violated in practice but, to preserve the convenience of a transfer-function analysis, the variable that causes either shift variance or nonlinearity is allowed to assume a set of discrete values, and a separate impulse response and transfer function are defined for each set. For instance, in an imaging system that has aberrations that depend on field angle, separate impulse responses are defined for different regions of the image plane. The image-forming portion of an infrared-imaging system is linear, but the detectors introduce a nonlinearity if the responsivity of the sensors is a function of flux level. In practice this nonlinearity is partially corrected by the signal processing, but particularly if the detectors are tested as a separate subsystem, the definition of an impulse response and MTF that are restricted to a given range of irradiance levels may be appropriate.

A transfer-function analysis considers the imaging of sinusoidal objects, rather than point objects. Using the convolution theorem for Fourier transforms (Gaskill, 1978), we can rewrite the convolution of Eq. (13.1) as a multiplication of the respective spectra:

$$F(\xi, \zeta) \times H(\xi, \zeta) = G(\xi, \zeta) \tag{13.2}$$

where the uppercase letters denote Fourier transforms, such that $F(\xi, \zeta)$ is the spectrum of the object, $G(\xi, \zeta)$ is the spectrum of the image, and $H(\xi, \zeta)$ is the transfer function, which is the Fourier transform of the impulse response. $H(\xi, \zeta)$ multiplies the object spectrum to yield the image spectrum. The variables ξ and ζ are spatial frequencies in the x and y directions. The transfer-function approach decomposes the object and image into a basis set of sinusoidal functions. The reciprocal of the crest-to-crest distance (the spatial period) of any such sinusoid is the corresponding spatial frequency. In two dimensions, a general sinusoidal basis function has an arbitrary orientation with respect to the x and y axes. There will be a spatial period along both the x and y axes and the reciprocals of these are the spatial frequencies ξ and ζ. Spatial frequency is typically specified in cycles/mm in the image plane, and in angular spatial frequency (cycles/milliradian) in object space. For an object located at infinity, these two spatial frequencies are related through the focal length of the image-forming optical system as seen in Fig. 13.1.

$$\xi_{ang,obj}[cycles/mrad] = \frac{\xi_{img}[cycles/mm] \times F\,[mm]}{1000} \tag{13.3}$$

For an imager system composed of m' subsystems, each subsystem with an impulse response $h_m(m, y)$ and transfer function $H_m(\xi, \zeta)$, we generalize Eq.

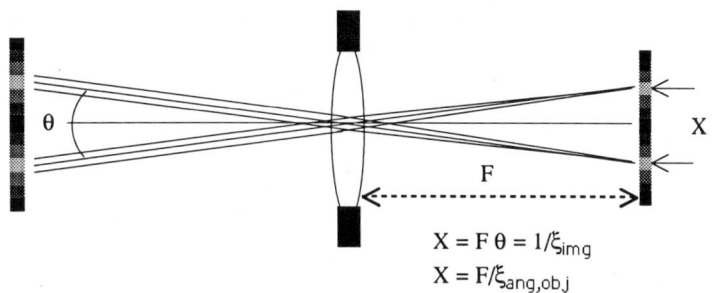

Figure 13.1. Relation of angular spatial frequency in object space to image spatial frequency.

(13.1) as

$$g(x, y) = f(x, y) ** h_1(x, y) ** \cdots ** h_{m'}(x, y) \tag{13.4}$$

and Eq. (13.2) as

$$G(\xi, \varsigma) = F(\xi, \varsigma) \times \prod_1^{m'} H_m(\xi, \varsigma) \tag{13.5}$$

The function $H(\xi, \varsigma)$ in Eq. (13.2) is usually normalized to have unit value at zero frequency (i.e., $H(0, 0) = 1$), which yields a relative transfer function that ignores frequency-independent attenuations, such as losses caused by Fresnel reflection or by obscurations. This normalization is appropriate for optical systems, because (Goodman, 1968) the transfer function of an incoherent optical system is proportional to the two-dimensional autocorrelation of the exit pupil, and an autocorrelation is necessarily maximum at the origin. However, in the context of a complete infrared-imager system, other subsystems, such as the electronics and signal processing, can have transfer functions that are not necessarily unity at the origin. These transfer functions are generally used for calculations in an unnormalized form, leaving the normalization until after the product in Eq. (13.5) has been taken.

13.3. OTF, MTF, AND PTF

In its normalized form, $H(\xi, \varsigma)$ is called the optical transfer function (OTF). The OTF is, in general, a complex function with both a magnitude and a phase portion:

$$\text{OTF}(\xi, \varsigma) = H(\xi, \varsigma) = |H(\xi, \varsigma)| \exp\{-j\theta(\xi, \varsigma)\} \tag{13.6}$$

The magnitude $|H(\xi, \varsigma)|$ is called the modulation transfer function (MTF), while the phase, $\theta(\xi, \zeta)$, is the phase transfer function (PTF). The MTF is the magnitude response of the imaging system to sinusoids of different spatial frequencies. This response is described in terms of the modulation depth $\mathfrak{M}$:

$$\mathfrak{M} = \frac{B_{max} - B_{min}}{B_{max} + B_{min}} \qquad (13.7)$$

where B_{max} and B_{min} refer to the maximum and minimum values of the brightness sinusoid in either the object waveform (exitance versus position) or the image waveform (irradiance versus position). The brightness in W/cm^2 is always nonnegative, so the sinusoids have a dc bias. Thus, from Eq. (13.7), modulation depth is between 0 and 1.

A linear system will yield a sinusoidal image of a sinusoidal object, and because the impulse response $h(x, y)$ is not a delta function, the modulation depth in the image sinusoid is lower than that in the object sinusoid. The reduction in $\mathfrak{M}$ is typically most severe for the high-spatial-frequency image detail. The MTF is the ratio of image-to-object modulation depth, as a function of spatial frequency:

$$\text{MTF}(\xi, \zeta) \equiv \frac{\mathfrak{M}_{img}(\xi, \zeta)}{\mathfrak{M}_{obj}(\xi, \zeta)} \qquad (13.8)$$

as seen in Fig. 13.2. The area under the MTF curve is indicative of overall modulation transfer from object to image, and is a common measure of overall image quality.

The PTF does not affect the modulation depth of the sinusoidal image components, but rather describes the relative phase with which the various components recombine. A linear PTF such as $\theta(\xi) = x_0\xi$ corresponds to an image shift by an amount x_0, where each frequency component is shifted by the amount required to reproduce the original waveform at the location x_0. For an impulse response $h(x, y)$ that is even about the ideal image point, the OTF will be real valued and the PTF will take a value of 0 or π radians as a function of spatial frequency. A general impulse response $h(x, y)$ yields a PTF that is a nonlinear function of frequency, resulting in image distortion. The degree of nonlinearity of optics-subsystem PTF is a sensitive test criterion for aberrations, such as coma, that produce asymmetric impulse responses. The PTF is also a consideration for the electronics subsystem in an infrared imager. Particularly near frequencies where sharp cutoffs are made to occur in the transfer function, the resulting nonlinearity of the PTF can affect the image quality. The PTF of the optics subsystem is not usually measured directly, but is calculated from the Fourier transform of the measured impulse response $h(x, y)$. The PTF of the electronics subsystem can be conveniently measured directly using a sinusoidal signal generator.

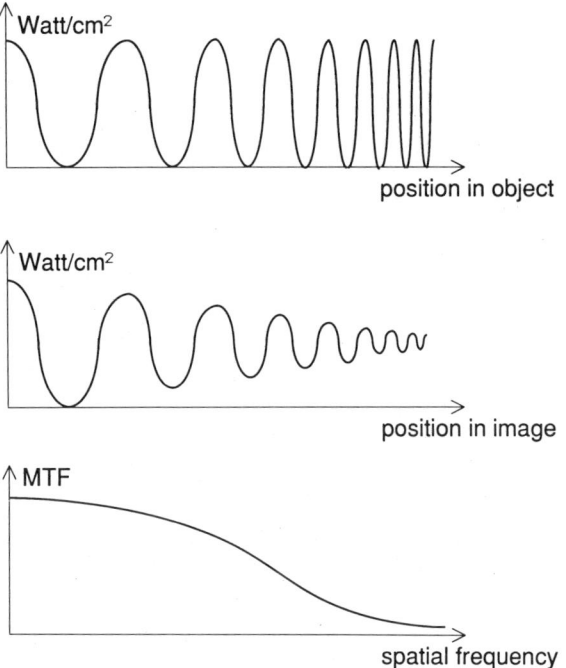

Figure 13.2. MTF is the ratio of the image-to-object modulation depth as a function of ξ.

13.4. CALCULATION OF OPTICAL-SUBSYSTEM MTF

The overall transfer function is the product of individual subsystem transfer functions, as seen in Eq. (13.5). It is convenient to consider two transfer functions to apply to the optics portion of an infrared-imager system, one that accounts for the effects of geometrical-optics aberrations, the other to account for the effects of diffraction. The overall transfer function of the optics subsystem can be calculated as the product of the two. This is equivalent to convolution of the impulse-response functions from diffraction effects and geometrical-aberration effects.

Ray-tracing procedures that produce a ray-intersection density (spot) diagram in the image plane generate the impulse response $h(x, y)$ of the geometrical-optics image blur arising from aberrations, for which the Fourier transform is the corresponding geometrical-optics component of the transfer function. Diffraction OTF is a wave-optics calculation. The diffraction OTF of an infrared imager is proportional to the two-dimensional autocorrelation of the exit pupil. A change of variables is needed to identify an autocorrelation (a function of position in the exit pupil) as a transfer function (a function of image-plane spatial frequency). The change of variables is

$$\xi \equiv \frac{x}{\lambda z_{\text{exp},i}} \tag{13.9}$$

where x is the autocorrelation shift distance in the pupil, λ is the wavelength, and $z_{\text{exp},i}$ is the distance from the exit pupil to the image. The pupil autocorrelation goes to zero at the cutoff frequency

$$\xi_{\text{cutoff}} \equiv \frac{1}{\lambda F/\#} \tag{13.10}$$

where the $F/\#$ is either the object-space or image-space $F/\#$ as defined in Chapter 1, corresponding to the object- or image-plane cutoff frequency.

The diffraction OTF represents the best performance that a system can achieve for a given $F/\#$ and λ, and accurately describes systems having negligible aberrations, where the size of the impulse response is dominated by diffraction. A diffraction-limited system with a circular exit pupil has a circularly symmetric MTF, with ξ profile (Gaskill, 1978)

$$\text{MTF}(\xi/\xi_{\text{cutoff}}) = \frac{2}{\pi} \left\{ \cos^{-1}(\xi/\xi_{\text{cutoff}}) - (\xi/\xi_{\text{cutoff}})[1 - (\xi/\xi_{\text{cutoff}})^2]^{1/2} \right\};$$

$$\xi \leq \xi_{\text{cutoff}}$$

$$= 0, \quad \text{if} \quad \xi > \xi_{\text{cutoff}} \tag{13.11}$$

Equation (13.11) is plotted in Fig. 13.3, along with MTF curves obtained for a range of annular pupils, which arise in obscured systems such as Cassegrain telescopes. The plots are functions of the obscuration ratio, and the emphasis at high frequencies has been obtained by an overall decrease in flux reaching the image, proportional to the obscured area. If the curves in Fig. 13.3 were plotted without normalization to 1 at $\xi = 0$, they would all be contained under the envelope of the unobscured diffraction-limited curve.

The MTF curve for a system with appreciable geometric aberrations is bounded by the diffraction-limited MTF curve as the upper envelope. Aberrations broaden the impulse response $h(x, y)$, resulting in a narrower and lower MTF, with less integrated area. Figure 13.4 shows the effect of defocus on the MTF, and Fig. 13.5 shows the MTF curves resulting from small amounts of third-order spherical aberration. Equation (13.11) is plotted as the limiting (no aberration) case in both figures. MTF curves for systems with higher-order spherical, coma, astigmatism, and chromatic are contained in Smith (1966).

For imager systems that employ a number of optical subsystems (for instance, an afocal telescope and a detector lens), the geometrical-aberration MTFs for each subsystem will cascade, and $H_{\text{geometrical}}$ can be expressed as a product of the form of Eq. (13.5), provided the subsystems do not correct for

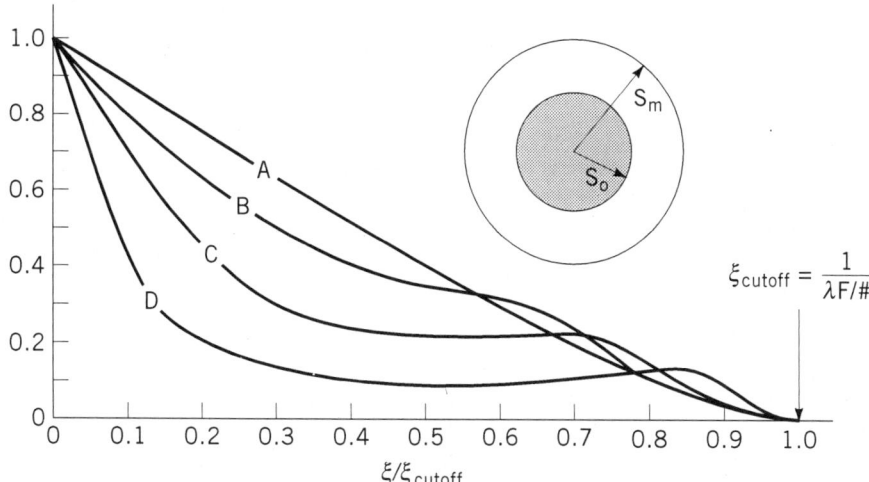

Figure 13.3. (A) Diffraction-limited MTF for circular-pupil system (no obscuration: $s_o/s_m = 0$. (B) through (D) are diffraction-limited MTF for an annular-pupil system: (B) $s_o/s_m = 0.25$; (C) $s_o/s_m = 0.5$; (D) $s_o/s_m = 0.75$. (Originally appeared in W. J. Smith, *Modern Optical Engineering*, 1966. Reprinted by permission of McGraw-Hill.)

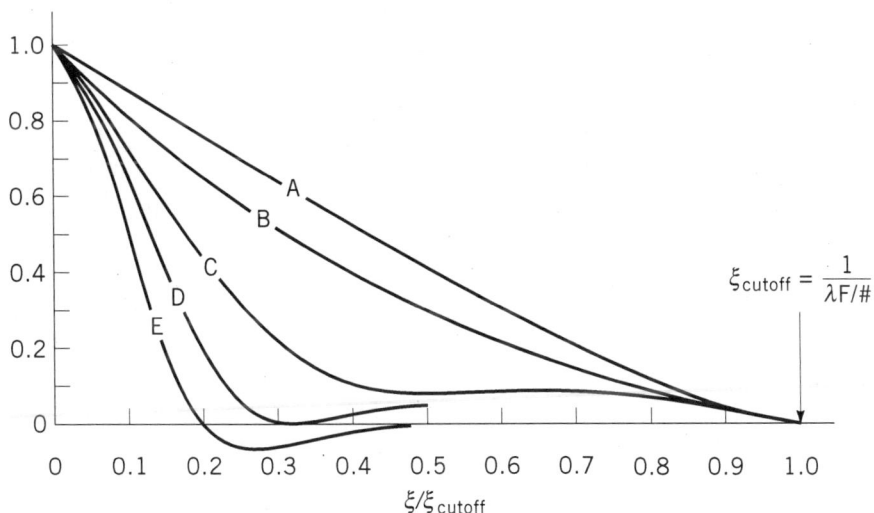

Figure 13.4. MTF for a defocused system, as a function of optical path difference (OPD): (A) in focus, OPD = 0.0; (B) defocus OPD = $\lambda/4$; (C) defocus OPD = $\lambda/2$; (D) defocus OPD = $3\lambda/4$; (E) defocus OPD = λ. (Originally appeared in W. J. Smith, *Modern Optical Engineering*, 1966. Reprinted by permission of McGraw-Hill.)

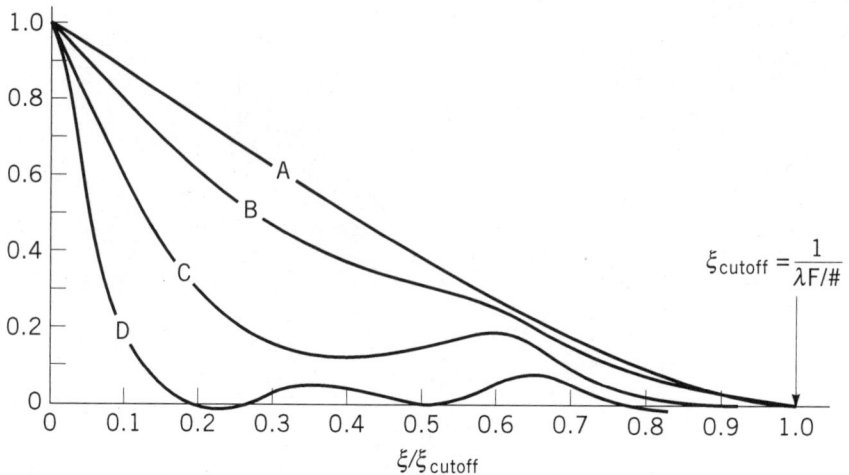

Figure 13.5. MTF for system with third-order spherical aberration (image plane midway between marginal and paraxial foci). (A) OPD = 0; (B) OPD = $\lambda/4$; (C) OPD = $\lambda/2$; (D) OPD = λ. (Originally appeared in W. J. Smith, *Modern Optical Engineering*, 1966. Reprinted by permission of McGraw-Hill.)

each other's aberrations or exhibit other partial-coherence effects (DeVelis and Parrent, 1967). The diffraction MTF does not multiply from one system to the next, and is calculated only once (according to Eq. (13.11)) for an entire optics train, at the limiting aperture of the system. The aggregate optics MTF at frequencies less than the cutoff is given by

$$\text{MTF}_{\text{optics}}(\xi) = \frac{2}{\pi} \{\cos^{-1}(\xi/\xi_{\text{cutoff}}) - (\xi/\xi_{\text{cutoff}})[1 - (\xi/\xi_{\text{cutoff}})^2]^{1/2}\}$$

$$\times \prod_{1}^{m'} H_{m,\text{geometrical}}(\xi) \qquad\qquad (13.12)$$

where $H_{m,\text{geometrical}}(\xi)$ are the individual geometrical-optics transfer functions arising from aberrations.

If two or more subsystems are designed to correct for each other's aberrations, the MTF of the combined system is better than the individual MTFs would indicate, and Eq. (13.12) does not apply as written. In this case, the product term of the subsystem geometrical-aberration MTFs is replaced by the geometrical MTF of the optical system as a single unit.

13.5. CALCULATION OF ELECTRONIC-SUBSYSTEM MTF

In infrared imagers, electronics subsystems perform both signal-handling and signal-processing functions. Characterization of electronic networks by transfer function is well established, where the frequency variable is the temporal frequency f [Hz]. To interpret the electronics transfer function as an MTF, the temporal frequencies are divided by the scan velocity (mm/s in the image or mrad/s in object space), converting the temporal frequencies to spatial frequencies. With this change of variables, the electronics subsystem will have its own MTF that will multiply the transfer functions of the other subsystems. In contrast to an optics MTF, an electronics transfer function is not necessarily maximum at the origin and can amplify certain frequencies and have sharp cutoffs at others.

An unavoidable characteristic of the electronics is the noise contribution, which limits the amount of electronic amplification that can be used in recovering modulation depth lost in other subsystems. A figure of merit that has been validated to correlate with image visibility (Snyder, 1973) is the area between the MTF curve and the power spectral density of the noise. The power spectrum is interpreted in terms of a noise-equivalent modulation depth, the modulation needed for unit signal-to-noise ratio as a function of spatial frequency.

13.6. DETECTOR MTF

The photosensitive area of the detectors is finite rather than a point sampler in any infrared imaging system. This results in spatial averaging (Boreman and Plogstedt, 1989) of the irradiance distribution incident on the detector. This spatial averaging has an MTF impact that multiplies the other transfer functions. This MTF is always present in a system with detectors. It may or may not be the limiting factor in determining the image quality, but the MTF contribution from the detector dimensions should always be considered. Large-area detectors exhibit more attenuation of high spatial frequencies than small detectors. For a detector of dimension ℓ_x in the x direction, the detector MTF is

$$\text{MTF}_{\text{detector}}(\xi) = \left| \frac{\sin \pi\xi\ell_x}{\pi\xi\ell_x} \right| \tag{13.13}$$

which has a zero at $\xi = 1/\ell_x$.

For scanning systems, the along-scan detector-footprint MTF is well represented by Eq. (13.13). For scanning systems in the cross-scan direction, and for staring systems, the effect of the location of the sampling sites must be considered to calculate the MTF. The MTF is not strictly defined for these cases, because the imaging process is no longer shift invariant (Wittenstein et al., 1982). A variety of impulse responses (and hence MTFs) will be obtained, depending on the exact alignment of the target with respect to the sampling sites. Park et al. (1984) calculate an MTF averaged over all possible target locations with respect to the sampling sites. The approximate result is that for a detector of dimension ℓ_x and a sampling interval of $\ell_{x,\text{samp}}$, the average spatial-averaging MTF will be the product of two sinc functions:

$$\text{MTF}_{\text{avg}}(\xi) = \left|\frac{\sin \pi\xi\ell_x}{\pi\xi\ell_x}\right|\left|\frac{\sin \pi\xi\ell_{x,\text{samp}}}{\pi\xi\ell_{x,\text{samp}}}\right| \qquad (13.14)$$

which is a poorer MTF than just the dimensions of the detector would indicate, particularly for arrays with substantial dead space between the detectors ($\ell_{x,\text{samp}} > \ell_x$).

13.7. MTF MEASUREMENTS

13.7.1. Overview

When characterizing the image quality of an infrared-imaging system, the MTF is assessed in a variety of ways, including measurement of the impulse response, line response, edge response, sine-wave response, bar-target response, and random-target response. Usually, infrared-imaging systems are designed for the case of an object at infinity. This impacts the MTF-test configuration. For any target referred to earlier, we assume the setup see in Fig. 13.6, where a collimator with the desired target at the focus simulates the infinite-conjugate condition. The MTF of the collimator is usually sufficiently high to avoid affecting the measurement over the spatial-frequency range of interest. If desired, the (geometrical) MTF of the collimator can be included as one of the subsystem MTFs in Eq. (13.5), and later divided out to yield just the MTF of the system under test (Alexander et al., 1993). The aperture of the collimator should be large enough to overfill the aperture of the system under test. This condition satisfies two requirements. First, it ensures that the system under test is measured at its full aperture (a measurement using less than the full aperture

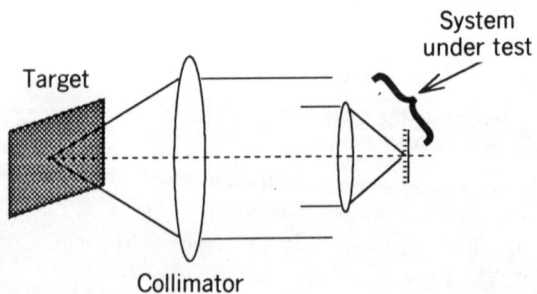

Figure 13.6. MTF measurement configuration for infinite-conjugate imager system.

does not necessarily contain all of the aberrations encountered in actual use). Second, the limiting aperture of the system is not at the collimator, ensuring that the diffraction MTF included in Eq. (13.12) represents the system under test and not the test setup.

Each measurement method has advantages and disadvantages, depending on the details of the system under test. All of the methods should yield the same result if the system is truly a linear shift-invariant system. Given that linearity and shift invariance are idealizations, the preferred measurement method uses a target similar to that which the system will encounter under operational conditions. This ensures good correlation between a laboratory MTF measurement and image quality in the field.

The measurement of MTF ultimately involves the detection of the image-plane flux with a finite-sized detector. Thus, one component of the measurement-system MTF is the result of the finite aperture of the detector, and can be accounted for in the calibration of the instrument by dividing out the detector MTF seen in Eq. (13.13).

It should also be noted that any MTF calculated from a measured profile is the product of a diffraction MTF and a geometrical-aberration MTF. When combining the separately measured MTFs of several optical subsystems, care should be taken to ensure that the diffraction MTF (determined by the aperture stop of the combined system) contributes only once to the calculation, as in Eq. (13.12).

13.7.2. Impulse Response

If a point-source object $\delta(x, y)$ were used as the test target, the object spectrum would be a constant, consistent with the Fourier transform of a delta function. The image formed by the system under test would then be the impulse response, the Fourier transform of the OTF. The impulse response $h(x, y)$ can be Fourier transformed in two dimensions to yield the complete two-dimensional OTF(ξ, ζ) in one measurement:

$$\mathcal{F}\{h(x, y)\} = \text{OTF}(\xi, \zeta) \tag{13.15}$$

The usual source for impulse-response measurements is an illuminated pinhole, which should be as small as possible, within the constraint of flux-detection considerations. The pinhole is small enough not to appreciably affect the MTF measurement if its angular subtense is less than $\approx 1/10$ of the angular subtense of the impulse response, when both are viewed from a common plane such as the aperture stop of the system. If it is necessary to use a pinhole of larger extent, a Fourier spectrum of the source can be calculated (because $F(\xi, \zeta)$ will no longer be a constant), and the OTF can be calculated using Eq. (13.2) over a range of spatial frequencies where $F(\xi, \zeta)$ is nonzero (Alexander et al., 1993).

13.7.3. Line Response

If a pinhole source does not provide a sufficient image-plane flux level to maintain a signal-to-noise ratio, a line response can be measured. The system is presented with an illuminated line source (or a hot current-carrying wire), which acts approximately as a delta function in one direction and a constant in the other, for instance $\delta(x)1(y)$. The image formed is the line response $\ell(x)$ of the system under test. The line response is a summation of vertically displaced impulse responses, and thus is not just a one-dimensional profile of the two-dimensional impulse response (in general $\ell(x) \neq h(x, 0)$), but has a different functional form (Gaskill, 1978). The line response only yields information about one profile of the two-dimensional OTF, and a one-dimensional Fourier-transform relationship applies between the line response and the corresponding profile of the OTF:

$$\mathcal{F}\{\ell(x)\} = \text{OTF}(\xi, 0) \tag{13.16}$$

To obtain other profiles of the OTF, the line source can be reoriented as desired.

The line response is usually measured by scanning across the narrow direction of the line-source image with a small detector. A correction for the finite width of the line source can be made using Eq. (13.2). The finite size of the detector will contribute its own MTF component to the measurement according to Eq. (13.13), and can also be corrected for. Line-response data can also be measured from the response of the system to a point source, using a scanning detector of long aspect ratio (scanned in the narrow direction, integrating in the long direction) that integrates the impulse response along the long direction of the detector.

13.7.4. Edge Response

Another measurement technique (Reichenbach et al., 1991) uses the edge response $e(x)$, which is the response of the system under test to an illuminated knife-edge target. Each line in the open part of the aperture produces a displaced line response, resulting in the following relationship between $e(x)$ and $\ell(x)$

$$e(x) = \int_{-\infty}^{x} \ell(x') \, dx' \tag{13.17}$$

and the inverse relationship

$$\frac{d}{dx}(e(x)) = \ell(x) \tag{13.18}$$

Measurement of $e(x)$ allows the calculation of the corresponding one-dimensional profile of the OTF using Eqs. (13.16) and (13.18). The derivative operation in Eq. (13.18) increases the effect of any detector or system noise present in the data, and a smoothing operation is often desirable. It should be noted that any digital filter used for data smoothing has its own impulse response, and hence its own OTF contribution, which can be corrected for in the data analysis. The edge response can be measured in a variety of equivalent ways. Using an edge source, the image-plane irradiance is scanned with a small discrete detector or a narrow linear detector. Alternatively, either a point source or a line source can be used, where a scanning knife edge is combined with a large-area detector in the image plane.

13.7.5. Sine-Wave Response

The MTF can be obtained by measuring the system's response to a sine-wave target, that is, the image modulation depth as a function of spatial frequency. The PTF can be found from the position of the waveform maxima as a function of frequency. Sine-wave targets are available as photographic prints or transparencies, which are suitable for testing visible-wavelength systems. Careful control in their manufacture is exercised (Lamberts, 1963) to avoid harmonic distortions, including a limitation to relatively small modulation depths. Sine-wave targets are difficult to fabricate for the testing of infrared systems, and require the use of either interferometric (Barnard et al., 1992) or half-tone-transparency (Daniels et al., 1995) techniques.

13.7.6. Bar-Target Response

Although the MTF is defined for sine-wave targets, the infrared-target fabrication difficulties noted earlier often dictate the use of bar targets with equal-sized lines and spaces, and binary transmission or reflection characteristics. A particular target can be specified by a fundamental spatial period $\mathfrak{X}$ of one cycle of the pattern. Infrared systems are typically characterized with four-bar targets; three-bar targets are more commonly used for visible-wavelength systems.

Targets with an infinite number of square-wave cycles are a simpler place to begin the analysis. For such targets, we define a contrast transfer function (CTF) as a function of fundamental spatial frequency $\xi_f = 1/\mathfrak{X}$:

$$\mathrm{CTF}(\xi_f) = \left[\frac{\mathfrak{M}_{\text{output}}(\xi_f)}{\mathfrak{M}_{\text{input-square-wave}}(\xi_f)} \right] \tag{13.19}$$

CTF is measured on the peak-to-valley variation of image irradiance. CTF is generally higher than the MTF at the same spatial frequency because of the contribution of the odd harmonics of the infinite-square-wave test pattern (which are absent from sine-wave targets) to the modulation depth in the image. The CTF at any frequency can be written as a sum of harmonic components, weighted by two multiplicative factors: the relative strength of the component in the input square wave and the MTF of the system under test at that frequency. This process yields an expression (Coltman, 1954) for the CTF in terms of MTF:

$$CTF(\xi_f) = \frac{4}{\pi} \left\{ MTF(\xi = \xi_f) - \frac{MTF(\xi = 3\xi_f)}{3} + \frac{MTF(\xi = 5\xi_f)}{5} \right.$$
$$\left. - \frac{MTF(\xi = 7\xi_f)}{7} + \frac{MTF(\xi = 9\xi_f)}{9} \cdots \right\} \tag{13.20}$$

Inversion of the series (Coltman, 1954) yields the MTF in terms of CTF:

$$MTF(\xi) = \frac{\pi}{4} \left\{ CTF(\xi_f = \xi) + \frac{CTF(\xi_f = 3\xi)}{3} - \frac{CTF(\xi_f = 5\xi)}{5} \right.$$
$$\left. + \frac{CTF(\xi_f = 7\xi)}{7} + \frac{CTF(\xi_f = 11\xi)}{11} \cdots \right\} \tag{13.21}$$

The series representations of Eqs. (13.20) and (13.21) are valid for infinite-square-wave targets because these targets have discrete harmonic components. However, the spectra of a finite-length bar target does not have discrete harmonic components (Kelly, 1965); thus a series representation cannot be exact. Bar-target data are interpreted in terms of image modulation depth as a function of fundamental spatial frequency, $IMD(\xi_f)$, according to Eq. (13.22):

$$IMD(\xi_f) = \left[\frac{\mathfrak{M}_{output}(\xi_f)}{\mathfrak{M}_{input\text{-}bar\text{-}target}(\xi_f)} \right] \tag{13.22}$$

The image modulation depth for three- and four-bar targets are slightly higher than the CTF curve for an infinite square wave. Figure 13.7 compares the MTF with modulation depths obtained for square-wave, three-bar, and four-bar targets, for the case of a diffraction-limited circular-aperture system with cutoff frequency ξ_0. Figure 13.8 makes the same comparison for the case of a diffraction-limited annular-aperture system (50% diameter obscuration) with cutoff frequency ξ_0, illustrating a more substantial difference between the bar-target response and the MTF (Boreman and Yang, 1995).

Conversion of bar-target image-modulation data to MTF using the series approach of Eq. (13.21) can yield biased results if there is substantial difference (as in Fig. 13.8) between the bar-target response and the CTF. In this case, narrow-band filtering of the image data, either electronic or digital, can isolate the fundamental (sine-wave) component, and allow conversion of bar-target image data to MTF directly, without the need for a series representation.

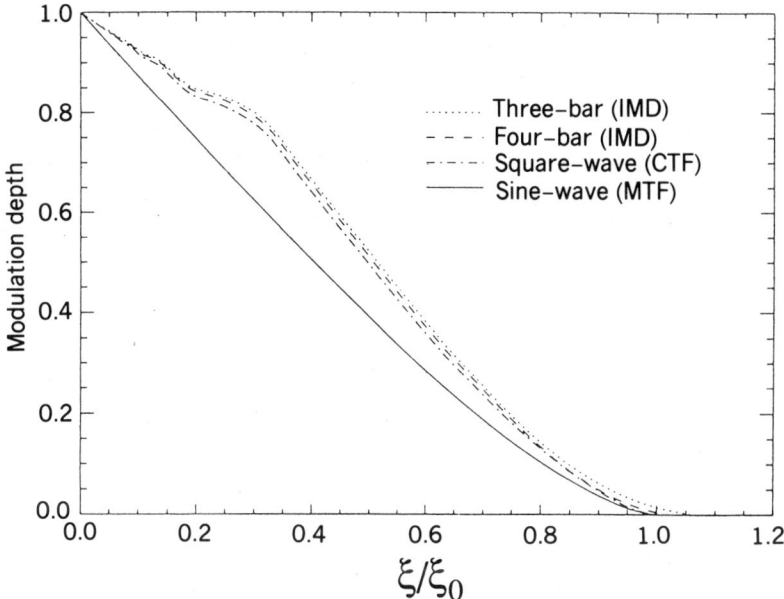

Figure 13.7. Comparison of MTF to image modulation depths obtained with infinite-square-wave, four-bar, and three-bar targets for a diffraction-limited system with circular pupil with cutoff frequency ξ_0.

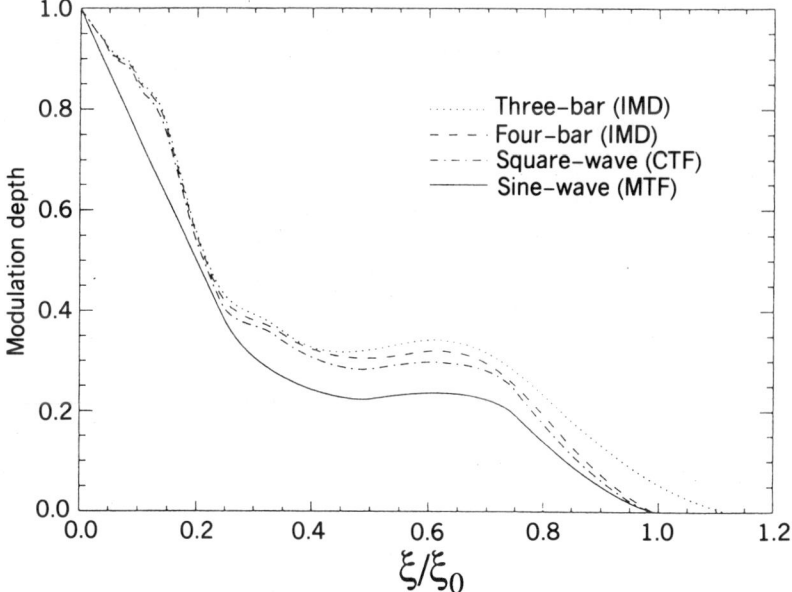

Figure 13.8. Comparison of MTF to image modulation depths obtained with infinite-square-wave, four-bar, and three-bar targets for a diffraction-limited annular-aperture MTF (50% diameter obscuration) with cutoff frequency ξ_0.

13.7.7. Random-Target Response

The MTF can also be measured by the response of the system to a random object of known spatial-frequency content. For situations where the MTF of an entire imager system is to be assessed (Fig. 13.6), the random target is made as a transparency (Daniels et al., 1995). Subsystem MTF measurements of a detector array alone are performed using laser-speckle techniques (Sensiper et al., 1993).

The main advantage of a random-target technique is that the MTF measurement is shift-invariant. Imager systems that contain a detector array typically have a shift-variant response (Park et al., 1984) for deterministic targets such as point sources, line sources, edges, and sine waves. A different impulse response is seen for different locations of the target, as in Fig. 13.9. A range of MTF results are seen, depending on the exact alignment of the image of the target with respect to the sampling sites of the detector array. Using noiselike test targets, as in Fig. 13.10, the target information has a random position with respect to the sampling sites. The resulting MTF measurement is shift invariant, in that small differences in the position of the target (in the x-y plane) do not change the measured MTF. The MTF produced is the position-averaged

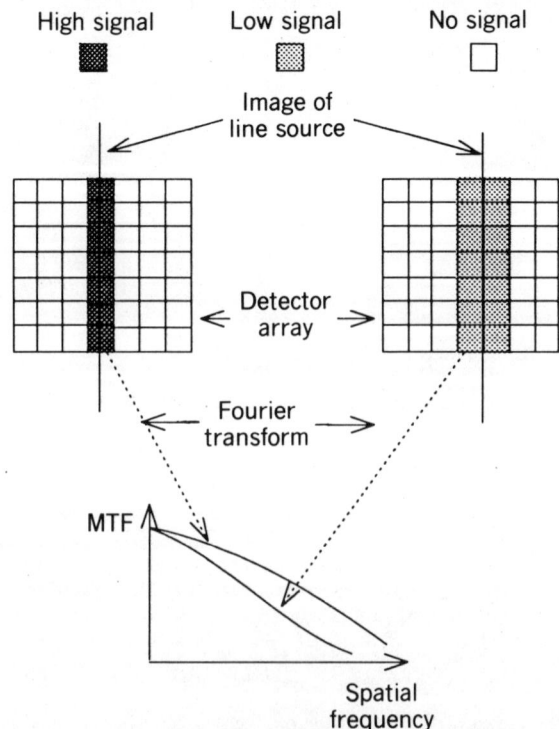

Figure 13.9. Shift-variant image formation with a detector array.

Figure 13.10. Band-limited white-noise random target.

MTF over the range of MTF curves seen in Fig. 13.9. This average MTF is more representative of the field performance of an imager system, where the target images of interest will be located randomly with respect to the rows and columns of the detector array.

Random targets are used that have a known input power spectral density (PSD) as a function of spatial frequency. The MTF relates the input and output PSDs:

$$PSD_{output}(\xi) = [MTF(\xi)]^2 \times PSD_{input}(\xi) \qquad (13.23)$$

The MTF can be calculated from a measured PSD_{output} as

$$MTF(\xi) = \sqrt{\frac{PSD_{output}(\xi)}{PSD_{input}(\xi)}} \qquad (13.24)$$

where the PSD_{output} is measured as an ensemble average of the squared finite-length Fourier transform of the detector-array voltage data

$$PSD_{output}(\xi) = \langle |\mathcal{F}\{v(x)\}|^2 \rangle \qquad (13.25)$$

Typically, these random targets are band-limited to avoid aliasing, so that the highest spatial frequency is less than the spatial Nyquist frequency of the detector array. This allows the MTF to be measured out to the Nyquist frequency. These targets have been used for infrared system testing (Daniels et al., 1995) in the 3- to 5-μm band (plastic-film transparencies) and in the 8- to 12-μm band (chrome-on-ZnSe transparencies).

REFERENCES

Alexander, T., G. D. Boreman, A. Ducharme, and R. Rapp, "Point Spread Function and MTF Characterization of the KHILS Infrared-Laser Scene Projector," *SPIE Proc.* **1969,** 270–284, (1993).

Barnard, K., G. D. Boreman, A. Plogstedt, and B. Anderson, "MTF Measurement of SPRITE Detectors: Sine Wave Response," *Appl. Opt.* **31,** 144–147 (1992).

Boreman, G. D., and A. E. Plogstedt, "Spatial Filtering by a Line-Scanned Nonrectangular Detector Application to SPRITE Readout MTF," *Appl. Opt.* **28,** 1165–1168 (1989).

Boreman, G. D., and S. Yang, "Modulation Transfer Function Measurement Using Three-Bar and Four-Bar Targets," *Appl. Opt.* **34,** 8050–8052 (1995).

Coltman, J. W., "The Specification of Imaging Properties by Response to a Sine Wave Input," *J. Opt. Soc. Am.* **44,** 468 (1954).

Daniels, A., G. D. Boreman, A. Ducharme, and E. Sapir, "Random Transparency Targets for Modulation-Transfer-Function Measurement in the Visible and Infrared," *Opt. Eng.* **34,** 860–868 (1995).

DeVelis, J. B., and G. B. Parrent, "Transfer Function for Cascaded Optical Systems," *J. Opt. Soc. Am.* **57,** 1486–1490 (1967).

Gaskill, J. D., *Linear Systems, Fourier Transforms, and Optics*, Wiley, New York, 1978.

Goodman, J. W. *Introduction to Fourier Optics*, McGraw-Hill, New York, 1968, pp. 116–120.

Kelly, D. H., "Spatial Frequency, Bandwidth, and Resolution," *Appl. Opt.* **4,** 435–437 (1965).

Lamberts, R. L., "The Production and Use of Variable-Transmittance Sinusoidal Test Objects," *Appl. Opt.* **2,** 273–276 (1963).

Park, S. K., R. Schowengerdt, and M. Kaczynski, "Modulation-Transfer-Function Analysis for Sampled Image Systems," *Appl. Opt.* **23,** 2572 (1984).

Reichenbach, S. E., S. K. Park, and R. Narayanswamy, "Characterizing Digital Image Acquisition Devices," *Opt. Eng.* **30,** 170–177 (1991).

Sensiper, M., G. D. Boreman, A. Ducharme, and D. Snyder, "Modulation Transfer Function Testing of Detector Arrays Using Narrow-Band Laser Speckle," *Opt. Eng.* **32,** 395–400 (1993).

Smith, W. J., *Modern Optical Engineering*, McGraw-Hill, New York, 1966.

Snyder, H. L., "Image Quality and Observer Performance," in *Perception of Displayed Information*, L. M. Biberman (ed.), Plenum, New York, 1973, 87–117.

Wittenstein, W., J. Fontanella, A. Newbery, and J. Baars, "The Definition of the OTF and the Measurement of Aliasing for Sampled-Imaging Systems," *Optica Acta* **29**, 41–50 (1982).

BIBLIOGRAPHY

Mallick, S., "Common-Path Interferometers, in *Optical Shop Testing* (D. Malacara, ed.), Wiley, New York, 1978, pp. 98–104.

Selected papers on Optical Transfer Function: Foundation and Theory, Lionel Baker, ed., SPIE Milestone Series, vol. MS 59, Bellingham, WA, 1992.

Selected papers on Optical Transfer Function: Measurement, Lionel Baker, ed., SPIE Milestone Series, vol. MS 60, Bellingham, WA, 1992.

Williams, C. S., and O. A. Becklund, *Introduction to the Optical Transfer Function*, Wiley-Interscience, New York, 1989.

PROBLEMS

1.31. For what spatial frequency (express in cycles per millimeter) will the MTF of a 50-μm-wide detector equal 50%?

13.2. A diffraction-limited, circular-aperture system operates at 10-μm wavelength. What $F/\#$ is required to have an MTF of at least 75% at 10 cy/mm? Is this a minimum or a maximum $F/\#$?

13.3 In what situation can the overall MTF of the system be better than the multiplication of the subsystem MTFs?

13.4. Given a measured MTF for each of two subsystems, an afocal telescope and a detector lens, and a measured MTF for both operating together as a system, will the measured system MTF be better, worse, or the same as the product of the two measured subsystem MTFs? Why?

13.5. What is the highest spatial frequency for which the MTF is at least 30% for a diffraction-limited $F/3$ system operating at 8 μm? Further assume that the image is scanned using a detector with an along-scan dimension 25 μm? What is the new 30%-MTF spatial frequency?

13.6. Given an imaging system that must fit completely within a standard 12-oz soft drink can. Estimate the best angular resolution (cycles/mrad) that the system could be expected to achieve, for operating wavelengths of 0.5 μm, 4 μm, and 10 μm.

13.7.   What is the image-plane spatial frequency if the object spatial frequency is 10 cycles/mrad, and the focal length of the system is 50 mm? 500 mm?

13.8.   If the collimator in Fig. 13.6 is diffraction-limited, and overfills the aperture of the system under test, must we correct the measured data for the MTF of the collimator? Why or why not?

13.9.   Assuming that the system under test has a circular aperture and is diffraction-limited, the impulse response is a circularly symmetric function proportional to $\{2J_1(r)/r\}^2$. Sketch the functional form of the line response. (*Hint:* You may want to consider the test configuration for which the impulse response is scanned by a long narrow detector.) Does the line response have nulls in the same manner in the Bessel function? Why or why not?

13.10. Sketch the orientation of the line source that gives a measurement of the ζ profile of the MTF. Under what conditions would two measurements (a ζ profile and an ξ profile) be sufficient to determine the two-dimensional MTF?

13.11. A measurement configuration similar to Fig. 13.6 is given. The system under test has a focal length of 50 mm and $F/\#$ of 2, and operates at a wavelength of 10 μm. The collimator has a focal length of 500 mm and a maximum aperture of 50 mm. Assume that both the collimator and the lens under test are diffraction-limited.

(a) How large is the impulse response formed in the image plane of the lens under test, assuming an infinitely narrow line source is used as the target? Sketch the measured MTF.

(b) How wide can the line source be and not affect the MTF measurement? Do not forget to include the two-lens magnification factor in your calculations.

(c) If the actual width of the line source used is 1 mm, sketch the Fourier transform of the measured image plane flux (uncorrected MTF).

(d) Along with the conditions of part (c), assume that the image-plane flux is scanned with a detector of dimension 25 μm. Sketch the Fourier transform of the measured image-plane flux (uncorrected MTF).

(e) Discuss the signal processing that you would use on the uncorrected MTF in part (d) to yield the MTF of the lens under test. What problems might arise in the signal processing because the detector is not noise free?

13.12. Find the cutoff frequency (express in cycles/mm) in both the object and the image, for a diffraction-limited, circular-aperture system that operates at a wavelength of 5 μm, and has a diameter of 2 cm. The object plane of the system is 5 cm away from the (thin) lens, and the image plane is at 10 cm. How are these spatial frequencies related to the geometrical magnification $\mathfrak{M}$ defined in Chapter 1?

13.13. Find the cutoff frequency (express in cycles/mrad in the object and cycles/mm in the image), for a diffraction-limited, circular-aperture system that operates at a wavelength of 5 μm, with a diameter of 2 cm, and a focal length of 4 cm.

14

Thermal-Imager Systems

14.1. INTRODUCTION

This chapter discusses concepts associated with the performance of thermal-imager systems, chiefly the noise-equivalent temperature difference (NETD) and the minimum resolvable temperature difference (MRTD). A system that is optimized for use as a thermal-imager system, as opposed to a search system, will operate on extended sources characterized by radiance, rather than point sources characterized by intensity. The thermal-imager system's function is to produce a picture that is a map of temperature differences related to spatial flux and emissivity differences across an extended target. We will consider the primary performance metrics relating to thermal-imager systems: thermal sensitivity and spatial resolution. Both attributes are necessary to produce good thermal imagery. Thermal sensitivity is concerned with the minimum temperature difference that can be discerned above the noise level, and spatial resolution is concerned with how small an object can be imaged by the system. We follow the general approach of Lloyd (1975) in the next section. As in Chapter 12, we derive the relevant design equations under the assumption that the $D*$ of the detector is independent of the $F/\#$ of the system that illuminates it; in other words, the system is not operating under background-limited infrared photodetector (BLIP) conditions. After carrying out the derivation under this assumption, we evaluate the BLIP case separately.

14.2. NETD DERIVATION

The configuration of the basic thermal-imager system is shown in Fig. 14.1. Using the methods of Chapter 2, we find the flux on the detector ϕ_d in terms of the radiance L at the location of the instantaneous field of view (IFOV):

$$\phi_d = L \times A_{\text{enp}} \times \Omega_d \tag{14.1}$$

$$\Omega_{\text{det}} = \frac{A_d}{F^2} = \frac{A_{\text{footprint}}}{r^2} \tag{14.2}$$

$$\phi_d = L \times \frac{\pi}{4} D_{\text{enp}}^2 \times \frac{A_d}{F^2} \tag{14.3}$$

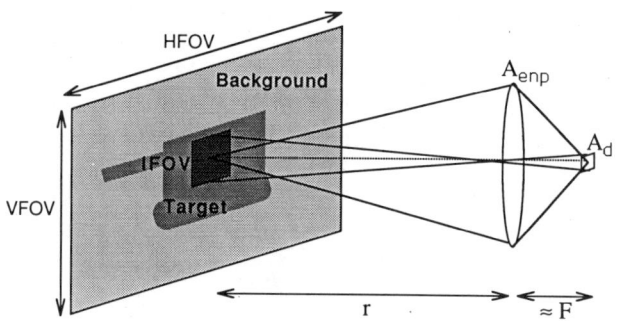

Figure 14.1. Thermal-imager system configuration.

$$\phi_d = L\,\frac{\pi}{4}\,\frac{A_d}{(F/\#)^2} \tag{14.4}$$

From Eq. (14.4), we see that the irradiance on the detector $E_d = \phi_d/A_d$ is independent of the range r, and depends only on the source radiance and the $F/\#$ of the optics. This assumes that the extended source fills (or overfills) the IFOV.

We obtain an expression for the signal voltage v_s produced by the detector by multiplying Eq. (14.4) by the responsivity, $\mathfrak{R}_v$:

$$v_s = \mathfrak{R}_v \times L\,\frac{\pi}{4}\,\frac{A_d}{(F/\#)^2} \tag{14.5}$$

where it is understood that detector responsivity and source radiance are functions of wavelength and that their product $\mathfrak{R}_v \times L$ denotes an integral over wavelength of the product of the spectral functions, over the passband of the system:

$$\mathfrak{R}_v \times L \equiv \int_{\lambda_1}^{\lambda_2} \mathfrak{R}_v(\lambda)\,L(\lambda)\,d\lambda \tag{14.6}$$

Because we are interested in a thermal mapping of the object, we write an expression for the change of the signal voltage for a change in temperature by taking $\partial/\partial T$ of each side of Eq. (14.5):

$$\frac{\Delta v_s}{\Delta T} = \mathfrak{R}_v \times \frac{\partial L}{\partial T} \times \frac{\pi}{4}\,\frac{A_d}{(F/\#)^2} \tag{14.7}$$

We substitute the following expression for responsivity into Eq. (14.7):

$$\Re_v \equiv \frac{v_n}{\text{NEP}} = \frac{v_n D^*}{\sqrt{A_d}\sqrt{\Delta f}} \tag{14.8}$$

realizing that the D^* is also a spectral quantity in the sense of Eq. (14.6). We obtain

$$\frac{\Delta v_s}{\Delta T} = \frac{v_n D^*}{\sqrt{A_d}\sqrt{\Delta f}} \times \frac{\partial L}{\partial T} \times \frac{\pi}{4} \frac{A_d}{(F/\#)^2} \tag{14.9}$$

Rearranging Eq. (14.9) in terms of a signal-to-noise ratio (SNR):

$$\text{SNR} = \frac{\Delta v_s}{v_n} = \Delta T \times \frac{D^*}{\sqrt{A_{\text{det}}}\sqrt{\Delta f}} \times \frac{\partial L}{\partial T} \times \frac{\pi}{4} \frac{A_{\text{det}}}{(F/\#)^2} \tag{14.10}$$

$$\text{SNR} = \Delta T \times \frac{D^*}{\sqrt{\Delta f}} \times \frac{\partial L}{\partial T} \times \frac{\pi}{4} \frac{\sqrt{A_d}}{(F/\#)^2} \tag{14.11}$$

We can now write Eq. (14.11) in terms of the NETD, which is the value of temperature difference that produces unity SNR, by setting SNR = 1 and solving for ΔT:

$$\text{NETD} = \frac{4}{\pi}\left[\frac{(F/\#)^2\sqrt{\Delta f}}{D^* \,\partial L/\partial T \,\sqrt{A_d}}\right] \tag{14.12}$$

where it should be noted that Eq. (14.12) applies specifically to a non-BLIP situation. The NETD characterizes the thermal sensitivity of an infrared system, that is, the amount of temperature difference required to produce a unity SNR. A smaller NETD indicates a better thermal sensitivity. Let us examine each term of Eq. (14.12) separately.

For the best sensitivity (lowest NETD), we want to maximize the spectral integral of the product of D^* and exitance contrast $\partial L/\partial T$, both being functions of wavelength

$$\int_{\lambda_1}^{\lambda_2} \frac{\partial L}{\partial T}(\lambda) \times D^*(\lambda)\,d\lambda \rightarrow \max \tag{14.13}$$

This will be approximately the case when the peak of the spectral responsivity (proportional to the spectral D^*) and the peak of the exitance contrast coincide. For a thermal-imager system, the spectral passband of the system should include the peak of the exitance contrast, at least from the point of view of thermal sensitivity. However, thermal-imager systems may not satisfy these conditions because of other constraints, such as atmospheric/obscurant transmittance effects or available detector technologies. For instance, many commercial thermal-imager systems are based on PtSi detector arrays, for which the system spectral passband is typically 3 to 5 μm. When used for imaging of room-temperature targets, this passband does not include the peak of the exitance-contrast curve, which is around 8 μm for a 300-K source. The choice of passband has been dictated by the availability of cost-effective detector arrays. The thermal sensitivity (NETD) is ultimately determined by Eq. (14.12), and while the NETD may be sufficient for a given imaging task, it is not as low as it might be, because the wavelength region where the source radiance changes the most for a given temperature change is not included in the detected flux.

NETD is proportional to the square root of the bandwidth, which is intuitive, given that the root-mean-square (rms) noise is proportional to the square root of the bandwidth. NETD is independent of the diameter of the optic as such, depending on the optic diameter only through the $F/\#$. According to Eq. (14.12), better (lower) NETDs result from lower $F/\#$s. This can be understood by consideration of Eq. (14.4) where, for an extended target (that at least fills the IFOV), a lower $F/\#$ results in more flux captured by the detector. This increases SNR for a given noise level.

The dependence of NETD on the area of the detector is critical. As seen in Eq. (14.10), the inverse-square-root dependence of NETD on detector area is the combined effect of two terms. The rms noise increases as the square root of the detector area. Also, for an extended source, the flux captured by the detector (proportional to the signal voltage produced) is proportional to the area of the detector. The net result is that NETD $\propto 1/(A_{det})^{1/2}$. A large detector thus gives a smaller (better) NETD. The fundamental drawback of NETD as a system-level performance descriptor is that while the thermal sensitivity of an infrared imager is better for larger detectors, the spatial resolution (image quality) is poorer for larger detectors. Thus, while NETD is a sufficient operational test, it cannot be applied as a design criterion. Another parameter, the MRTD, considers both thermal sensitivity and spatial resolution and is more appropriate for design.

Equation (14.12) must be modified when the system operates under BLIP conditions. As seen in Eq. (12.47), D^*_{BLIP} is proportional to the $F/\#$:

$$D^*_{BLIP}(\lambda, f) = (F/\#) \frac{\lambda}{hc} \sqrt{\frac{2\eta}{\pi L_{q,bkg}}} \qquad (14.14)$$

again assuming a cooled photovoltaic detector. Substituting Eq. (14.14) into Eq. (14.12), we find the following equation for NETD under BLIP conditions:

$$NETD = \frac{2\sqrt{2}}{\sqrt{\pi}} \frac{hc}{\lambda} \left[\frac{(F/\#)\sqrt{\Delta f}\sqrt{L_{q,bkg}}}{\partial L/\partial T \sqrt{A_d}\sqrt{\eta}} \right] \qquad (14.15)$$

Equation (14.15) has only a linear dependence on $F/\#$, rather than a squared dependence because in a BLIP system, the lower $F/\#$ brings in a higher background irradiance on the detector in addition to a higher signal irradiance. Other changes for the BLIP condition include that the NETD is inversely proportional to the square root of the quantum efficiency and that the NETD is proportional to the square root of the in-band background flux.

14.3. MEASUREMENT AND USE OF NETD

Measurement of NETD calibrates the thermal sensitivity of the imager system, and is a useful diagnostic tool to verify that full operational performance is available. It is an advantage that it is a relatively easy test to set up and perform. However, it is not a suitable design criterion, because it ignores image-resolution issues. The NETD of a system will get better (smaller) if larger detectors are used. The better thermal sensitivity obtained is a consequence of Eq. (14.12) and Eq. (14.15). However, for larger IFOVs, the image quality will be poorer because of the spatial averaging that occurs across the surface of the detector.

A typical NETD test target and the corresponding system response are shown in Fig. 14.2, consisting of a region of higher temperature superimposed on a background of similar but lower temperature. The dimensions of the higher-temperature region should be such that several IFOVs (projected to the target plane) will fit within it, to ensure that the spatial response of the system does not affect the measurement. The goal is to have the target overfill the IFOV.

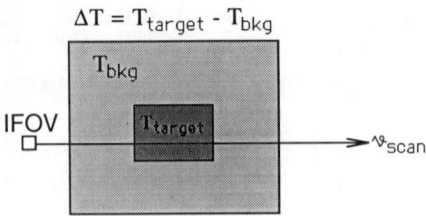

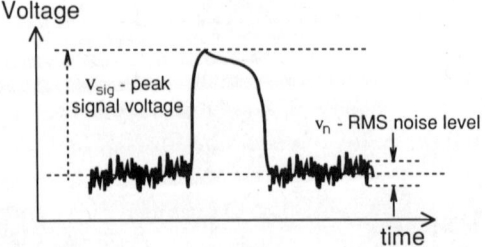

Figure 14.2. An NETD test pattern and the resulting waveform output from the system.

The temperature difference between target and background should be several times the expected NETD to ensure a response that is easily visible above the system noise. The target and background regions usually approximate black-bodies.

From the data (peak signal and rms noise) in Fig. 14.2, the NETD can be calculated as the temperature difference that would produce an SNR of 1:

$$\text{NETD} = \frac{\Delta T}{(v_s/v_n)} \tag{14.16}$$

For example, if an SNR of 50 is obtained for a ΔT of 2 K, the NETD is 40 mK. Some practical issues associated with the measurement of NETD are that the source must have a precise control for temperature difference, particularly if systems with low NETD are to be measured. Also, the calculation of the rms noise must exclude the hot-target region. The measurement is conceptually easy on an oscilloscope, but care must be taken when performing automatic computations on digitized data. Because of the dependence of noise on band-with, the NETD must be measured with the system running at its full operational scan rate, to obtain the proper dwell time and bandwidth.

The NETD (and also the MRTD) is defined in terms of a temperature difference, rather than the more fundamental quantity, the flux difference. It is a common practice to specify system performance in terms of temperature, but the amount of flux difference corresponding to a given temperature difference depends on the background or reference temperature used. Usually NETD and MRTD are measured with an ambient background temperature, but to extrapolate from this laboratory measurement to predict field performance requires a correction factor if the temperature of the background in the field is different from the temperature of the laboratory background (Driggers et al., 1992). In the measurement of high-performance systems with very small NETDs, the ambient temperature uncertainty can introduce variability into the NETD measurement.

14.4. JOHNSON CRITERIA

A more complete figure of merit for a thermal-imaging system will account for the spatial resolution as well as the thermal sensitivity. We explore the process of determining the spatial-resolution requirement for a thermal imager by consideration of the Johnson criteria (Johnson, 1958). The Johnson criteria provide a way of describing real targets in terms of simpler visual patterns. In the original tests, a variety of military targets were placed in front of a television camera. Alongside these targets were placed sets of square-wave patterns, with a selection of bar widths. When it was determined that a particular target could be detected (or recognized or identified), the corresponding just-resolved bar pattern of the set was noted also. The result of these tests for a variety of observers was that it was possible to characterize the performance of the average observer for any particular decision task in terms of the number of just-resolved cycles that would fit across the minimum dimension of the target. By normalizing the number of cycles in that bar pattern by the minimum dimension of the target, it was found that the spatial resolution required for a particular decision task was approximately constant.

The more complex decision tasks required a higher level of detail, and hence required better image resolution. For instance, better resolution is required for the identification task than for the detection task. Detection (presence discerned) was found to require 1 cycle per minimum dimension, recognition (such as, tank versus truck) required 4 cycles per minimum target dimension, while identification (such as, what kind of tank) required 6.4 cycles per minimum target dimension. These resolutions are based on a 50% probability of correct detection (or recognition or identification). This is the criterion generally used in design work. A higher degree of decision certainty (probabilities of correct decision greater than 50%) requires finer spatial resolution than the numbers quoted.

This information allows us to set the resolution requirements for the system, and begins to solidify the parameter choices made in the design process. The mission of the system (the desired decision task) must be consistent with the relationship of the target size, target distance, optics focal length, and detector dimension. Once the required number of cycles per minimum target dimension (n_{cy}) is determined from the Johnson criterion, the required angular spatial frequency (ξ in cycles/rad) is given by

$$\xi = \frac{n_{cy}}{x_{min}/r} \tag{14.17}$$

where x_{min} is the minimum target dimension and r is the distance to the target (in the same units). The denominator of Eq. (14.17) is the angular subtense of the target. For Nyquist sampling, two IFOVs are needed per cycle of the highest spatial frequency required for the decision task of interest. This determines the IFOV in terms of the other parameters:

$$\frac{1}{2 \, \text{IFOV}} = \frac{F}{2\sqrt{A_d}} = \frac{n_{cy}}{x_{min}/r} \tag{14.18}$$

$$\frac{F}{\sqrt{A_d}} = \frac{2n_{cy}r}{x_{min}} \tag{14.19}$$

As an example of the use of Eq. (14.12) and Eq. (14.19) in first-order design, we consider the recognition of a tank (2.3 m × 2.3 m) at a distance of 2 km. We have a single-element detector, of dimension 50 μm × 50 μm with a band-averaged D^* of 2×10^{10} cm Hz$^{-1/2}$ W^{-1}. Suppose that the required NETD is 0.10 K. What is the required front-end diameter of the optics,

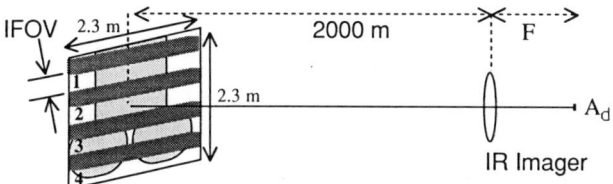

Figure 14.3. Use of the Johnson criterion to determine the required IFOV.

such that the system can operate at 30 frames/s? Assume operation in the 8- to 12-μm band, and a 1-degree field. First, we use Eq. (14.19) to determine the focal length of the system as 34.78 cm, with the requirement of 4 cycles resolved on target (Fig. 14.3). The IFOV $= (A_d)^{1/2}/F$ is 144 μrad. We have approximately 128 144-μrad-subtense pixels across the 1-degree FOV. The dwell time is then $(1/30)/128^2 \approx 2$ μs, with a resulting bandwidth $\Delta f = 246$ kHz. We then use Eq. (14.12), with a value of $\partial M_\lambda(\lambda = 10 \ \mu\text{m})/\partial T = \pi \partial L_\lambda/\partial T \approx 5 \times 10^{-5}$ W cm^{-2} μm^{-1} K^{-1} calculated from the Planck equation, and a spectral bandwidth of 4 μm (across the 8- to 12-μm band). This yields an $F/\#$ of 1.0, giving a front-end optics diameter of 34.78 cm ($\approx$ 14''). If, instead of a single detector, we use a 128 $\times$ 128 focal-plane array (with the same D^*), the bandwidth is reduced to 15 Hz and the allowed $F/\#$ increases to 11, yielding a smaller required optics diameter of 3.2 cm ($\approx$ 1.25'') to achieve the same NETD. This purely radiometric analysis has ignored diffraction effects. We have assumed that the front-end aperture diameter is such that the diffraction modulation transfer function (MTF) yields an impulse response that is smaller than the detector. This allows a geometrical-optics calculation of the IFOV. The 3.2-cm-diameter case above violates this assumption, yielding a differential blur of around 600 μrad.

14.5. MINIMUM RESOLVABLE TEMPERATURE DIFFERENCE

In thermal-imager systems, a combination of both spatial resolution and thermal sensitivity determines the final performance of the system. At low spatial frequencies, thermal sensitivity is most important, while at high spatial frequencies, spatial resolution is the dominant effect. The MRTD combines thermal sensitivity and spatial resolution. It is useful as a summary measure of performance and a design criterion. MRTD answers the question: What temperature difference is required for various-sized bar targets to be visible on the display? MRTD is the temperature difference required between bars and spaces of a test target (having fundamental spatial frequency ξ_f) so that the bars can be ''just discerned'' by a trained observer with an unlimited viewing time. Smaller is better for MRTD, as is the case for NETD. MRTD includes the effects of the display and the observer, which makes it a good end-to-end metric, but this makes it difficult to quantitatively measure because of inconsis-

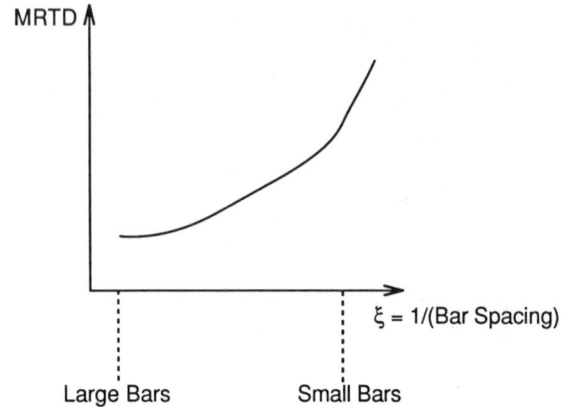

Figure 14.4.   MRTD is an increasing function of spatial frequency.

tencies and variabilities of the human observer, and the number of factors involved. MRTD is the observed thermal sensitivity of the thermal imager as a function of spatial frequency.

MRTD is a better system-performance descriptor than MTF alone because MTF is just an attenuation of modulation depth, without regard for a noise level. MRTD is better than NETD because it accounts for both resolution and noise level. MRTD includes the effect of spatial integration, temporal integration, noise, and resolution. It correlates well with target recognition in imagery corrupted by temporal noise, but does not have a one-to-one correspondence with field performance because of complex issues such as background clutter and spatial noise, the effects of which are not included in the MRTD. Smaller bar targets will require larger temperature differences to be visible (Fig. 14.4), because of the falloff in MTF at higher spatial frequencies.

14.6. MEASUREMENT OF MRTD

MRTD is measured using the setup shown schematically in Fig. 14.5. The MRTD test target, which may be either a single bar target or a set of various-sized bar targets, is backlit by a uniform extended-blackbody source. The target is placed at the focus of a high-quality collimator (typically reflective) and thus appears to be at infinity, simulating the infinite-conjugate case for which the thermal imager is typically designed. The collimator MTF should be of sufficient quality not to affect the MRTD measurement at the spatial frequencies of interest. The collimator should be of sufficient aperture to overfill the aperture of the system under test, so that all optical aberrations in the system are included in the measurement. This also ensures that the limiting aperture, from a diffraction point of view, is that of the system under test. The target is then imaged into the system under test and is displayed to the observer. The controls on the display are typically set so that the noise of the display is just

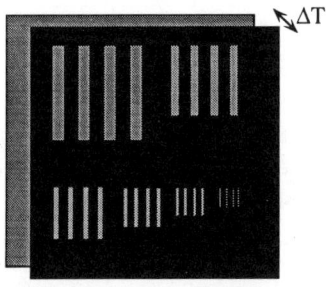

MRTD test target, backlit by an
extended blackbody source.

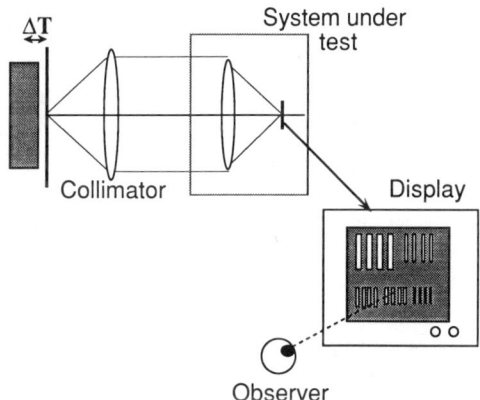

Figure 14.5. Schematic of apparatus for MRTD measurement.

visible, because the best sensitivity is obtained under noise-limited rather than contrast-limited conditions. For largest bar target (lowest spatial frequency), begin at $\Delta T = 0$ and increase the temperature difference slowly until the bars are seen by the observer. Repeat the process for each higher spatial frequency until the spatial frequency is sufficiently high that the bars cannot be seen for any value of ΔT.

14.7. ANALYTICAL MODEL FOR MRTD

The use of MRTD as a system-level descriptor overcomes the main deficiency inherent in a specification of NETD, in that MRTD accounts for resolution. When specified in terms of MRTD, the system does not appear to have better performance as the IFOV increases in size, as it would if NEDT were used. Another difference is the inclusion of the human-observer effects into the performance model. The derivation of an exact analytical expression for MRTD (Lloyd, 1975) is complex because of the number of variables that contribute. We can gain insight to the behavior of MRTD through the proportionality of Eq. (14.20), which includes the main variables of interest for our purposes:

$$MRTD(\xi_f) \propto \frac{NETD \times \xi_f \times \sqrt{HIFOV \times VIFOV}}{MTF(\xi_f) \times \sqrt{\tau_{int}/\tau_f}} \qquad (14.20)$$

where ξ_f is the fundamental spatial frequency of the bar target, τ_{int} is the integration time of the eye, and τ_f is the frame time. Substituting in for a (non-BLIP) NETD (Eq. (14.12)), we obtain

$$MRTD(\xi_f) \propto \frac{\xi_f \times \sqrt{HIFOV \times VIFOV}}{MTF(\xi_f) \times \sqrt{\tau_{int}/\tau_f}} \left[\frac{(F/\#)^2 \sqrt{\Delta f}}{D^* \, \partial L/\partial T \, \sqrt{A_d}} \right] \qquad (14.21)$$

From Eq. (14.21) we can see that MRTD has the same dependence as NETD on $F/\#$, bandwidth, D^*, and radiance contrast. However, it is no longer possible to increase the apparent performance of the system by increasing the IFOVs. The MRTD is not improved by an increase in detector area or IFOV. The MTF term is the denominator of Eq. (14.21), and as the MTF decreases, the ΔT required for the bars to be seen will increase. Actually MRTD will increase faster than MTF drops off because of the extra term of spatial frequency in the numerator. Equation (14.21) also shows that there is benefit to be gained from the temporal integration of the eye. If the frame rate is fast enough that several samples are received by the eye during its integration time, then the eye–brain system will tend to average out some of the noise, leading to a lower MRTD.

We also see from Eq. (14.21) that the MRTD curve will have a vertical asymptote at the spatial frequency where the MTF goes to zero. Because all thermal-imager systems operate with a finite detector size, this asymptote frequency cannot be greater than the inverse of the detector angular subtense (the zero of the corresponding sinc function) seen in Fig. 14.6. Actually, the MRTD will have large values at a somewhat lower spatial frequency than 1/IFOV because there are other MTF terms than just the detector footprint, and because of the extra term of ξ_f in the numerator of Eq. (14.21).

MRTD is a design criterion that flows down from computer-aided performance models (for example, FLIR92). The detection and recognition ranges can be predicted from an MRTD curve for a given target size, target temperature, and background temperature, given an atmospheric attenuation. It is of practical interest to measure MRTD without the need for a human observer, particularly from the point of view of system acceptance testing. This "objective MRTD" measurement is faster and more cost effective than measurements that include the observer each time. In this context, Eq. (14.21) can be written as

$$MRTD(\xi_f) = K(\xi_f) \frac{NETD}{MTF(\xi_f)} \qquad (14.22)$$

where the constant of proportionality and any spatial-frequency-dependent terms (including the effect of the observer) are taken up into the function $K(\xi_f)$.

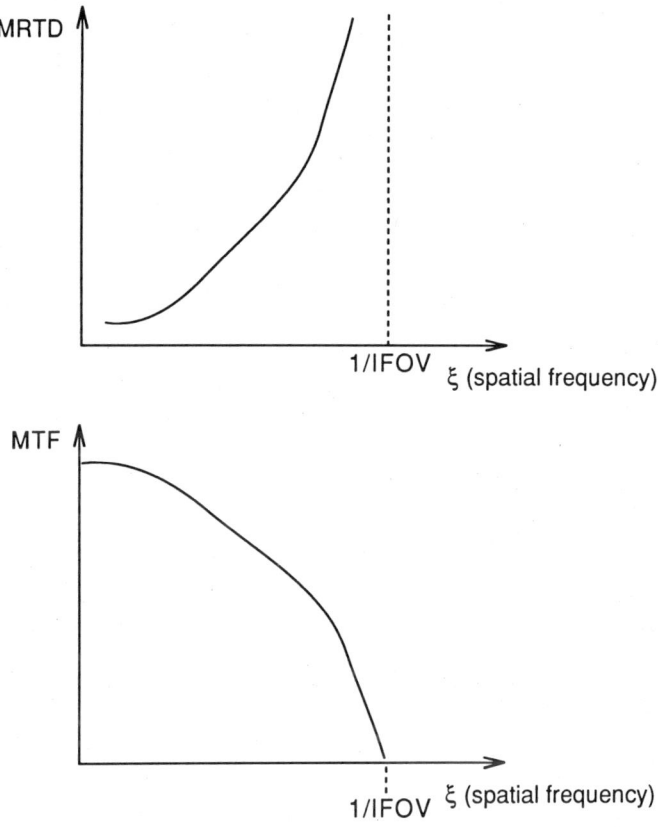

Figure 14.6. MRTD has a vertical asymptote where MTF goes to zero.

To characterize the average effects of the observer, for a given display and viewing geometry an MRTD curve is measured for a representative sample of the system under test. Along with the MRTD data, the NETD and MTF are recorded for this system. From these data, the function $K(\xi_f)$ can be determined, and subsequent tests of similar systems can be performed without the observer. NETD and MTF can be measured on the analog video before the display, and thus can be automated. The resulting matrix of acceptable NETD values and MTF cuves will produce an MRTD curve that will satisfy a given specification, and acceptance testing can proceed without the need for observers for each system test.

REFERENCES

Driggers, R. G., G. L. Boylston, and G. T. Edwards, "Equivalent Temperature Differences with Respect to Ambient Temperature Difference as a Function of Background Temperature," *Opt. Eng.* **31**(6), 1357–1361 (1992).

Johnson, J., "Analysis of Image Forming Systems,"*Proceedings of the Image Intensifier Symposium*, Ft. Belvoir, VA, 1958. (Reprinted in *SPIE Proc.* vol. 513, *Selected Papers on Infrared Design*, R. B. Johnson and W. L. Wolfe (eds.), Bellingham, WA, 1985, pp. 761–781.)

Lloyd, J. M., *Thermal Imaging Systems*, Plenum, New York, 1975.

BIBLIOGRAPHY

Campana, S. B., ed., *The Infrared and Electro-Optical Systems Handbook*, vol. 5, SPIE, Bellingham, WA, 1993.

Johnson, R. B., and W. L. Wolfe, eds., *SPIE Proc.* vol. 513, *Selected Papers on Infrared Design*, Bellingham, WA, 1985.

Seyrafi, K., *Electro-Optical Systems Analysis*, Electro-Optical Research Company, Los Angeles, 1985.

Spiro, I. J., and M. Schlessinger, *Infrared Technology Fundamentals*, Dekker, New York, 1989.

PROBLEMS

14.1. Write an expression analogous to Eq. (14.21) for a BLIP MRTD.

14.2. The amount of flux difference corresponding to a 3-K temperature difference at a background temperature of 300 K corresponds to what quantity temperature difference if the background is at (a) 250 K? (b) 325 K?

14.3. What is the impact of making the system bandpass wider on the NETD in Eq. (14.12)? Is this a realistic way to design the system? Why or why not?

14.4. If the NETD of a system were measured to be 0.25 K, and the scan rate were then slowed by a factor of 2, what would the new NETD be?

14.5. Discuss the uncertainty in the NETD, as a function of the length of the digitized data record used to calculate it.

14.6. If a target is imaged with sufficient resolution that it can be recognized with a 50% probability, what can be said about the corresponding probabilities of detection and identification?

14.7. (a) What is the spatial-frequency cutoff of a system that can detect a 2.3-m-by-2.3-m target at a distance of 10 km?

   (b) For a given detector size, does this imply a longer or shorter focal length than for use against the same target at closer range?

   (c) If this system operated in the 8- to 12-μm band, what would be the minimum front-end aperture size, from a diffraction point of view?

14.8. Ignoring diffraction effects, plot the required aperture diameter versus $D*$ for a non-BLIP system to have a given NETD. Repeat for a BLIP system.

14.9. Ignoring diffraction effects, we want to design a system that will identify a 1-m² target at a distance of 600 m. The required NETD is 0.025 K. The field is 5 degrees by 5 degrees. Assume non-BLIP conditions apply.

(a) What is the maximum frame rate, if the system has a single-element detector of $D* = 10^{10}$, and an aperture diameter of 5 cm?

(b) Recalculate the maximum frame rate if the required NETD is 0.5 K.

(c) For the NETD of 0.025 K, what aperture diameter would be required if a staring array were available with sufficient resolution? Assume $D*$ for each detector is 5×10^{10}, and a frame rate of 1/30 s. Is the resulting aperture diameter reasonable from a diffraction point of view if the system is to operate in the 8- to 12-μm band?

Appendix A

Blackbody Integrals for Photon Exitance and Radiant Exitance

Temperature = 28 K

Lambda Cutoff (μm)	Photon Exitance (photons/s-cm^2)	Blackbody Exitance (W/cm^2)
1	0.000E+00	0.000E+00
2	0.000E+00	0.000E+00
3	0.000E+00	0.000E+00
4	0.000E+00	0.000E+00
5	0.000E+00	0.000E+00
6	0.000E+00	0.000E+00
7	0.000E+00	0.000E+00
8	0.000E+00	0.000E+00
9	0.000E+00	0.000E+00
10	0.000E+00	0.000E+00
11	1.628E−02	3.010E−22
12	6.738E−01	1.144E−20
13	1.553E+01	2.440E−19
14	2.264E+02	3.308E−18
15	2.287E+03	3.126E−17
16	1.717E+04	2.204E−16
17	1.010E+05	1.223E−15
18	4.849E+05	5.557E−15
19	1.963E+06	2.136E−14
20	6.876E+06	7.122E−14
21	2.128E+07	2.104E−13
22	5.920E+07	5.598E−13
23	1.501E+08	1.361E−12
24	3.511E+08	3.057E−12
25	7.650E+08	6.407E−12
26	1.565E+09	1.263E−11
27	3.030E+09	2.360E−11
28	5.581E+09	4.201E−11
29	9.834E+09	7.163E−11
30	1.665E+10	1.175E−10
31	2.721E+10	1.862E−10
32	4.303E+10	2.859E−10
33	6.607E+10	4.267E−10
34	9.878E+10	6.206E−10
35	1.441E+11	8.816E−10
36	2.056E+11	1.226E−09
37	2.874E+11	1.671E−09
38	3.942E+11	2.236E−09
39	5.314E+11	2.944E−09
40	7.049E+11	3.817E−09
41	9.214E+11	4.879E−09
42	1.188E+12	6.154E−09
43	1.512E+12	7.670E−09
44	1.902E+12	9.451E−09
45	2.366E+12	1.152E−08
46	2.914E+12	1.391E−08
47	3.554E+12	1.665E−08
48	4.295E+12	1.975E−08
49	5.148E+12	2.324E−08
50	6.122E+12	2.715E−08

Temperature = 77 K

Lambda Cutoff (μm)	Photon Exitance (photons/s-cm²)	Blackbody Exitance (W/cm²)
1	0.000E+00	0.000E+00
2	0.000E+00	0.000E+00
3	0.000E+00	0.000E+00
4	3.386E−01	1.722E−20
5	2.501E+03	1.023E−16
6	8.904E+05	3.051E−14
7	5.656E+07	1.671E−12
8	1.231E+09	3.202E−11
9	1.318E+10	3.064E−10
10	8.602E+10	1.811E−09
11	3.928E+11	7.564E−09
12	1.374E+12	2.441E−08
13	3.920E+12	6.468E−08
14	9.538E+12	1.470E−07
15	2.044E+13	2.960E−07
16	3.956E+13	5.405E−07
17	7.039E+13	9.111E−07
18	1.169E+14	1.438E−06
19	1.831E+14	2.148E−06
20	2.730E+14	3.064E−06
21	3.904E+14	4.201E−06
22	5.387E+14	5.571E−06
23	7.206E+14	7.176E−06
24	9.382E+14	9.015E−06
25	1.193E+15	1.108E−05
26	1.486E+15	1.336E−05
27	1.817E+15	1.584E−05
28	2.186E+15	1.851E−05
29	2.592E+15	2.134E−05
30	3.034E+15	2.432E−05
31	3.511E+15	2.742E−05
32	4.020E+15	3.063E−05
33	4.559E+15	3.393E−05
34	5.126E+15	3.729E−05
35	5.720E+15	4.071E−05
36	6.337E+15	4.416E−05
37	6.975E+15	4.764E−05
38	7.632E+15	5.112E−05
39	8.306E+15	5.460E−05
40	8.995E+15	5.806E−05
41	9.695E+15	6.150E−05
42	1.041E+16	6.490E−05
43	1.113E+16	6.827E−05
44	1.185E+16	7.159E−05
45	1.259E+16	7.486E−05
46	1.332E+16	7.807E−05
47	1.406E+16	8.122E−05
48	1.480E+16	8.431E−05
49	1.553E+16	8.733E−05
50	1.627E+16	9.029E−05

Temperature = 197 K

Lambda Cutoff (μm)	Photon Exitance (photons/s-cm^2)	Blackbody Exitance (W/cm^2)
1	0.000E+00	0.000E+00
2	9.405E+04	9.622E−15
3	8.311E+09	5.752E−10
4	2.113E+12	1.114E−07
5	5.356E+13	2.295E−06
6	4.361E+14	1.582E−05
7	1.874E+15	5.925E−05
8	5.430E+15	1.528E−04
9	1.215E+16	3.094E−04
10	2.277E+16	5.308E−04
11	3.754E+16	8.100E−04
12	5.635E+16	1.135E−03
13	7.877E+16	1.491E−03
14	1.042E+17	1.865E−03
15	1.320E+17	2.246E−03
16	1.616E+17	2.625E−03
17	1.923E+17	2.995E−03
18	2.237E+17	3.352E−03
19	2.553E+17	3.692E−03
20	2.869E+17	4.013E−03
21	3.180E+17	4.314E−03
22	3.485E+17	4.596E−03
23	3.783E+17	4.895E−03
24	4.071E+17	5.104E−03
25	4.351E+17	5.330E−03
26	4.620E+17	5.540E−03
27	4.879E+17	5.735E−03
28	5.128E+17	5.914E−03
29	5.367E+17	6.081E−03
30	5.595E+17	6.234E−03
31	5.814E+17	6.377E−03
32	6.022E+17	6.508E−03
33	6.222E+17	6.630E−03
34	6.413E+17	6.744E−03
35	6.595E+17	6.848E−03
36	6.768E+17	6.946E−03
37	6.934E+17	7.036E−03
38	7.093E+17	7.120E−03
39	7.244E+17	7.198E−03
40	7.389E+17	7.271E−03
41	7.527E+17	7.338E−03
42	7.659E+17	7.402E−03
43	7.785E+17	7.461E−03
44	7.905E+17	7.516E−03
45	8.021E+17	7.567E−03
46	8.131E+17	7.615E−03
47	8.237E+17	7.661E−03
48	8.338E+17	7.703E−03
49	8.435E+17	7.743E−03
50	8.528E+17	7.780E−03

Temperature = 300 K

Lambda Cutoff (μm)	Photon Exitance (photons/s-cm^2)	Blackbody Exitance (W/cm^2)
1	6.067E+00	1.233E−18
2	4.108E+10	4.268E−09
3	5.636E+13	3.997E−06
4	1.798E+15	9.801E−05
5	1.319E+16	5.902E−04
6	4.721E+16	1.807E−03
7	1.132E+17	3.814E−03
8	2.126E+17	6.442E−03
9	3.406E+17	9.432E−03
10	4.897E+17	1.255E−02
11	6.519E+17	1.562E−02
12	8.207E+17	1.854E−02
13	9.905E+17	2.124E−02
14	1.158E+18	2.370E−02
15	1.319E+18	2.592E−02
16	1.474E+18	2.790E−02
17	1.621E+187	2.967E−02
18	1.760E+18	3.125E−02
19	1.890E+18	3.264E−02
20	2.012E+18	3.389E−02
21	2.125E+18	3.499E−02
22	2.232E+18	3.597E−02
23	2.330E+18	3.684E−02
24	2.423E+18	3.762E−02
25	2.509E+18	3.832E−02
26	2.589E+18	3.895E−02
27	2.664E+18	3.951E−02
28	2.733E+18	4.001E−02
29	2.799E+18	4.047E−02
30	2.860E+18	4.088E−02
31	2.917E+18	4.125E−02
32	2.970E+18	4.158E−02
33	3.020E+18	4.189E−02
34	3.067E+18	4.217E−02
35	3.111E+18	4.242E−02
36	3.153E+18	4.266E−02
37	3.192E+18	4.287E−02
38	3.228E+18	4.306E−02
39	3.263E+18	4.324E−02
40	3.296E+18	4.341E−02
41	3.326E+18	4.356E−02
42	3.356E+18	4.370E−02
43	3.383E+18	4.383E−02
44	3.409E+18	4.395E−02
45	3.434E+18	4.406E−02
46	3.458E+18	4.416E−02
47	3.480E+18	4.425E−02
48	3.501E+18	4.434E−02
49	3.521E+18	4.443E−02
50	3.541E+18	4.450E−02

Temperature = 500 K

Lambda Cutoff (μm)	Photon Exitance (photons/s-cm^2)	Blackbody Exitance (W/cm^2)
1	2.231E+09	4.601E−10
2	1.060E+15	1.137E−04
3	6.108E+16	4.554E−03
4	4.046E+17	2.365E−02
5	1.169E+18	5.718E−02
6	2.267E+18	9.683E−02
7	3.537E+18	1.357E−01
8	4.846E+18	1.704E−01
9	6.109E+18	2.000E−01
10	7.283E+18	2.246E−01
11	8.352E+18	2.448E−01
12	9.313E+18	2.615E−01
13	1.017E+19	2.751E−01
14	1.093E+19	2.864E−01
15	1.161E+19	2.957E−01
16	1.222E+19	3.034E−01
17	1.276E+19	3.099E−01
18	1.324E+19	3.154E−01
19	1.367E+19	3.200E−01
20	1.406E+19	3.240E−01
21	1.440E+19	3.273E−01
22	1.472E+19	3.302E−01
23	1.500E+19	3.327E−01
24	1.526E+19	3.349E−01
25	1.549E+19	3.368E−01
26	1.570E+19	3.385E−01
27	1.590E+19	3.399E−01
28	1.608E+19	3.412E−01
29	1.624E+19	3.424E−01
30	1.639E+19	3.434E−01
31	1.653E+19	3.443E−01
32	1.666E+19	3.451E−01
33	1.678E+19	3.458E−01
34	1.689E+19	3.465E−01
35	1.699E+19	3.471E−01
36	1.708E+19	3.476E−01
37	1.717E+19	3.481E−01
38	1.726E+19	3.485E−01
39	1.733E+19	3.489E−01
40	1.740E+19	3.493E−01
41	1.747E+19	3.496E−01
42	1.754E+19	3.499E−01
43	1.760E+19	3.502E−01
44	1.765E+19	3.504E−01
45	1.770E+19	3.507E−01
46	1.775E+19	3.509E−01
47	1.780E+19	3.511E−01
48	1.785E+19	3.513E−01
49	1.789E+19	3.515E−01
50	1.793E+19	3.516E−01

Temperature = 1000 K

Lambda Cutoff (μm)	Photon Exitance (photons/s-cm²)	Blackbody Exitance (W/cm²)
1	8.480E+15	1.819E−03
2	3.237E+18	3.784E−01
3	1.814E+19	1.549E+00
4	3.877E+19	2.727E+00
5	5.827E+19	3.593E+00
6	7.450E+19	4.183E+00
7	8.748E+19	4.582E+00
8	9.775E+19	4.855E+00
9	1.059E+20	5.046E+00
10	1.124E+20	5.183E+00
11	1.177E+20	5.284E+00
12	1.221E+20	5.359E+00
13	1.256E+20	5.416E+00
14	1.286E+20	5.460E+00
15	1.311E+20	5.494E+00
16	1.333E+20	5.521E+00
17	1.351E+20	5.544E+00
18	1.367E+20	5.561E+00
19	1.380E+20	5.576E+00
20	1.392E+20	5.588E+00
21	1.403E+20	5.599E+00
22	1.412E+20	5.607E+00
23	1.420E+20	5.614E+00
24	1.428E+20	5.621E+00
25	1.434E+20	5.626E+00
26	1.440E+20	5.630E+00
27	1.445E+20	5.634E+00
28	1.450E+20	5.638E+00
29	1.455E+20	5.641E+00
30	1.458E+20	5.644E+00
31	1.462E+20	5.646E+00
32	1.465E+20	5.648E+00
33	1.468E+20	5.650E+00
34	1.471E+20	5.651E+00
35	1.474E+20	5.653E+00
36	1.476E+20	5.654E+00
37	1.478E+20	5.655E+00
38	1.480E+20	5.657E+00
39	1.482E+20	5.658E+00
40	1.484E+20	5.658E+00
41	1.486E+20	5.659E+00
42	1.487E+20	5.660E+00
43	1.489E+20	5.661E+00
44	1.490E+20	5.661E+00
45	1.491E+20	5.662E+00
46	1.492E+20	5.662E+00
47	1.494E+20	5.663E+00
48	1.495E+20	5.663E+00
49	1.496E+20	5.664E+00
50	1.497E+20	5.664E+00

Temperature = 2856 K

Lambda Cutoff (μm)	Photon Exitance (photons/s-cm^2)	Blackbody Exitance (W/cm^2)
1	3.589E+20	9.076E+01
2	1.635E+21	2.686E+02
3	2.390E+21	3.308E+02
4	2.788E+21	3.538E+02
5	3.014E+21	3.639E+02
6	3.153E+21	3.690E+02
7	3.243E+21	3.718E+02
8	3.306E+21	3.735E+02
9	3.351E+21	3.745E+02
10	3.384E+21	3.752E+02
11	3.409E+21	3.757E+02
12	3.429E+21	3.760E+02
13	3.445E+21	3.763E+02
14	3.457E+21	3.765E+02
15	3.467E+21	3.766E+02
16	3.476E+21	3.767E+02
17	3.483E+21	3.768E+02
18	3.489E+21	3.769E+02
19	3.494E+21	3.769E+02
20	3.499E+21	3.770E+02
21	3.502E+21	3.770E+02
22	3.506E+21	3.770E+02
23	3.509E+21	3.771E+02
24	3.511E+21	3.771E+02
25	3.514E+21	3.771E+02
26	3.516E+21	3.771E+02
27	3.517E+21	3.771E+02
28	3.519E+21	3.771E+02
29	3.521E+21	3.772E+02
30	3.522E+21	3.772E+02
31	3.523E+21	3.772E+02
32	3.524E+21	3.772E+02
33	3.525E+21	3.772E+02
34	3.526E+21	3.772E+02
35	3.527E+21	3.772E+02
36	3.528E+21	3.772E+02
37	3.529E+21	3.772E+02
38	3.529E+21	3.772E+02
39	3.530E+21	3.772E+02
40	3.530E+21	3.772E+02
41	3.531E+21	3.772E+02
42	3.531E+21	3.772E+02
43	3.532E+21	3.772E+02
44	3.532E+21	3.772E+02
45	3.533E+21	3.772E+02
46	3.533E+21	3.772E+02
47	3.533E+21	3.772E+02
48	3.534E+21	3.772E+02
49	3.534E+21	3.772E+02
50	3.534E+21	3.772E+02

Appendix B

Basic Program for Integrating Planck Equation

```
100 DIM STOREPH(5, 20), STOREL(5, 20)
110 Y = 1 : Z = 1
120 KEY OFF : CLS
130 INPUT "TEMPERATURE(K), SHORT, LONG WAVELENGTHS(um)"; T, W1,
       W2
140 INPUT "PHOTONS OR ENERGY: 1 OR 2"; A
150 C2 = 1.4388 : C = 3E+10 : H = 6.626E-34
160 ON A GOTO 170, 300
170 W = W1 : GOSUB 430
180 Q1 = Q : W = W2 : GOSUB 430
190 Q2 = Q : QB = Q2 - Q1 : W = W1 : GOSUB 510
200 Q3 = DQ : W = W2 : GOSUB 510
210 Q4 = DQ : QD = Q4 - Q3
220 PRINT : PRINT "WAVEBAND", "TEMPERATURE",
       "PHOTANCE(ph/s-cm-2-sr)", "CONTRAST(ph/s-cm-2-sr-k)"
230 STOREPH(1, Y) = W1 : STOREPH(2, Y) = W2 : STOREPH(3, Y) = T :
       STOREPH(4, Y) = QB : STOREPH(5, Y) = QD
240 FOR J = 1 TO Y
250 PRINT STOREPH(1, J); "-"; STOREPH(2, J), STOREPH(3, J), STOREPH(4,
       J), STOREPH(5, J) : NEXT J
270 PRINT : INPUT "RUN AGAIN OR QUIT: 1 OR 2"; V
280 IF V = 1 THEN Y = Y + 1 : GOTO 120
290 END
300 W = W1 : GOSUB 590
310 L1 = L : W = W2 : GOSUB 590
320 L2 = L : LB = L2 - L1 : W = W1 : GOSUB 670
330 D1 = D : W = W2 : GOSUB 670
340 D2 = D : DB = D2 - D1
350 PRINT : PRINT "WAVEBAND", "TEMPERATURE", "RADIANCE
       (W/cm-2-sr)", "CONTRAST (W/cm-2-sr-K)"
360 STOREL(1, Z) = W1 : STOREL(2, Z) = W2 : STOREL (3, Z) = T : STOREL(4, Z)
       = LB : STOREL(5, Z) = DB
370 FOR B = 1 TO Z
```

```
380 PRINT STOREL(1, B); "-"; STOREL(2, B), STOREL(3, B), STOREL(4, B),
       STOREL(5, B): NEXT B
400 PRINT : INPUT "RUN AGAIN OR QUIT: 1 OR 2"; V
410 IF V = 1 THEN Z = Z + 1: GOTO 120
420 END
430 IF W = 0 THEN Q = 0: RETURN
440 X = 14388 / W / T: M = 0: S = 0
450 M = M + 1: U = M * X
460 S0 = (U * U + 2 * U + 2) / M / M / M: S1 = S0 * EXP(-U)
470 S = S + S1: IF S1 = 0 AND S = 0 THEN W = 14388 / T / (38 * LOG(10) +
       LOG(S0)) + .1: GOTO 440
480 IF S1 / S > 1E-09 THEN 450
490 Q = 2 * C * (T / C2) ^ 3 * S
500 RETURN
510 IF W = 0 THEN DQ = 0: RETURN
520 X = 14388 / W / T: M = 0: S = 0
530 M = M + 1: U = M * X
540 S0 = (U ^ 3 + 3 * U * U + 6 * U + 6) / M / M / M: S1 = S0 * EXP(-U)
550 S = S + S1: IF S1 = 0 AND S = 0 THEN W = 14388 / T / (38 * LOG(10) +
       LOG(S0)) + .1: GOTO 520
560 IF S1 / S > 1E-09 THEN 530
570 DQ = 2 * C * (T / C2) ^ 3 / T * S
580 RETURN
590 IF W = 0 THEN L = 0: RETURN
600 X = 14388 / W / T: S = 0: M = 0
610 M = M + 1: U = M * X
620 S0 = (U * U * U + 3 * U * U + 6 * U + 6) / M / M / M / M: S1 = S0 * EXP(-U)
630 S = S + S1: IF S1 = 0 AND S = 0 THEN W = 14388 / T / (38 * LOG(10) +
       LOG(S0)) + .1: GOTO 600
640 IF S1 / S > 1E-09 THEN 610
650 L = 2 * C * C * H * (T / C2) ^ 4 * S
660 RETURN
670 IF W = 0 THEN D = 0: RETURN
680 X = 14388 / W / T: S = 0: M = 0
690 M = M + 1: U = M * X
700 S0 = (U ^ 4 + 4 * U ^ 3 + 12 * U * U + 24 * U + 24) / M / M / M / M: S1 = S0 *
       EXP(-U)
710 S = S + S1: IF S1 = 0 AND S = 0 THEN W = 14388 / T / (38 * LOG(10) +
       LOG(s0)) + .1: GOTO 680
720 IF S1 / S > 1E-09 THEN 690
730 D = 2 * C * C * H * (T / C2) ^ 4 / T * S
740 RETURN
```

Appendix C

RMS Evaluation

Consider a root-mean-square (rms) voltage

$$\Delta V_{rms} = \sqrt{\frac{1}{T} \int_0^T (V(t) - \overline{V})^2 \, dt}$$

that is, a sine wave; since we can divide any waveform into a Fourier series:

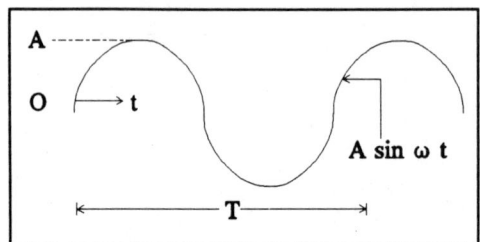

$$\Delta V_{rms} = \sqrt{\frac{1}{T} \int V(t)^2 \, dt}$$

$$= \sqrt{\frac{1}{T} \int_0^T (A^2 \sin^2 \omega t - 0) \, dt}$$

$$= \sqrt{\frac{A^2}{2T} \int_0^T (1 - \cos 2\omega t) \, dt}$$

$$= \left(\frac{A^2}{2T} \left[t - \frac{1}{2\omega} \sin 2\omega t \right]_0^T \right)^{1/2}$$

$$\Delta V_{rms} = \frac{A}{\sqrt{2}} = 0.707 \, A$$

Appendix D

10% to 90% Rise Time Related to Time Constant

Relationship of Time Constant (τ) to t_R^{10-90} for First-Order Low-Pass Filters

First-order low-pass filter response:

$$f(t) = f_0(1 - e^{-t/\tau})$$

$$f(t_R^{10}) = 0.10 f_0$$

$$1 - \exp(-t_R^{10}) = 0.10$$

$$t_R^{10} = -\tau \ln(0.90) = 0.1054\tau$$

Similarly, for the 90% value:

$$f(t_R^{90}) = 0.90 f_0$$

$$1 - \exp(-t_R^{90}) = 0.90$$

$$t_R^{90} = -\tau \ln(0.10) = 2.3026\tau$$

$$t_R^{10-90} = t_R^{90} - t_R^{10} = 2.1972\tau$$

$$f_{-3\,\mathrm{dB}} = \frac{1}{2\pi\tau} = \frac{1}{2\pi(t_R^{10-90}/2.1972)} = \frac{0.350}{t_R^{10-90}}$$

Index